Mathematical Theory of Elasticity and Generalized Dynamics of Quasicrystals and Its Applications

Mathematical Theory of Elasticity and Generalized Dynamics of Quasicrystals and Its Applications

Tian-You Fan
Beijing Institute of Technology, China

Xian-Fang Li
Central South University, China

Xiao-Hong Sun
Zheng Zhou University, China

Ming-Jun Huang
South China University of Technology, China

Yu-Chu Liu
South China University of Technology, China

World Scientific

NEW JERSEY • LONDON • SINGAPORE • BEIJING • SHANGHAI • HONG KONG • TAIPEI • CHENNAI • TOKYO

Published by

World Scientific Publishing Co. Pte. Ltd.
5 Toh Tuck Link, Singapore 596224
USA office: 27 Warren Street, Suite 401-402, Hackensack, NJ 07601
UK office: 57 Shelton Street, Covent Garden, London WC2H 9HE

British Library Cataloguing-in-Publication Data
A catalogue record for this book is available from the British Library.

MATHEMATICAL THEORY OF ELASTICITY AND GENERALIZED DYNAMICS OF QUASICRYSTALS AND ITS APPLICATIONS

ISBN 978-981-127-909-6 (hardcover)
ISBN 978-981-127-910-2 (ebook for institutions)
ISBN 978-981-127-911-9 (ebook for individuals)

For any available supplementary material, please visit
https://www.worldscientific.com/worldscibooks/10.1142/13492#t=suppl

Typeset by Stallion Press
Email: enquiries@stallionpress.com

Preface

The book titled *Mathematical Theory of Elasticity of Quasicrystals and Its Applications* was first published in Chinese by Beijing Institute of Technology Press in 1999. Then, the first two English editions of the book were jointly published by Science Press, Beijing, and Springer-Verlag, Heidelberg, in 2010 and 2016, respectively. In the second English edition of the book, some new contents are added, including phonon–phason dynamics, defect dynamics and hydrodynamics of solid quasicrystals, which belong to generalized dynamics of quasicrystals. Therefore, the title in this edition will be changed to *Mathematical Theory of Elasticity and Generalized Dynamics of Quasicrystals and Their Applications.*

In the meantime, the studies on the theory and applications of quasicrystals have been going on over the past decades. In some areas, lots of achievements have been obtained. Especially, the discovery of soft-matter quasicrystals in 2004 is an important event of 21st century chemistry, which enlarged the scope of quasicrystals. In this new branch, the metastable state and metastability in thermodynamics and the determination of dynamic effects in various sizes and time-scales in dynamics are very important, which may belong to one of generalized dynamics of quasicrystals as well as soft matter science. This increases the complexity and difficulty of studying the new phase. In the new edition, we will give a preliminary discussion only on the formation mechanism, metastability and phase transitions of the phase.

In addition, the applications of quasicrystal study are emphasized in different parts of the book, showing its importance in science and engineering.

The first author of this book thanks the support of National Natural Science Foundation of China and the Alexander von Humboldt Foundation of Germany over the latest decades. We thank also Professors U Messerschmidt in Max-Planck Institute for Microstructure Physics (Halle), H-R Trebin in the Institute for Theoretical Physics, the University of Stuttgart in Germany, T C Lubensky in the University of Pennsylvania and Stephen Z D Cheng in the University of Akron in the USA for their discussions and help.

Due to the important contributions for the new edition of the book, I invite Professors Xian-Fang Li, Xiao-Hong Sun, Ming-Jun Huang and Yu-Chu Liu to join me as co-authors of the present book and take charge of the book with me.

I also thank Professors Zhong-Qi Ma in Institute of High Energy Physics of Chinese Academy of Science, Zi-Tong Li and Zhu-Feng Sun in Beijing Institute of Technology for their kind help during the manuscript preparation of the book.

Tian-You Fan
March 29, 2023, Beijing

Contents

Authors

Tian-You Fan
School of Physics
Beijing Institute of Technology
Beijing, China

Xian-Fang Li
Central South University
Changsha, China

Xiao-Hong Sun
Zhengzhou University
Zhengzhou, China

Ming-Jun Huang
South China University of Technology
Guangzhou, China

Yu-Chu Liu
South China University of Technology
Guangzhou, China

Chapter 1

Crystals

This book discusses some structures and properties of quasicrystals, in which the solids and soft-matter are included. The former, i.e., solid quasicrystals, have inherent connections with crystals. This chapter provides the basic knowledge of crystals which may be beneficial to study solid quasicrystals and relevant topics.

1.1 Periodicity of Crystal Structure, Crystal Cell

Based on X-ray diffraction patterns, it is known that a crystal consists of particles (i.e., collections of ions, atoms and molecules) which are arranged regularly in space. The arrangement is a repetition of the smallest unit, called a unit cell, resulting in the periodicity of a complete crystal. The frame of the periodic arrangement of centres of particles is called a lattice. Thus, the properties of corresponding points of different cells in a crystal will be the same. The positions of these points can be defined by radius vectors $\mathbf{r}$ and $\mathbf{r}'$ in a coordinate frame $\mathbf{e}_1, \mathbf{e}_2, \mathbf{e}_3$, and $\mathbf{a}$, $\mathbf{b}$ and $\mathbf{c}$ are three non-mutually co-linear vectors, respectively (the general concept on vector referring to Chapter 2). Hence, we have

$$\mathbf{r}' = \mathbf{r} + l\mathbf{a} + m\mathbf{b} + n\mathbf{c} \tag{1.1-1}$$

in which $\mathbf{a}$, $\mathbf{b}$ and $\mathbf{c}$ are the basic translational vectors describing the particle arrangement in the complete crystal, and l, m and n are arbitrary integers. If the physical properties are described by function $f(\mathbf{r})$, then the above-mentioned invariance may be expressed mathematically as

$$f(\mathbf{r}') = f(\mathbf{r} + l\mathbf{a} + m\mathbf{b} + n\mathbf{c}) = f(\mathbf{r}). \qquad (1.1\text{-}2)$$

This is called the translational symmetry or long-range translational order of a crystal, because the symmetry is realized by the operation of translation.

Equation (1.1-1) represents a kind of translational transform, while (1.1-2) shows that the lattice is invariant under transformation (1.1-1). The collection of all translational transforms of the remaining lattice invariants constitute the translational group.

1.2 Three-dimensional Lattice Types

Cells of a lattice may be described by a parallel hexahedron having lengths of its three sides **a**, **b** and **c** and angles α, β and γ between the sides. According to the relationship between length of sides and angles, there are seven different forms observed for the cells, which form seven crystal systems given by Table 1.1.

Among each crystal system, there are some classes of crystals that are classified based on the configuration such as whether the face centre or body centre contains lattice points. For example, the cubic system can be classified into three classes: the simple cubic, body centre cubic and face centre cubic. According to this classification, the seven crystal systems contain 14 different lattice cells, which are called Bravais cells, as shown in Fig. 1.1.

Table 1.1. Crystals and the relationship of length of sides and angles.

Crystal system	Characters of cell
Triclinic	$a \neq b \neq c, \alpha \neq \beta \neq \gamma$
Monoclinic	$a \neq b \neq c, \alpha = \gamma = 90^\circ \neq \beta$
Orthorhombic	$a \neq b \neq c, \alpha = \beta = \gamma = 90^\circ$
Rhombohedral	$a = b = c, \alpha = \beta = \gamma \neq 90^\circ$
Tetragonal	$a = b \neq c, \alpha = \beta = \gamma = 90^\circ$
Hexagonal	$a = b \neq c, \alpha = \beta = 90^\circ, \gamma = 120^\circ$
Cubic	$a = b = c, \alpha = \beta = \gamma = 90^\circ$

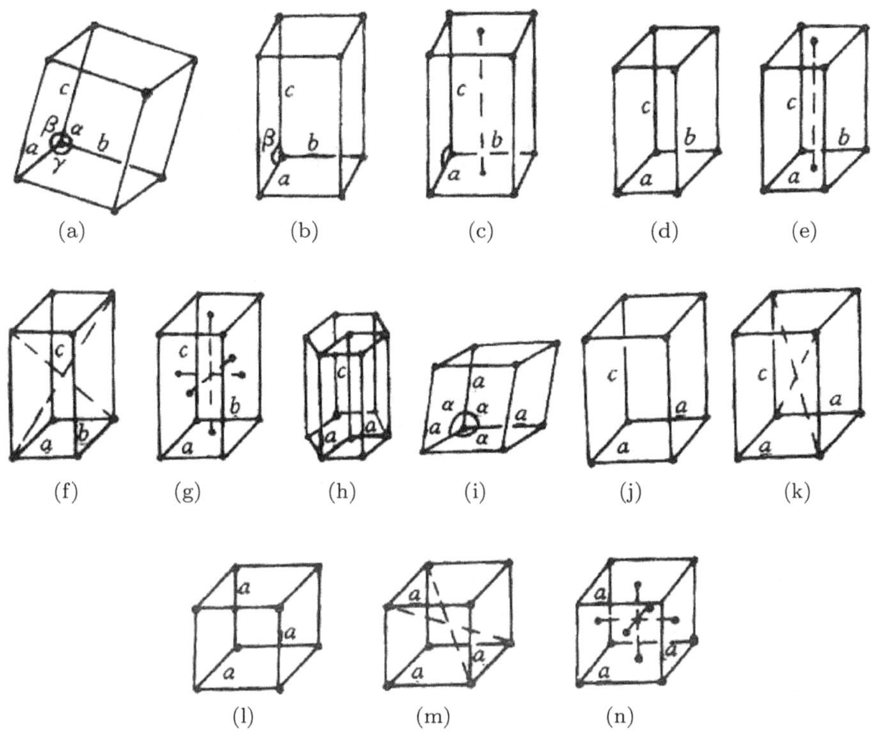

Fig. 1.1. The 14 crystal cells of three-dimensional lattices. (a) Simple triclinic, (b) simple monoclinic, (c) button centre monoclinic, (d) simple orthorhombic, (e) button centre orthorhombic, (f) body centre orthorhombic, (g) face centre orthorhombic, (h) hexagonal, (i) rhombohedral, (j) simple tetragonal, (k) body centre tetragonal, (l) simple cubic, (m) body centre cubic, (n) face centre cubic.

Apart from the above-mentioned 14 Bravais cells with three-dimensional lattice, there are five Bravais cells of two-dimensional lattices, we do not give any others.

1.3 Symmetry and Point Groups

In Section 1.1, we have discussed the translational symmetry of crystals. Here, we point out that the symmetry reveals invariance of crystals under translational transformation

$$\mathbf{T} = l\mathbf{a} + m\mathbf{b} + n\mathbf{c} \tag{1.3-1}$$

Fig. 1.2. There is no fivefold rotational symmetry in crystals.

Equation (1.3-1) is referred to as an operation of symmetry, which is a translational operation. Apart from this, there are rotation operation and reflection (or mapping) operation, which belong to the so-called point operation. A brief introduction of the rotation operation and orientational symmetry of crystals follows.

By rotating about an axis through a lattice, the crystal can always return to the original state since the rotational angles are $2\pi/1$, $2\pi/2$, $2\pi/3$, $2\pi/4$ and $2\pi/6$ or integer times of these values. This is the orientational symmetry or the long-range orientational order of a crystal. Because of the constraint of translational symmetry, the orientational symmetry holds for $n = 1, 2, 3, 4$ and 6 only, which is neither equal to 5 nor greater than 6, where n is the denominator of $2\pi/n$ (e.g., a molecule can have fivefold rotation symmetry, but a crystal cannot have this symmetry because the cells either overlap or have gaps between the cells when $n = 5$, Fig. 1.2 is an example). The fact constitutes the following fundamental law of crystallography:

Law of symmetry of crystals. Under rotation operation, n-fold symmetry axis is marked by n. Due to the constraint of translational symmetry, axes $n = 1, 2, 3, 4$ and 6 exist only, neither 5 nor a number greater than 6 exists.

In contrast to translational symmetry, rotation is a point symmetry. Other point symmetries are as follows: plane of symmetry, the corresponding operation is mapping, marked by m; centre of symmetry, the corresponding operation is inversion, marked by I; rotation–inversion axis, the corresponding operation is composition of rotation and inversion, when the inversion after rotation is $2\pi/n$, marked by $\bar{n}$.

For crystals, the point operation consists of eight independent ones only, i.e.,

$$1, 2, 3, 4, 6, I = (\bar{I}), \quad m = \bar{2}, \bar{4} \tag{1.3-2}$$

which are basic symmetric elements of point symmetry. The rotation operation is also denoted by C_n, $n = 1, 2, 3, 4, 6$.

The mapping operation can also be expressed by σ. The horizontal mapping is by m_h and S_h, and vertical one is by S_v. The mapping–rotation is a composite operation, denoted by S_n, which can be understood as

$$S_n = C_n\sigma_h = \sigma_h C_n$$

The inversion mentioned previously can be understood as

$$I = S_2 = C_2\sigma_h = \sigma_h C_2$$

Another composite operation — rotation–inversion — $\bar{n}$ is related with S_n, e.g., $\bar{1} = S_2 = I$, $\bar{2} = S_1 = \sigma$, $\bar{3} = S_6$, $\bar{4} = S_4$, $\bar{6} = S_3$.

So that (1.3-2) can also be redescribed as

$$C_1, C_2, C_3, C_4, C_6, I, \sigma, S_4 \tag{1.3-3}$$

The collection of each symmetric operation among these eight basic operations constitutes a point group, the collection of their composition forms 32 point groups, listed in Table 1.2.

1.4 Reciprocal Lattice

The concept of the reciprocal lattice will be our focus in the following chapters, here is a brief introduction.

Table 1.2. 32 Point groups of crystals.

Sign	Meaning of sign	Point group	Number
C_n	Having n-fold axis	C_1, C_2, C_3, C_4, C_6	5
I	Symmetry centre	$I(i)$	1
$\sigma(m)$	Mapping	$\sigma(m)$	1
C_{nh}	Having n-fold axis and horizontal symmetry plane	$C_{2h}, C_{3h}, C_{4h}, C_{6h}$	4
$C_{n\nu}$	Having n-fold axis and vertical symmetry plane	$C_{2\nu}, C_{3\nu}, C_{4\nu}, C_{6\nu}$	4
D_n	Having n-fold axis and n twofold axis, they are perpendicular to each other	D_2, D_3, D_4, D_6	4
D_{nh}	Meaning of h is the same as before	$D_{2h}, D_{3h}, D_{4h}, D_{6h}$	4
D_{nd}	d means in D_n there is a symmetry plane dividing the angle between 2 twofold axes	D_{2d}, D_{3d}	2
S_n	Having n-fold mapping rotation axis	$S_4, S_6 = C_{3i}$	2
T	Having 4 threefold axes and 3 twofold axes	T	1
T_h	Meaning of h is the same as before	T_h	1
T_d	Meaning of d is the same as previous	T_d	1
O	Having 3 fourfold axes which are perpendicular to each other and 6 twofold axes and 4 threefold axes	O, O_h	2

Note:
$T = C_{3'}D_2$ means the composition between operations $C_{3'}$ and D_2, where suffix $3'$ denotes a threefold axis.
$O = C_{3'}C_4C_{2''}$ means the composition between operations $C_{3'}$, C_4 and $C_{2''}$ where $3'$ represents a threefold axis, $2''$ a twofold axis.
The concept and sign of point groups will be used in the subsequent chapters.

Assume there is a relation between base vectors $\mathbf{a_1}$, $\mathbf{a_2}$ and $\mathbf{a_3}$ of a lattice (L) and base vectors $\mathbf{b_1}$, $\mathbf{b_2}$ and $\mathbf{b_3}$ for another lattice (L_R)

$$\mathbf{b}_i \cdot \mathbf{a}_j = \delta_{ij} = \begin{cases} 1, & i = j \\ 0, & i \neq j \end{cases} \quad (i, j = 1, 2, 3) \tag{1.4-1}$$

such that the lattice with base vectors $\mathbf{b_1}$, $\mathbf{b_2}$, $\mathbf{b_3}$ is the reciprocal lattice L_R of crystal lattice L, which has base vectors $\mathbf{a_1}$, $\mathbf{a_2}$, $\mathbf{a_3}$.

Between $\mathbf{b_i}$ and $\mathbf{a_j}$ there exist the following relationships:

$$\mathbf{b}_1 = \frac{\mathbf{a}_2 \times \mathbf{a}_3}{\Omega}, \quad \mathbf{b}_2 = \frac{\mathbf{a}_3 \times \mathbf{a}_1}{\Omega}, \quad \mathbf{b}_3 = \frac{\mathbf{a}_1 \times \mathbf{a}_2}{\Omega} \tag{1.4-2}$$

where $\Omega = \mathbf{a_1} \cdot (\mathbf{a_2} \times \mathbf{a_3})$ is the volume of lattice cell.

Denote

$$\Omega^* = \mathbf{b_1} \cdot (\mathbf{b_2} \times \mathbf{b_3})$$

then

$$\Omega^* = \frac{1}{\Omega}$$

The position of any point in the reciprocal lattice can be expressed by

$$\mathbf{G} = h_1\mathbf{b_1} + h_2\mathbf{b_2} + +h_3\mathbf{b_3} \tag{1.4-3}$$

in which

$$h_1, h_2, h_3 = \pm 1, \pm 2, \ldots$$

Points in a lattice can be described by $\mathbf{a_1}$, $\mathbf{a_2}$, $\mathbf{a_3}$ as well as by $\mathbf{b_1}$, $\mathbf{b_2}$, $\mathbf{b_3}$.

Similarly, the concept of reciprocal lattice can be extended to higher-dimensional space, e.g., six-dimensional space, which will be discussed in Chapter 4.

The brief introduction above provides a basic knowledge for reading the subsequent text of the book. Further information on crystals, diffraction theory and point group can be found in the book [1] and monograph [2]. We will recall the concepts in the following sections.

1.5 Appendix: Some Basic Concepts

Some basic concepts will be described in the following chapters, with which most physicists are familiar. For the readers who are non-physicists, a simple introduction is provided as follows, the details can be found in the relevant references.

1.5.1 *Concept of phonon*

In general, the course of crystallography does not contain the contents given in this section. Because the discussion here is dependent on quasicrystals, especially with the elasticity of solid quasicrystals, we have to introduce some of the simplest relevant arguments.

In 1900, Planck put forward quantum theory. Soon after, Einstein developed the theory and created the concept of the photon and explained the photo-electric effect, which leads to the photon concept. Einstein also studied the specific heat c_v of crystals arising from lattice vibration by using the Planck quantum theory. There are some unsatisfactory points in the work of Einstein on specific heat, though his formula successfully explained the phenomenon of $c_v = 0$ at $T = 0$, where T denotes the absolute temperature (or Kelvin temperature). To improve Einstein's work, Debye [3] and Born *et al.* [4,5] applied quantum theory to study specific heat in 1912 and 1913, respectively, and saw great success. The theoretical prediction is in excellent agreement with the experimental results, at least for specific heat of the atom crystals.

The propagation of the lattice vibration is called lattice wave. Under the long-wavelength approximation, lattice vibration can be seen as continuum elastic vibration, i.e., the lattice wave can be approximately seen as continuum elastic wave. The motion is a mechanical motion, but Debye and Born assumed that the energy can be quantized based on Planck's hypothesis. With the elastic wave approximation and quantization, Debye and Born successfully explained the specific heat of crystals at low temperature, and the theoretical prediction is consistent with experimental results in all ranges of temperature, at least for atom crystals. The quanta of the elastic vibration, or the smallest unit of energy of the elastic wave, is named phonon, because the elastic wave is an acoustic wave. Unlike photon, the phonon is not an elementary particle, but in the sense of quantization, the phonon presents natural similarity to that of photon and other elementary particles, thus it can be named quasiparticle. The concept created by Debye and Born opened the study on lattice dynamics, an important branch of solid-state physics. Yet, according to the view point at present, the Debye and

Born theory of solids belongs to a phenomenological theory, though they used the classical quantum theory. For the quantum mechanics theory, see Born and Huang [5] and Landau and Lifshitz [6].

Landau [6] further developed the phenomenological theory and put forward the concept of elementary excitation. According to the concept, photon, phonon, etc. belong to elementary excitations. In general, one elementary excitation corresponds to a certain field, e.g., photon corresponds to electro-magnetic wave, phonon corresponds to elastic wave, etc. The phonon concept is further extended by Born [5] and other scientists, who pointed out that the phonon theory given by Debye, in general, is not suitable for compounds. As atoms in a compound constitute a molecular collection in solid state, the vibration of atoms can be approximately classified into two cases: a vibration of whole body of molecule, and the other is relative vibration among atoms within a molecule. The first type of vibration is the same as that described by Debye theory, called phonetic frequency vibration mode, or the phonetic branch of phonon. In this case, the physical quantity phonon (under long-wavelength approximation) describing displacement field deviated from the equilibrium position of particles at lattice is also called as phonon-type displacement, or phonon field, or briefly phonon. Macroscopically, it is the displacement vector $\mathbf{u}$ of an elastic body (if the crystal is regarded as an elastic body). And the second type of vibration, i.e., the relative vibration among atoms within a molecule, is called photonic frequency vibration mode, or photonic branch of phonon. For this branch, the phonon cannot be simply understood as macroscopic displacement field. But our discussion in this book is confined to the framework of continuum medium, with no concern for the photonic branch, so the phonon field is the macro-displacement field in the consideration.

In many physical systems (classical or quantum systems), motion presents the discrete spectrum (the energy spectrum or frequency spectrum, which corresponds to the discrete spectrum of an eigen-value problem of a certain operator from the mathematical point of view). The lowest energy (frequency) level state is called ground state and that beyond the ground state is named excited state.

The so-called elementary excitation induces a transfer from the ground state to the state with the smallest non-zero energy (or frequency). Strictly speaking, it should be named lowest energy (or frequency) elementary excitation.

Solid-state physics was intensively developed in 1960s–1970s, and then evolved into condensed matter physics. Condensed matter physics not only extends the scope of solid-state physics by considering liquid state and micro-powder structure, it also develops basic concepts and principles. Modern condensed matter physics is established as a result of the construction of its paradigm, in which the symmetry-breaking occurs in a central place, which was contributed by Landau [6] and Anderson [7] and other scientists.

Considering the importance of the concept and principle of symmetry-breaking in the development of elasticity of solid and soft-matter quasicrystals, we discuss it briefly here.

It is well known that for a system with a constant volume, the equilibrium state thermodynamically requires that the free energy of the system

$$F = E - TS \tag{1.5-1}$$

be minimum, in which E is the internal energy, S the entropy and T the absolute temperature, respectively.

Landau proposed the so-called second-order phase transition theory by introducing a macroscopic order parameter η (to describe order–disorder) phase transition, i.e., assuming that the free energy can be expanded as a power series of η

$$F(\eta, T) = F_0(T) + A(T)\eta^2 + B(T)\eta^4 + \cdots \tag{1.5-2}$$

in which according to the requirement of the stability condition of phase transition (i.e., the variational condition $\delta F = 0$ or $\partial F/\partial \eta = 0$), the coefficients of odd terms should be taken to zero, and $B(T) > 0$. At high temperature the system is in disorder state, so $A(T) > 0$, too; as temperature reduces, $A(T)$ will change its sign; at the critical temperature T_C there exists $A(T_C) = 0$. The simplest choice to satisfy these conditions is

$$A(T) = \alpha(T - T_C), \quad B(T) = B(T_C) \tag{1.5-3}$$

in which α is a constant. Without concerning with concrete micro-mechanism, the Landau theory has the merits of simplicity and generality; it can be used in many systems and has achieved successes, especially for the study of superconductivity, liquid crystals, high energy physics (to the author's understanding, the quasicrystals study is another area that has been achieved following the line of the symmetry-breaking principle). Applying the above principle to periodic crystals, we have [7]

$$F = \frac{1}{2}\alpha(|\mathbf{G}|)(T - T_C(\mathbf{G}))\eta^2 + \text{higher-order terms} \tag{1.5-4}$$

where the constant α is related with reciprocal vector $\mathbf{G}$ (for the concepts of the reciprocal vector and reciprocal lattice, refer to Section 1.4). Further, Anderson [7] proved for crystals that if the density of periodic crystals can be expressed by Fourier series (the expansion exists due to the periodicity of the structure in three-dimensional lattice or reciprocal lattice)

$$\rho(\mathbf{r}) = \sum_{\mathbf{G}\in L_R} \rho_{\mathbf{G}} \exp\{i\mathbf{G}\cdot\mathbf{r}\} = \sum_{\mathbf{G}\in L_R} |\rho_{\mathbf{G}}| \exp\{-i\Phi_{\mathbf{G}} + i\mathbf{G}\cdot\mathbf{r}\} \tag{1.5-5}$$

where $\mathbf{G}$ is a reciprocal vector just mentioned above, and L_R is the reciprocal lattice in three-dimensional space, $\rho_{\mathbf{G}}$ is a complex number

$$\rho_{\mathbf{G}} = |\rho_{\mathbf{G}}|e^{i\Phi_{\mathbf{G}}} \tag{1.5-6}$$

with the amplitude $|\rho_{\mathbf{G}}|$ and phase angle $\Phi_{\mathbf{G}}$, due to $\rho(r)$ being real, $|\rho_{\mathbf{G}}| = |\rho_{-\mathbf{G}}|$ and $\Phi_{\mathbf{G}} = -\Phi_{-\mathbf{G}}$, then the order parameter is

$$\eta = |\rho_{\mathbf{G}}| \tag{1.5-7}$$

Anderson pointed out further that for crystals the phase angle $\Phi_{\mathbf{G}}$ contains the phonon $\mathbf{u}$, i.e.,

$$\Phi_{\mathbf{G}} = \mathbf{G}\cdot\mathbf{u} \tag{1.5-8}$$

in which both $\mathbf{G}$ and $\mathbf{u}$ are in three-dimensional physical space. If we consider only the phonetic branch of phonon, then $\mathbf{u}$ can be understood as phonon-type displacement field. So the displacement field in periodic crystals can be understood as phonon field from the Landau symmetry-breaking hypothesis, though it possesses an intuitive physical meaning under the approximation of long-wavelength (refer to Chapter 2). The description based on the symmetry-breaking, physical quantity $\mathbf{u}$ is connected with reciprocal vector $\mathbf{G}$ and reciprocal lattice L_R of crystals, so it presents more profound insights (a result of symmetry-breaking of the crystals) than that of the intuitive description of displacement field $\mathbf{u}$, though the explanation here is still phenomenological (because the Landau theory is a phenomenological theory), rather than that from the first-principles.

The concept of phonon originated from Debye [3] and Born [4,5], who describe the mechanical vibration of lattice mass points (atoms, or ions, or molecules) deviating from their equilibrium position. The propagation of the vibration leads to the lattice wave, and the motion can be quantized, the quanta is the phonon. This is an elementary excitation in condensed matter. The symmetry-breaking leads to the appearance of new ordered phase (e.g., crystals), new order parameter (e.g., $\eta = |\rho_{\mathbf{G}}|$, the wave amplitude of mass density wave), new elementary excitation (e.g., phonon) and new conservation law (e.g., the crystal symmetry law given in Section 1.3). The above description on phonon from the symmetry-breaking point of view helps us to understand in depth the physical nature of phonons.

Following this scheme, some elementary excitations (e.g., phason) temporarily without complete intuitive meaning can also be explained by the Landau theory, and one can further find out their physical meaning from the point of view of symmetry rather than from the point of view of simple intuition, because for some complex phenomena, the simple intuition cannot give a complete and correct explanation.

We recall that the elementary excitations are related with broken symmetry or symmetry-breaking, e.g., a liquid behaves with arbitrary translational symmetry and arbitrary orientational symmetry, then the periodicity of crystals (i.e., the lattice) breaks the translational symmetry and orientational symmetry of the liquid, the phonon is the elementary excitation, resulting from the symmetry-breaking. We can say the quasicrystal is a result of symmetry-breaking of crystal, which will be discussed in Chapter 4.

If it is not necessary to give a description on phason concept in depth, this section can be omitted. The readers are advised to jump over the section if they are not interested in.

More profound discussions on the concept of phonon can be found in Born and Huang's classical monograph on lattice dynamics [5] and vol V of *Course of Theoretical Physics* by Landau and Lifshitz [6]. They give both phenomenological and quantum mechanical descriptions, the previous introduction is only a phenomenological description.

1.5.2 *Incommensurate crystals*

In this book, we do not discuss incommensurate phases. Considering that quasicrystals are related with the so-called incommensurate structure, we have to mention it in brief.

Since the 1960s, incommensurate crystals have been studied by many physicists, see, e.g., [8]. The term incommensurate phase means that it is added with an additional incommensurate modulate to the basic lattice, in which the modulated ones may be displacements or compounds of atoms or arrangement of spin, etc. As an example, if a modulated displacement λ is added to a lattice with period a, and if λ/a is a rational number, the crystal becomes a super-structure with a long period (which is the integer times of a); and if λ/a is an irrational number, the crystal becomes an incommensurate structure. In this case, along the modulation direction, the periodicity is lost. The modulation can be one-dimensional, e.g., Na_2CO_3, $NaNO_2$, etc., or two-dimensional, e.g., $TaSe_2$, quartz, etc., or three-dimensional, e.g., $Fe_{1-x}O$, etc. In incommensurate phases, the modulation is only a "perturbation" of another period in the basic lattice, the diffraction pattern of the basic lattice holds, i.e., the crystallography symmetry holds still, so one calls the structure incommensurate crystals.

It is noted that there are new degrees of freedom in the phases, named phasons. Here the phason modes present long-wavelength propagation similar to that of phonon modes. In Chapter 4, we shall discuss how the phason modes in quasicrystals present quite different natures, i.e., the motion of atoms exhibits discontinuous jumps rather than the long-wavelength propagation. In addition, in quasicrystals there is non-crystallographic orientational symmetry, which is essentially different from that of general incommensurate crystals.

1.5.3 *Glassy structure*

The crystals are solid with long-range order due to regular arrangement of atoms. In contrast, there is a solid without order, but it has short-range order in scale within the atom size. This material is the glassy structure, and is studied as a branch of the condensed matter physics.

1.5.4 *Mathematical aspect of group*

1.5.4.1 *Mathematical definition of group*

In previous sections, we introduced group concept through the symmetry operations, this is intuitive and easy understood. Under a transformation when a system maintains invariant, we can call the transformation a symmetry transformation. In succeeding taking two transformations is defined as a product of the two transformations, it is evident, which is the symmetry transformation of the system, too. The product of three symmetry transformations satisfies the associate law. The identical transformation is also a symmetry transformation, the product between it and any symmetry transformation is the transformation. The inversion of a symmetry transformation is still a symmetry transformation. The connection of transformations of a system forms a transformation group. We have defined the concept of product, the mathematical definition of group can be given as follows:

(1) Closeness of connection: The product between any two elements g_i and g_j in a connection G belongs to the connection, i.e., $g_i \in G$, $g_j \in G$, $g_i g_j \in G$.
(2) Associate law: If $g_i, g_j, g_k \in G$, then $g_i(g_j, g_k) = (g_i, g_j)g_k$.
(3) There is an identical element E: If $g_i \in G$, then $Eg_i = g_i$.
(4) There is an inversion element g_i^{-1}: If $g_i \in G$, then $g_i g_i^{-1} = E$.

The definition of group is also called axiom of group, which is valid for all groups, including the point groups. In this book, we mainly concern ourselves not only with the point groups, but also Lie groups in Chapter 16 and Appendix III in Major Appendices of the book.

1.5.4.2 *The linear representation of group*

Assume element g_i belongs to group G, which corresponds to matrix A_i, and assume all matrixes have the same order, their determinates are not equal to zero, if product $g_i g_j$ corresponds to product $A_i A_j$, then we can say that matrix A_i is a linear representation of group G. Assume a linear expression of group G corresponds to a matrix A_i of n order, and a linear expression of the same group that corresponds to matrix B_i of m order constitutes quasidiagonal matrix of $(n+m)$ order

$$[A_i, B_i] = \begin{bmatrix} A_i, & 0 \\ 0, & B_i \end{bmatrix} \equiv [C_i]$$

If we assume transforming to the equivalent expression according to matrix XC_iX^{-1}, then in general, the character of quasi-diagonal matrix will be lost. If this character can be maintained, we say this expression is reducible, otherwise, it is irreducible.

References

[1] Kittel C, 1976, *Introduction to the Solid State Physics*, Wiley John & Sons, Inc., New York.
[2] Wybourne B G, 1974, *Classical Group Theory for Physicists*, Wiley John & Sons, Inc., New York; Ma Zhong-Qi, 2019, *Group Theory for Physicists*, 2nd edition, World Scientific, Singapore.
[3] Debye P, 1912, Die Eigentuemlichkeit der spezifischen Waermen bei tiefen Temperaturen, *Arch de Genéve*, 33(4), 256–258.

[4] Born M and von Kármán Th, 1913, Zur Theorie der spezifischen Waermen, *Physikalische Zeitschrift*, 14(1), 15–19.
[5] Born M and Huang K, 1954, *Dynamic Theory of Crystal Lattices*, Clarendon Press, Oxford.
[6] Landau L D and Lifshitz M E, 1980, *Theoretical Physics V: Statistical Physics*, 3rd edition, Pergamon Press, Oxford.
[7] Anderson P W, 1984, *Basic Notations of Condensed Matter Physics*, Benjamin-Cummings, Menlo Park.
[8] Blinc B and Lavanyuk A P, 1986, *Incommensurate Phases in Dielectrics I, II*, North Holland, Amsterdam.

Chapter 2

Framework of Crystal Elasticity

As the knowledge of crystals is beneficial for understanding quasicrystals, it is worthwhile having a concise review of crystal elasticity or classical elasticity before learning about the elasticity of solid quasicrystals. Here is a brief description of the theory. The detailed material for this theory can be found in many monographs and textbooks, e.g., Landau and Lifshitz [1]. Though the discussion here is limited within the framework of continuum medium mechanics, there are still connections to the physical nature of the elasticity of crystals reflected by the concept of phonons (discussed in Section 1.5). The readers are advised to refer to the relevant chapters and sections of monographs of Born and Huang [2] and Anderson [3] which would help us understanding the phonon concept and thus the phason concept and elasticity of solid quasicrystals, which will be presented in the following chapters. The practice shows that it would be hard to understand phasons and phason elasticity if we limited our knowledge only to the classical continuum medium and our intuition.

For simplicity, tensor algebra will be used in the text.

2.1 Review of Some Basic Concepts

2.1.1 *Vector*

A quantity with both magnitude and direction is named vector, denoted by $\mathbf{a}$, and $a = |\mathbf{a}|$ represents its magnitude. The scalar product of two vectors $\mathbf{a}$ and $\mathbf{b}$ is $\mathbf{a} \cdot \mathbf{b} = ab\cos(\mathbf{a}, \mathbf{b})$. The vector product $\mathbf{a} \times \mathbf{b} = \mathbf{n}\, ab\sin(\mathbf{a}, \mathbf{b})$, in which $\mathbf{n}$ is the unit vector perpendicular to both $\mathbf{a}$ and $\mathbf{b}$, so $|\mathbf{n}| = 1$. A more general definition of vector is given later.

2.1.2 *Coordinate frame*

To describe vector and tensor, it is convenient to introduce the coordinate frame. We will consider the orthogonal frame. Assume that $\mathbf{e}_1, \mathbf{e}_2$ and $\mathbf{e}_3$ are three unit vectors and mutually perpendicular, i.e., $\mathbf{e}_1 \cdot \mathbf{e}_2 = 0$, $\mathbf{e}_2 \cdot \mathbf{e}_3 = 0$, $\mathbf{e}_3 \cdot \mathbf{e}_1 = 0$, and $\mathbf{e}_3 = \mathbf{e}_1 \times \mathbf{e}_2$, $\mathbf{e}_2 = \mathbf{e}_3 \times \mathbf{e}_1$, $\mathbf{e}_1 = \mathbf{e}_2 \times \mathbf{e}_3$, then they are base vectors of an orthogonal coordinate frame, and often called base vectors briefly.

In the orthogonal coordinate frame $\mathbf{e}_1, \mathbf{e}_2, \mathbf{e}_3$, any vector $\mathbf{a}$ can be expressed by

$$\mathbf{a} = a_1\mathbf{e}_1 + a_2\mathbf{e}_2 + a_3\mathbf{e}_3 \tag{2.1-1}$$

Or

$$\mathbf{a} = (a_1, a_2, a_3)$$

2.1.3 *Coordinate transformation*

Consider another orthogonal frame $\mathbf{e}'_1, \mathbf{e}'_2, \mathbf{e}'_3$ which can be expressed in terms of $\mathbf{e}_1, \mathbf{e}_2, \mathbf{e}_3$. From (2.1-1), there are

$$\begin{aligned}
\mathbf{e}'_1 &= c_{11}\mathbf{e}_1 + c_{12}\mathbf{e}_2 + c_{13}\mathbf{e}_3 \\
\mathbf{e}'_2 &= c_{21}\mathbf{e}_1 + c_{22}\mathbf{e}_2 + c_{23}\mathbf{e}_3 \\
\mathbf{e}'_3 &= c_{31}\mathbf{e}_1 + c_{32}\mathbf{e}_2 + c_{33}\mathbf{e}_3
\end{aligned} \tag{2.1-2}$$

where $c_{11}, c_{12}, \ldots, c_{33}$ are some scalar constants. The relation (2.1-2) is named coordinate transformation too, which can also be denoted by the matrix

$$\begin{bmatrix} \mathbf{e}'_1 \\ \mathbf{e}'_2 \\ \mathbf{e}'_3 \end{bmatrix} = [C] \begin{bmatrix} \mathbf{e}_1 \\ \mathbf{e}_2 \\ \mathbf{e}_3 \end{bmatrix} \tag{2.1-3}$$

where

$$[C] = \begin{bmatrix} c_{11} & c_{12} & c_{13} \\ c_{21} & c_{22} & c_{23} \\ c_{31} & c_{32} & c_{33} \end{bmatrix}$$

which is an orthogonal matrix, consequently

$$[C]^T = [C]^{-1} \tag{2.1-4}$$

Here notation "T" marks transpose operation, and "−1", the inversion operation. It is natural that

$$\begin{bmatrix} \mathbf{e}_1 \\ \mathbf{e}_2 \\ \mathbf{e}_3 \end{bmatrix} = [C]^T \begin{bmatrix} \mathbf{e}'_1 \\ \mathbf{e}'_2 \\ \mathbf{e}'_3 \end{bmatrix} = [C]^{-1} \begin{bmatrix} \mathbf{e}'_1 \\ \mathbf{e}'_2 \\ \mathbf{e}'_3 \end{bmatrix} \tag{2.1-5}$$

Based on (2.1-1), if in the frame $\mathbf{e}'_1, \mathbf{e}'_2, \mathbf{e}'_3$, then

$$\mathbf{a} = a_1\mathbf{e}'_1 + a_2\mathbf{e}'_2 + a_3\mathbf{e}'_3 \tag{2.1-6}$$

Substituting (2.1-5) into (2.1-1) yields

$$\begin{aligned} \mathbf{a} = {} & (c_{11}a_1 + c_{12}a_2 + c_{13}a_3)\mathbf{e}'_1 + (c_{21}a_1 + c_{22}a_2 + c_{23}a_3)\mathbf{e}'_2 \\ & + (c_{31}a_1 + c_{32}a_2 + c_{33}a_3)\mathbf{e}'_3 \end{aligned} \tag{2.1-7}$$

It follows that by comparison between (2.1-6) and (2.1-7):

$$\begin{aligned} a'_1 &= c_{11}a_1 + c_{12}a_2 + c_{13}a_3 \\ a'_2 &= c_{21}a_1 + c_{22}a_2 + c_{23}a_3 \\ a'_3 &= c_{31}a_1 + c_{32}a_2 + c_{33}a_3 \end{aligned} \tag{2.1-8}$$

or by the matrix expression, i.e.,

$$\begin{bmatrix} a'_1 \\ a'_2 \\ a'_3 \end{bmatrix} = [C] \begin{bmatrix} a_1 \\ a_2 \\ a_3 \end{bmatrix} \tag{2.1-8$'$}$$

Whether (2.1-8) or (2.1-8′), there is

$$a'_i = \sum_{j=1}^{3} c_{ij}a_j = c_{ij}a_j \tag{2.1-9}$$

The summation sign in the right-hand side of (2.1-9) is omitted, when the repeated indexes in $c_{ij}a_j$ represent summing. Henceforth, the summation convention will be used throughout.

A set of numbers (a_1, a_2, a_3) satisfying the relation (2.1-9) under linear transformation (2.1-2) will be called a vector regardless of its physical meaning. This is an algebraic definition of vector; it is more general than saying that the vector has both magnitude and direction.

2.1.4 *Tensor*

Let us define nine numbers in the orthogonal frame $\mathbf{e}_1, \mathbf{e}_2, \mathbf{e}_3$ as $\mathbf{A}$:

$$\mathbf{A} = \begin{bmatrix} A_{11} & A_{12} & A_{13} \\ A_{21} & A_{22} & A_{23} \\ A_{31} & A_{32} & A_{33} \end{bmatrix} \tag{2.1-10}$$

in which the components satisfy the relation

$$A'_{kl} = \sum_{i,j=1}^{3} c_{ki}c_{lj}A_{ij} = c_{ki}c_{lj}A_{ij} \tag{2.1-11}$$

under the linear transformation, then $\mathbf{A}$ is a tensor of rank 2, where c_{ij} are given by (2.1-3), and the summing sign is omitted in the right-hand side of (2.1-11). It is evident that the concept of tensor is an extension of that of vector. According to the definition A_{ij} represents a tensor where $i = 1, 2, 3$, $j = 1, 2, 3$. It is understood that it represents a component with the indexes i and j of the tensor.

2.1.5 *Algebraic operations of tensor*

(i) Unit tensor:

$$I = \delta_{ij} = \begin{cases} 0 & i \neq j \\ 1 & i = j \end{cases} \tag{2.1-12}$$

which is named the Kronecker sign conventionally.

(ii) Transpose of tensor:

$$A^T = \begin{bmatrix} A_{11} & A_{21} & A_{31} \\ A_{12} & A_{22} & A_{32} \\ A_{13} & A_{23} & A_{33} \end{bmatrix} \tag{2.1-13}$$

(iii) Algebraic sum of tensors:

$$A \pm B = A_{ij} \pm B_{ij} \tag{2.1-14}$$

(iv) Product of scalar and tensor:

$$mA = mA_{ij} \tag{2.1-15}$$

(v) Product of tensors:

$$AB = A_{ij}B_{kl} \tag{2.1-16}$$

Other operations related to tensors will be provided in the description of the subsequent text.

2.2 Basic Assumptions of Theory of Elasticity

The theory of elasticity is a branch of continuum mechanics, it follows certain basic assumptions:

(i) **Continuity:** In the theory one assumes that the medium fills the full space that it occupies, and this means the medium is continuous. Connected with this, the field variables concerning the medium are continuous and differentiable functions of coordinates.
(ii) **Homogeneity:** Physical constants describing the medium are independent from coordinates, so the medium is homogeneous.
(iii) **Small deformation:** Assume that displacements u_i are small and $\partial u_i/\partial x_j$ less than a unit. Due to small deformation, the boundary conditions are written at the boundaries before deformation even though those boundaries have taken some deformation. This makes the problem linearized and simplifies the solution procedure.

2.3 Displacement and Deformation

That an elastic body exhibits deformation is connected to the relative displacement between points on it. So we first look for the displacement field.

Consider a region R in an elastic body, refer to Fig. 2.1, it becomes another region R' after deformation. The point O with radius vector $\mathbf{r}$, before deformation, which becomes point O' with

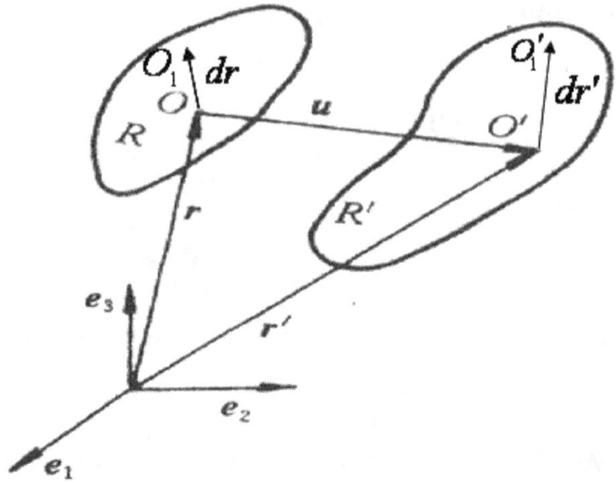

Fig. 2.1. Displacement of a point on an elastic body.

radius vector $\mathbf{r}'$ after deformation, and $\mathbf{u}$ is the displacement vector of point O during the deformation process (see Fig. 2.1), i.e.,

$$\mathbf{r}' = \mathbf{r} + \mathbf{u} \tag{2.3-1}$$

or

$$\mathbf{u} = \mathbf{r}' - \mathbf{r} = x_i' - x_i \tag{2.3-1$'$}$$

In Fig. 2.1, frame $\mathbf{e_1}$, $\mathbf{e_2}$, $\mathbf{e_3}$ depicts any orthogonal coordinate system, especially we use the rectilinear coordinate system (x_1, x_2, x_3) or (x, y, z). Assume that O_1 in R is a point near the point O, the radius vector joining them is $d\mathbf{r} = dx_i$. The point O_1 becomes point O_1' in R' after deformation. The vector radius joining points O_1 and point O_1' is $d\mathbf{r}' = dx_i' = dx_i + du_i$. The displacement of point O_1 is $\mathbf{u}'$, there is

$$\mathbf{u}' = \mathbf{u} + d\mathbf{u} \tag{2.3-2}$$

i.e.,

$$du_i = u_i' - u_i \tag{2.3-3}$$

and

$$du_i = \frac{\partial u_i}{\partial x_j} dx_j \tag{2.3-4}$$

Equation (2.3-4) expresses the Taylor expansion at point O and takes the first-order term only. Under the small deformation assumption, this reaches a very high accuracy. It denotes

$$\frac{\partial u_i}{\partial x_j} = \varepsilon_{ij} + \omega_{ij} \tag{2.3-5}$$

in which

$$\varepsilon_{ij} = \frac{1}{2}\left(\frac{\partial u_i}{\partial x_j} + \frac{\partial u_j}{\partial x_i}\right) \tag{2.3-6}$$

$$\omega_{ij} = \frac{1}{2}\left(\frac{\partial u_i}{\partial x_j} - \frac{\partial u_j}{\partial x_i}\right) \tag{2.3-7}$$

Here ε_{ij} is a symmetric tensor

$$\varepsilon_{ij} = \varepsilon_{ji} \tag{2.3-8}$$

and called the strain tensor, while ω_{ij} is an asymmetric tensor, which has only three independent components, as follows:

$$\begin{aligned}
\Omega_x &= \omega_{yz} = \frac{1}{2}\left(\frac{\partial u_y}{\partial z} - \frac{\partial u_z}{\partial y}\right) \\
\Omega_y &= \omega_{zx} = \frac{1}{2}\left(\frac{\partial u_z}{\partial x} - \frac{\partial u_x}{\partial z}\right) \\
\Omega_z &= \omega_{xy} = \frac{1}{2}\left(\frac{\partial u_x}{\partial y} - \frac{\partial u_y}{\partial x}\right)
\end{aligned} \tag{2.3-9}$$

The physical meaning of ε_{ij} describes volume and shape changes of a cell, and that of ω_{ij}, the rigid rotation, which is independent of deformation, so only ε_{ij} is considered afterward.

The components $\varepsilon_{11}, \varepsilon_{22}$ and ε_{33} (if we denote $x = x_1, y = x_2, z = x_3$, then we have $\varepsilon_{xx}, \varepsilon_{yy}$ and ε_{zz}) represent normal strains describing volume change of a cell, while $\varepsilon_{32} = \varepsilon_{23}$, $\varepsilon_{13} = \varepsilon_{31}$ and $\varepsilon_{12} = \varepsilon_{21}$ (or $\varepsilon_{yz} = \varepsilon_{zy}$, $\varepsilon_{zx} = \varepsilon_{xz}$ and $\varepsilon_{xy} = \varepsilon_{yx}$) represent shear strains describing shape change of a cell.

2.4 Stress Analysis

The internal forces per unit area due to deformation are called stresses, and denoted by σ_{ij}, which will be zero if there is no deformation of a body. When the body is in static equilibrium according to the law of momentum conservation, we have

$$\frac{\partial \sigma_{ij}}{\partial x_i} + f_j = 0 \tag{2.4-1}$$

in which the equation holds for any infinitesimal volume element of the body, σ_{ij} represents the components of the stress tensor as mentioned above, suffix j is the acting direction of the component, i is the direction of outward normal vector of the surface element that the component exerted, and f_i is the body force density vector. Among all the components of σ_{ij}, σ_{xx}, σ_{yy} and σ_{zz} are normal to the surface elements which they exerted, and $\sigma_{yz}, \sigma_{zy}, \sigma_{zx}, \sigma_{xz}, \sigma_{xy}$ and σ_{yx} are along the tangent directions of the surface elements. The former are called normal stresses, and the latter shear stresses.

According to the angular momentum conservation, one finds that

$$\sigma_{ij} = \sigma_{ji} \tag{2.4-2}$$

This means the stress tensor is a symmetric tensor, and Equation (2.4-2) is named as the shear stress mutual equal law.

External surface forces' density (tractions) T_i subjected to the surface of a body should be balanced with the internal stresses, this leads to

$$\sigma_{ij} n_j = T_i \tag{2.4-3}$$

where n_j is the unit vector along the outward normal to the surface element. People also call T_i area force density.

Equation (2.4-3) describes the stress boundary conditions which play a very important role in elasticity.

2.5 Generalized Hooke's Law

Between stresses σ_{ij} and strains ε_{ij}, there exists a certain relationship depending upon the material behaviour of the body. Hereafter we consider only the linear elastic behaviour of materials, and the state without initial stresses. In the case, the classical experimental law — Hooke's law — can be extended as

$$\sigma_{ij} = \frac{\partial U}{\partial \varepsilon_{ij}} = C_{ijkl}\varepsilon_{kl} \tag{2.5-1}$$

in which U denotes the free energy density, or the strain energy density, i.e.,

$$U = F = \frac{1}{2}C_{ijkl}\varepsilon_{ij}\varepsilon_{kl} \tag{2.5-2}$$

and C_{ijkl} is the elastic constant tensor, consisting of 81 components. Due to the symmetry of σ_{ij} and ε_{ij}, each of them has six independent components only, such that the independent components of C_{ijkl} reduce to 36. Equation (2.5-2) shows that U is a homogeneous quantity of ε_{ij} of rank two, considering the symmetry of ε_{ij}, then we have

$$C_{ijkl} = C_{klij} \tag{2.5-3}$$

so the independent components among 36 reduce to 21.

The relation (2.5-1) with 21 independent elastic constants is named as generalized Hooke's law.

The generalized Hooke's law describes anisotropic elastic bodies including crystals. Stress and strain tensors can also be expressed by corresponding vectors with six independent elements, then can be denoted by the corresponding elastic constants matrix $[B_{ijkl}]$ as follows:

$$\begin{bmatrix}\sigma_{11}\\ \sigma_{22}\\ \sigma_{33}\\ \sigma_{23}\\ \sigma_{31}\\ \sigma_{12}\end{bmatrix} = \begin{bmatrix} B_{1111} & B_{1122} & B_{1133} & B_{1123} & B_{1131} & B_{1112}\\ & B_{2222} & B_{2233} & B_{2223} & B_{2231} & B_{2212}\\ & & B_{3333} & B_{3323} & B_{3331} & B_{3312}\\ & & & B_{2323} & B_{2331} & B_{2312}\\ & (symmetry) & & & B_{3131} & B_{3112}\\ & & & & & B_{1212}\end{bmatrix} \begin{bmatrix}\varepsilon_{11}\\ \varepsilon_{22}\\ \varepsilon_{33}\\ \varepsilon_{23}\\ \varepsilon_{31}\\ \varepsilon_{12}\end{bmatrix} \tag{2.5-4}$$

Applying Equation (2.5-4) to crystals, between the elements C_{ijkl} (or B_{ijkl}) we obtain some relations by considering certain symmetry of the crystals, so that the resulting number of the elastic constants for certain individual crystal systems may be less than 21. In the following we give a brief discussion on the argument.

(1) Triclinic system (classes 1 or C_1 and C_i)

The triclinic symmetry does not add any restrictions to the components of tensor C_{ijkl} (or B_{ijkl} in (2.5-4)), however, appropriate choice of the coordinate system enables us to reduce the number of non-zero independent elastic constants. Because the orientation of the coordinate system is determined by three rotation angles, this provides three conditions to restrict some components in C_{ijkl} (or B_{ijkl} in (2.5-4)); for example, one can take three of them to be zero, such that the triclinic crystal system has 18 components of elastic moduli.

(2) Monoclinic system (classes C_s, C_2, and C_{2h})

In the class C_s, there is a plane of symmetry, we take it as $x_3 = 0$ $(z = 0)$ in the coordinate frame $\mathbf{e_1}, \mathbf{e_2}, \mathbf{e_3}$. Making a coordinate transformation with this plane of symmetry, one can obtain a new coordinate frame $\mathbf{e}'_1, \mathbf{e}'_2, \mathbf{e}'_3$. Between these two coordinate frames, we obtain the following relation:

$$\mathbf{e}'_1 = \mathbf{e_1}, \quad \mathbf{e}'_2 = \mathbf{e_2}, \quad \mathbf{e}'_3 = -\mathbf{e_3} \tag{2.5-5}$$

This operation is the reflection or mapping. In addition, we know that between σ'_{ij} in $\mathbf{e}'_1, \mathbf{e}'_2, \mathbf{e}'_3$ and σ_{ij} in $\mathbf{e_1}, \mathbf{e_2}, \mathbf{e_3}$ there are (refer to Section 2.1)

$$\sigma'_{kl} = \alpha_{kj}\alpha_{li}\sigma_{ji} \tag{2.5-6}$$

in which α_{ij} are the coefficients of linear transformation, i.e.,

$$\mathbf{e}'_i = \alpha_{ij}\mathbf{e}_j \tag{2.5-7}$$

Under the transformation (2.5-5), there are

$$\alpha_{11} = 1, \quad \alpha_{22} = 1, \quad \alpha_{33} = -1, \quad \text{others} = 0 \tag{2.5-8}$$

Therefore, under the transformation, for C_{ijkl} in (2.5-1) (or B_{ijkl} in (2.5-4)) the suffixes containing 3 with an odd number of times (1 or 3) will change sign, while the others will remain invariant. Considering the symmetry of the crystal, however, the physical properties including C_{ijkl} (or B_{ijkl} in (2.5-4)) should remain unchanged under symmetric operation (including the reflection). It is obvious that all components with an odd number of suffixes 3 must vanish, i.e.,

$$B_{1123} = B_{1131} = B_{2223} = B_{2231} = B_{3323} = B_{3331} = B_{2312} = B_{3112} = 0 \tag{2.5-9}$$

Consequently, there are only 13 independent elastic constants.

A similar discussion can be presented for the classes C_2 and C_{2h}.

(3) Orthorhombic system (classes C_{2v}, D_2, and D_{2h})

This crystal system has a macroscopic corresponding, i.e., the orthotropic materials, in which there exist two planes of symmetry perpendicular to each other. Let us take $x_3 = 0$ and $x_1 = 0$ as the planes. If on reflection in plane $x_3 = 0$, it is just the case of the monoclinic system mentioned previously. Subsequently, considering the mapping in plane $x_1 = 0$, between the new and old coordinate systems, there is the following relation:

$$\begin{bmatrix} e_1' \\ e_2' \\ e_3' \end{bmatrix} = \begin{bmatrix} -1 & 0 & 0 \\ 0 & 1 & 0 \\ 0 & 0 & 1 \end{bmatrix} \begin{bmatrix} e_1 \\ e_2 \\ e_3 \end{bmatrix}$$

By a similar description to that of a monoclinic system, one finds that

$$B_{1112} = B_{2212} = B_{3312} = B_{2331} = 0 \tag{2.5-10}$$

Collecting to (2.5-9), the system contains nine independent elastic constants.

(4) Tetragonal system (classes C_{4v}, D_{2d}, D_4, and D_{4h})

This crystal system has four axes of symmetry. Similar to the previous discussion, independent elastic moduli are

$$B_{1111}, B_{3333}, B_{1122}, B_{1212}, B_{1133}, B_{1313},$$

the total number of these is six.

(5) Rhombohedral system (classes C_{3v}, 3 or C_3, D_3, D_{3d} and S_6)

In this system there is a third-order axis of symmetry (or three-fold symmetric axis). We can take the axis of symmetry as the axis $\mathbf{e_3}$, after a lengthy description that six independent elastic constants are as follows:

$$B_{3333}, B_{\xi\eta\xi\eta}, B_{\xi\xi\eta\eta}, B_{\xi\eta 33}, B_{\xi 3\eta 3}, B_{\xi\xi\xi 3}$$

with

$$\xi = x_1 + ix_2, \quad \eta = x_1 - ix_2$$

The moduli can also be written in conventional version as

$$B_{3333}, B_{1212}, B_{1122}, B_{1233}, B_{1323}, B_{1113}$$

(6) Hexagonal system (class C_6)

The crystal system has a macroscopic correspondence — the transverse isotropic material — whose elasticity presents fundamental importance to the elasticity of one- and two-dimensional quasi-crystals.

There is a sixth-order axis of symmetry (or say six-fold symmetric axis) in the system. By taking this axis as x_3-axis, and using the coordinate substitution $\xi = x_1 + ix_2$, $\eta = x_1 - ix_2$. In a rotation with angle $2\pi/6$ about the x_3-axis, the coordinates ξ and η experience a transformation, $\xi \to \xi e^{i2\pi/6}$, $\eta \to \eta e^{-i2\pi/6}$. Then one can see that only those components C_{ijkl} do not vanish which have the same number of suffixes ξ and η. These are

$$B_{3333}, B_{\xi\eta\xi\eta}, B_{\xi\xi\eta\eta}, B_{\xi\eta 33}, B_{\xi 3\eta\xi}$$

or in conventional expressions

$$\begin{aligned} &C_{1111} = C_{2222}, C_{3333}, C_{2323} = C_{3131}, C_{1122}, \\ &C_{1133} = C_{2233}, C_{1212} \end{aligned}$$

in which $2C_{1212} = C_{1111} - C_{1122}$, so the number of independent elastic constants is five.

(7) Cubic system

For this system there are three four-fold symmetric axes, in which there is tetragonal symmetry. If we take the four-fold symmetric axis of the tetragonal symmetry in the x_3-direction, the number of independent components of C_{ijkl} (or B_{ijkl} in (2.5-4)) are $B_{1111}, B_{1122}, B_{1212}$

(8) Isotropic body

In this case there are two elastic moduli, e.g., Young's modulus and Poisson's ratio

$$E, \quad \nu,$$

respectively, or the Lamé constants

$$\lambda = \frac{\nu E}{(1+\nu)(1-2\nu)}, \quad \mu = \frac{E}{2(1+\nu)} \tag{2.5-11}$$

or the bulk modulus of compression and shear modulus

$$K = \frac{E}{3(1-2\nu)}, \quad \mu = \frac{E}{2(1+\nu)} = G$$

In this case, the generalized Hooke's law presents a very simple form, i.e.,

$$\sigma_{ij} = 2\mu\varepsilon_{ij} + \lambda\varepsilon_{kk}\delta_{ij} \tag{2.5-12}$$

where $\varepsilon_{kk} = \varepsilon_{11} + \varepsilon_{22} + \varepsilon_{33} = \varepsilon_{xx} + \varepsilon_{yy} + \varepsilon_{zz}$, δ_{ij} is the unit tensor. An equivalent form of (2.5-12) is

$$\varepsilon_{ij} = \frac{1+\nu}{E}\sigma_{ij} - \frac{\nu}{E}\sigma_{kk}\delta_{ij} \tag{2.5-13}$$

in which $\sigma_{kk} = \sigma_{11} + \sigma_{22} + \sigma_{33} = \sigma_{xx} + \sigma_{yy} + \sigma_{zz}$

2.6 Elastodynamics, Wave Motion

When the inertia effect is considered in Equation (2.4-1), then it becomes

$$\frac{\partial \sigma_{ij}}{\partial x_j} + f_i = \rho\frac{\partial^2 u_i}{\partial t^2} \tag{2.6-1}$$

where ρ is the mass density of the material.

Considering isotropic medium and omitting body forces, from (2.6-1), (2.3-6) and (2.5-12), the equation of wave motion is obtained as

$$(c_1^2 - c_2^2)\frac{\partial^2 u_i}{\partial x_i \partial x_j} + c_2^2 \frac{\partial^2 u_j}{\partial x_i^2} = \frac{\partial^2 u_j}{\partial t^2} \tag{2.6-2}$$

where c_1 and c_2 are defined by

$$c_1 = \left(\frac{\lambda + 2\mu}{\rho}\right)^{\frac{1}{2}}, \quad c_2 = \left(\frac{\mu}{\rho}\right)^{\frac{1}{2}} \tag{2.6-3}$$

which are speeds of elastic longitudinal and transverse waves, respectively. If we let

$$\mathbf{u} = \nabla\phi + \nabla \times \boldsymbol{\psi}, \tag{2.6-4}$$

then Equation (2.6-2) can be reduced to

$$\nabla^2\phi = \frac{1}{c_1^2}\frac{\partial^2 \phi}{\partial t^2}, \quad \nabla^2\boldsymbol{\psi} = \frac{1}{c_2^2}\frac{\partial^2 \boldsymbol{\psi}}{\partial t^2} \tag{2.6-5}$$

where ϕ is scalar potential, $\boldsymbol{\psi}$, the vector potential, and $\nabla^2 = \frac{\partial^2}{\partial x^2} + \frac{\partial^2}{\partial y^2} + \frac{\partial^2}{\partial z^2}$. Equation (2.6-5) are typical wave equations of mathematical physics. To solve the problem, apart from the boundary conditions one needs initial conditions, i.e.,

$$\left.\begin{aligned} u_i(x_i, 0) &= u_{i0}(x_i) \\ \dot{u}_i(x_i, 0) &= \dot{u}_{i0}(x_i), \end{aligned}\right. \quad x_i \in D$$

2.7 Summary

The classical theory of elasticity is concluded to solve the following initial-boundary value problem:

$$\begin{aligned} &\varepsilon_{ij} = \frac{1}{2}\left(\frac{\partial u_i}{\partial x_j} + \frac{\partial u_j}{\partial x_i}\right) \\ &\frac{\partial \sigma_{ij}}{\partial x_i} = \rho\frac{\partial^2 u_j}{\partial t^2} - f_j, \quad (t > 0, x_i \in D) \\ &\sigma_{ij} = C_{ijkl}\varepsilon_{ijkl} \end{aligned}$$

$$\begin{aligned}
&u_i(x_i, 0) = u_{i0}(x_i) \\
&\dot{u}_i(x_i, 0) = \dot{u}_{i0}(x_i), \quad (x_i \in D) \\
&\sigma_{ij} n_j = T_i, \quad t > 0, \quad x_i \in S_t \\
&u_i = \bar{u}_i, \quad t > 0, \quad x_i \in S_u
\end{aligned}$$

where $u_{i0}(x_i), \dot{u}_{i0}(x_i), T_i$, and $\bar{u}_i$ are known functions, D denotes the region of materials we studied, S_t and S_u are parts of boundary S on which the tractions and displacements are prescribed, respectively, and $S = S_t + S_u$. If $\frac{\partial^2 u_j}{\partial t^2} = 0$, the problem reduces to a static problem as for a pure boundary value problem, there are no initial conditions at all.

References

[1] Landau L D and Lifshitz E M, 1986, *Theoretical Physics V: Theory of Elasticity*, Pergamon Press, Oxford.

[2] Born M and Huang K, 1954, *Dynamic Theory of Crystal Lattices*, Clarendon Press, Oxford.

[3] Anderson P W, 1984, *Basic Notations of Condensed Matter Physics*, Benjamin-Cummings, Menlo Park.

Chapter 3

Solid Quasicrystals and Their Properties

3.1 Discovery of Solid Quasicrystal

The first observation of metallic alloy quasicrystal was made in April 1982 by D. Shechtman as guest scholar working in the Bureau of Standards in the USA. He observed from electronic microscopy that a rapidly cooled Al-Mn alloy exhibits fivefold orientational symmetry by the diffraction patterns with bright diffraction spots and sharp Bragg reflections, as shown in Fig. 3.1. Because the fivefold orientational symmetry is in contradiction to the basic law of symmetry of crystals (refer to Chapter 1), the result could not be understood within the first two years since the discovery. A colleague of Shechtman, I. Blech in Israel, supported him powerfully, and explained it might be an icosahedral glass. They drafted a paper concerning the experimental results and sent it off to a journal, but it was rejected. Then they submitted it to another journal, which did not publish it either. J. W. Cahn, the hosting scientist in the Bureau of Standards recommended streamlining the paper; leaving out details of the model and experiment, and limiting it solely to the experimental findings. After consulting with D. Gratias, a mathematical crystallographer at the Centre National de la Recherche Scientifique in France, the group submitted an abbreviated paper to *Physical Review Letters* (PRL) in October 1984, more than 2 years after Shechtman's initial experiment. The paper was published several weeks later. This is the Ref. [1] (PRL Top 10 articles: #8 of APS).

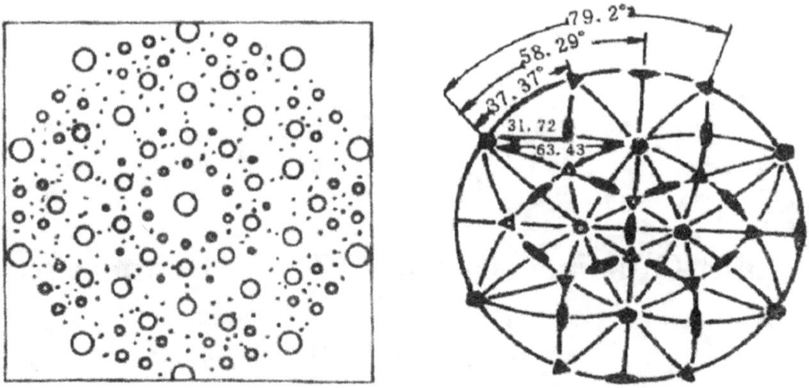

(a) (b)

Fig. 3.1. The patterns of diffraction of icosahedral quasicrystal. (a) The fivefold symmetry; (b) the stereographic structure of the quasicrystal.

More recently, the natural quasicrystals, apart from alloys, were observed by Bindi *et al.*[2].

After 4 weeks, Levine and Steinhardt [3] published their work by PRL to introduce ordered structure with quasiperiodicity and called the novel alloy "quasicrystal" formally, in which their theoretical (computed) diffraction pattern is in excellent agreement with that of the experimental observation. Soon after, other groups, e.g., the group of Guo [4] also found similar structures with fivefold symmetry and icosahedral quasicrystal in Ni–V and Ni–Ti alloys.

In December 1984, Bak [5] sent out his first paper, which might be the third completed manuscript after Shechtman *et al.* [1] and Levine and Steinhardt [3] published in English. He studied the icosahedral quasicrystals, but from the incommensurate phase point of view, which is very interesting. He pointed out his discussion is based on the phenomenological Landau symmetry breaking principle, and this opened an important direction in quasicrystals. In his second paper, he gave a further development, and first claimed that there are phason elementary excitations apart from phonon elementary excitations. He pointed out the first-principle computation for the material is difficult. The readers can get a hint from Bak's work that due to the complexity of quasicrystal structure, people cannot obtain a similar principle like Bloch theorem with great power. The difficulty has not been solved so far yet. The two papers of Bak have fruitful and profound scientific connotations, and greatly promote the development of quasicrystal study. We will further discuss them in the subsequent chapters.

The icosahedral quasicrystals are one kind of three-dimensional quasicrystals, in which atomic arrangement is quasiperiodic in three directions. Another kind of three-dimensional quasicrystal is cubic quasicrystal observed by Fung *et al.* [6] later.

Successively, two-dimensional quasicrystals were observed. Here, the atomic arrangement is quasiperiodic along two directions and periodic along the third direction, which is just the directions of 5-, 8-, 10- and 12-fold symmetrical axis of the two-dimensional quasicrystals observed to date, such that one finds four kinds of two-dimensional quasicrystals with 5-, 8-, 10- and 12-fold rotation symmetries, which are also called pentagonal, octagonal, decagonal and dodecagonal quasicrystals, respectively (see [7–12]).

There is another class of quasicrystals, the one-dimensional quasicrystals, in which the atomic arrangement is quasiperiodic along one axis, and periodic along the plane perpendicular to the axis (see [13–18]).

The discovery of this novel matter with long-range order and non-crystallographic symmetry changes the traditional concept of classifying solids into two classes: crystal and non-crystal, and gives a strong impact on traditional crystallography, bringing a profound new physical idea into the matter structure and symmetry.

The unusual structure of quasicrystals lends it a series of properties that is different from those of crystals, which has drawn a great deal of attention from researchers in a range of fields, such as physics, crystallography, chemistry, etc.

Two decades before the discovery of the physical quasicrystals, mathematicians [19] suggested aperiodic tiling, which may be considered as mathematical quasicrystals. One among them is the well-known Penrose tiling, in which a tiling without an overlap or a gap in two different rhombohedra can result in quasiperiodic structure. After the discovery of quasicrystals, the Penrose tiling has become

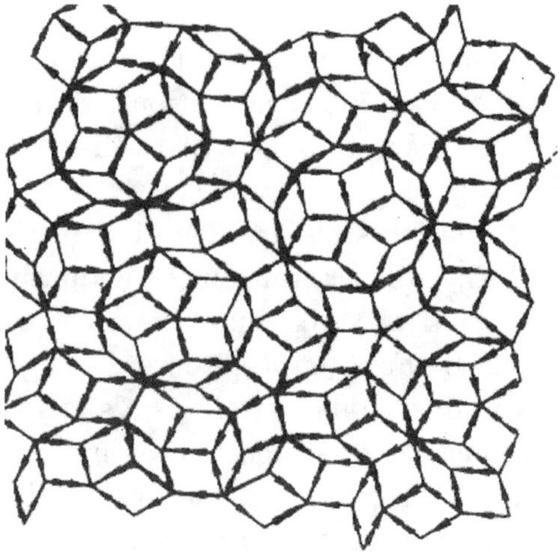

Fig. 3.2. Penrose tiling of two-dimensional quasicrystal with 5-fold symmetry.

a geometry tool for the new solid phase. As an example, a Penrose tiling for describing a two-dimensional 5-fold symmetry is shown in Fig. 3.2, in which the local structure of the tiling is similar, and this is called local isomorphism (LI). The quasiperiodic symmetry and local isomorphism of the Penrose tiling present significance for describing the new solid phase — quasicrystals. The discovery of physical quasicrystals promotes the development of the relevant discrete geometry, ergodic theory, group theory, mathematical physics, partial differential equations, complex analysis, etc. Of course, these pure and applied mathematics promote the development of quasicrystal study, too.

Quasicrystals with thermodynamical stability are becoming a new class of functional and structural materials, which have many prospective engineering applications. The study of physical and mechanical properties of the material is put forward. Among the mechanical behaviour of quasicrystals, the elasticity and defects are important topics. These provide many new challenges as well as opportunities for the continuum mechanics.

However, there was a different point of view, e.g., Pauling [20] claimed that the icosahedral quasicrystal is a cubic crystal rather than a quasicrystal, but the various quasicrystals are observed experimentally, the argument is ended.

3.2 Structure and Symmetry of Quasicrystals

Quasicrystals are different from periodic crystals, but with certain symmetry, so they are one kind of aperiodic crystals. The unusual characters of quasicrystals originate from their special atomic constitution. The character of this structure is explored by diffraction patterns. Just through these diffraction patterns people discovered differences between crystals and quasicrystals, and realized the discovery of quasicrystals.

Similar to other aperiodic crystals, quasiperiodicity induces new degrees of freedom, which can be explained as follows. In crystallography and solid-state physics, the Miller indices (h, k, l) are often used to describe the structure of crystals. These indices can explain the spectra of diffraction patterns of all crystals. In Chapter 1, we mentioned that the number of base vectors for crystal N is identical to the number of the dimensions d of the crystal, i.e., $N = d$. However, because quasicrystals have quasiperiodic symmetry (including both or either quasiperiodic translational and orientational symmetries disallowed by the rule of crystallography), the Miller indices cannot be used and instead we need to employ six indices $(n_1, n_2, n_3, n_4, n_5, n_6)$. This feature implies that, to characterize the symmetry of quasicrystals it is necessary to introduce higher-dimensional (four-, five- or six-dimensional) spaces. This idea is identical to that of group theory, i.e., the quasiperiodic structure is periodic in higher-dimensional space (four-, five- or six-dimensional space). Quasicrystals in real three-dimensional space (physical space) may be seen as a projection of a periodic lattice in higher-dimensional space (mathematical space). The projection of periodic lattice at four-, five- and six-dimensional space to physical space generates one-, two- and three-dimensional quasicrystals, respectively. The six-dimensional space is denoted by E^6, which consists of two sub-spaces, one is physical space, called the parallel space and denoted by $E^3_{\|}$, another is the complementary space, also called vertical space and

Table 3.1. The systems, Laue classes and point groups of one-dimensional quasicrystals.

System	Laue class	Point group
Triclinic	1	$1, \overline{1}$
Monoclinic	2	$2,\ m_h,\ 2/m_h$
	3	$2_h,\ m,\ 2_h/m$
Orthorhombic	4	$2_h\ 2_h\ 2,\ mm2,\ 2_h\ mm_h,\ mmm_h$
Tetragonal	5	$4, \overline{4}, 4/m_h,$
	6	$42_h2_h,\ 4mm,\ 4/m_h,\ mm$
Rhombohedral (trigonal)	7	$3, \overline{3}$
	8	$32_h, 3m, \overline{3}m$
Hexagonal	9	$6,\ \overline{6},\ 6/m_h$
	10	$62_h2_h,\ 6mm, \overline{6}m2_h,\ 6/m_hmm$

denoted by $E_\perp^3$ so that

$$E^6 = E_\parallel^3 \oplus E_\perp^3 \tag{3.2-1}$$

where $\oplus$ denotes the direct sum.

For one-, two- and three-dimensional quasicrystals, the number of the base vectors is $N = 4, 5, 6$ and the number of the realistic dimension of the material (in physical space) is $d = 3$, so $N > d$, this is different from that of crystals.

The method of group theory is the most appropriate method to describe symmetry of quasicrystals.

The one-dimensional quasicrystals have 31 point groups, consisting of 6 quasicrystal systems and 10 Laue classes, respectively, in which all point groups are crystallographic point groups, listed in Table 3.1.

The two-dimensional quasicrystals contain two kinds of matter: the first kind and the second kind, respectively. In this chapter, the discussion is only on the first kind of ones, and the second kind of ones will be discussed in Chapter 17.

The first kind of two-dimensional quasicrystals have 57 point groups, in which 31 are crystallographic point groups listed in Table 1.1, and other 26 are non-crystallographic point groups listed in Table 3.2. They belong to four quasicrystal systems and eight Laue classes, respectively.

Table 3.2. The systems, Laue classes and point groups of non-crystallography of the first kind of two-dimensional quasicrystals.

System	Laue class	Point groups
Pentagonal	11	$5, \bar{5}$
	12	$5m, 52, \bar{5}m$
Octagonal	13	$8, \bar{8}, 8/m$
	14	$8mm$, 822, $\bar{8}m2$, $8/mmm$
Decagonal	15	10, $\overline{10}$, $10/m$
	16	$10mm, 1022$, $\overline{10}m2, 10/mmm$
Dodecagonal	17	$12, \overline{12}, 12/m$
	18	$12mm, 1222, \overline{12}m2, 12/mmm$

The three-dimensional quasicrystals have 60 point groups. They are: (1) 32 crystallographic point groups and 28 non-crystallographic point groups i.e., icosahedral point groups $\left(235, \frac{2}{m}\overline{35}\right)$ and 26 point groups with 5-, 8-, 10- and 12-fold symmetries (5, $\bar{5}$, 52, $\bar{5}m$, $5m$, and N, $\overline{N}$, N/m, $N22$, Nmm, $\overline{N}\ m2$, N/mmm, $N = 8, 10, 12$), the latter have been listed in Table 3.2.

3.3 A Brief Introduction of the Physical Properties of Solid Quasicrystals

The unusual structure of quasicrystals lends them some new physical properties.

The mechanical behaviour of quasicrystals, especially, the distinguishing features of elasticity to those of crystals, have aroused a great deal of interest of researchers. These will be discussed starting from Chapter 4 in detail, so we do not talk about them here.

The thermal properties of quasicrystals are a field that attracts attention of many scientists [21–29]. The thermal conductivity of quasicrystals is lower than that of conventional metals.

Among the profoundly studied properties of quasicrystals, the first are the structural and elastic properties, and the second may be the properties of electricity of the material. The electric conductivity of quasicrystals is lower. The Hall effect [30–34] was well studied. The absolute number of the Hall coefficient RH is two orders of magnitude greater than that of conventional metals. In addition, the pressure-resistance properties of quasicrystals are also discussed by some references, see, e.g., [35].

The light conductivity rate of quasicrystals is quite different from that of conventional metals. For the people who are interested in the singularity, see [36–38]. Recently, the quasicrystals-based photonic crystal study has become a focus [39–41], and there is a trend of further developing the study [42–44]. Detailed discussions will be given in Chapter 19 of this book.

The electronic structure of quasicrystals and relevant topics have also been of concern after the discovery [45–51]. Due to lack of periodicity, the Bloch theorem and Brillouin zone concept in solid physics cannot be used for quasicrystals. By some simple models, for example, based on the so-called approximate phase of quasicrystals, because the approximate phase has certain kinds of crystals, the Bloch theorem can be used and through numerical computations, one can obtain results on electronic energy spectra. Some results of wave functions exhibit behaviour neither in extending state nor in localization state. For some quasicrystalline materials, e.g., AlCuLi, AlFe, there are pseudo-gaps when energy is over the Fermi energy. Quilichini and Janssen [52] have obtained some lattice dynamics results in computational and experimental results.

Some mathematicians developed several mathematical models to describe possible electronic energy spectrum of quasicrystals, readers can refer to Refs. [53–60] if interested in.

The magnetic properties of quasicrystals as pointed out by Fukamichi [61] have been actively studied since the discovery of Shechtman *et al.*, since they are very sensitive to the local atomic structure. The results obtained up to 1990 were reviewed by O'Handley *et al.* [62]. The review of Fukamichi is very interesting to researchers lies in the correlation between magnetic properties of quasicrystals and phason field. Of course, the magnetic properties of quasicrystals are an important correlation with spin, which requires the understanding of the electronic structure mentioned above. At present, this is a difficult problem due to the lack of important basis in quasicrystals similar to the Bloch theorem like that in crystals.

The properties discussed above are only limited to solid quasicrystals. After 20 years since 1984, the quasicrystals are observed in soft matter too, which can be called soft-matter quasicrystals and are quite different from the metallic alloy quasicrystals, whose properties are not touched upon here but will be discussed in Chapter 18.

Because the present book discusses elasticity, dynamics of solid quasicrystals and some chemistry problems on soft-matter quasicrystals only, the other physical properties will not be dealt with. The above introduction is very simple and in brief.

3.4 One-, Two- and Three-dimensional Quasicrystals

It is needed to recall the concept of one-, two- and three-dimensional quasicrystals. The one-dimensional quasicrystals are the ones in which the atom arrangement is quasiperiodic in one direction, and periodic in the other two directions. The two-dimensional quasicrystals belong to ones in which the atom arrangement is quasiperiodic in two directions and periodic in another one. The three-dimensional quasicrystals behave such that the arrangement presents quasiperiodicity in all three directions.

There exist many solid quasicrystals observed to date, among them the halves are icosahedral in the first place and decagonal quasicrystals in the second one, respectively. These two kinds of quasicrystal systems present major importance in the material. Recently, the two-dimensional quasicrystals with 10-, 12- and 18-fold symmetries in soft matter have been observed and studied [63–68], as well as two-dimensional quasicrystals with higher fold symmetry through nanomaterial manufacturing.

3.5 Two-dimensional Quasicrystals and Planar Quasicrystals

The two-dimensional quasicrystals and planar quasicrystals are different concepts. The two-dimensional quasicrystals are introduced in the previous section, which represent a three-dimensional structure with two-dimensional quasiperiodic planes stacked along the third direction, in which the atom arrangement is periodic. While the planar quasicrystals belong to a two-dimensional structure within the plane the atom arrangement is quasiperiodic, and there is no third dimension.

3.6 The First and Second Kind of Two-dimensional Quasicrystals

Hu *et al.* [69] predicted there will exist 7-, 9-, 14- and 18-fold symmetry quasicrystals in solid. Nevertheless, they have not been observed so far, except that the soft-matter quasicrystals of 18-fold symmetry were observed in colloids in 2011. Therefore, the prediction of Hu *et al.* is meaningful. We call the 7-, 9-, 14- and 18-fold symmetry quasicrystals as the second kind two-dimensional quasicrystals, which will be specially discussed in Chapter 17. Consider this reason, we can call the 5-, 8-, 10- and 12-fold symmetry quasicrystals mentioned above as the first-kind two-dimensional quasicrystals.

References

[1] Shechtman D, Blech I, Gratias D and Cahn J W, 1984, Metallic phase with long-range orientational order and no translational symmetry, *Phys Rev Lett*, **53**(20), 1951–1953.

[2] Bindi L, Steinhardt P J, Yao N and Lu P J, 2009, Natural quasicrystals, *Science*, **324**, 1306–1309.

[3] Levine D and Steinhardt P J, 1984, Quasicrystals: A new class of ordered structure, *Phys Rev Lett*, **53**(26), 2477–2480.

[4] Ye H Q, Wang D and Kuo K H, 1985, Five-fold symmetry in real and reciprocal space, *Ultramicrossopy*, **16**(2), 273–277; Zhang Z, Ye H Q and Kuo K H, 1985, A new icosahedral phase with m35 symmetry, *Phil Mag* A, **52**(6), L49–L52.

[5] Bak P, 1985, Phenomenological theory of icosahedral incommensurate ("quaisiperiodic") order in Mn-Al alloys, *Phys. Rev. Lett.*, **54**(8), 1517–1519; Bak P, 1985, Symmetry, stability and elastic properties of icosahedral incommensurate crystals, *Phys. Rev. B*, **32**(9), 5764–5772.

[6] Feng Y C, Lu G and Witers R I, 1989, An incommensurate structure with cubic point group symmetry in rapidly solidified V-Vi-Si alloy, *J Phys. Condens. Matter*, **1**(23), 3695–3700.

[7] Bendersky L, 1985, Quasicrystal with one-dimensional translational symmetry and a tenfold rotation axis, *Phys. Rev. Lett.*, **55**(14), 1461–1463.

[8] Chattopadhyay K, Lele S and Thangarai N *et al.*, 1987, Vacancy ordered phases and one-dimensional quasiperiodicity, *Acta Metall.*, **35**(3), 727–733.

[9] Fung K K, Yang C Y and Zhou Y Q *et al.*, 1986, Icosahedrally related decagonal quasicrystal in rapidly cooled Al-14-at.%-Fe alloy, *Phys. Rev. Lett.*, **56**(19), 2060–2063.

[10] Urban K, Mayer J, Rapp M, Wilkens M, Csanady A, and Fidler J, 1986, Studies on aperiodic crystals in Al-Mn and Al-V alloys by means of transmission electron microscopy, *Journal de Physique Colloque*, **47C**(3), 465–475.

[11] Wang N, Chen H and Kuo K H, 1987, Two-dimensional quasicrystal with eightfold rotational symmetry, *Phys. Rev. Lett.*, **59**(9), 1010–1013.

[12] Li X Z and Guo K H, 1988, Decagonal quasicrystals with different periodicities along the tenfold axis in rapidly solidified Al–Ni alloys, *Phil. Mag. Lett.*, **58**(3), 167–171.

[13] Merdin R, Bajema K, Clarke R, Juang F-Y and Bhattacharya P K, 1985, Quasiperiodic GaAs-AlAs heterostructures, *Phys. Rev. Lett.*, **55**(17), 1768–1770.

[14] Hu A, Tien C and Li X *et al.*, 1986, X-ray diffraction pattern of quasiperiodic (Fibonacci) Nb-Cu superlattices, *Phys. Lett.* A, **119**(6), 313–314.

[15] Feng D, Hu A, Chen K J *et al.*, 1987, Research on quasiperiodic superlattice, *Mater. Sci. Forum*, **22**(24), 489–498.

[16] Terauchi H, Noda Y, Kamigami K, Matsunaka S, Nkayama M, Kato H, Naokatsu Sano and Yasusada Yamada, 1988, X-Ray diffraction patterns of configuration Fibonacci lattices, *J. Phys. Jpn.*, **57**(7), 2416–2424.

[17] Chen K J, Mao G M and Fend D *et al.*, 1987, Quasiperiodic a-Si: H/a-SiNx: H multilayer structures, *J. Non-cryst. Solids*, **97**(1), 341–344.

[18] Yang W G, Wang R H and Gui J, 1996, Some new stable one-dimensional quasicrystals in $Al_{65}\ Cu_{20}\ Fe_{10}\ Mn_5$ alloy, *Phil. Mag. Lett.*, **74**(5), 357–366.

[19] Berger R, 1966, The undecidability of the domino problem, Mem Amer Math Soc, no 66; Penrose R, 1974, The role of aesthetics in pure and applied mathematical research, in *Bulletin of the Institute of Mathematics and Its Applications* (*Bull. Inst. Math. Appl.*). Southend-on-Sea, 10, 266–271, ISSN 0146-3942;

Pentaplexity — A Class of non-periodic tilings of the plane, in *The Mathematical Intelligencer*, Vol. 2, No. 1, Springer, New York, 1979, pp. 32–37, ISSN 0343-6993.

[20] Pauling L, 1985, Apparent icosahedral symmetry is due to directed multiple twinning of cubic crystals, *Nature*, **317**, 512–514.

[21] Biham O, Mukamel D and Shtrikman, 1986, Symmetry and stability of icosahedral and other quasicrystalline phases, *Phys. Rev. Lett.*, **56**(20), 2191–2194.

[22] Scheater R J and Bendersky L A, 1988, *Introduction to Quasicrystals*, ed. By Jaric M. V., Academic Press, Boston, MA.

[23] Widom M, Deng D P and Henleg C L, 1989, Transfer-matrix analysis of a two-dimensional quasicrystal, *Phys. Rev. Lett.*, **63**(3), 310–313.

[24] Yang W G, Ding D H, Wang R H and Hu C Z, 1996, Thermodynamics of equilibrium properties of quasicrystals, *Z Phys.* B, **100**(3), 447–454.

[25] Hu C Z, Yang W G, Wang R H and Ding D H, 1997, Quasicrystal Symmetry and Physical Properties. *Progress in Phys.* (in Chinese), **17** (4), 345–367.

[26] Fan T Y, 1999, A study on specific heat of one-dimensional hexagonal quasicrystal, *J. Phys.: Condens. Matter*, **11**(45), L513–517.

[27] Fan T Y and Mai Y W, 2003, Partition function and state equation of point group 12mm two-dimensional dodecagonal quasicrystals, *Euro. Phys. J.B*, **31**(2), 17–21.

[28] Li C Y and Liu Y Y, 2001, Phason-strain influence on low-temperature specific heat of the decagonal Al-Ni-Co quasicrystal, *Chin. Phys. Lett.*, **18**(4), 570–573.

[29] Li C Y and Liu Y Y, 2001, Low-temperature lattice excitation of icosahedral Al-Mn-Pd quasicrystals, *Phys. Rev.* B, **63**(6), 064203–064211.

[30] Biggs B D, Li Y and Poon S J, 1991, Electronic properties of icosahedral, approximant, and amorphous phases of an Al-Cu-Fe alloy, *Phys. Rev.* B, **43**(10), 8747–8750.

[31] Lindqvist P, Berger C, Klein T *et al.*, 1993, Role of Fe and sign reversal of the Hall coefficient in quasicrystalline Al-Cu-Fe, *Phys. Rev.* B, **48**(1), 630–633.

[32] Klein T, Gozlen A, Berger C, Cyrot-Lackmann F, Calvayrac Y and Quivy A, 1990, Anomalous transport properties in pure AlCuFe icosahedral phases of high structural quality, *Euro. Phys. Lett.*, **13**(2), 129–134.

[33] Pierce F S, Guo Q and Poon S J, 1994, Enhanced insulatorlike electron transport behavior of thermally tuned quasicrystalline states of Al-Pd-Re alloys, *Phys. Rev. Lett.*, **73**(16), 2220–2223.

[34] Wang A J, Zhou X, Hu C Z and Miao L, 2005, Properties of nonlinear elasticity of quasicrystals with five-fold symmetry, Wuhan University Journal, *Nat. Sci.*, **51**(5), 536–566 (in Chinese).

[35] Zhou X, Hu C Z, Gong P and Qiu S D, 2004, Piezoresistance properties of quasicrystals, *J. Phys.: Condens. Matter*, **16**(30), 5419–5425.

[36] Homes C C, Timusk T, Wu X, Altounian Z, Sahnoune A and Stroem-Olsen J O, 1991, Optical conductivity of the stable icosahedral quasicrystal $Al_{63.5}Cu_{24.5}Fe_{12}$, *Phy. Rev. Lett.*, **67**(19), 2694–2696.

[37] Burkov S E, 1992, Optical conductivity of icosahedral quasi-crystals, *J. Phys.: Condens. Matter*, **4**(47), 9447–9458.

[38] Basov D N, Timusk T, Barakat F, Greedan J and Grushko B, 1994, Anisotropic optical conductivity of decagonal quasicrystals, *Phys. Rev. Lett.*, **72**(12), 1937–1940.

[39] Villa D, Enoch S and Tayeb G *et al.*, 2005, Band gap formation and multiple scattering in photonic quasicrystals with a Penrose-type lattice, *Phys. Rev. Lett.*, **94**(18), 183903–183907.

[40] Didier M, 2000, Generalized Drude Formula for the Optical Conductivity of Quasicrystals, *Phys. Rev. Lett.*, **85**(6), 1290–1293.

[41] Notomi M, Suzuk H and Tamamura T *et al.*, 2004, Lasing action due to the two-dimensional quasiperiodicity of photonic quasicrystals with a Penrose lattice, *Phys. Rev. Lett.*, **90**(12), 123906–123910.

[42] Feng Z F, Zhang X D and Wang Y Q, 2005, Negative refraction and imaging using 12-fold-symmetry quasicrystals, *Phys. Rev. Lett.*, **94**(24), 247402–247406.

[43] Lifshitz R, Arie A and Bahabad A, 2005, Photonic quasicrystals for nonlinear optical frequency conversion, *Phys. Rev. Lett.*, **95**(13), 13390.

[44] Marn W and Megens M, 2005, Experimental measurement of the photonic properties of icosahedral quasicrystals, *Nature*, **436**(7053), 993–996.

[45] Kohmoto M and Sutherland B, 1986, Electronic states on a Penrose lattice, *Phys. Rev. Lett.*, **56**(25), 2740–2743.

[46] Sutherland B, 1986, Simple system with quasiperiodic dynamics: a spin in a magnetic field, *Phys. Rev.* B, **34**(8), 5208–5211.

[47] Fujiwara T and Yokokawa T, 1991, Universal pseudo-gap at Fermi energy in quasicrystals, *Phys. Rev. Lett.*, **66**(3), 333–336.

[48] Fujiwara T and Tsunetsugu H, 1991, Electronic structure and transport, in *Quasicrystals, The States of the Art*, ed by Di-Vincenzo D P and Steinhardt, P J, World Scientific, Singapore, pp. 343–360.

[49] Fujiwara T, 1999, Theory of electronic structure in quasicrystals, in *Physical Properties of Quasicrystals*, ed by Stadnik Z M, Springer-Verlag, Berlin, pp. 169–208.

[50] Ishii Y and Fujiwara T, 2008, Electronic structures and stability, mechanism of quasicrystals, in *Quasicrystals — Handbook of Metal Physics*, ed by Fujiwara T and Ishii Y, Elsevier, Amsterdam, pp. 170–208.

[51] Stadnik Z M, 1999, Spectroscopic studies of the electronic structure, in *Physical Properties of Quasicrystals*, ed by Stadnik Z M, Springer-Verlag, Berlin, pp. 257–293.

[52] Quilichini M and Janssen T, 1997, Phonon excitations in quasicrystals, *Rev. Mod. Phys.*, **69**(1), 277–314.

[53] Casdagli M, 1986, Symbolic dynamics for the renormalization map of a quasiperiodic Schroedinger equation, *Commun. Math. Phys.* **107**(2), 295–318.

[54] Sueto A, 1987, The spectrum of a quasiperiodic Schroedinger operator, *Comm. Math. Phys.*, **111**(3), 409–415.

[55] Kotani S, 1989, Jacobi matrices with random potentials taking finitely many values, *Rev. Math. Phys.*, **1**(1), 129–133.

[56] Bellissard J, Lochum B, Scoppola E and Testart D, 1989, Spectral properties of one dimensional quasicrystals, *Commun. Math. Phys.*, **125**(3), 527–543.

[57] Bovier A and Ghez J-M, 1993, Spectrum properties of one-dimensional Schroedinger operators with potentials generated by substitutions, *Commun. Math. Phys.*, **158**(1), 45–66.

[58] Liu Q H, Tan B, Wen Z X and Wu J, 2002, Measure zero spectrum of a class of Schroedinger operators, *J. Statist. Phys.*, **106**(3–4), 681–691.

[59] Lenz D, 2002, Singular spectrum of Lebesgue measure zero for one-dimensional quasicrystals, *Comm. Math. Phys.*, **227**(1), 119–130.

[60] Furman A, 1997, On the multiplicative ergodic theorem for uniquely ergodic systems, *Ann. Inst. H. Poincare Probab. Statist.*, **33**(6), 797–815.

[61] Fukamichi K, 1999, Magnetic properties of quasicrystals, in *Physical Properties of Quasicrystals*, ed by Stadnik Z M, Springer-Verlag, Heidelberg, Chapter 9, pp. 295–326.

[62] O'handley R C, Dunlap R A and McHenry M E, 1991, *Handbook of Magnetic Materials*, Elsevier, Amsterdam, Vol. 6, pp. 543–510.

[63] Zeng X, Ungar G, Liu Y, Percec V, Dulcey A E and Hobbs J K, 2004, Supermolecular dentritic liquid quasicrystals, *Nature*, **428**, 157–160.

[64] Talapin V D, Shevechenko E V, Bodnarchuk M I, Ye X C, Chen J and Murray C B, 2009, Quasicrystalline order in self-assembled binary nanoparticle superlattices, *Nature*, **461**, 964–967.

[65] Fischer S, Exner A, Zielske K, Perlich J, Deloudi S, Steuer W, Linder P and Foestor S, 2011, Colloidal quasicrystals with 12-fold and 18-fold symmetry, *Proc. Nat. Ac. Sci.*, **108**, 1810–1814.

[66] Takano K, 2005, A mesoscopic Archimedian tiling having a complexity in polymeric stars, *J. Polym. Sci. Pol. Phys.*, **43**, 2427–2432.

[67] Yue K, Huang M J, Marson R, He J L, Huang J H, Zhou Z, Liu C, Yan X S, Wu K, Wang J, Guo Z H, Liu H, Zhang W, Ni P H, Wesdemiotis C, Wen-Bin Zhang W B, Sharon, Glotzer S C and Cheng S Z D, 2016, Geometry induced sequence of nanoscale Frank-Kasper and quasicrystal mesophases in giant surfactants, *Proc. Nat. Ac. Sci.*, **113**, 1392–1400.

[68] Liu Y, Liu T, Yan X Y, Guo Q Y, Lei H, Huang Z, Zhang R, Wang Y, Wang J, Liu F, Bian F, Meijer E W, Aida T, Huang M and Cheng S Z D, 2022, Expanding quasiperiodicity in soft matter: Supramolecular decagonal quasicrystals by binary giant molecule blends, *Proc. Nat. Ac. Sci.*, **119**, e2115304119.

[69] Hu C Z, Ding D H, Yang W G and Wang R H, 1994, Possible two-dimensional quasicrystal structures, *Phys. Rev. B.*, **49**, 9423–9427.

Chapter 4

The Physical Basis of Elasticity of Solid Quasicrystals

4.1 Physical Basis of Elasticity of Quasicrystals

Solid quasicrystal is becoming a kind of functional and structural material, presenting prospective engineering applications. As a kind of material, it can be deformed under the action of applied forces, thermal field or inner effects, etc. In Chapter 2, the deformation of crystals has been discussed. Then, what is the rule for the deformation of solid quasicrystals? Which characters will be exhibited in the deformation process? And so on. To answer these questions, it is necessary to search a physical background of elasticity of quasicrystals. The study about this was put forward immediately after the discovery of the new solid phase.

Because quasicrystal observed in binary and ternary alloys is a new structure of solid, to its elasticity theoretical physicists put forward some different descriptions. The majority of the scientists confirm that the physical basis of elasticity of quasicrystals is the Landau density wave theory (refer to Refs. [1–25]), in which Refs. [1, 2] were published by outstanding theoretical physicist P Bak, immediately after the work of Shechtman *et al.* and Levine and Steinhardt. Bak's work is significant not only to the elasticity of quasicrystals, but also for the whole field of the new phase, since he pointed out and emphasized that the Landau symmetry breaking and elementary excitation principle are the key point of the study on quasicrystals. He was an expert studying incommensurate phase and phason physics. The quasicrystals are one type of incommensurate phase, and physical quantity phason plays a very important role in the phase. He introduced these important concepts into the quasicrystals study for the first time. The phason becomes one of the elementary excitations in quasicrystals, and he developed the Landau theory in this respect. Due to some difficulties, the relevant theory will be introduced in the Appendix of this chapter (i.e., Section 4.9). The results of the description widely acknowledged are that there are two displacement fields $\mathbf{u}$ and $\mathbf{w}$, in which the former can be understood to be similar to the one appearing in crystals, named phonon field according to the physical terminology, and its macro-mechanical behaviour is discussed in Chapter 2, and the latter is a new displacement field named phason field. The total displacement field for quasicrystals can be expressed by

$$\bar{\mathbf{u}} = \mathbf{u}^{\parallel} \oplus \mathbf{u}^{\perp} = \mathbf{u} \oplus \mathbf{w} \tag{4.1-1}$$

where $\oplus$ represents the direct sum. According to the explanation of physicists, $\mathbf{u}$ is in the physical space, or the parallel space $E_{\parallel}^{3}$, while $\mathbf{w}$ is in the complement space, or perpendicular space $E_{\perp}^{3}$, which is an internal space.

One further indicates that these two displacement vectors are dependent only upon the coordinate vector $\mathbf{r}^{\parallel}$ in physical space, i.e.,

$$\mathbf{u} = \mathbf{u}(\mathbf{r}^{\parallel}), \quad \mathbf{w} = \mathbf{w}(\mathbf{r}^{\parallel}) \tag{4.1-2}$$

For simplicity, the superscript of $\mathbf{r}^{\parallel}$ will be removed afterwards. From the angle of mathematical theory of elasticity of quasicrystals and its technological applications, formulas (4.1-1) and (4.1-2) are enough for understanding the following contents within Chapters 1–15. If readers are interested in further physical background on the phonon and phason fields in quasicrystals, we suggest that they could read the Appendix of this chapter (i.e., Section 4.9).

With basic formulas (4.1-1) and (4.1-2) and some fundamental conservation laws well known in physics, the macroscopic basis of the continuous medium model of elasticity of solid quasicrystals can be set up, to some extent, the discussion is an extension to that in Chapter 2, which will be done in the following sections.

4.2 Deformation Tensors

In Chapter 2, we introduced that the deformation of phonon field lies in the relative displacement (i.e., the rigid translation and rotation do not result in deformation), which can be expressed by

$$d\mathbf{u} = \mathbf{u}' - \mathbf{u}$$

If we set up an orthogonal coordinate system (x_1, x_2, x_3) or (x, y, z), then we have

$$\mathbf{u} = (u_x, u_y, u_z) = (u_1, u_2, u_3)$$

and

$$du_i = \frac{\partial u_i}{\partial x_j} dx_j \tag{4.2-1}$$

in which $\partial u_i / \partial x_j$ has the meaning of the gradient of vector $\mathbf{u}$. In some publications, one denotes

$$\nabla \mathbf{u} = \frac{\partial u_i}{\partial x_j} = \begin{bmatrix} \dfrac{\partial u_x}{\partial x} & \dfrac{\partial u_x}{\partial y} & \dfrac{\partial u_x}{\partial z} \\ \dfrac{\partial u_y}{\partial x} & \dfrac{\partial u_y}{\partial y} & \dfrac{\partial u_y}{\partial z} \\ \dfrac{\partial u_z}{\partial x} & \dfrac{\partial u_y}{\partial y} & \dfrac{\partial u_y}{\partial z} \end{bmatrix} \tag{4.2-2}$$

and

$$
\begin{bmatrix}
\frac{\partial u_x}{\partial x} & \frac{\partial u_x}{\partial y} & \frac{\partial u_x}{\partial z} \\
\frac{\partial u_y}{\partial x} & \frac{\partial u_y}{\partial y} & \frac{\partial u_y}{\partial z} \\
\frac{\partial u_z}{\partial x} & \frac{\partial u_z}{\partial y} & \frac{\partial u_z}{\partial z}
\end{bmatrix}
$$

$$
= \begin{bmatrix}
\frac{\partial u_x}{\partial x} & \frac{1}{2}\left(\frac{\partial u_x}{\partial y} + \frac{\partial u_y}{\partial x}\right) & \frac{1}{2}\left(\frac{\partial u_x}{\partial z} + \frac{\partial u_z}{\partial x}\right) \\
\frac{1}{2}\left(\frac{\partial u_x}{\partial y} + \frac{\partial u_y}{\partial x}\right) & \frac{\partial u_y}{\partial y} & \frac{1}{2}\left(\frac{\partial u_y}{\partial z} + \frac{\partial u_z}{\partial y}\right) \\
\frac{1}{2}\left(\frac{\partial u_x}{\partial z} + \frac{\partial u_z}{\partial x}\right) & \frac{1}{2}\left(\frac{\partial u_y}{\partial z} + \frac{\partial u_z}{\partial y}\right) & \frac{\partial u_z}{\partial z}
\end{bmatrix}
$$

$$
+ \begin{bmatrix}
0 & -\frac{1}{2}\left(\frac{\partial u_y}{\partial x} - \frac{\partial u_x}{\partial y}\right) & -\frac{1}{2}\left(\frac{\partial u_z}{\partial x} - \frac{\partial u_x}{\partial z}\right) \\
-\frac{1}{2}\left(\frac{\partial u_x}{\partial y} - \frac{\partial u_y}{\partial x}\right) & 0 & -\frac{1}{2}\left(\frac{\partial u_z}{\partial y} - \frac{\partial u_y}{\partial z}\right) \\
-\frac{1}{2}\left(\frac{\partial u_x}{\partial z} - \frac{\partial u_z}{\partial x}\right) & -\frac{1}{2}\left(\frac{\partial u_y}{\partial z} - \frac{\partial u_z}{\partial y}\right) & 0
\end{bmatrix}
$$

$$
= \frac{1}{2}\left(\frac{\partial u_i}{\partial x_j} + \frac{\partial u_j}{\partial x_i}\right) - \frac{1}{2}\left(\frac{\partial u_j}{\partial x_i} - \frac{\partial u_i}{\partial x_j}\right) = \varepsilon_{ij} + \omega_{ij}
$$

$$
\varepsilon_{ij} = \frac{1}{2}\left(\frac{\partial u_i}{\partial x_j} + \frac{\partial u_j}{\partial x_i}\right) \tag{4.2-3}
$$

$$
\omega_{ij} = \frac{1}{2}\left(\frac{\partial u_i}{\partial x_j} - \frac{\partial u_j}{\partial x_i}\right) \tag{4.2-4}
$$

this means the gradient of phonon vector **u** can be decomposed into two parts ε_{ij} and ω_{ij}, in which ε_{ij} has contribution to deformation energy and ω_{ij} represents a kind of rigid rotations. We consider only ε_{ij}, which is the phonon deformation tensor, called strain tensor, and a symmetric tensor: $\varepsilon_{ij} = \varepsilon_{ji}$.

For one-, the first kind of two- and three-dimensional quasicrystals, there is one phason field **w**, and for the second kind of two-dimensional quasicrystals, there are two phason fields **w** and **v**, respectively, as pointed out in Chapter 3. In this chapter, we discuss the problem of one phason field, i.e., the field **w**, while the problem of two phason fields **w** and **v** will be discussed in Chapter 17.

Similarly, for phason field **w**, we have

$$dw_i = \frac{\partial w_i}{\partial x_j} dx_j \tag{4.2-5}$$

and

$$\nabla \mathbf{w} = \frac{\partial w_i}{\partial x_j} = \begin{bmatrix} \frac{\partial w_x}{\partial x} & \frac{\partial w_x}{\partial y} & \frac{\partial w_x}{\partial z} \\ \frac{\partial w_y}{\partial x} & \frac{\partial w_y}{\partial y} & \frac{\partial w_y}{\partial z} \\ \frac{\partial w_z}{\partial x} & \frac{\partial w_z}{\partial y} & \frac{\partial w_z}{\partial z} \end{bmatrix} \tag{4.2-6}$$

Though it can be decomposed into symmetric and asymmetric parts, all components $\frac{\partial w_i}{\partial x_j}$ contribute to the deformation of quasicrystals. The phason deformation tensor, or phason strain tensor, is defined by

$$w_{ij} = \frac{\partial w_i}{\partial x_j} \tag{4.2-7}$$

which describes the local rearrangement of atoms in a cell, and is a asymmetric tensor $w_{ij} \neq w_{ji}$.

The difference between ε_{ij} and w_{ij} given by (4.2-3) and (4.2-7) originated from the physical properties of phonon modes and phason modes. This can also be explained by group theory, i.e., they follow different irreducible representations for some symmetry transformations for most quasicrystal systems, except the three-dimensional cubic quasicrystal system. The detail about this is omitted here.

For the three-dimensional cubic quasicrystals, the phason modes exhibit the same behaviour like that of phonon modes, which will be particularly discussed in Chapter 9.

4.3 Stress Tensors and Equations of Motion

The gradient of displacement field $\mathbf{w}$ figures out the local rearrangement of atoms in a cell in quasicrystals. It needs external forces to drive the atoms through barriers when they make the local rearrangement in a cell such that there are another kind of body forces and tractions apart from the conventional body forces $\mathbf{f}$ and tractions $\mathbf{T}$ for deformed quasicrystals, which are named the generalized body forces (density) $\mathbf{g}$ and generalized tractions (the generalized area forces density) $\mathbf{h}$.

At first, we consider the static case.

Denoting the stress tensor corresponding to ε_{ij} by σ_{ij}, called the phonon stress tensor, and that to w_{ij} by H_{ij}, the phason stress tensor, we have the following equilibrium equations:

$$\left.\begin{aligned} \frac{\partial \sigma_{ij}}{\partial x_j} + f_i = 0 \\ \frac{\partial H_{ij}}{\partial x_j} + g_i = 0 \end{aligned}\right\} (x, y, z) \in \Omega \tag{4.3-1}$$

based on the momentum conservation law.

From the angular momentum conservation law to the phonon field

$$\frac{d}{dt}\int_{\Omega} \mathbf{r}^{\|} \times \rho \dot{\mathbf{u}} d\Omega = \int_{\Omega} \mathbf{r}^{\|} \times \mathbf{f} d\Omega + \int_{\Omega} \mathbf{r}^{\|} \times \mathbf{T} d\Gamma \tag{4.3-2}$$

and by using the Gauss theorem, it follows that

$$\sigma_{ij} = \sigma_{ji} \tag{4.3-3}$$

this indicates that the phonon stress tensor is symmetric.

Since $\mathbf{r}^{\|}$ and $\mathbf{w}(\mathbf{g}, \mathbf{h})$ transform under different representations of the point groups, more precisely that the former transforms like a vector but the latter does not, the product representation $\mathbf{r}^{\|} \times \mathbf{w}$, $\mathbf{r}^{\|} \times \mathbf{g}$ and $\mathbf{r}^{\|} \times \mathbf{h}$ do not contain any vector representations. This implies that for the phason field, there is no equation analogous to (4.3-2), from which it follows that, generally,

$$H_{ij} \neq H_{ji} \tag{4.3-4}$$

The result holds for all quasicrystal systems except the case for three-dimensional cubic quasicrystals.

In dynamic case, the deformation process is quite complicated and there are different arguments. Lubensky *et al.* [5] claimed that phonon modes and phason modes are different based on their role in six-dimensional hydrodynamics, and phonons are wave propagation while phasons are diffusive with very large diffusive time. Physically, the phason modes represent a relative motion of the constituent density waves. Dolinsek *et al.* [22, 23] further developed the point of view of Lubensky *et al.* and argued the atom flip or atom hopping concept for the phason dynamics. But according to Bak [1, 2], the phason describes particular structural disorders or structure fluctuations in quasicrystals, and it can be formulated based on a six-dimensional space description. Since there are six continuous symmetries, there exist six hydrodynamic vibration modes. In the following, we give a brief introduction on elastodynamics based on Bak's argument as well as the argument of Lubensky *et al.*

Ding *et al.* [26] and Hu *et al.* [16] derived that

$$\left.\begin{aligned}\frac{\partial \sigma_{ij}}{\partial x_j} + f_i = \rho\frac{\partial^2 u_i}{\partial t^2}\\ \frac{\partial H_{ij}}{\partial x_j} + g_i = \rho\frac{\partial^2 w_i}{\partial t^2}\end{aligned}\right\}(x, y, z) \in \Omega, \quad t > 0 \tag{4.3-5}$$

based on the momentum conservation law. We believe that the derivation is carried out by Bak's argument, in which ρ is the mass density of quasicrystals.

If according to the argument of Lubensky *et al.* people cannot obtain (4.3-5), instead

$$\left.\begin{aligned}\frac{\partial \sigma_{ij}}{\partial x_j} + f_i = \rho\frac{\partial^2 u_i}{\partial t^2}\\ \frac{\partial H_{ij}}{\partial x_j} + g_i = \kappa\frac{\partial w_i}{\partial t}\end{aligned}\right\}(x, y, z) \in \Omega, \quad t > 0 \tag{4.3-6}$$

in which $\kappa = 1/\Gamma_w$, where Γ_w is the kinetic coefficient of phason field. The equations are given by Fan *et al.* [27] and Rochal and Norman [28], which are identical to those given by Lubensky *et al.* [5] for linear case and omitting velocity field due to the viscosity field. Lubensky *et al.* gave their hydrodynamics formulation based on conservation law and symmetry breaking principle, so Eq. (4.3-6) may be seen as the elasto-/hydro-dynamic equation of quasicrystals. In particular, the second equation of (4.3-6) presents the dissipation feature of motion of phason degrees in dynamic process, and it is irreversible thermodynamically.

The dynamics of quasicrystals faces a great challenge, see, e.g., [28–30], and we discuss it in Chapters 8, 10 and 16 in the text and Appendix III in the Major Appendices in detail.

4.4 Free Energy and Elastic Constants

Consider the free energy or the strain energy density of a quasicrystal $F(\varepsilon_{ij}, w_{ij})$ whose general expression is difficult to obtain. We take a Taylor expansion in the neighbourhood of $\varepsilon_{ij} = 0$ and $w_{ij} = 0$, and remain up to the second-order term, then

$$\begin{aligned} F(\varepsilon_{ij}, w_{ij}) &= \frac{1}{2}\left[\frac{\partial^2 F}{\partial\varepsilon_{ij}\partial\varepsilon_{kl}}\right]_0 \varepsilon_{ij}\varepsilon_{kl} + \frac{1}{2}\left[\frac{\partial^2 F}{\partial\varepsilon_{ij}\partial w_{kl}}\right]_0 \varepsilon_{ij}w_{kl} \\ &\quad + \frac{1}{2}\left[\frac{\partial^2 F}{\partial w_{ij}\partial w_{kl}}\right]_0 w_{ij}w_{kl} + \frac{1}{2}\left[\frac{\partial^2 F}{\partial w_{ij}\partial\varepsilon_{kl}}\right]_0 w_{ij}\varepsilon_{kl} \\ &= \frac{1}{2}C_{ijkl}\varepsilon_{ij}\varepsilon_{kl} + \frac{1}{2}R_{ijkl}\varepsilon_{ij}w_{kl} + \frac{1}{2}K_{ijkl}w_{ij}w_{kl} \\ &\quad + \frac{1}{2}R'_{ijkl}w_{ij}\varepsilon_{kl} \\ &= F_u + F_w + F_{uw} \end{aligned} \tag{4.4-1}$$

where F_u, F_w and F_{uw} denote the parts contributed by phonon, phason and phonon–phason coupling respectively and

$$C_{ijkl} = \left[\frac{\partial^2 F}{\partial\varepsilon_{ij}\partial\varepsilon_{kl}}\right]_0 \tag{4.4-2}$$

is the phonon elastic tensor, discussed in Chapter 2 already, and

$$C_{ijkl} = C_{klij} = C_{jikl} = C_{ijlk} \tag{4.4-3}$$

the tensor can be expressed by a symmetric matrix

$$[C]_{9\times 9}$$

In (4.4-1), another elastic constant tensor

$$K_{ijkl} = \left[\frac{\partial^2 F}{\partial w_{ij} \partial w_{kl}} \right]_0 \tag{4.4-4}$$

in which the suffixes j, l belong to space $E_{\|}^3$ and i, k to space $E_{\perp}^3$, and

$$K_{ijkl} = K_{klij} \tag{4.4-5}$$

All components of K_{ijkl} can also be expressed by symmetric matrix

$$[K]_{9\times9}$$

In addition,

$$R_{ijkl} = \left[\frac{\partial^2 F}{\partial \varepsilon_{ij} \partial w_{kl}} \right]_0 \tag{4.4-6}$$

$$R'_{ijkl} = \left[\frac{\partial^2 F}{\partial w_{ij} \partial \varepsilon_{kl}} \right]_0 \tag{4.4-7}$$

are the elastic constants of phonon–phason coupling, and it is to be noted that the suffixes i, j, l belong to space $E_{\|}^3$ and k belongs to space $E_{\perp}^3$, and

$$R_{ijkl} = R_{jikl}, \quad R'_{ijkl} = R_{klij}, \quad R'_{klij} = R_{ijkl} \tag{4.4-8}$$

but

$$R_{ijkl} \neq R_{klij}, \quad R'_{ijkl} \neq R'_{klij} \tag{4.4-9}$$

all components of which can be expressed in symmetric matrices

$$[R]_{9\times9}, \quad [R']_{9\times9}$$

and

$$[R]^T = [R'] \tag{4.4-10}$$

where the superscript T denotes the transpose operator. The composition of four matrixes $[C], [K], [R]$ and $[R']$ forms a matrix with

18×18

$$[C, K, R] = \begin{bmatrix} [C] & [R] \\ [R'] & [K] \end{bmatrix} = \begin{bmatrix} [C] & [R] \\ [R]^T & [K] \end{bmatrix} \tag{4.4-11}$$

If the strain tensor is expressed by a row vector with 18 elements, i.e.,

$$[\varepsilon_{ij}, w_{ij}] = \begin{bmatrix} \varepsilon_{11}, \varepsilon_{22}, \varepsilon_{33}, \varepsilon_{23}, \varepsilon_{31}, \varepsilon_{12}, \varepsilon_{32}, \varepsilon_{13}, \varepsilon_{21}, \\ w_{11}, w_{22}, w_{33}, w_{23}, w_{31}, w_{12}, w_{32}, w_{13}, w_{21} \end{bmatrix} \tag{4.4-12}$$

the transpose of which denotes the array vector, then the free energy (or strain energy density) may be expressed by

$$F = \frac{1}{2}[\varepsilon_{ij}, w_{ij}] \begin{bmatrix} [C] & [R] \\ [R]^T & [K] \end{bmatrix} [\varepsilon_{ij}, w_{ij}]^T \tag{4.4-13}$$

which is identical to that given by (4.4-1)

4.5 Generalized Hooke's Law

For the application of theory of elasticity of quasicrystals to any science or engineering problem, one must determine the displacement field and stress field. This requires that we need to set up the relationship between strains and stresses, which are called the generalized Hooke's law of quasicrystalline material. From the free energy (4.4-1) or (4.4-13), we have

$$\begin{aligned} \sigma_{ij} &= \frac{\partial F}{\partial \varepsilon_{ij}} = C_{ijkl}\varepsilon_{kl} + R_{ijkl}w_{kl} \\ H_{ij} &= \frac{\partial F}{\partial w_{ij}} = K_{ijkl}w_{kl} + R_{klij}\varepsilon_{kl} \end{aligned} \tag{4.5-1}$$

or in the form of matrices

$$\begin{bmatrix} \sigma_{ij} \\ H_{ij} \end{bmatrix} = \begin{bmatrix} [C] & [R] \\ [R]^T & [K] \end{bmatrix} \begin{bmatrix} \varepsilon_{ij} \\ w_{ij} \end{bmatrix} \tag{4.5-2}$$

where

$$\begin{aligned} \begin{bmatrix} \sigma_{ij} \\ H_{ij} \end{bmatrix} &= [\sigma_{ij}, H_{ij}]^T \\ \begin{bmatrix} \varepsilon_{ij} \\ w_{ij} \end{bmatrix} &= [\varepsilon_{ij}, w_{ij}]^T \end{aligned} \tag{4.5-3}$$

4.6 Boundary Conditions and Initial Conditions

The above general formulae give a description of the basic law of elasticity of quasicrystals and provide a key to solve those problems in application for academic research and engineering practice. The formulae hold in any interior of the body, i.e., $(x, y, z) \in \Omega$ where (x, y, z) denote the coordinates of any point of the interior and Ω the body. The formulae are concluded as some partial differential equations. To solve them, it is necessary to know the situation of the field variables at the boundary Γ of Ω. Without appropriate information at the boundary, the solution has no physical meaning. According to practical case, the boundary Γ consists of two parts Γ_t and Γ_u, i.e., $\Gamma = \Gamma_t + \Gamma_u$; at Γ_t, the tractions are given and at Γ_u, the displacements are prescribed. For the former case,

$$\left. \begin{aligned} \sigma_{ij} n_j &= T_i \\ H_{ij} n_j &= h_i \end{aligned} \right\} (x, y, z) \in \Gamma_t \tag{4.6-1}$$

where n_j represents the unit outward normal vector at any point at Γ and T_i and h_i the traction and generalized traction vectors, which are given functions at the boundary. Formula (4.6-1) is called the stress boundary conditions. And for the latter case,

$$\left. \begin{aligned} u_i &= \bar{u}_i \\ w_i &= \bar{w}_i \end{aligned} \right\} (x, y, z) \in \Gamma_u \tag{4.6-2}$$

where $\bar{u}_i$ and $\bar{w}_i$ are known functions at the boundary. Formula (4.6-2) is named the displacement boundary conditions.

If $\Gamma = \Gamma_t$ (i.e., $\Gamma_u = 0$), the problem for solving Eqs. (4.2-3), (4.2-7), (4.3-1) and (4.5-1) under boundary conditions (4.6-1) is called stress boundary value problem. While $\Gamma = \Gamma_u$ (i.e., $\Gamma_t = 0$), the problem for solving Eqs. (4.2-3), (4.2-7), (4.3-1) and (4.5-1) under boundary conditions (4.6-2) is called displacement boundary value problem.

If $\Gamma = \Gamma_u + \Gamma_t$ and both $\Gamma_t \neq 0$ and $\Gamma_u \neq 0$, the problem for solving Eqs. (4.2-3), (4.2-7), (4.3-1) and (4.5-1) under boundary conditions (4.6-1) and (4.6-2) is called mixed boundary value problem.

For dynamic problem, if taking wave equations (4.3-5) collaborating Eqs. (4.2-3), (4.2-7) and (4.5-1), apart from boundary conditions (4.6-1) and (4.6-2), we must give relevant initial value conditions, i.e.,

$$\left.\begin{array}{l} u_i(x,y,z,0) = u_{i0}(x,y,z), \dot{u}_i(x,y,z,0) = \dot{u}_{i0}(x,y,z) \\ w_i(x,y,z,0) = w_{i0}(x,y,z), \dot{w}_i(x,y,z,0) = \dot{w}_{i0}(x,y,z) \end{array}\right\} (x,y,z) \in \Omega \tag{4.6-3}$$

in which $u_{i0}(x,y,z,0)$, $\dot{u}_{i0}(x,y,z,0)$, $w_{i0}(x,y,z,0)$ and $\dot{w}_{i0}(x,y,z,0)$ are known functions and $\dot{u}_i = \frac{\partial u_i}{\partial t}$, etc. In this case, the problem is called initial-boundary value problem.

But if taking wave equations coupling diffusion equations (4.3-6) collaborating (4.2-3) and (4.5-1), then the initial value conditions will be

$$\left.\begin{array}{l} u_i(x,y,z,0) = u_{i0}(x,y,z), \dot{u}_i(x,y,z,0) = \dot{u}_{i0}(x,y,z) \\ w_i(x,y,z,0) = w_{i0}(x,y,z) \end{array}\right\} (x,y,z) \in \Omega \tag{4.6-4}$$

this is also called initial-boundary value problem but is different from the previous one.

4.7 A Brief Introduction on Relevant Material Constants of Quasicrystals

In the above discussion, we find that the quasicrystals present different nature from those of crystals. Connection with this, the material constants of the solid should be different from those of crystals and other conventional structural materials, in which the constants that appeared in the above basic equations are interesting at least. We here give a brief introduction to help readers in the conceptual point of view. A detailed introduction will be given in Chapters 6, 9 and 10.

The measurement of material constants of quasicrystals is difficult, but the experimental technique is progressed, especially in recent years. Due to the majority of icosahedral and decagonal quasicrystals in the material, the measured data are most for these two kinds of the solid phase.

For icosahedral quasicrystals, the independent non-zero components of phonon elastic constants C_{ij} are only λ and μ, phason elastic constants K_{ij} only K_1 and K_2, and phonon–phason coupling elastic constants R_{ij} only R. For the most important icosahedral Al–Pd–Mn quasicrystal, the measured data including the mass density and kinetic coefficient of phason are as follows [31, 32], respectively:

$$\begin{aligned}
&\rho = 5.1\,\mathrm{g/cm^3}, \quad \lambda = 74.9, \quad \mu = 72.4(\mathrm{GPa}), \quad K_1 = 72,\\
&K_2 = -37(\mathrm{MPa}), \quad R \approx 0.01\mu,\\
&\Gamma_w = 4.8 \times 10^{-19}\mathrm{m^3 \cdot s/kg} = 4.8 \times 10^{-10}\mathrm{cm^3 \cdot \mu s/g}
\end{aligned}$$

and for two-dimensional quasicrystals, the independent non-zero components of the phonon elastic constants are only C_{11}, C_{33}, C_{44}, C_{12}, C_{13} and $C_{66} = (C_{11} - C_{12})/2$, the phason ones only K_1, K_2, K_3, and the phonon–phason coupling ones only R_1, R_2. For decagonal Al–Ni–Co quasicrystal, the measured data are as follows [31], respectively:

$$\begin{aligned}
&\rho = 4.186\mathrm{g/cm^3}, \quad C_{11} = 234.3, \quad C_{33} = 232.22, \quad C_{44} = 70.19,\\
&C_{12} = 57.41, \quad C_{13} = 66.63\ (\mathrm{GPa})\\
&R_1 = -1.1, |R_2| < 0.2\ (\mathrm{GPa})
\end{aligned}$$

and there are no measured data for K_1, K_2 (but we can use those obtained by the Monte Carlo simulation), and Γ_w can approximately be taken as the value of icosahedral quasicrystal. In addition, the tensile strength $\sigma_c = 450\,\mathrm{MPa}$ for decagonal Al–Cu–Co quasicrystals before annealing and $\sigma_c = 550\,\mathrm{MPa}$ after annealing. The hardness for decagonal A–Cu–Co quasicrystals is 4.10 GPa [33, 34], the fracture toughness of which is 1.0–1.2 $\mathrm{MPa}\sqrt{m}$ [33].

With these basic data, the computation for stress analysis for statics and dynamics can be undertaken.

4.8 Summary and Mathematical Solvability of Boundary Value or Initial-boundary Value Problem

For static equilibrium problems, the mathematical formulation is

$$\frac{\partial \sigma_{ij}}{\partial x_j} + f_i = 0, \quad \frac{\partial H_{ij}}{\partial x_j} + h_i = 0, \quad x_i \in \Omega \tag{4.8-1}$$

$$\varepsilon_{ij} = \frac{1}{2}\left(\frac{\partial u_i}{\partial x_j} + \frac{\partial u_j}{\partial x_i}\right), \quad w_{ij} = \frac{\partial w_i}{\partial x_j}, \quad x_i \in \Omega \tag{4.8-2}$$

$$\left.\begin{aligned} \sigma_{ij} &= C_{ijkl}\varepsilon_{kl} + R_{ijkl}w_{kl} \\ H_{ij} &= K_{ijkl}w_{kl} + R_{klij}\varepsilon_{kl} \end{aligned}\right\}, \quad x_i \in \Omega \tag{4.8-3}$$

$$\sigma_{ij}n_j = T_i, \quad H_{ij}n_j = h_i, \quad x_i \in \Gamma_t \tag{4.8-4}$$

$$u_i = \bar{u}_i, \quad w_i = \bar{w}_i, \quad x_i \in \Gamma_u \tag{4.8-5}$$

For dynamic problems, based on Bak's argument, the mathematical formulation is

$$\frac{\partial \sigma_{ij}}{\partial x_j} + f_i = \rho\frac{\partial^2 u_i}{\partial t^2}, \quad \frac{\partial H_{ij}}{\partial x_j} + g_i = \rho\frac{\partial^2 w_i}{\partial t^2}, \quad x_i \in \Omega, \quad t > 0 \tag{4.8-6}$$

$$\varepsilon_{ij} = \frac{1}{2}\left(\frac{\partial u_i}{\partial x_j} + \frac{\partial u_j}{\partial x_i}\right), \quad w_{ij} = \frac{\partial w_i}{\partial x_j}, \quad x_i \in \Omega, \quad t > 0 \tag{4.8-7}$$

$$\left.\begin{aligned} \sigma_{ij} &= C_{ijkl}\varepsilon_{kl} + R_{ijkl}w_{kl} \\ H_{ij} &= K_{ijkl}w_{kl} + R_{klij}\varepsilon_{kl} \end{aligned}\right\}, \quad x_i \in \Omega, \quad t > 0 \tag{4.8-8}$$

$$\sigma_{ij}n_j = T_i, \quad H_{ij}n_j = h_i, \quad x_i \in \Gamma_t, \quad t > 0 \tag{4.8-9}$$

$$u_i = \bar{u}_i, \quad w_i = \bar{w}_i, \quad x_i \in \Gamma_u, \quad t > 0 \tag{4.8-10}$$

$$u_i|_{t=0} = u_{i0}, \quad \dot{u}_i|_{t=0} = \dot{u}_{i0}, \quad w_i|_{t=0} = w_{i0}, \quad \dot{w}_i|_{t=0} = \dot{w}_{i0}, \quad x_i \in \Omega \tag{4.8-11}$$

For a dynamic problem, based on the argument of Lubensky *et al.*, the mathematical formulation is

$$\frac{\partial \sigma_{ij}}{\partial x_j} + f_i = \rho\frac{\partial^2 u_i}{\partial^2 t}, \quad \frac{\partial H_{ij}}{\partial x_j} + h_i = \kappa\frac{\partial w_i}{\partial t},$$

$$\kappa = \frac{1}{\Gamma_w}, \quad x_i \in \Omega, \quad t > 0 \tag{4.8-12}$$

$$\varepsilon_{ij} = \frac{1}{2}\left(\frac{\partial u_i}{\partial x_j} + \frac{\partial u_j}{\partial x_i}\right), \quad w_{ij} = \frac{\partial w_i}{\partial x_j}, \quad x_i \in \Omega, \quad t > 0 \quad (4.8\text{-}13)$$

$$\left.\begin{array}{l} \sigma_{ij} = C_{ijkl}\varepsilon_{kl} + R_{ijkl}w_{kl} \\ H_{ij} = K_{ijkl}w_{kl} + R_{klij}\varepsilon_{kl} \end{array}\right\}, \quad x_i \in \Omega, \quad t > 0 \quad (4.8\text{-}14)$$

$$\sigma_{ij}n_j = T_j, \quad H_{ij}n_j = h_i, \quad x_i \in \Gamma_t, \quad t > 0 \quad (4.8\text{-}15)$$

$$u_i = \bar{u}_i, \quad w_i = \bar{w}_i, \quad x_i \in \Gamma_u, \quad t > 0 \quad (4.8\text{-}16)$$

$$u_i|_{t=0} = u_{i0}, \quad \dot{u}_i|_{t=0} = \dot{u}_{i0}, \quad w_i|_{t=0} = w_{i0}, \quad x_i \in \Omega \quad (4.8\text{-}17)$$

The solution satisfying all equations and corresponding initial conditions and boundary conditions is just the realistic solution of elasticity of quasicrystals mathematically and has physical meaning.

The existence and uniqueness of solutions of elasticity of quasicrystals will be further discussed in Chapter 13.

4.9 Appendix A: Description on Physical Basis of Elasticity of Quasicrystals Based on the Landau Density Wave Theory

In Section 4.1, we gave formula (4.1-1) as the physical basis of elasticity of quasicrystals and did not discuss its profound physical source because the discussion is concerned with quite complicated background, which is not the most necessary part for a beginner before reading Sections 4.2–4.8. After reading the previous sections, it may be a benefit to turn considering the profound physical background. We suggest that the reader can read Section 1.5 (Appendix of Chapter 4) first.

Chapter 3 shows quasicrystals that belong to the subject of condensed matter physics rather than traditional solid-state physics, though the former is evolved from the latter. In the development, the symmetry breaking (or broken symmetry) forms the core concept and principle of the condensed matter physics. According to the understanding of physicists that the Landau density wave description on the elasticity of quasicrystals is a natural choice, though there are some other descriptions, e.g., the unit cell description based on the Penrose tiling. Now, the difficulty lies in that the readers in other disciplines are not so familiar with the Landau theory and the relevant topics. For this reason, we have introduced the Landau theory in Appendix of Chapter 1 (i.e., Section 1.5, in which the concept of incommensurate crystals is also introduced in brief, which will be concerned though it is not related to Landau theory) and the Penrose tiling in Section 3.1. These important physical and mathematical results can help us understand the elasticity of quasicrystals.

Immediately after the discovery of quasicrystals, Bak [1] published the theory of elasticity in which he used the three important results in physics and mathematics mentioned above, but the core is the Landau theory on symmetry breaking and elementary excitation of condensed matter. Bak [1, 2] pointed out too, ideally, one would like to explain the structure from first-principle calculations taking into account the actual electronic properties of constituent atoms. Such a calculation is hardly possible to date. So, he suggested that Landau's phenomenological theory [3] on structural transition can be used, i.e., the condensed phase is described by a symmetry-breaking order parameter which transforms as an irreducible representation of the symmetry group of a liquid with full translational and rotational symmetry. According to the Landau theory, the order parameter of quasicrystals is the density wave. For the density of the ordered, a low-temperature d-dimensional quasicrystal can be expressed as a Fourier series by extended formula (1.5-5) (the expansion exists due to the periodicity in lattice or reciprocal lattice of higher-dimensional space)

$$\rho(\mathbf{r}) = \sum_{\mathbf{G}\in L_R} \rho_{\mathbf{G}} \exp\{i\mathbf{G}\cdot\mathbf{r}\} = \sum_{\mathbf{G}\in L_R} |\rho_{\mathbf{G}}| \exp\{-i\Phi_{\mathbf{G}} + i\mathbf{G}\cdot\mathbf{r}\} \tag{4.9-1}$$

where $\mathbf{G}$ is a reciprocal vector L_R is the reciprocal lattice (the concepts on the reciprocal vector and reciprocal lattice, referring to Chapter 1) and $\rho_{\mathbf{G}}$ is a complex number

$$\rho_{\mathbf{G}} = |\rho_{\mathbf{G}}| e^{-i\Phi_{\mathbf{G}}} \tag{4.9-2}$$

with an amplitude $|\rho_{\mathbf{G}}|$ and phase angle $\Phi_{\mathbf{G}}$, due to $\rho(r)$ being real, $|\rho_{\mathbf{G}}| = |\rho_{-\mathbf{G}}|$ and $\Phi_{\mathbf{G}} = -\Phi_{-\mathbf{G}}$.

There exists a set of N base vectors, $\{\mathbf{G}_n\}$, so that each $\mathbf{G} \in L_R$ can be written as $\sum m_n \mathbf{G}_n$ for integers m_n. Furthermore, $N = kd$, where k is the number of the mutually incommensurate vectors in the d-dimensional quasicrystal. In general, $k = 2$. A convenient parametrization of the phase angle is given by

$$\Phi_n = \mathbf{G}_n^{\|} \cdot \mathbf{u} + \mathbf{G}_n^{\perp} \cdot \mathbf{w} \tag{4.9-3}$$

in which $\mathbf{u}$ can be understood similar to the phonon like that in conventional crystals, while $\mathbf{w}$ can be understood as the phason degrees of freedom in quasicrystals, which describe the local rearrangement of unit cell description based on the Penrose tiling. Both are functions of the position vector in the physical space only, where $\mathbf{G}_n^{\|}$ is the reciprocal vector in the physical space $E_{\|}^3$ just mentioned and $\mathbf{G}_n^{\perp}$ is the conjugate vector in the perpendicular space $E_{\perp}^3$. People can realize that the above-mentioned Bak's hypothesis is a natural development of Anderson's theory introduced in Section 1.5.

Almost in the same time, Levine *et al.* [4], Lubensky *et al.* [5–8], Kalugin *et al.* [9], Torian and Mermin [10], Jaric [11], Duneau and Katz [12], Socolar *et al.* [13] and Gahler and Phyner [14] carried out the study on elasticity of quasicrystals. Though the researchers studied the elasticity from different descriptions, e.g., the unit-cell description based on the Penrose tiling is adopted too, but the density wave description based on the Landau phenomenological theory on symmetry breaking of condensed matter has played the central role and been widely acknowledged. This means there are two elementary excitations of low energy, phonon $\mathbf{u}$ and phason $\mathbf{w}$ for quasicrystals, in which vector $\mathbf{u}$ is in the parallel space $E_{\|}^3$ and vector $\mathbf{w}$ is in the perpendicular space $E_{\perp}^3$, respectively, so that total displacement field for quasicrystals is

$$\bar{\mathbf{u}} = \mathbf{u}^{\|} \oplus \mathbf{u}^{\perp} = \mathbf{u} \oplus \mathbf{w}$$

which is formula (4.1-1), where $\oplus$ represents the direct sum.

According to the argument of Bak, etc.,

$$\mathbf{u} = \mathbf{u}(\mathbf{r}^{\|}), \ \ \mathbf{w} = \mathbf{w}(\mathbf{r}^{\|})$$

i.e., $\mathbf{u}$ and $\mathbf{w}$ depend upon special radius vector $\mathbf{r}^{\parallel}$ in parallel space $E_{\parallel}^3$ only, this is formula (4.1-2). For simplicity, the superscript of $\mathbf{r}^{\parallel}$ is removed in the presentation in Sections 4.2–4.8.

Even if introducing $\mathbf{u}$ and $\mathbf{w}$ in such a way, the concept of phason is hard to be accepted by some readers. We would like to give some additional explanation in terms of projection concept as the following.

We previously said that the quasicrystal in three-dimensional space may be seen as a projection of periodic structure in higher-dimensional space. For example, one-dimensional quasicrystals in physical space may be seen as a projection of "periodic crystals" in four-dimensional space, while in the one-dimensional quasicrystals, the atom arrangement is quasiperiodic only in one direction, say z-axis direction, and periodic in the other two directions. The atom arrangement quasiperiodic axis may be seen as a projection of a two-dimensional periodic crystal shown in Fig. 4.1(a), in which dots form

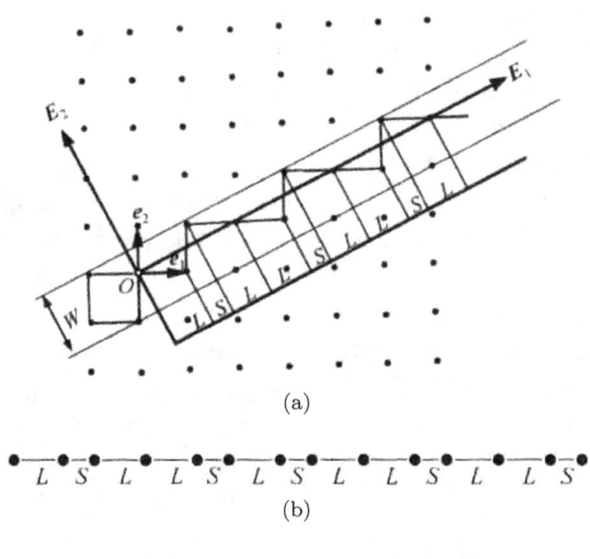

Fig. 4.1. A geometrical representation for one-dimensional quasiperiodic structure. (a) A projection of two-dimensional crystal can generate a one-dimensional quasiperiodic structure. (b) The Fibonacci sequence.

the two-dimensional, e.g., right square crystal, and lines with the slop of irrational numbers can correspond to quasiperiodic structures (by contrast, if the slope is a rational number, it corresponds to the periodic structure). For this purpose, we can use the so-called Fibonacci sequence formed by a longer segment L and a shorter segment S according to the recurrence (whose geometry depiction is shown in Fig. 4.1(b))

$$F_{n+1} = F_n + F_{n-1}$$

and

$$\begin{aligned} &F_0 : S \\ &F_1 : L \\ &F_2 : LS \\ &F_3 : LSL \\ &F_4 : LSLLS \end{aligned}$$

The geometric expression of the sequence can be shown in the axis E_1, i.e., $E_{\|}$ in Fig. 4.1(a), and $E_{\|}$ is the so-called parallel space and that perpendicular to which is the so-called vertical space E_2, i.e., $E_{\perp}$. The Fibonacci sequence is a useful tool to describe the geometry of one-dimensional quasiperiodic structures, like that of the Penrose tiling to describe the geometry of two- and three-dimensional quasicrystals. Figure 4.1 may help us to understand the internal space $E_{\perp}$. For one-dimensional quasicrystals, the figure can give an explicit description, while for two- and three-dimensional quasicrystals, there is no such explicit graph.

Since quasicrystals belong to one of the incommensurate phases, and there are phason modes in the incommensurate crystals, denoted by $\mathbf{w}(\mathbf{r}^{\|})$, which may be understood as the corresponding new displacement field, if people have knowledge on incommensurate phases, then they may easily understand the origin of phason modes in quasicrystal, though conventional incommensurate crystals are not certainly the actual quasicrystals.

The phonon variables $\mathbf{u}(\mathbf{r}^{\|})$ appear in the physical space $E_{\|}^3$, vector $\mathbf{u}$ represents the displacement of lattice point deviated from its equilibrium position due to the vibration of the lattice, and the propagation of this vibration is sound waves in solids. Though vibration is a mechanical motion, which can be quantized, the quanta of this motion are named phonon. So, the **u** field is called phonon field in physical terminology. The gradient of the **u** field characterizes the changes in volume and shape of cells — this is identical to that in the classical elasticity (see, e.g., Chapter 2 and previous sections of this chapter).

As mentioned before, the phason variables are substantively related with structural transitions of alloys, some among which can be observed from the characteristics of diffraction patterns. Lubensky *et al.* [5, 7] and Horn *et al.* [15] discussed the connection between the phenomena and phason strain. These profound observations could not be discussed here, so readers can refer to the review given by Hu *et al.* [16]. This makes us know that the phason modes exist truly. The physical meaning of phason variables can be explained as a quantity to describe the local rearrangement of atoms in a cell. We know that the phase transition in crystalline materials is just arisen by the atomic local rearrangement. The unit cell description on quasicrystals mentioned above predicts that **w** describes the local arrangement of Penrose tiling. These findings may help us to understand the meaning of the unusual field variables. After experimental investigations by neutron scattering [17–20], Mossbauer spectroscopy [21], nuclear magnetic resonance (NMR) [22, 23] and specific heat measurements [24, 25], the concept of thermal-induced phason flips is suggested. This is identical to the diffusive essentiality of phasons, but note that the so-called diffusion here is quite different from that appeared in metallic periodic crystals (which mainly results from the presence of vacancies in the lattice, and the vacancies are not necessary for atomic motion in the quasicrystal structures), which will be discussed in Chapter 10.

We must point out that vector **u** and vector **w** present different behaviour under some symmetry operations. This can be explained by group theory. The discussion is omitted here.

References

[1] Bak P, 1985, Phenomenological theory of icosahedral incommensurate ("quaisiperiodic") order in Mn-Al alloys, *Phys. Rev. Lett.*, **54**(8), 1517–1519.

[2] Bak P, 1985, Symmetry, stability and elastic properties of icosahedral incommensurate crystals, *Phys. Rev.* B, **32**(9), 5764–5772.

[3] Landau L D and Lifshitz E M, 1980, *Theoretical Physics V: Statistical Physics*, 3rd edition, Pregamon Press, New York.

[4] Levine D, Lubensky T C, Ostlund S, Ramaswamy S, Steinhardt P J and Toner J, 1985, Elasticity and dislocations in pentagonal and icosahedral quasicrystals, *Phys. Rev. Lett.*, **54**(8), 1520–1523.

[5] Lubensky T C, Ramaswamy S and Nad Toner J, 1985, Hydrodynamics of icosahedral quasicrystals, *Phys. Rev.* B, **32**(11), 7444–7452.

[6] Lubensky T C, Ramaswamy S and Toner J, 1986, Dislocation motion in quasicrystals and implications for macroscopic properties, *Phys. Rev.* B, **33**(11), 7715–7719.

[7] Lubensky T C, Socolar J E S, Steinhardt P J, Bancel P A and Heiney P A, 1986, Distortion and peak broadening in quasicrystal diffraction patterns, *Phys. Rev. Lett.*, **57**(12), 1440–1443.

[8] Lubensky T C, 1988, *Introduction to Quasicrystals*, ed by Jaric M V, Academic Press, Boston.

[9] Kalugin P A, Kitaev A and Levitov L S, 1985, 6-dimensional properties of $Al_{0.86}Mn_{0.14}$alloy, *J. Phys. Lett.*, **46**(13), 601–607.

[10] Torian S M and Mermin D, 1985, Mean-field theory of quasicrystalline order, *Phys. Rev. Lett.*, **54**(14), 1524–1527.

[11] Jaric M V, 1985, Long-range icosahedral orientational order and quasicrystals, *Phys. Rev. Lett.*, **55**(6), 607–610.

[12] Duneau M and Katz A, 1985, Quasiperiodic patterns, *Phys. Rev. Lett.*, **54**(25), 2688–2691.

[13] Socolar J E S, Lubensky T C and Steinhardt P J, 1986, Phonons, phasons, and dislocations in quasicrystals, *Phys. Rev.* B, **34**(5), 3345–3360.

[14] Gahler F and Rhyner J, 1986, Equivalence of the generalised grid and projection methods for the construction of quasiperiodic tilings, *J. Phys. A: Math. Gen.*, **19**(2), 267–277.

[15] Horn P M, Melzfeldt W, Di Vincenzo D P, Toner J and Gambine R, 1986, Systematics of disorder in quasiperiodic material, *Phys. Rev. Lett.*, **57**(12), 1444–1447.

[16] Hu C Z, Wang R H and Ding D H, 2000, Symmetry groups, physical property tensors, elasticity and dislocations in quasicrystals, *Rep. Prog. Phys.*, **63**(1), 1–39.

[17] Coddens G, Bellissent R, Calvayrac Y *et al.*, 1991, Evidence for phason hopping in icosahedral AlFeCu quasi-crystals, *Euro. phys. Lett.*, **16**(3), 271–276.

[18] Coddens G and Sturer W, 1999, Time-of-flight neutron-scattering study of phason hopping in decagonal Al-Co-Ni quasicrystals, *Phys. Rev. B*, **60**(1), 270–276.

[19] Coddens G, Lyonnard S, Hennion B *et al.*, 2000, Triple-axis neutron-scattering study of phason dynamics in Al-Mn-Pd quasicrystals, *Phys. Rev.* B, **62**(10), 6268–6295.

[20] Coddens G, Lyonnard S, Calvayrac Y *et al.*, 1996, Atomic (phason) hopping in perfect icosahedral quasicrystals Al70.3Pd21.4Mn8.3 by time-of-flight quasielastic neutron scattering, *Phys. Rev.* B, **53**(6), 3150–3160.

[21] Coddens G, Lyonnard S, Sepilo B *et al.*, 1995, Evidence for atomic hopping of Fe in perfectly icosahedral AlFeCu quasicrystals by ^{57}Fe Moessbauer spectroscopy, *J. Phys.*, **5**(7), 771–776.

[22] Dolisek J, Ambrosini B, Vonlanthen P *et al.*, 1998, Atomic motion in quasicrystalline $Al_{70}Re_{8.6}Pd_{21.4}$: A two-dimensional exchange NMR study, *Phys. Rev. Lett.*, **81**(17), 3671–3674.

[23] Dolisek J, Apih T, Simsic M *et al.*, 1999, Self-diffusion in icosahedral $Al_{72.4}Pd_{20.5}Mn_{7.1}$ and phason percolation at low temperatures studied by ^{27}Al NMR, *Phys. Rev. Lett.*, **82**(3), 572–575.

[24] Edagawa K and Kajiyama K, 2000, High temperature specific heat of Al-Pd-Mn and Al-Cu-Co quasicrystals, *Mater. Sci. and Eng.* A, **294–296**(5), 646–649.

[25] Edagawa K, Kajiyama K and Tamura R *et al.*, 2001, High-temperature specific heat of quasicrystals and a crystal approximant, *Mater. Sci. and Eng.* A, **312**(1–2), 293–298.

[26] Ding D H, Yang W G, Hu C Z *et al.*, 1993, Generalized elasticity theory of quasicrystals, *Phys. Rev.* B, **48**(10), 7003–7010.

[27] Fan T Y, Wang X F and Li W *et al.*, 2009, Elasto-/hydro-dynamics of quasicrystals, *Phil. Mag.*, **89**(6), 501–512.

[28] Rochal S B and Lorman V L, 2002, Minimal model of the phonon-phason dynamics on icosahedral quasicrystals and its application for the problem of internal friction in the i-AIPdMn alloys, *Phys. Rev.* B, **66**(14), 144204.

[29] Francoual S, Levit F, de Boussieu M *et al.*, 2003, Dynamics of phason fluctuations in the $i-$Al-Pd-Mn quasicrystals, *Phys. Rev. Lett.*, **91**(22), 225501.

[30] Coddens G, 2006, On the problem of the relation between phason elasticity and phason dynamics in quasicrystals, *Eur. Phys. J.* B, **54**(1), 37–65.

[31] Edagawa K and Takeuchi S, 2006, Elasticity, dislocations and their motion in quasicrystals, in *Dislocation in Solids*, Chapter 76, ed. by Nabarro E R N and Hirth J P, pp. 367–417.
[32] Edagawa K and Giso Y, 2007, Experimental evaluation of phonon-phason coupling in icosahedral quasicrystals, *Phil. Mag.*, **87**(1), 77–95.
[33] Meng X M, Tong B Y and Wu Y K, 1994, Mechanical properties of quasicrystal $Al_{65}Cu_{20}Co_{15}$, *Acta Metallurgica Sinica*, **30**(2), 61–64 (in Chinese).
[34] Takeuchi S, Iwanhaga H and Shibuya T, 1991, Hardness of quasicrystals, *Japanese J. Appl. Phys.*, **30**(3), 561–562.

Chapter 5

Elasticity Theory of One-dimensional Quasicrystals and Simplification

As mentioned in Chapter 4, there are three kinds of one-, two- and three-dimensional quasicrystals observed experimentally, each kind of them can be further divided into certain subclasses with respect to symmetry consideration.

In the class of one-dimensional quasicrystals, the atom arrangement is quasiperiodic only in one direction (say, the z-direction), while that in the plane perpendicular to the direction (i.e., xy-plane) is periodic. Even if for the one-dimensional quasicrystals, the structure is three-dimensional, i.e., it is generated in a three-dimensional body. Strictly speaking, one-dimensional quasicrystals may be seen as a projection of periodic crystals in four-dimensional space to three-dimensional physical space. So that the problem substantively is a four-dimensional problem, there are four non-zero displacements, i.e., u_x, u_y, u_z and w_z (and $w_x = w_y = 0$).

Here we briefly list the systems and Laue classes of one-dimensional quasicrystals in which the concept of point group must be concerned, and we do not concern with the concept of space group.

5.1 Elasticity of Hexagonal Quasicrystals

As pointed out previously that for one-dimensional quasicrystals, there are phonon displacements u_x, u_y, u_z and phason displacement

w_z (and $w_x = w_y = 0$), the corresponding strains are

$$\varepsilon_{xx} = \frac{\partial u_x}{\partial x} \quad \varepsilon_{yy} = \frac{\partial u_y}{\partial y}, \quad \varepsilon_{zz} = \frac{\partial u_z}{\partial z},$$

$$\varepsilon_{yz} = \varepsilon_{zy} = \frac{1}{2}\left(\frac{\partial u_z}{\partial y} + \frac{\partial u_y}{\partial z}\right), \quad \varepsilon_{zx} = \varepsilon_{xz} = \frac{1}{2}\left(\frac{\partial u_z}{\partial x} + \frac{\partial u_x}{\partial z}\right),$$

$$\varepsilon_{xy} = \varepsilon_{yx} = \frac{1}{2}\left(\frac{\partial u_x}{\partial y} + \frac{\partial u_y}{\partial x}\right) \tag{5.1-1}$$

$$w_{zx} = \frac{\partial w_z}{\partial x}, \quad w_{zy} = \frac{\partial w_z}{\partial y}, \quad w_{zz} = \frac{\partial w_z}{\partial z} \tag{5.1-2}$$

and other $w_{ij} = 0$. Formulas (5.1-1) and (5.1-2) hold for all one-dimensional quasicrystals. In this section, we only discuss one-dimensional hexagonal quasicrystals.

If put the strains given by (5.1-1) and (5.1-2) as a vector with nine components, i.e.,

$$[\varepsilon_{11}, \varepsilon_{22}, \varepsilon_{33}, 2\varepsilon_{23}, 2\varepsilon_{31}, 2\varepsilon_{12}, w_{33}, w_{31}, w_{32}] \tag{5.1-3}$$

or

$$[\varepsilon_{xx}, \varepsilon_{yy}, \varepsilon_{zz}, 2\varepsilon_{yz}, 2\varepsilon_{zx}, 2\varepsilon_{xy}, w_{zz}, w_{zx}, w_{zy}] \tag{5.1-4}$$

corresponding stresses can be arranged as

$$[\sigma_{xx}, \sigma_{yy}, \sigma_{zz}, \sigma_{yz}, \sigma_{zx}, \sigma_{xy}, H_{zz}, H_{zx}, H_{zy}] \tag{5.1-5}$$

and we have the elastic constant matrix such as

$$[CKR] = \begin{bmatrix} C_{11} & C_{12} & C_{13} & 0 & 0 & 0 & R_1 & 0 & 0 \\ C_{12} & C_{11} & C_{13} & 0 & 0 & 0 & R_1 & 0 & 0 \\ C_{13} & C_{13} & C_{33} & 0 & 0 & 0 & R_2 & 0 & 0 \\ 0 & 0 & 0 & C_{44} & 0 & 0 & 0 & 0 & R_3 \\ 0 & 0 & 0 & 0 & C_{44} & 0 & 0 & R_3 & 0 \\ 0 & 0 & 0 & 0 & 0 & C_{66} & 0 & 0 & 0 \\ R_1 & R_1 & R_2 & 0 & 0 & 0 & K_1 & 0 & 0 \\ 0 & 0 & 0 & 0 & R_3 & 0 & 0 & K_2 & 0 \\ 0 & 0 & 0 & R_3 & 0 & 0 & 0 & 0 & K_2 \end{bmatrix}$$

where for the phonon elastic constant tensor, a brief notation is used, i.e., index 11 → 1, 22 → 2, 33 → 3, 23 → 4, 31 → 5, 12 → 6, and C_{ijkl} is denoted as C_{pq} accordingly,

$$C_{11} = C_{1111} = C_{2222}, C_{12} = C_{1122}, C_{33} = C_{3333}, C_{44} = C_{2323} = C_{3131}$$
$$C_{13} = C_{1133} = C_{2233}, C_{66} = (C_{11} - C_{12})/2 = (C_{1111} - C_{1122})/2$$

This shows the independent phonon elastic constants are five; in addition, $K_1 = K_{3333}$, $K_2 = K_{3131} = K_{3232}$, i.e., the independent phason elastic constants are two, and $R_1 = R_{1133} = R_{2233}$, $R_2 = R_{3333}$, $R_3 = R_{2332} = R_{3131}$, i.e., phonon–phason coupling elastic constants are three.

From the elastic constant matrix given above, it is easy to find that the corresponding stress–strain relations (or so-called constitutive equations, or the generalized Hooke's law) are as follows:

$$\begin{aligned}
\sigma_{xx} &= C_{11}\varepsilon_{xx} + C_{12}\varepsilon_{yy} + C_{13}\varepsilon_{zz} + R_1 w_{zz} \\
\sigma_{yy} &= C_{12}\varepsilon_{xx} + C_{11}\varepsilon_{yy} + C_{13}\varepsilon_{zz} + R_1 w_{zz} \\
\sigma_{zz} &= C_{13}\varepsilon_{xx} + C_{13}\varepsilon_{yy} + C_{33}\varepsilon_{zz} + R_2 w_{zz} \\
\sigma_{yz} &= \sigma_{zy} = 2C_{44}\varepsilon_{yz} + R_3 w_{zy} \\
\sigma_{zx} &= \sigma_{xz} = 2C_{44}\varepsilon_{zx} + R_3 w_{zx} \\
\sigma_{xy} &= \sigma_{yx} = 2C_{66}\varepsilon_{xy} \\
H_{zz} &= R_1(\varepsilon_{xx} + \varepsilon_{yy}) + R_2\varepsilon_{zz} + K_1 w_{zz} \\
H_{zx} &= 2R_3\varepsilon_{zx} + K_2 w_{zx} \\
H_{zy} &= 2R_3\varepsilon_{yz} + K_2 w_{zy}
\end{aligned} \tag{5.1-6}$$

and other $H_{ij} = 0$.

The stresses satisfy the following equilibrium equations:

$$\frac{\partial \sigma_{xx}}{\partial x} + \frac{\partial \sigma_{xy}}{\partial y} + \frac{\partial \sigma_{xz}}{\partial z} = 0$$
$$\frac{\partial \sigma_{yx}}{\partial x} + \frac{\partial \sigma_{yy}}{\partial y} + \frac{\partial \sigma_{yz}}{\partial z} = 0$$

$$\begin{aligned}\frac{\partial\sigma_{zx}}{\partial x}+\frac{\partial\sigma_{zy}}{\partial y}+\frac{\partial\sigma_{zz}}{\partial z}=0\\ \frac{\partial H_{zx}}{\partial x}+\frac{\partial H_{zy}}{\partial y}+\frac{\partial H_{zz}}{\partial z}=0\end{aligned} \tag{5.1-7}$$

The above results are obtained by Wang *et al.* [1].

It is evident that the elastic equilibrium problem of one-dimensional hexagonal quasicrystals is more complicated than one of three-dimensional classical elasticity. Here, there are 4 displacements, 9 strains and 9 stresses, which added up to get 22 field variables. The corresponding field equations are also 22, in which 4 is for equilibrium equations and 9 for equations of deformation geometry and 9 for stress–strain relations. The complete solution of the problem will be given later. In the following, we give a simplified treatment of the problem, which will be easily solved and provides experiences for solving more complicated problems.

5.2 Decomposition of the Elasticity into a Superposition of Plane and Anti-plane Elasticity

If there is a straight dislocation or a Griffith crack along the direction of atom quasiperiodic arrangement, i.e., the deformation is independent from the direction, or say

$$\frac{\partial}{\partial z}=0 \tag{5.2-1}$$

then we have

$$\frac{\partial u_i}{\partial z}=0,\ (i=1,2,3),\quad \frac{\partial w_z}{\partial z}=0 \tag{5.2-2}$$

so that

$$\varepsilon_{zz}=w_{zz}=0,\quad \varepsilon_{yz}=\varepsilon_{zy}=\frac{1}{2}\frac{\partial u_z}{\partial y},\quad \varepsilon_{zx}=\varepsilon_{xz}=\frac{1}{2}\frac{\partial u_z}{\partial x} \tag{5.2-3}$$

$$\frac{\partial\sigma_{ij}}{\partial z}=0,\quad \frac{\partial H_{ij}}{\partial z}=0 \tag{5.2-4}$$

In this case, the generalized Hooke's law is simplified as

$$
\begin{aligned}
\sigma_{xx} &= C_{11}\varepsilon_{xx} + C_{12}\varepsilon_{yy} \\
\sigma_{yy} &= C_{12}\varepsilon_{xx} + C_{11}\varepsilon_{yy} \\
\sigma_{xy} &= \sigma_{yx} = 2C_{66}\varepsilon_{xy} \\
\sigma_{zz} &= C_{13}(\varepsilon_{xx} + \varepsilon_{yy}) \\
\sigma_{yz} &= \sigma_{zy} = 2C_{44}\varepsilon_{yz} + R_3 w_{zy} \\
\sigma_{zx} &= \sigma_{xz} = 2C_{44}\varepsilon_{zx} + R_3 w_{zx} \\
H_{zz} &= R_1(\varepsilon_{xx} + \varepsilon_{yy}) \\
H_{zx} &= 2R_3\varepsilon_{zx} + K_2 w_{zx} \\
H_{zy} &= 2R_3\varepsilon_{yz} + K_2 w_{zy}
\end{aligned} \tag{5.2-5}
$$

The equilibrium equations are (if the body force and generalized body force are omitted)

$$
\frac{\partial \sigma_{xx}}{\partial x} + \frac{\partial \sigma_{xy}}{\partial y} = 0, \quad \frac{\partial \sigma_{yx}}{\partial x} + \frac{\partial \sigma_{yy}}{\partial y} = 0, \quad \frac{\partial \sigma_{zx}}{\partial x} + \frac{\partial \sigma_{zy}}{\partial y} = 0 \tag{5.2-6}
$$

$$
\frac{\partial H_{zx}}{\partial x} + \frac{\partial H_{zy}}{\partial y} = 0 \tag{5.2-7}
$$

Equations (5.1-2), (5.1-3) and (5.2-5)–(5.2-7) define two decoupled problems [2]; the first of them is

$$
\begin{aligned}
\sigma_{xx} &= C_{11}\varepsilon_{xx} + C_{12}\varepsilon_{yy} \\
\sigma_{yy} &= C_{12}\varepsilon_{xx} + C_{11}\varepsilon_{yy} \\
\sigma_{xy} &= (C_{11} - C_{12})\varepsilon_{xy} \\
\sigma_{zz} &= C_{13}(\varepsilon_{xx} + \varepsilon_{yy}) \\
H_{zz} &= R_1(\varepsilon_{xx} + \varepsilon_{yy})
\end{aligned} \tag{5.2-8}
$$

$$\frac{\partial \sigma_{xx}}{\partial x} + \frac{\partial \sigma_{xy}}{\partial y} = 0, \quad \frac{\partial \sigma_{yx}}{\partial x} + \frac{\partial \sigma_{yy}}{\partial y} = 0$$

$$\varepsilon_{xx} = \frac{\partial u_x}{\partial x}, \quad \varepsilon_{yy} = \frac{\partial u_y}{\partial y}, \quad \varepsilon_{xy} = \frac{1}{2}\left(\frac{\partial u_y}{\partial x} + \frac{\partial u_x}{\partial y}\right)$$

this is plane elasticity of conventional hexagonal crystals. The second is

$$\begin{aligned}
&\sigma_{yz} = \sigma_{zy} = 2C_{44}\varepsilon_{yz} + R_3 w_{zy} \\
&\sigma_{zx} = \sigma_{xz} = 2C_{44}\varepsilon_{zx} + R_3 w_{zx} \\
&H_{zx} = 2R_3\varepsilon_{zx} + K_2 w_{zx} \\
&H_{zy} = 2R_3\varepsilon_{yz} + K_2 w_{zy} \\
&\frac{\partial \sigma_{zx}}{\partial x} + \frac{\partial \sigma_{zy}}{\partial y} = 0, \quad \frac{\partial H_{zx}}{\partial x} + \frac{\partial \sigma_{zy}}{\partial y} = 0 \\
&\varepsilon_{zx} = \frac{1}{2}\frac{\partial u_z}{\partial x} = \varepsilon_{xz}, \quad \varepsilon_{zy} = \frac{1}{2}\frac{\partial u_z}{\partial y} = \varepsilon_{yz} \\
&w_{zx} = \frac{\partial w_z}{\partial x}, \quad w_{zy} = \frac{\partial w_z}{\partial y}
\end{aligned} \tag{5.2-9}$$

which is a phonon–phason coupling elasticity problem. If there are only displacements u_z and w_z, it is an anti-plane elasticity problem.

The plane elasticity described by (5.2-8) belongs to the pure classical elasticity, which has well been studied, e.g., it introduces

$$\sigma_{xx} = \frac{\partial^2 U}{\partial y^2}, \quad \sigma_{yy} = \frac{\partial^2 U}{\partial x^2}, \quad \sigma_{xy} = -\frac{\partial^2 U}{\partial x \partial y}$$

then Eqs. (5.2-8) are reduced to solve

$$\nabla^2\nabla^2 U = 0$$

The problem is considerably discussed in crystal (or classical) elasticity, here we need not discuss it any more.

We are interested in the phonon–phason coupling anti-plane elasticity described by (5.2-9), which may bring some new insight into the scope of elasticity of quasicrystals.

Substituting deformation geometry relations into the stress–strain relations and then into the equilibrium equations yields the final governing equations such as

$$\begin{aligned} C_{44}\nabla^2 u_z + R_3\nabla^2 w_z = 0 \\ R_3\nabla^2 u_z + K_2\nabla^2 w_z = 0 \end{aligned} \tag{5.2-10}$$

Because $C_{44}K_2 - R_3^2 \neq 0$, we have

$$\nabla^2 u_z = 0, \quad \nabla^2 w_z = 0 \tag{5.2-11}$$

where $\nabla^2 = \frac{\partial}{\partial x^2} + \frac{\partial}{\partial y^2}$, so u_z and w_z are harmonic functions.

It is well known that the two-dimensional harmonic functions u_z and w_z can be a real part or an imaginary part of any analytic functions $\varphi(t)$ and $\psi(t)$ of complex variable $t = x + iy$, $i = \sqrt{-1}$, respectively, i.e.,

$$\begin{aligned} u_z(x, y) = \text{Re}\varphi(t) \\ w_z(x, y) = \text{Re}\psi(t) \end{aligned} \tag{5.2-12}$$

In this version, Eqs. (5.2-11) should be automatically satisfied. The determination of $\phi(t)$ and $\psi(t)$ depends upon appropriate boundary conditions, which will be discussed in detail in Chapters 7 and 8. The complex variable function method for solving elasticity of one-, two- and three-dimensional quasicrystals will be summarized fully in Chapter 11.

5.3 Elasticity of Monoclinic Quasicrystals

The decomposition and superposition procedure suggested by the author see, e.g., in Ref. [2, 3] is applicable not only for hexagonal quasicrystals but also for other one-dimensional quasicrystals. We here discuss monoclinic quasicrystals.

For this kind of one-dimensional quasicrystals, there are 25 non-zero elastic constants in total, namely, $C_{1111}, C_{2222}, C_{3333}, C_{1122}$, $C_{1133}, C_{1112}, C_{2233}, C_{2212}, C_{3312}, C_{3232}, C_{3231}, C_{3131}, C_{1212}$ for phonon field, $K_{3333}, K_{3131}, K_{3232}, K_{3132}$ for phason field and R_{1133}, R_{2233}, $R_{3333}, R_{1233}, R_{2331}$, R_{2332}, R_{3132}, R_{1233} for phonon–phason coupling.

The corresponding generalized Hooke's law for the case is given as follows [1]:

$$
\begin{aligned}
\sigma_{xx} &= C_{11}\varepsilon_{xx} + C_{12}\varepsilon_{yy} + C_{13}\varepsilon_{zz} + 2C_{16}\varepsilon_{xy} + R_1 w_{zz} \\
\sigma_{yy} &= C_{12}\varepsilon_{xx} + C_{22}\varepsilon_{yy} + C_{23}\varepsilon_{zz} + 2C_{26}\varepsilon_{xy} + R_2 w_{zz} \\
\sigma_{zz} &= C_{13}\varepsilon_{xx} + C_{23}\varepsilon_{yy} + C_{33}\varepsilon_{zz} + 2C_{36}\varepsilon_{xy} + R_3 w_{zz} \\
\sigma_{yz} &= \sigma_{zy} = 2C_{44}\varepsilon_{yz} + 2C_{45}\varepsilon_{zx} + R_4 w_{zx} + R_5 w_{zy} \\
\sigma_{zx} &= \sigma_{xz} = 2C_{45}\varepsilon_{yz} + 2C_{55}\varepsilon_{zx} + R_6 w_{zx} + R_7 w_{zy} \\
\sigma_{xy} &= \sigma_{yx} = C_{16}\varepsilon_{xx} + C_{26}\varepsilon_{yy} + C_{36}\varepsilon_{zz} + 2C_{66}\varepsilon_{xy} + R_8 w_{zz} \\
H_{zx} &= 2R_4\varepsilon_{yz} + 2R_6\varepsilon_{zx} + K_1 w_{zx} + K_4 w_{zy} \\
H_{zy} &= 2R_5\varepsilon_{yz} + 2R_7\varepsilon_{zx} + K_4 w_{zx} + K_2 w_{zy} \\
H_{zz} &= R_1\varepsilon_{xx} + R_2\varepsilon_{yy} + R_3\varepsilon_{zz} + 2R_8\varepsilon_{xy} + K_3 w_{zz}
\end{aligned} \tag{5.3-1}
$$

where recalling that for the phonon elastic constant tensor, a brief notation is used, i.e., index 11 → 1, 22 → 2, 33 → 3, 23 → 4, 31 → 5, 12 → 6, and C_{ijkl} is denoted as C_{pq}. In addition, for the phason elastic constants, $K_{3131} = K_1$, $K_{3232} = K_2$, $K_{3333} = K_3$, $K_{3132} = K_4$, and for the phonon–phason coupling elastic constants, $R_{1133} = R_1$, $R_{2233} = R_2$, $R_{3333} = R_3$, $R_{2331} = R_4$, $R_{2332} = R_5$, $R_{3131} = R_6$, $R_{3132} = R_7$, $R_{1233} = R_8$.

Under the assumption (5.2-1), the problem will be decomposed into two separate problems as

$$
\begin{aligned}
\sigma_{xx} &= C_{11}\varepsilon_{xx} + C_{12}\varepsilon_{yy} + 2C_{16}\varepsilon_{xy} \\
\sigma_{yy} &= C_{12}\varepsilon_{xx} + C_{22}\varepsilon_{yy} + 2C_{26}\varepsilon_{xy} \\
\sigma_{xy} &= \sigma_{yx} = C_{16}\varepsilon_{xx} + C_{26}\varepsilon_{yy} + 2C_{66}\varepsilon_{xy} \\
\sigma_{zz} &= C_{13}\varepsilon_{xx} + C_{23}\varepsilon_{yy} + 2C_{36}\varepsilon_{xy} \\
H_{zz} &= R_1\varepsilon_{xx} + R_2\varepsilon_{yy} + R_3\varepsilon_{zz} + 2R_8\varepsilon_{xy}
\end{aligned} \tag{5.3-2}
$$

and

$$
\begin{aligned}
\sigma_{yz} &= \sigma_{zy} = 2C_{44}\varepsilon_{yz} + 2C_{45}\varepsilon_{zx} + R_4 w_{zx} + R_5 w_{zy} \\
\sigma_{zx} &= \sigma_{xz} = 2C_{45}\varepsilon_{yz} + 2C_{55}\varepsilon_{zx} + R_6 w_{zx} + R_7 w_{zy} \\
H_{zx} &= 2R_4\varepsilon_{yz} + 2R_6\varepsilon_{zx} + K_1 w_{zx} + K_4 w_{zy} \\
H_{zy} &= 2R_5\varepsilon_{yz} + 2R_7\varepsilon_{zx} + K_4 w_{zx} + K_2 w_{zy}
\end{aligned}
\tag{5.3-3}
$$

in which the problem described by Eqs. (5.3-2) is plane elasticity of monocline crystals, by introducing displacement potential $G(x, y)$

$$
\begin{aligned}
u_x &= \left[C_{16}\frac{\partial^2}{\partial x^2} + C_{26}\frac{\partial^2}{\partial y^2} + (C_{12} + C_{66})\frac{\partial^2}{\partial x \partial y}\right] G \\
u_y &= -\left[C_{11}\frac{\partial^2}{\partial x^2} + C_{66}\frac{\partial^2}{\partial y^2} + 2C_{16}\frac{\partial^2}{\partial x \partial y}\right] G
\end{aligned}
$$

The elasticity equations are reduced at last to

$$
\left(c_1\frac{\partial^4}{\partial x^4} + c_2\frac{\partial^4}{\partial x^3 \partial y} + c_3\frac{\partial^4}{\partial x^2 \partial y^2} + c_4\frac{\partial^4}{\partial x \partial y^3} + c_5\frac{\partial^4}{\partial y^4}\right) G = 0
$$

with constants

$$
\begin{aligned}
c_1 &= C_{16}^2 - C_{11}C_{66}, \quad c_2 = 2(C_{16}C_{12} - C_{11}C_{26}) \\
c_3 &= C_{12}^2 - 2C_{16}C_{26} + 2C_{12}C_{66} - C_{11}C_{22} \\
c_4 &= 2(C_{26}C_{12} - C_{16}C_{22}), \quad c_5 = C_{26}^2 - C_{22}C_{66}
\end{aligned}
$$

Because this is classical elasticity and does not have direct connection to phason elasticity of one-dimensional quasicrystal, we do not discuss it here further.

We are interested in the problem described by Eq. (5.3-3), which is a phonon–phason coupling problem. Substituting the equations of deformation geometry into the stress–strain relations and then into the equilibrium equations, we obtain the final governing equation as

$$
\left(a_1\frac{\partial^4}{\partial x^4} + a_2\frac{\partial^4}{\partial x^3 \partial y} + a_3\frac{\partial^4}{\partial x^2 \partial y^2} + a_4\frac{\partial^4}{\partial x \partial y^3} + a_5\frac{\partial^4}{\partial y^4}\right) F = 0
\tag{5.3-4}
$$

where

$$\left.\begin{aligned} u_z &= \left[R_6\frac{\partial^2}{\partial x^2} + R_5\frac{\partial^2}{\partial y^2} + (R_4 + R_7)\frac{\partial^2}{\partial x \partial y}\right] F \\ w_z &= -\left[C_{55}\frac{\partial^2}{\partial x^2} + C_{44}\frac{\partial^2}{\partial y^2} + 2C_{45}\frac{\partial^2}{\partial x \partial y}\right] F \end{aligned}\right\} \tag{5.3-5}$$

and

$$\begin{aligned} &a_1 = R_6^2 - K_1 C_{56}, \quad a_2 = 2(R_6(R_4 + R_7) - K_1 C_{45} - K_4 C_{55}) \\ &a_3 = 2R_5 R_6 + (R_4 + R_7)^2 - K_1 C_{44} - K_2 C_{55} - 4K_4 C_{45} \\ &a_4 = 2[R_5(R_4 + R_7) - K_2 C_{45} - K_4 C_{44}], \quad a_5 = R_5^2 - K_2 C_{44} \end{aligned} \tag{5.3-6}$$

In the subsequent contents of this book, we discuss only the anti-plane elasticity of monocline quasicrystals, which has the complex representation of solution as

$$F(x, y) = 2Re\sum_{k=1}^{2} F_k(z_k), \quad z_k = x + \mu_k y \tag{5.3-7}$$

where $F_k(z_k)$ are analytic functions of z_k and

$$\mu_k = \alpha_k + i\beta_k \tag{5.3-8}$$

are the distinct complex parameters to be determined by the characteristic equation (or call the eigenvalue equation)

$$a_5\mu^4 + a_4\mu^3 + a_3\mu^2 + a_2\mu + a_1 = 0 \tag{5.3-9}$$

and $\mu_1 \neq \mu_2$

If the roots of Eq. (5.3-9) are multi-roots, i.e., $\mu_1 = \mu_2$, then

$$F(x, y) = 2Re[F_1(z_1) + \bar{z}_1 F_2(z_1)], \quad z_1 = x + \mu_1 y \tag{5.3-10}$$

Substituting formula (5.3-7) into (5.3-5) and then into (5.3-3), we can get the complex representation of the displacements and stresses

$$
\begin{aligned}
u_z &= 2Re\sum_{k=1}^{2}[R_6 + (R_4 + R_7)\mu_k + R_5\mu_k^2]f_k(z_k) \\
w_z &= -2Re\sum_{k=1}^{2}(C_{55} + 2C_{45}\mu_k + C_{44}\mu_k^2)f_k(z_k) \\
\sigma_{zy} = \sigma_{yz} &= 2Re\sum_{k=1}^{2}[R_6C_{45} - R_4C_{55} + (R_6C_{44} - R_4C_{45} \\
&\quad + R_7C_{45} - R_5C_{55})\mu_k + (R_7C_{44} - R_5C_{45})\mu_k^2]f_k'(z_k) \\
\sigma_{zx} = \sigma_{xz} &= 2Re\sum_{k=1}^{2}[R_4C_{55} - R_6C_{45} + (R_5C_{55} + R_4C_{45} - R_6C_{44} \\
&\quad - R_5C_{55})\mu_k + (R_3C_{45} - R_7C_{44})\mu_k^2]\mu_k f_k'(z_k) \\
H_{zx} &= 2Re\sum_{k=1}^{2}[(R_7 + R_5\mu_k)(R_6 + R_4\mu_k + R_7\mu_k + R_5\mu_k^2) \\
&\quad - (K_4 + K_2\mu_k)(C_{55} + 2C_{45}\mu_k + C_{44}\mu_k^2)]f_k'(z_k) \\
H_{zy} &= 2Re\sum_{k=1}^{2}[(R_6 + R_4\mu_k)(R_6 + R_4\mu_k + R_7\mu_k + R_5\mu_k^2) \\
&\quad - (K_4 + K_2\mu_k)(C_{55} + 2C_{45}\mu_k + C_{44}\mu_k^2)]f_k'(z_k)
\end{aligned}
\tag{5.3-11}
$$

where $f_k(z_k) \equiv \partial^2 F_k(z_k)/\partial z_k^2 = F_k''(z_k)$.

The determination of analytic functions $F_k(z_k)$ depends on the boundary conditions of concrete problems, which will be investigated later in Chapters 7 and 8.

5.4 Elasticity of Orthorhombic Quasicrystals

In Table 5.1, we know that orthorhombic quasicrystals contain the points $2_h2_h2, mm2, 2_hmm_h$ and mmm_h, which belong to one Laue

Table 5.1. System, Laue classes and point groups of one-dimensional quasicrystals.

Systems	Laue classes	Point groups
Triclinic	1	$1\ \bar{1}$
Monoclinic	2	$2\ \ m_h\ 2/m_h$
	3	$2_h\ m\ 2_h/m$
Orthorhombic	4	$2_h 2_h 2\ 2mm\ 2_h mm_h m_h mm$
Tetragonal	5	$4\ \bar{4}\ 4/m_h$
	6	$42_h\ 2_h\ 4mm\ \bar{4}2_h m\ 4/m_h mm$
Rhombohedral (or trigonal)	7	$3\ \bar{3}$
	8	$32_h\ 3m\ \bar{3}m$
Hexgonal	9	$6\ \ \bar{6}6/m_h$
	10	$62_h 2_h\ \ \ 6mm$ $\bar{6}2_h m\ \ \ 6/m_h mm$

class 4. Due to the increase in symmetric elements of this quasicrystal system in comparison with monocline quasicrystal system, one has

$$C_{16} = C_{26} = C_{36} = C_{45} = K_4 = R_4 = R_7 = R_8 = 0 \tag{5.4-1}$$

Therefore, the total number of non-zero elastic constants in the case reduces to 17, i.e., $C_{11}, C_{22}, C_{33}, C_{12}, C_{13,}C_{23}, C_{44}, C_{55}, C_{66}$ for the phonon fields, K_1, K_2, K_3 for the phason fields and R_1, R_2, R_3, R_5, R_6 for the phonon–phason coupling fields.

Considering results (5.4-1), then (5.3-6) can be simplified to

$$a_2 = a_4 = 0, \quad a_1 = R_6^2 - K_1 C_{55}, \quad a_3 = 2R_5 R_6 - K_1 C_{44} - K_2 C_{55},$$
$$a_5 = R_5^2 - K_2 C_{66} \tag{5.4-2}$$

and a_1 and a_5 no change from (5.3-6) so that Eq. (5.3-4) is reduced to

$$\left(a_1 \frac{\partial^4}{\partial x^4} + a_3 \frac{\partial^4}{\partial x^2 \partial y^2} + a_5 \frac{\partial^4}{\partial y^4} \right) F = 0 \tag{5.4-3}$$

we have the expression of solution

$$
\begin{aligned}
u_z &= 2Re\sum_{k=1}^{2}[R_6 + R_5\mu_k^2]f_k(z_k)\\
w_z &= -2Re\sum_{k=1}^{2}(C_{55} + C_{44}\mu_k^2)f_k(z_k)\\
\sigma_{zy} &= \sigma_{yz} = 2Re\sum_{k=1}^{2}(R_6C_{44} - R_5C_{55})\mu_k f_k'(z_k)\\
\sigma_{zx} &= \sigma_{xz} = 2Re\sum_{k=1}^{2}(R_5C_{55} - R_6C_{44})\mu_k^2 f_k'(z_k)\\
H_{zy} &= 2Re\sum_{k=1}^{2}[R_5R_6 - K_2C_{55} + (R_5^2 - K_2C_{44})\mu_k^2)\\
&\quad - (K_4 + K_2\mu_k)]\mu_k f_k'(z_k)\\
H_{zx} &= 2Re\sum_{k=1}^{2}[R_6^2 - K_1C_{55} + (R_5R_6 - K_1C_{44})\mu_k^2]f_k'(z_k)
\end{aligned}
\tag{5.4-4}
$$

in which $f_k(z_k)$ represents analytic function of z_k.

5.5 Tetragonal Quasicrystals

From Table 5.1, we know that one-dimensional tetragonal quasicrystals have seven point groups, in which groups $\bar{4}2_hm, 4mm, 42_h2_h$ and $4/m_hmm$ belong to Laue class 6, besides (5.4-1), we have also

$$
\begin{gathered}
C_{11} = C_{22}, \quad C_{13} = C_{23}, \quad C_{44} = C_{55}, \quad K_1 = K_2,\\
R_1 = R_2, \quad R_5 = R_6
\end{gathered}
\tag{5.5-1}
$$

Therefore, the total number of non-zero elastic constants reduces to 11.

With (5.5-1) and (5.4-2), one can find the complex representation of solution for the anti-plane elasticity of quasicrystals which belong to Laue class 6.

The point groups 4, $\bar{4}$ and $4/m_h$ belong to Laue class 5, in which the solution of the anti-plane elasticity can also be expressed in a similar manner.

The final governing equation of anti-plane elasticity for this kind of quasicrystals is

$$\nabla^2\nabla^2 F = 0 \tag{5.5-2}$$

5.6 The Space Elasticity of Hexagonal Quasicrystals

The decomposition and superposition procedure proposed by Ref. [2] has been developed for simplifying the elasticity of various systems of one-dimensional quasicrystals, in the previous sections. The main feature of the procedure lies in the decomposition of a space (three-dimensional) elasticity into a superposition of a plane elasticity and an anti-plane elasticity for the studied quasicrystalline material. This often simplifies the solution process, whose worth will be shown in Chapters 7 and 8 and other chapters.

In some cases, when the procedure cannot be used, we have to solve space elasticity. In this section, taking an example, we discuss the solution of space elasticity hexagonal quasicrystals, which has been studied by many authors, see, e.g., [6,9,10]. Here, the derivation of Ref. [6] is listed.

Substituting (5.5-1) into (5.1-6) and then into (5.1-7) yields the equilibrium equations expressed by the displacements:

$$\left(C_{11}\frac{\partial^2}{\partial x^2} + C_{66}\frac{\partial^2}{\partial y^2} + C_{44}\frac{\partial^2}{\partial z^2}\right)u_x + (C_{11} - C_{66})\frac{\partial^2 u_y}{\partial x \partial y}$$

$$+ (C_{13} + C_{44})\frac{\partial^2 u_z}{\partial x \partial z} + (R_1 + R_3)\frac{\partial^2 w_z}{\partial x \partial z} = 0$$

$$(C_{11} - C_{66})\frac{\partial^2 u_x}{\partial x \partial y} + \left(C_{66}\frac{\partial^2}{\partial x^2} + C_{11}\frac{\partial^2}{\partial y^2} + C_{44}\frac{\partial^2}{\partial z^2}\right)u_y$$

$$+ (C_{13} + C_{44})\frac{\partial^2 u_z}{\partial y \partial z} + (R_1 + R_3)\frac{\partial^2 w_z}{\partial y \partial z} = 0$$

$$
\begin{aligned}
&(C_{13}+C_{44})\left(\frac{\partial^2 u_x}{\partial x\partial z}+\frac{\partial^2 u_y}{\partial y\partial z}\right)+\left(C_{44}\frac{\partial^2}{\partial x^2}+C_{44}\frac{\partial^2}{\partial y^2}+C_{33}\frac{\partial^2}{\partial z^2}\right)u_z\\
&\quad+\left[R_3\left(\frac{\partial^2}{\partial x^2}+\frac{\partial^2}{\partial y^2}\right)+R_2\frac{\partial^2}{\partial z^2}\right]w_z=0\\
&(R_1+R_3)\left(\frac{\partial^2 u_x}{\partial x\partial z}+\frac{\partial^2 u_y}{\partial y\partial z}\right)+\left[R_3\left(\frac{\partial^2}{\partial x^2}+\frac{\partial^2}{\partial y^2}\right)+R_2\frac{\partial^2}{\partial z^2}\right]u_z\\
&\quad+\left[K_2\left(\frac{\partial^2}{\partial x^2}+\frac{\partial^2}{\partial y^2}\right)+K_1\frac{\partial^2}{\partial z^2}\right]w_z=0
\end{aligned}
\tag{5.6-1}
$$

If we introduce four displacement functions such as

$$
\begin{aligned}
&u_x=\frac{\partial}{\partial x}(F_1+F_2+F_3)-\frac{\partial F_4}{\partial y},\quad u_y=\frac{\partial}{\partial y}(F_1+F_2+F_3)+\frac{\partial F_4}{\partial x}\\
&u_z=\frac{\partial}{\partial z}(m_1F_1+m_2F_2+m_3F_3),\quad w_z=\frac{\partial}{\partial z}(l_1F_1+l_2F_2+l_3F_3)
\end{aligned}
\tag{5.6-2}
$$

and

$$\nabla_i^2 F_i=0\quad (i=1,2,3,4) \tag{5.6-3}$$

$$\nabla_i^2=\frac{\partial^2}{\partial x^2}+\frac{\partial^2}{\partial y^2}+\gamma_i^2\frac{\partial^2}{\partial z^2}\quad i=1,2,3,4 \tag{5.6-4}$$

with m_i, l_i and γ_i defined by

$$
\begin{aligned}
&\frac{C_{44}+(C_{13}+C_{44})m_i+(R_1+R_3)l_i}{C_{11}}=\frac{C_{33}m_i+R_2l_i}{C_{13}+C_{44}+C_{44}m_i+R_3l_i}\\
&=\frac{R_2m_i+K_1l_i}{R_1+R_2+R_3m_i+K_2l_i}\quad i=1,2,3,\ C_{44}/C_{66}=\gamma_4^2
\end{aligned}
\tag{5.6-5}
$$

then Eqs. (5.6-1) will be automatically satisfied.

It is obvious that the final governing Eqs. (5.6-3) are three-dimensional harmonic equations; this greatly simplifies the solution.

By substituting (5.6-2) into (5.1-1), (5.1-2) then into (5.1-6), one can find the stresses expressed in terms of F_1, F_2, F_3 and F_4

$$
\begin{aligned}
\sigma_{xx} &= \left[C_{11}\frac{\partial^2}{\partial x^2} + (C_{11} - 2C_{66})\frac{\partial^2}{\partial y^2}\right](F_1 + F_2 + F_3) \\
&\quad - 2C_{66}\frac{\partial^2 F_4}{\partial x \partial y} + C_{13}\frac{\partial^2}{\partial z^2}(m_1 F_1 + m_2 F_2 + m_3 F_3) \\
&\quad + R_1\frac{\partial^2}{\partial z^2}(l_1 F_1 + l_2 F_2 + l_3 F_3) \\
\sigma_{yy} &= \left[(C_{11} - 2C_{66})\frac{\partial^2}{\partial x^2} + C_{11}\frac{\partial^2}{\partial y^2}\right](F_1 + F_2 + F_3) + 2C_{66}\frac{\partial^2 F_4}{\partial x \partial y} \\
&\quad + C_{13}\frac{\partial^2}{\partial z^2}(m_1 F_1 + m_2 F_2 + m_3 F_3) + R_1\frac{\partial^2}{\partial z^2}(l_1 F_1 + l_2 F_2 + l_3 F_3) \\
\sigma_{zz} &= -C_{13}\frac{\partial^2}{\partial z^2}(\gamma_1^2 F_1 + \gamma_2^2 F_2 + \gamma_3^2 F_3) + C_{33}\frac{\partial^2}{\partial z^2}(m_1 F_1 + m_2 F_2 + m_3 F_3) \\
&\quad + R_2\frac{\partial^2}{\partial z^2}(l_1 F_1 + l_2 F_2 + l_3 F_3) \\
\sigma_{xy} &= \sigma_{yx} = 2C_{66}\frac{\partial^2}{\partial x \partial y}(F_1 + F_2 + F_3) + C_{66}\left(\frac{\partial^2}{\partial x^2} - \frac{\partial^2}{\partial y^2}\right)F_4 \\
\sigma_{yz} &= \sigma_{zy} = C_{44}\frac{\partial^2}{\partial y \partial z}[(m_1 + 1)F_1 + (m_2 + 1)F_2 + (m_3 + 1)F_3] \\
&\qquad + C_{44}\frac{\partial^2 F_4}{\partial x \partial z} + R_3\frac{\partial^2}{\partial y \partial z}(l_1 F_1 + l_2 F_2 + l_3 F_3) \\
\sigma_{zx} &= \sigma_{xz} = C_{44}\frac{\partial^2}{\partial x \partial z}[(m_1 + 1)F_1 + (m_2 + 1)F_2 + (m_3 + 1)F_3] \\
&\qquad - C_{44}\frac{\partial^2 F_4}{\partial y \partial z} + R_3\frac{\partial^2}{\partial x \partial z}(l_1 F_1 + l_2 F_2 + l_3 F_3) \\
H_{zz} &= -R_1\frac{\partial^2}{\partial z^2}(\gamma_1^2 F_1 + \gamma_2^2 F_2 + \gamma_3^2 F_3) + R_2\frac{\partial^2}{\partial z^2}(m_1 F_1 + m_2 F_2 + m_3 F_3) \\
&\quad + K_1\frac{\partial^2}{\partial z^2}(l_1 F_1 + l_2 F_2 + l_3 F_3) \\
H_{zx} &= R_3\frac{\partial^2}{\partial x \partial z}[(m_1 + 1)F_1 + (m_2 + 1)F_2 + (m_3 + 1)F_3] \\
&\quad - R_3\frac{\partial^2 F_4}{\partial y \partial z} + K_2\frac{\partial^2}{\partial x \partial z}(l_1 F_1 + l_2 F_2 + l_3 F_3)
\end{aligned}
$$

$$H_{zy} = R_3 \frac{\partial^2}{\partial y \partial z}[(m_1+1)F_1 + (m_2+1)F_2 + (m_3+1)F_3] + R_3 \frac{\partial^2 F_4}{\partial y \partial z} + K_2 \frac{\partial^2}{\partial y \partial z}(l_1 F_1 + l_2 F_2 + l_3 F_3)$$

Harmonic equations (5.6-3) will be solved under appropriate boundary conditions, which are mentioned in Chapter 8.

5.7 Other Results of Elasticity of One-dimensional Quasicrystals

There are many other results of elasticity of one-dimensional quasicrystals, e.g., Fan *et al.* [8] on elasticity of one-dimensional crystal–crystal coexisting phase (a briefed discussion referring to Chapter 7), Chen *et al.* [9], Wang *et al.* [10], Gao *et al.* [11], Li *et al.* [12], etc. On the three-dimensional elasticity of hexagonal quasicrystals, they carried out considerable research in the area and have got quite a lot of achievements. Due to the limitation of the space, the details of their work could not be handled here and readers can refer to the original literature.

References

[1] Wang R H, Yang W G, Hu C Z and Ding D H, 1997, Point and space groups and elastic behaviour of one-dimensional quasicrystals, *J. Phys.: Condens. Matter*, **9**(11), 2411–2422.

[2] Fan T Y, 2000, Mathematical theory of elasticity and defects of quasicrystals, *Advances in Mechanics*, **30**(2), 161–174 (in Chinese).

[3] Fan T Y and Mai Y W, 2004, Elasticity theory, fracture mechanics and some relevant thermal properties of quasicrystalline materials, *Appl. Mech. Rev.*, **57**(5), 325–344.

[4] Liu G T, Fan TY and Guo R P, 2004, Governing equations and general solutions of plane elasticity of one-dimensional quasicrystals, *Int. J. Solids Struct.*, **41**(14), 3949–3959.

[5] Liu G T, 2004, The complex variable function method of the elastic theory of quasicrystals and defects and auxiliary equation method for solving some nonlinear evolution equations, Dissertation, Beijing Institute of Technology (in Chinese).

[6] Peng Y Z and Fan T Y, 2000, Elastic theory of 1-D quasiperiodic stacking 2-D crystals, *J. Phys.: Condens. Matter*, **12**(45), 9381–9387.
[7] Peng Y Z, 2001, Study on elastic three-dimensional problems of cracks for quasicrystals, Dissertation, Beijing Institute of Technology (in Chinese).
[8] Fan T Y, Xie L Y, Fan L and Wang Q Z, 2011, Study on interface of quasicrystal-crystal, *Chin. Phys. B*, **20**(7), 076102.
[9] Chen W Q, Ma Y L and Ding H J, 2004, On three-dimensional elastic problems of one-dimensional hexagonal quasicrystal bodies, *Mech. Res. Commun.*, **31**(5), 633–641.
[10] Wang X, 2006, The general solution of one-dimensional hexagonal quasicrystal, *Appl. Math. Mech.*, **33**(4), 576–580.
[11] Gao Y, Zhao Y T and Zhao B S, 2007, Boundary value problems of holomorphic vector functions in one-dimensional hexagonal quasicrystals, *Physica* B, **394**, 56–61.
[12] Li X Y, 2013, Fundamental solutions of a penny-shaped embedded crack and half-infinite plane crack in finite space of one-dimensional hexagonal quasicrystals under thermal loading, *Proc. Roy. Soc. A*, **469**, 20130023.

Chapter 6

Elasticity Theory of Two-dimensional Solid Quasicrystals of First Kind and Simplification

As has been shown in Chapter 5, in one-dimensional quasicrystals, elasticity can be decomposed into the superposition of plane elasticity and anti-plane elasticity in case the configuration is independent of the quasi-periodic axis. In this case, plane elasticity is a classical elasticity problem that need not be considered here, whereas the latter is a problem concerned with the quasiperiodic structure which we are just interested in. This decomposition leads to great simplification for the solution.

We now consider elasticity of two-dimensional quasicrystals, which is mathematically much more complicated than that of one-dimensional quasicrystals. However, the text of this chapter is limited to discuss the first kind of two-dimensional quasicrystals only. According to the introduction in Chapter 3, the concept of two-dimensional quasicrystals is developed thanks to the advances of the research work in the field. The discussion on the second kind of two-dimensional quasicrystals is given in Chapter 17, which is newly added in the present edition of this book.

The decomposition and superposition procedure developed in Chapter 5 hints that the elasticity of two-dimensional quasicrystals may somehow also be made of a decomposition and superposition for a wide range of applications. In this way, the problem can be greatly simplified and it is helpful to solve the boundary value problems by the analytic method.

Two-dimensional quasicrystals of the first kind so far observed cover four systems, i.e., those involving 5-, 8-, 10- and 12-fold symmetries, named the pentagonal, octagonal, decagonal and dodecagonal, respectively, and among them there are different Laue classes. The importance of two-dimensional quasicrystals is only less than that of three-dimensional icosahedral quasicrystals. To date, among solid quasicrystals, half of them are icosahedral quasicrystals, while the second important type belongs to two-dimensional decagonal quasicrystals. These two types constitute the majority of the quasicrystal materials. The elasticity of icosahedral quasicrystals will be discussed in Chapter 9.

We will, at first, give a brief description of the point groups and Laue classes of the two-dimensional quasicrystals. There are 31 kinds of crystallographic point groups and 26 kinds of non-crystallographic point groups of the quasicrystals. The former will not be discussed here. We focus on the latter, which is further divided into eight Laue classes and four quasicrystal systems observed so far, as listed in Table 6.1.

Like that in the one-dimensional quasicrystals, the phonon field of two-dimensional quasicrystals is transversally isotropic. If we take the x–y plane (or the x_1–x_2 plane) as the quasiperiodic plane, and the z-axis (or x_3) as the periodic axis, then the x–y plane is the elasticity isotropic plane, within which the elastic constants are

$$
\begin{aligned}
&C_{1111} = C_{2222} = C_{11} \\
&C_{1122} = C_{12} \\
&C_{1212} = C_{1111} - C_{1122} = C_{11} - C_{12} = 2C_{66}
\end{aligned}
$$

Table 6.1. Systems, Laue classes and point groups of two-dimensional quasicrystals.

Systems	Laue classes	Point groups
Pentagonal	11	$5,\ \bar{5}$
	12	$5m,\ 52,\ \bar{5}m$
Decagonal	13	$10,\ \overline{10}, 10/m$
	14	$10mm,\ 1022,\ \overline{10}m2,\ 10/mmm$
Octagonal	15	$8,\ \bar{8},\ 8/m$
	16	$8mm,\ 822,\ \bar{8}m2,\ 8/mmm$
Dodecagonal	17	$12,\ \overline{12},\ 12/m$
	18	$12mm,\ 1222,\ \overline{12}m2,\ 12/mmm$

This shows that C_{66} is not independent. Other independent elastic constants are out of the plane, i.e.,

$$C_{2323} = C_{3131} = C_{44}$$
$$C_{1133} = C_{2233} = C_{13}$$
$$C_{3333} = C_{33}$$

which are listed in Table 6.2.

The relevant phason elastic constants and phonon–phason coupling elastic constants are listed in Tables 6.3–6.6.

For Laue class 12:

$$\textbf{If } 2//x_1,\ m\perp x_1\textbf{:}\quad K_6 = R_2 = R_3 = R_6 = 0$$
$$\textbf{If } 2//x_2,\ m\perp x_2\textbf{:}\quad K_7 = R_2 = R_4 = R_6 = 0$$

Table 6.2. Phonon elastic constants in two-dimensional quasicrystals (C_{ijkl}).

	11	22	33	23	31	12
11	C_{11}	C_{12}	C_{13}	0	0	0
22	C_{12}	C_{11}	C_{13}	0	0	0
33	C_{13}	C_{13}	C_{33}	0	0	0
23	0	0	0	C_{44}	0	0
31	0	0	0	0	C_{66}	0
12	0	0	0	0	0	C_{66}

Table 6.3. Phason elastic constants for Laue class 11 (K_{ijkl}).

	11	22	23	12	13	21
11	K_1	K_2	K_7	0	K_6	0
22	K_2	K_1	K_7	0	K_6	0
23	K_7	K_7	K_4	K_6	0	$-K_6$
12	0	0	K_6	K_1	$-K_7$	$-K_2$
13	K_6	K_6	0	$-K_7$	K_4	K_7
21	0	0	$-K_6$	$-K_2$	K_7	K_1

Table 6.4. Phonon–phason coupling elastic constants for Laue class 11 (R_{ijkl}).

$\varepsilon_{ij}\backslash w_{ij}$	11	22	23	12	13	21
11	R_1	R_1	R_6	R_2	R_5	$-R_2$
22	$-R_1$	$-R_1$	$-R_6$	$-R_2$	$-R_5$	R_2
33	0	0	0	0	0	0
23	R_4	$-R_4$	0	R_3	0	R_3
31	$-R_3$	R_3	0	R_4	0	R_4
12	R_2	R_2	$-R_5$	$-R_1$	R_6	R_1

Table 6.5. The phason elastic constants for Laue class 15 (K_{ijkl}).

	11	22	23	12	13	21
11	K_1	K_2	0	K_5	0	K_5
22	K_2	K_1	0	$-K_5$	0	$-K_5$
23	0	0	K_4	0	0	0
12	K_5	$-K_5$	0	K	0	K_3
13	0	0	0	0	K_4	0
21	K_5	$-K_5$	0	K_3	0	K

Table 6.6. The phonon–phason coupling elastic constants for Laue class 15 (R_{ijkl}).

$\varepsilon_{ij}\backslash w_{ij}$	11	22	23	12	13	21
11	R_1	R_1	0	R_2	0	$-R_2$
22	$-R_1$	$-R_1$	0	$-R_2$	0	R_2
33	0	0	0	0	0	0
23	0	0	0	0	0	0
31	0	0	0	0	0	0
12	R_2	R_2	0	$-R_1$	0	R_1

For Laure class 13:

$$K_6 = K_7 = R_3 = R_4 = R_5 = R_6 = 0$$

For Laue class 14:

$$\begin{aligned} &K_6 = K_7 = R_2 = R_3 = R_4 = R_5 = R_6 = 0 \\ &K = K_1 + K_2 + K_3 \end{aligned}$$

For Laue class 16: $K_6 = R_2 = 0$.

For Laue class 17: the constants K_{ijkl} are the same as Table 6.5 and $R_{ijkl} = 0$.

For Laue class 18: the constants K_{ijkl} are the same as those of Laue 16 and $R_{ijkl} = 0$.

The experimental measurement of the above data for different quasicrystals is very important. Since putting forward of elasticity of quasicrystals, the study has been carried out. We here list some data for decagonal quasicrystals in Tables 6.7–6.9, respectively.

For an Al–Ni–Co decagonal quasicrystal, anisotropic diffuse scattering has been observed in synchrotron X-ray diffraction measurements [2]. It has been shown that the measurement can attribute the phason elastic constants, although no quantitative evaluation on K_1 and K_2 is provided. The Monte Carlo simulation was used to evaluate the phason elastic constants, e.g., given in Table 6.8 [3], where $1\,\mathrm{GPa} = 10^{10}\mathrm{dyn/cm}^2$. But the accuracy of the values given by Monte Carlo simulation should be verified.

Recently, the experimental measurement of phonon–phason coupling elastic constants for decagonal quasicrystals has been achieved. The results are listed in Table 6.9.

Tables 6.2–6.6 indicate that the phonon elasticity of two-dimensional quasicrystals is three-dimensional and that the phason

Table 6.7. Phonon elastic constants of a decagonal quasicrystal (the unit of C_{ij}, B, G is GPa) by experimental measurement [1].

Alloy	C_{11}	C_{33}	C_{44}	C_{12}	C_{13}	B	G	v
Al–Ni–Co	234.33	232.22	70.19	57.41	66.63	120.25	79.98	0.228

Note: B: bulk modulus; G: shear modulus; ν: Poisson's ratio.

Table 6.8. Phason elastic constants of an Al–Ni–Co decagonal quasicrystal by Monte Carlo simulation [3].

Alloy	K_1 ($10^{12}\mathrm{dyn/cm}^2$)	K_2 ($10^{12}\mathrm{dyn/cm}^2$)
Al–Ni–Co	1.22	0.24

Table 6.9. Coupling elastic constants for Al–Ni–Co decagonal quasicrystals with point group$10, \overline{10}$. [4].

Alloy	R_1 (GPa)	$\|R_2\|$ (GPa)
Al–Ni–Co	−1.1	<0.2

and phonon–phason coupling elasticity are also three-dimensional. In general, they cannot be reduced to two-dimensional problems. In this case, there are 29 field variables and 29 field equations, making their solution being very hard to obtain.

In practice, there is often a case in which the configuration of a system is uniform along a periodic axis, say the z-axis, both physically and geometrically, so that the field variables are free from the z-coordinate, i.e.,

$$\frac{\partial}{\partial z} = 0. \tag{6.0-1}$$

In this case, the elasticity of two-dimensional quasicrystals can be decomposed into a superposition of a plane elasticity of quasicrystal and an anti-plane elasticity of crystal. The latter is a pure phonon or classical elasticity, whose governing equation and boundary conditions are decoupled with the plane elasticity of quasiperiodic structure, which can be treated separately.

Uniform configuration along the z-axis physically and geometrically represents several meaningful physical problems, e.g., a straight dislocation line or a Griffith crack along the direction, and so on. This shows that the decomposition and superposition procedure presents important physical meaning.

In what follows, final governing equations for the plane and anti-plane elasticity of the four different systems will be derived by introducing some displacement or stress potential functions. We will see that the field equations are dramatically simplified. This is greatly helpful to solve them by separate variable method; furthermore, we use the Fourier analysis or complex analysis method to construct analytic solutions of some boundary value problems of the equations. The mathematical methods and numerous exact solutions for different boundary value problems will be performed in Chapters 7 and 8 in detail. They reveal the exact mathematical structure of dislocations and cracks in two-dimensional quasicrystalline materials. In addition, they develop further the potential theory of classical elasticity as well as the classical methods of solution in mathematical physics.

6.1 Basic Equations of Plane Elasticity in Two-dimensional Quasicrystals: Point Groups $5m$ and $10mm$ in Five- and 10-fold Symmetries

Following the introduction given in Chapter 1, it is understood that the sign 10 in the symbol $10mm$ represents a 10-fold rotation symmetry and m signifies a mirror symmetry. The sign $10mm$ therefore means a combined operation of one rotation symmetry and two mirror symmetries. For other signs, the explanation is similar.

The point groups $5m$ and $10mm$ quasicrystals are quasicrystals with five- and 10-fold symmetries, respectively, whose plane elasticity has been studied earlier, see, e.g., in reviews given by Fan and Mai [5] or by Hu *et al.* [6] or by Bohsung *et al.* [7]. The first solution of dislocation of pentagonal quasicrystals was given by De and Pelcovis [8] in a point group $5m$ two-dimensional quasicrystal. The formulation and the solution method developed here and hereafter are quite different from those given by Ref. [8].

The diffraction patterns and relevant Penrose tilings are shown in Figs. 6.1 and 6.2. The Penrose tiling of pentagonal quasicrystals was shown in Chapter 3.

Quasicrystals with these symmetries belong to two-dimensional quasicrystals, i.e., the atomic arrangement is quasiperiodic in a plane and periodic in the third direction. For clarity, the quasicrystals can be defined as being generated by stacking from planar quasiperiodic structures of five-fold or 10-fold symmetry along the third symmetry axis. Here, the quasiperiodic plane is the x–y plane, and the five-fold or 10-fold rotational axis is the z-axis, which is the only axis of symmetry. Due to the presence of scaling of incommensurate length in the quasiperiodic plane, it leads to the additional degree of freedom that does not hold in conventional crystals, named the phason field $\mathbf{w}$.

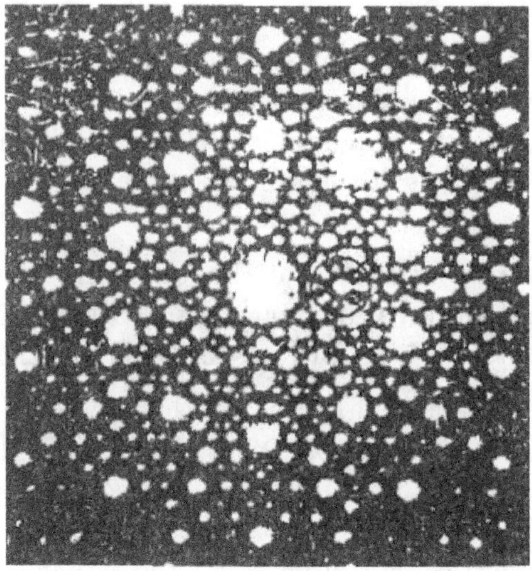

Fig. 6.1. Diffraction pattern of two-dimensional quasicrystal with 10-fold symmetry.

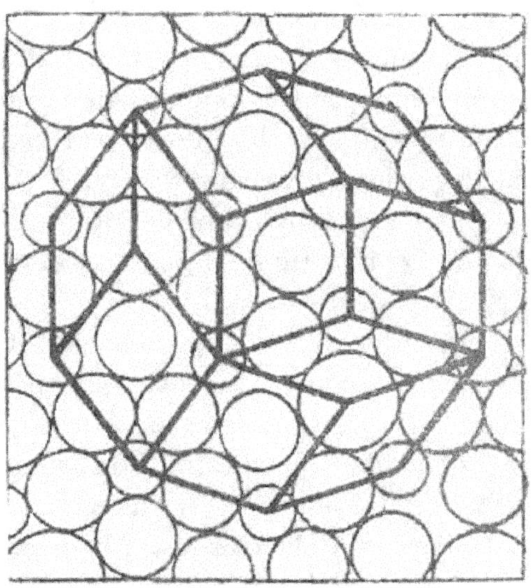

Fig. 6.2. Penrose tiling of point group $10mm$ of 10-fold symmetry quasicrystals.

Although two-dimensional quasicrystals can be understood as a stacked planar quasiperiodic structure along the periodic symmetrical axis, it is three-dimensional in elasticity and is different from the plane problem in classical elasticity. It can be decomposed into a superposition of plane and anti-plane elasticity only in particular cases. We here consider only the plane elasticity because the anti-plane elasticity is a classical one and is independent of phason variables.

Tables 6.2–6.6 list all the elastic constants. Considering a plane in two-dimensional quasicrystals, and assuming that it is perpendicular to the periodic symmetrical axis (e.g., z-axis),

$$\mathbf{u} = (u_x, u_y, u_z), \quad \mathbf{w} = (w_x, w_y, 0)$$

so the strains $w_{zz} = w_{zx} = w_{xz} = w_{zy} = w_{yz} = 0$, and assumption (6.0-1) leads to $\varepsilon_{zz} = \varepsilon_{xz} = \varepsilon_{yz} = 0$, then Table 6.2 is simplified to Table 6.1 for phonon elastic constants for the plane elasticity.

The phonon elastic constants listed in Table 6.10 can be expressed in tensors of four rank

$$C_{ijkl} = L\delta_{ij}\delta_{kl} + M(\delta_{jk}\delta_{jl} + \delta_{il}\delta_{jk}) \quad (i, j, k, l = 1, 2) \tag{6.1-1}$$

$$L = C_{12}, \quad M = (C_{11} - C_{12})/2 = C_{66} \tag{6.1-2}$$

There are only two independent phonon elastic constants.

The data given by Table 6.10 and relation by Eqs. (6.1-1) and (6.1-2) hold for all two-dimensional quasicrystals for the plane elasticity. In the plane elasticity for two-dimensional quasicrystals

Table 6.10. Phonon elastic constants for plane elasticity of two-dimensional quasicrystals.

	11	22	12	21
11	C_{11}	C_{12}	0	0
22	C_{12}	C_{11}	0	0
12	0	0	C_{66}	C_{66}
21	0	0	C_{66}	C_{66}

Table 6.11. Phason elastic constants for plane elasticity for point group $10mm$ quasicrystals.

	11	22	12	21
11	K_1	K_2	0	0
22	K_2	K_1	0	0
12	0	0	K_1	$-K_2$
21	0	0	$-K_2$	K_1

Table 6.12. Phonon–phason coupling elastic constants for plane elasticity.

$\varepsilon_{ij}\backslash w_{ij}$	11	22	12	21
11	R	R	0	0
22	$-R$	$-R$	0	0
12	0	0	$-R$	R
21	0	0	$-R$	R

with point groups $5m$, $52, \bar{5}m$, $10mm$, 1022, $\overline{10}m2$ and $10/mmm$, Table 6.3 is simplified to Table 6.11.

This means that

$$\begin{aligned} &K_{1111} = K_{2222} = K_{2121} = K_1 \\ &K_{1122} = K_{2211} = -K_{2112} = -K_{1221} = K_2 \end{aligned} \tag{6.1-3}$$

and other $K_{ijkl} = 0$, and the expression of them by tensors of four rank is

$$K_{ijkl} = K_1\delta_{ik}\delta_{jl} + K_2(\delta_{ij}\delta_{kl} - \delta_{il}\delta_{jk}) \quad (i, j, k, l = 1, 2) \tag{6.1-4}$$

Table 6.4 in this case is simplified into Table 6.12 as follows. This shows that

$$\begin{aligned} R_{1111} &= R_{1122} = -R_{2222} = R_{1221} = R_{2121} \\ &= -R_{1212} = -R_{2211} = -R_{2112} = R \end{aligned} \tag{6.1-5}$$

or

$$R_{ijkl} = R(\delta_{i1} - \delta_{i2})(\delta_{ij}\delta_{kl} - \delta_{il}\delta_{jk} + \delta_{iljk})(i, j, k, l = 1, 2) \tag{6.1-6}$$

One can find that for plane elasticity, point group $5m$ has the same elastic constants with point group $10mm$, so they can be collectively discussed here.

The definition of strain tensor is as given in Chapter 4, i.e.,

$$\varepsilon_{ij} = \frac{1}{2}\left(\frac{\partial u_i}{\partial x_j} + \frac{\partial u_j}{\partial x_i}\right), \quad w_{ij} = \frac{\partial w_i}{\partial x_j} \tag{6.1-7}$$

In Chapter 4, it was seen that the stress, strain and elastic constant tensors can be expressed by matrices. The above-mentioned elastic constants may be denoted by matrix $[CKR]$. For this case, the strain vector defined by (4.4-12) is simplified into

$$[\varepsilon_{ij} \; w_{ij}] = [\varepsilon_{11} \; \varepsilon_{22} \; \varepsilon_{12} \; \varepsilon_{21} \; w_{11} \; w_{22} \; w_{12} \; w_{21}] \tag{6.1-8}$$

and $[CKR]$ is

$$[CKR] = \begin{bmatrix} L+2M & L & 0 & 0 & R & R & 0 & 0 \\ L & L+2M & 0 & 0 & -R & -R & 0 & 0 \\ 0 & 0 & M & M & 0 & 0 & -R & R \\ 0 & 0 & M & M & 0 & 0 & -R & R \\ R & -R & 0 & 0 & K_1 & K_2 & 0 & 0 \\ R & -R & 0 & 0 & K_2 & K_1 & 0 & 0 \\ 0 & 0 & -R & -R & 0 & 0 & K_1 & -K_2 \\ 0 & 0 & R & R & 0 & 0 & -K_2 & K_1 \end{bmatrix}$$

$$= \begin{bmatrix} L+2M & L & 0 & 0 & R & R & 0 & 0 \\ & L+2M & 0 & 0 & -R & -R & 0 & 0 \\ & & M & M & 0 & 0 & -R & R \\ & & & M & 0 & 0 & -R & R \\ & & & & K_1 & K_2 & 0 & 0 \\ & symmetry & & & & K_1 & 0 & 0 \\ & & & & & & K_1 & -K_2 \\ & & & & & & & K_1 \end{bmatrix} \tag{6.1-9}$$

The free energy density (or strain energy density)

$$\begin{aligned}F &= \frac{1}{2}L\varepsilon_{ii}\varepsilon_{ii} + M\varepsilon_{ij}\varepsilon_{ij} + \frac{1}{2}K_1 w_{ij}w_{ij} + K_2(w_{xx}w_{yy} - w_{xy}w_{xy}) \\ &\quad + R\left[(\varepsilon_{xx} - \varepsilon_{yy})(w_{xx} + w_{yy}) + 2\varepsilon_{xy}(w_{xy} - w_{yx})\right]\end{aligned} \tag{6.1-10}$$

From (6.1-9) and (4.5-3) of Chapter 4, or from (6.1-10) and (4.5-1) of Chapter 4, the generalized Hooke's law for plane elasticity of point group $10mm$ quasicrystals of 10-fold symmetry is as

$$\left.\begin{aligned}\sigma_{xx} &= L(\varepsilon_{xx} + \varepsilon_{yy}) + 2M\varepsilon_{xx} + R(w_{xx} + w_{yy}) \\ \sigma_{yy} &= L(\varepsilon_{xx} + \varepsilon_{yy}) + 2M\varepsilon_{yy} - R(w_{xx} + w_{yy}) \\ \sigma_{xy} &= \sigma_{yx} = 2M\varepsilon_{xy} + R(w_{yx} - w_{xy}) \\ H_{xx} &= K_1 w_{xx} + K_2 w_{yy} + R(\varepsilon_{xx} - \varepsilon_{yy}) \\ H_{yy} &= K_1 w_{yy} + K_2 w_{xx} + R(\varepsilon_{xx} - \varepsilon_{yy}) \\ H_{xy} &= K_1 w_{xy} - K_2 w_{yx} - 2R\varepsilon_{xy} \\ H_{yx} &= K_1 w_{yx} - K_2 w_{xy} + 2R\varepsilon_{xy}\end{aligned}\right\} \tag{6.1-11}$$

In addition, for anti-plane elasticity, there is

$$\left.\begin{aligned}\sigma_{xz} &= 2M\varepsilon_{xz} \\ \sigma_{yz} &= 2M\varepsilon_{yz}\end{aligned}\right\} \tag{6.1-12}$$

Monograph [9] pointed out first that (6.1-9)–(6.1-11) hold for the plane elasticity for both point group 5m and point group $10mm$; this is identical to the argument of Ref [6].

Equation (6.1-11) is the physical basis of elasticity of point groups 5m and 10mm quasicrystals. The geometry (or kinetics) basis of the subject is equation (6.1-7). Another necessary basis comes from statics, i.e.,

$$\left.\begin{aligned}\frac{\partial \sigma_{xx}}{\partial x} + \frac{\partial \sigma_{xy}}{\partial y} = 0, \quad \frac{\partial \sigma_{yx}}{\partial x} + \frac{\partial \sigma_{yy}}{\partial y} = 0 \\ \frac{\partial H_{xx}}{\partial x} + \frac{\partial H_{xy}}{\partial y} = 0, \quad \frac{\partial H_{yx}}{\partial x} + \frac{\partial H_{yy}}{\partial y} = 0\end{aligned}\right\} \tag{6.1-13}$$

In addition, for the anti-plane elasticity, there is

$$\frac{\partial \sigma_{zx}}{\partial x} + \frac{\partial \sigma_{zy}}{\partial y} = 0. \tag{6.1-14}$$

Here, the body force density is omitted.

From (6.1-7), (6.1-11) and (6.1-13), we found that there are eighteen field variables, i.e., four displacements u_x, u_y, w_x, w_y, seven strains $\varepsilon_{xx}, \varepsilon_{yy}, \varepsilon_{xy} = \varepsilon_{yx}, w_{xx}, w_{yy}, w_{xy}, w_{yx}$ and seven stresses $\sigma_{xx}, \sigma_{yy}, \sigma_{xy} = \sigma_{yx}, H_{xx}, H_{yy}, H_{xy}, H_{yx}$. And the number of corresponding field equations is also eighteen, including four equations of statics, seven stress–strain relations and seven equations of deformation geometry. The number of field equations is equal to that of field variables. This means that the mathematical presentation of the problem is consistent and solvable under appropriate boundary conditions. In addition, (6.1-7), (6.1-12) and (6.1-14) give the description of anti-plane pure classical (phonon) elasticity. De and Pelcovits [10] solved the equations for the dislocation problem by using the Green function method and iteration procedure. Ding *et al.* [11] solved a similar problem in terms of Fourier transforms and the Green function method. Taking a different way from either De and Pelcovits or Ding *et al.*, we at first reduce the 18 equations into a single partial differential of higher order by introducing some displacement potentials or stress potentials. This will be very helpful to solve by separate variable procedures and to develop systematic and straightforward methods, in which the Fourier transform method and complex analysis method are the best ones. In the following, we derive the final governing equations and fundamental solutions for different quasicrystal systems. The applications of the theory and methodology will be given in detail in Chapters 7 and 8 for dislocation and crack problems of quasicrystals, respectively.

6.2 Simplification of the Basic Equation Set: Displacement Potential Function Method

The number of basic equations as given in the previous section is too large and it is very hard to solve them directly. In mathematical physics, a conventional procedure is to reduce the number of field equations. In the classical theory of elasticity, there is a way by introducing the so-called displacement or stress potential functions to realize the purpose, and the procedure is called the displacement potential method or stress potential method, respectively. In this section, we eliminate the stress and strain components in the basic equations and obtain some equilibrium equations expressed by displacement components, thus the strain compatibility is automatically satisfied. Furthermore, by introducing a displacement potential function, the final governing equations become a single quadruple harmonic equation concerning the displacement potential function, thus this is the so-called displacement potential method. In the following section, we will introduce the stress potential method; there the governing equation becomes a single quadruple harmonic equation concerning the stress potential function that would be the so-called stress potential method. In this way, the huge number of equations involving elasticity can be simplified very much.

Substituting (6.1-7) into (6.1-11) and then into (6.1-12), we have

$$\left.\begin{aligned}
&M\nabla^2 u_x + (L+M)\frac{\partial}{\partial x}\nabla\cdot\mathbf{u} + R\left(\frac{\partial^2 w_x}{\partial x^2} + 2\frac{\partial^2 w_y}{\partial x\partial y} - \frac{\partial w_x}{\partial y^2}\right) = 0\\
&M\nabla^2 u_y + (L+M)\frac{\partial}{\partial y}\nabla\cdot\mathbf{u} + R\left(\frac{\partial^2 w_y}{\partial x^2} - 2\frac{\partial^2 w_x}{\partial x\partial y} - \frac{\partial^2 w_y}{\partial y^2}\right) = 0\\
&K_1\nabla^2 w_x + R\left(\frac{\partial^2 u_x}{\partial x^2} - 2\frac{\partial^2 u_y}{\partial x\partial y} - \frac{\partial^2 u_x}{\partial y^2}\right) = 0\\
&K_1\nabla^2 w_y + R\left(\frac{\partial^2 u_y}{\partial x^2} + 2\frac{\partial^2 u_x}{\partial x\partial y} - \frac{\partial^2 u_y}{\partial y^2}\right) = 0
\end{aligned}\right\} \tag{6.2-1}$$

in which

$$\nabla^2 = \frac{\partial^2}{\partial x^2} + \frac{\partial^2}{\partial y^2}, \quad \nabla\cdot\mathbf{u} = \frac{\partial u_x}{\partial x} + \frac{\partial u_y}{\partial y}$$

Equations (6.2-1) are actually the displacement equilibrium equations of the plane elasticity of point group $5m$ with 5 fold symmetry or point group $10mm$ with 10 fold symmetry, as mentioned in the previous section. Here, there are only four displacement components u_x, u_y, w_x, w_y, i.e., the number of field variables is only four, and the number of order of equations is eight.

Observation for the first two equations of (6.2-1) suggests that by introducing the unknown functions $\varphi(x, y)$ and $\psi(x, y)$ as follows:

$$\left.\begin{aligned}
u_x &= (L+M)\frac{\partial^2\varphi}{\partial x\partial y} + M\frac{\partial^2\psi}{\partial x^2} + (L+2M)\frac{\partial^2\psi}{\partial y^2}\\
u_y &= -\left[(L+2M)\frac{\partial^2\varphi}{\partial x^2} + M\frac{\partial^2\varphi}{\partial y^2} + (L+M)\frac{\partial^2\psi}{\partial x\partial y}\right]\\
w_x &= -\frac{M(L+2M)}{R}\left[2\frac{\partial^2\varphi}{\partial x\partial y} + \frac{\partial^2\psi}{\partial x^2} - \frac{\partial^2\psi}{\partial y^2}\right]\\
w_y &= \frac{M(L+2M)}{R}\left[\frac{\partial^2\varphi}{\partial x^2} - \frac{\partial^2\varphi}{\partial y^2} - 2\frac{\partial^2\psi}{\partial x\partial y}\right]
\end{aligned}\right\} \tag{6.2-2}$$

then the first two equations are automatically satisfied already. Substituting (6.2-2) into the last two equations in (6.2-1), there follow the equations:

$$\left.\begin{aligned}
(\alpha\Pi_1 + \beta\Pi_2)\frac{\partial^2\varphi}{\partial x\partial y} + \left(\alpha\Pi_1\frac{\partial^2}{\partial y^2} - \beta\Pi_2\frac{\partial^2}{\partial x^2}\right)\psi = 0\\
\left(\alpha\Pi_2\frac{\partial^2}{\partial x^2} - \beta\Pi_1\frac{\partial^2}{\partial y^2}\right)\varphi + (\alpha\Pi_2 + \beta\Pi_1)\frac{\partial^2\psi}{\partial x\partial y} = 0
\end{aligned}\right\} \tag{6.2-3}$$

where

$$\Pi_1 = 3\frac{\partial^2}{\partial x^2} - \frac{\partial^2}{\partial y^2},\quad \Pi_2 = 3\frac{\partial^2}{\partial y^2} - \frac{\partial^2}{\partial x^2} \tag{6.2-4}$$

$$\left.\begin{aligned}
\alpha = R(L+2M) - \omega K_1,\quad \beta = RM - \omega K_1\\
\delta = RM - \omega K_1,\quad \omega = M(L+2M)/R
\end{aligned}\right\} \tag{6.2-5}$$

in which δ will be used in formula (6.2-6).

Equation (6.2-3) is simpler enough compared with (6.2-1), but it is not the simplest yet. Based on an observation of the first one in Eq. (6.2-3), if we assume

$$\varphi = \left(\beta\Pi_2\frac{\partial^2}{\partial x^2} - \alpha\Pi_1\frac{\partial^2}{\partial y^2}\right)F,\quad \psi = (\alpha\Pi_1 + \delta\Pi_2)\frac{\partial^2 F}{\partial x\partial y} \tag{6.2-6}$$

in which $F(x, y)$ is any function, then the equation has been automatically satisfied. Substituting (6.2-6) into the second one of Eq. (6.2-3), after some simple manipulation, it reduces to

$$\nabla^2\nabla^2\nabla^2\nabla^2 F = 0 \tag{6.2-7}$$

This is the final governing equation of the plane elasticity of point groups $5m$ and $10mm$ quasicrystals. We call $F(x, y)$ a displacement potential function or displacement potential briefly. Equation (6.2-7) is a quadruple harmonic equation, the order of which is much higher than that in the classical elasticity, where that is a biharmonic equation.

All of displacement and stress components can be expressed by a potential function $F(x, y)$ as follows:

$$\left.\begin{aligned} u_x &= [M\alpha\Pi_1 + (L+2M)\beta\Pi_2]\frac{\partial^2}{\partial x\partial y}\nabla^2 F \\ u_y &= \left[M\alpha\Pi_1\frac{\partial^2}{\partial y^2} - (L+2M)\beta\Pi_2\frac{\partial^2}{\partial x^2}\right]\nabla^2 F \end{aligned}\right\} \tag{6.2-8}$$

$$\left.\begin{aligned} w_x &= \omega(\alpha-\beta)\Pi_1\Pi_2\frac{\partial^2}{\partial x\partial y}F \\ w_y &= -\omega\left[\alpha\Pi_1^2\frac{\partial^2}{\partial y^2} + \beta\Pi_2^2\frac{\partial^2}{\partial x^2}\right]F \end{aligned}\right\} \tag{6.2-9}$$

$$\left.\begin{aligned} \sigma_{xx} &= 2M(L+M)\alpha\Pi_1\frac{\partial^3}{\partial y^3}\nabla^2 F \\ \sigma_{yy} &= 2M(L+M)\alpha\Pi_1\frac{\partial^3}{\partial x^2\partial y}\nabla^2 F \\ \sigma_{xy} &= \sigma_{yx} = -2M(L+M)\alpha\Pi_1\frac{\partial^3}{\partial x\partial y^2}\nabla^2 F \end{aligned}\right\} \tag{6.2-10}$$

$$\left.\begin{aligned}
H_{xx} &= \alpha\beta\frac{\partial}{\partial y}\nabla^2\nabla^2\nabla^2 F \\
&\quad +\omega(K_1-K_2)\frac{\partial}{\partial y}\left(\alpha\Pi_1^2\frac{\partial^2}{\partial y^2}+\beta\Pi_2^2\frac{\partial^2}{\partial x^2}\right)F \\
H_{yy} &= \alpha\beta\frac{\partial}{\partial y}\nabla^2\nabla^2\nabla^2 F \\
&\quad -\omega(K_1-K_2)(\alpha-\beta)\Pi_1\Pi_2\frac{\partial^3}{\partial x^2\partial y}F \\
H_{xy} &= -\alpha\beta\frac{\partial}{\partial x}\nabla^2\nabla^2\nabla^2 F \\
&\quad -\omega(K_1-K_2)\frac{\partial}{\partial x}\left(\alpha\Pi_1^2\frac{\partial^2}{\partial y^2}+\beta\Pi_2^2\frac{\partial^2}{\partial x^2}\right)F \\
H_{yx} &= \alpha\beta\frac{\partial}{\partial x}\nabla^2\nabla^2\nabla^2 F \\
&\quad +\omega(K_1-K_2)(\alpha-\beta)\Pi_1\Pi_2\frac{\partial^3}{\partial x\partial y^2}F
\end{aligned}\right\} \tag{6.2-11}$$

Li and Fan [12] have suggested the approach, and the practice shows that it is very effective. In the following two chapters, many applications of the approach will be given.

In addition, (6.1-12) and (6.1-14) yield the final governing equation for anti-plane elasticity

$$\nabla^2 u_z = 0. \tag{6.2-12}$$

It is obvious that the anti-plane problem is decoupled from the plane one.

6.3 Simplification of Basic Equations Set: Stress Potential Function Method

The stress potential method has widely been used in classical theory of elasticity. The first author of the book and his students extended it to the study of elasticity of quasicrystals, refer to [13, 14].

From (6.1-7), the strain compatibility equations

$$\left.\begin{array}{l}\dfrac{\partial^2 \varepsilon_{xx}}{\partial y^2}+\dfrac{\partial^2 \varepsilon_{yy}}{\partial x^2}=2\dfrac{\partial^2 \varepsilon_{xy}}{\partial x \partial y}\\[2ex]\dfrac{\partial w_{xy}}{\partial x}=\dfrac{\partial w_{xx}}{\partial y}, \quad \dfrac{\partial w_{yx}}{\partial y}=\dfrac{\partial w_{yy}}{\partial x}\end{array}\right\} \tag{6.3-1}$$

The strain components ε_{ij} and w_{ij} can be expressed by the stress components σ_{ij} and H_{ij}, i.e.,

$$\left.\begin{array}{l}\varepsilon_{xx}=\dfrac{1}{4(L+M)}(\sigma_{xx}+\sigma_{yy})\\[1ex]\qquad +\dfrac{1}{4C}[(K_1+K_2)(\sigma_{xx}+\sigma_{yy})-2R(H_{xx}+H_{yy})]\\[2ex]\varepsilon_{yy}=\dfrac{1}{4(L+M)}(\sigma_{xx}+\sigma_{yy})\\[1ex]\qquad -\dfrac{1}{4C}[(K_1+K_2)(\sigma_{xx}+\sigma_{yy})-2R(H_{xx}+H_{yy})]\\[2ex]\varepsilon_{xy}=\varepsilon_{yx}=\dfrac{1}{C}[(K_1+K_2)\sigma_{xy}-R(H_{xy}+H_{yx})]\\[2ex]w_{xx}=\dfrac{1}{2(K_1-K_2)}(H_{xx}-H_{yy})\\[1ex]\qquad +\dfrac{1}{2C}[M(H_{xx}+H_{yy})-R(\sigma_{xx}-\sigma_{yy})]\\[2ex]w_{yy}=-\dfrac{1}{2(K_1-K_2)}(H_{xx}-H_{yy})\\[1ex]\qquad +\dfrac{1}{2C}[M(H_{xx}+H_{yy})-R(\sigma_{xx}-\sigma_{yy})]\\[2ex]w_{xy}=\dfrac{1}{C}\left[R\sigma_{xy}-\dfrac{(R^2-MK_1)H_{xy}+(R^2-MK_2)H_{yx}}{K_1-K_2}\right]\\[2ex]w_{yx}=\dfrac{1}{C}\left[-R\sigma_{xy}-\dfrac{(R^2-MK_1)H_{xy}+(R^2-MK_2)H_{yx}}{K_1-K_2}\right]\end{array}\right\} \tag{6.3-2}$$

with

$$C=M(K_1+K_2)-2R^2 \tag{6.3-3}$$

Substituting (6.3-2) into (6.3-1), one can find the strain compatibility equations expressed by stress components (because they are too long to be listed, but we will give them in Section 6.7). In addition, there are equilibrium equations

$$\left.\begin{aligned}&\frac{\partial\sigma_{xx}}{\partial x}+\frac{\partial\sigma_{xy}}{\partial y}=0, \quad \frac{\partial\sigma_{yx}}{\partial x}+\frac{\partial\sigma_{yy}}{\partial y}=0\\&\frac{\partial H_{xx}}{\partial x}+\frac{\partial H_{xy}}{\partial y}=0, \quad \frac{\partial H_{yx}}{\partial x}+\frac{\partial H_{yy}}{\partial y}=0\end{aligned}\right\} \tag{6.3-4}$$

so that we have seven equations in total in which there are three compatibility equations expressed by stress components and four equilibrium equations, and the number of the unknown functions is also 7, i.e., $\sigma_{xx}, \sigma_{yy}, \sigma_{xy} = \sigma_{yx}, H_{xx}, H_{yy}, H_{xy}, H_{yx}$. The equation set is closed and is solvable.

If introducing the stress functions φ, ψ_1 and ψ_2 such as

$$\left.\begin{aligned}&\sigma_{xx}=\frac{\partial^2\varphi}{\partial y^2}, \quad \sigma_{yy}=\frac{\partial^2\varphi}{\partial x^2}, \quad \sigma_{xy}=\sigma_{yx}=-\frac{\partial^2\varphi}{\partial x\partial y}\\&H_{xx}=\frac{\partial\psi_1}{\partial y}, \quad H_{xy}=-\frac{\partial\psi_1}{\partial x}, \quad H_{yx}=-\frac{\partial\psi_2}{\partial y}, \quad H_{yy}=\frac{\partial\psi_2}{\partial x}\end{aligned}\right\} \tag{6.3-5}$$

then the equilibrium equations (6.3-4) are automatically satisfied. Substituting (6.3-5) into deformation compatibility equations expressed by stress components yields

$$\left.\begin{aligned}&\frac{1}{2C(L+M)}\nabla^2\nabla^2\varphi+\frac{K_1+K_2}{2C}\nabla^2\nabla^2\varphi\\&\qquad+\frac{R}{C}\left(\frac{\partial}{\partial y}\Pi_1\psi_1-\frac{\partial}{\partial x}\Pi_2\psi_2\right)=0\\&\left(\frac{C}{K_1-K_2}+M\right)\nabla^2\psi_1+R\frac{\partial}{\partial y}\Pi_1\varphi=0\\&\left(\frac{C}{K_1-K_2}+M\right)\nabla^2\psi_2-R\frac{\partial}{\partial x}\Pi_2\varphi=0\end{aligned}\right\} \tag{6.3-6}$$

in which Π_1 and Π_2 are defined by (6.2-4) and C is given by (6.3-3). By now, the number of equations and unknown functions are reduced to 3.

Now, introducing a new unknown function $G(x,y)$ such as

$$\left.\begin{aligned}&\varphi = D\frac{\partial}{\partial y}\Pi_1\nabla^2 G\\&\psi_1 = -\frac{1}{R}(MK_1 - R_2)[(L+2M)(K_1-K_2) - 2R^2]\nabla^2\nabla^2\nabla^2 G\\&\qquad + (L+M)(K_1-K_2)R\frac{\partial^2}{\partial x\partial y}\Pi_1\Pi_2 G\\&\psi_2 = (L+M)(K_1-K_2)R\frac{\partial^2}{\partial x\partial y}\Pi_1\Pi_2 G\end{aligned}\right\} \tag{6.3-7}$$

If

$$\nabla^2\nabla^2\nabla^2\nabla^2 G = 0 \tag{6.3-8}$$

then Eq. (6.3-6) are satisfied, in which

$$D = 2(MK_1 - R^2)(L+M) \tag{6.3-9}$$

At the same time, Eq. (6.2-12) holds for anti-plane elasticity, too.

6.4 Plane Elasticity of Point Group $5,\bar{5}$ and $10,\overline{10}$ Pentagonal and Decagonal Quasicrystals

For plane elasticity, the fivefold and ten-fold symmetry quasicrystals of point groups $5,\bar{5}$ and $10,\overline{10}$ are different in elasticity with that of point groups $5m$ and $10mm$; the difference lies in only the phonon–phason coupling elastic constants, in which the former has two coupling elastic constants R_1 and R_2 rather than one R, So, we have

$$\begin{aligned}R_{ijkl} &= R_1(\delta_{i1}-\delta_{i2})(\delta_{ij}\delta_{kl} - \delta_{ik}\delta_{jl} + \delta_{il}\delta_{jk})\\&\quad + R_2[(1-\delta_{ij})\delta_{kl} + \delta_{ij}(\delta_{i1}-\delta_{i2})\\&\quad \times(\delta_{k1}\delta_{l2} - \delta_{k2}\delta_{l1})] \quad i,j,k,l = 1,2\end{aligned} \tag{6.4-1}$$

Apart from this, the phonon and phason elastic constants of point group $5,\bar{5}$ and point group $10,\overline{10}$ quasicrystals are the same as those of point groups $5m$ and $10mm$. The corresponding elastic constant matrix is

$$[CKR] = \begin{bmatrix} L+2M & L & 0 & 0 & R_1 & R_1 & R_2 & -R_2 \\ L & L+2M & 0 & 0 & -R_1 & -R_1 & -R_2 & R_2 \\ 0 & 0 & M & M & R_2 & R_2 & -R_1 & R_1 \\ 0 & 0 & M & M & R_2 & R_2 & -R_1 & R_1 \\ R_1 & -R_1 & R_2 & R_2 & K_1 & K_2 & 0 & 0 \\ R_1 & -R_1 & R_2 & R_2 & K_2 & K_1 & 0 & 0 \\ R_2 & -R_2 & -R_1 & -R_1 & 0 & 0 & K_1 & -K_2 \\ -R_2 & R_2 & R_1 & R_1 & 0 & 0 & -K_2 & K_1 \end{bmatrix}$$

$$= \begin{bmatrix} L+2M & L & 0 & 0 & R_1 & R_1 & R_2 & -R_2 \\ & L+2M & 0 & 0 & -R_1 & -R_1 & -R_2 & R_2 \\ & & M & M & R_2 & R_2 & -R_1 & R_1 \\ & & & M & R_2 & R_2 & -R_1 & R_1 \\ & symmetry & & & K_1 & K_2 & 0 & 0 \\ & & & & & K_1 & 0 & 0 \\ & & & & & & K_1 & -K_2 \\ & & & & & & & K_1 \end{bmatrix} \tag{6.4-2}$$

According to this elastic constant matrix, the stress–strain relation can be written as

$$\left.\begin{aligned} \sigma_{xx} &= L(\varepsilon_{xx}+\varepsilon_{yy}) + 2M\varepsilon_{xx} + R_1(w_{xx}+w_{yy}) + R_2(w_{xy}-w_{yx}) \\ \sigma_{yy} &= L(\varepsilon_{xx}+\varepsilon_{yy}) + 2M\varepsilon_{yy} - R_1(w_{xx}+w_{yy}) - R_2(w_{xy}-w_{yx}) \\ \sigma_{xy} &= \sigma_{yx} = 2M\varepsilon_{xy} + R_1(w_{yx}-w_{xy}) + R_2(w_{xx}+w_{yy}) \\ H_{xx} &= K_1 w_{xx} + K_2 w_{yy} + R_1(\varepsilon_{xx}-\varepsilon_{yy}) + 2R_2\varepsilon_{xy} \\ H_{yy} &= K_1 w_{yy} + K_2 w_{xx} + R_1(\varepsilon_{xx}-\varepsilon_{yy}) + 2R_2\varepsilon_{xy} \\ H_{xy} &= K_1 w_{xy} - K_2 w_{yx} - 2R_1\varepsilon_{xy} + R_2(\varepsilon_{xx}-\varepsilon_{yy}) \\ H_{yx} &= K_1 w_{yx} - K_2 w_{xy} + 2R_1\varepsilon_{xy} - R_2(\varepsilon_{xx}-\varepsilon_{yy}) \end{aligned}\right\} \tag{6.4-3}$$

In addition, the stresses σ_{ij} and H_{ij} satisfy the same equilibrium equations as (6.1-13).

Substituting (6.1-7) into (6.4-3) and then into (6.1-13) leads to the equilibrium equations expressed by displacements such as

$$
\begin{aligned}
&M\nabla^2 u_x + (L+M)\frac{\partial}{\partial x}\nabla\cdot\mathbf{u} + R_1\left(\frac{\partial^2 w_x}{\partial x^2} + 2\frac{\partial^2 w_y}{\partial x\partial y} - \frac{\partial^2 w_x}{\partial y^2}\right)\\
&\quad - R_2\left(\frac{\partial^2 w_y}{\partial x^2} - 2\frac{\partial^2 w_x}{\partial x\partial y} - \frac{\partial^2 w_y}{\partial y^2}\right) = 0\\
&M\nabla^2 u_y + (L+M)\frac{\partial}{\partial y}\nabla\cdot\mathbf{u} + R_1\left(\frac{\partial^2 w_y}{\partial x^2} - 2\frac{\partial^2 w_x}{\partial x\partial y} - \frac{\partial^2 w_y}{\partial y^2}\right)\\
&\quad + R_2\left(\frac{\partial^2 w_x}{\partial x^2} + 2\frac{\partial^2 w_y}{\partial x\partial y} - \frac{\partial^2 w_x}{\partial y^2}\right) = 0\\
&K_1\nabla^2 w_x + R_1\left(\frac{\partial^2 u_x}{\partial x^2} - 2\frac{\partial^2 u_y}{\partial x\partial y} - \frac{\partial^2 u_x}{\partial y^2}\right)\\
&\quad + R_2\left(\frac{\partial^2 u_y}{\partial x^2} + 2\frac{\partial^2 u_x}{\partial x\partial y} - \frac{\partial^2 u_y}{\partial y^2}\right) = 0\\
&K_1\nabla^2 w_y + R_1\left(\frac{\partial^2 u_y}{\partial x^2} + 2\frac{\partial^2 u_x}{\partial x\partial y} - \frac{\partial^2 u_y}{\partial y^2}\right)\\
&\quad - R_2\left(\frac{\partial^2 u_x}{\partial x^2} - 2\frac{\partial^2 u_y}{\partial x\partial y} - \frac{\partial^2 u_x}{\partial y^2}\right) = 0
\end{aligned}
\tag{6.4-4}
$$

The equations are similar to those of (6.2-1), and the definitions of operators ∇^2 and $\nabla\cdot$ are the same there. It is obvious that Eq. (6.4-4) are more complex than those of (6.2-1). We now simplify the equations in terms of the displacement potential function method.

Introduce displacement potentials $\varphi(x,y)$ and $\psi(x,y)$ as follows:

$$\left.\begin{aligned}
u_x &= (L+M)\frac{\partial^2\varphi}{\partial x\partial y} + M\frac{\partial^2\psi}{\partial x^2} + (L+2M)\frac{\partial^2\psi}{\partial y^2} \\
u_y &= -\left[(L+2M)\frac{\partial^2\varphi}{\partial x^2} + M\frac{\partial^2\varphi}{\partial y^2} + (L+M)\frac{\partial^2\psi}{\partial x\partial y}\right] \\
w_x &= -\omega\left\{\left[2R_1\frac{\partial^2}{\partial x\partial y} - R_2\left(\frac{\partial^2}{\partial x^2} - \frac{\partial^2}{\partial y^2}\right)\right]\varphi\right. \\
&\quad \left. + \left[R_1\left(\frac{\partial^2}{\partial x^2} - \frac{\partial^2}{\partial y^2}\right) + 2R_2\frac{\partial^2}{\partial x\partial y}\right]\psi\right\} \\
w_y &= \omega\left\{\left[R_1\left(\frac{\partial^2}{\partial x^2} - \frac{\partial^2}{\partial y^2}\right) + 2R_2\frac{\partial^2}{\partial x\partial y}\right]\varphi\right. \\
&\quad \left. - \left[2R_1\frac{\partial^2}{\partial x\partial y} - R_2\left(\frac{\partial^2}{\partial x^2} - \frac{\partial^2}{\partial y^2}\right)\right]\psi\right\}
\end{aligned}\right\} \tag{6.4-5}$$

in which

$$\omega = \frac{M(L+2M)}{R^2}, \quad R^2 = R_1^2 + R_2^2 \tag{6.4-6}$$

The functions $\varphi(x,y)$ and $\psi(x,y)$ defined by (6.4-6) automatically satisfy the first two of Eq. (6.4-4), and the substitution of formulas (6.4-6) into the second two of Eq. (6.4-4) results that

$$\left.\begin{aligned}
&(L+2M)c_2\frac{\partial}{\partial x}\Lambda_1\varphi + Mc_1\frac{\partial}{\partial y}\Lambda_2\varphi \\
&\quad + (L+2M)c_2\frac{\partial}{\partial y}\Lambda_1\psi - Mc_1\frac{\partial}{\partial x}\Lambda_2\psi = 0 \\
&(L+2M)c_2\frac{\partial}{\partial x}\Lambda_2\varphi - Mc_1\frac{\partial}{\partial y}\Lambda_1\varphi \\
&\quad + (L+2M)c_2\frac{\partial}{\partial y}\Lambda_2\psi + Mc_1\frac{\partial}{\partial x}\Lambda_1\psi = 0
\end{aligned}\right\} \tag{6.4-7}$$

with

$$\left.\begin{aligned}\Lambda_1 &= R_1\frac{\partial}{\partial y}\Pi_1 + R_2\frac{\partial}{\partial x}\Pi_2\\ \Lambda_2 &= R_1\frac{\partial}{\partial x}\Pi_2 - R_2\frac{\partial}{\partial y}\Pi_1\end{aligned}\right\} \tag{6.4-8}$$

$$c_1 = (L+2M)K_1 - R^2, \quad c_2 = MK_1 - R^2 \tag{6.4-9}$$

If we introduce a new function $F(x, y)$ by means of

$$\left.\begin{aligned}\varphi &= -(L+2M)c_2R\frac{\partial}{\partial y}\Lambda_1F + Mc_1R\frac{\partial}{\partial x}\Lambda_2F\\ \psi &= (L+2M)c_2R\frac{\partial}{\partial x}\Lambda_1F + Mc_1R\frac{\partial}{\partial y}\Lambda_2F\end{aligned}\right\} \tag{6.4-10}$$

then the first one of Eq. (6.4-7) is automatically satisfied, and from the second one, we find that

$$\nabla^2\nabla^2\nabla^2\nabla^2F = 0 \tag{6.4-11}$$

The definitions of operators $\nabla^2$and $\nabla\cdot$ are the same as before.

All components of displacement vectors and stress tensors can be expressed by the potential function $F(x, y)$, e.g.,

$$u_x = R\left[c_2\frac{\partial}{\partial x}\Lambda_1 + c_1\frac{\partial}{\partial y}\Lambda_2\right]\nabla^2F \tag{6.4-12a}$$

$$u_y = R\left[c_2\frac{\partial}{\partial y}\Lambda_1 - c_1\frac{\partial}{\partial x}\Lambda_2\right]\nabla^2F \tag{6.4-12b}$$

$$w_x = -c_0\Lambda_1\Lambda_2F \tag{6.4-12c}$$

$$w_y = -R^{-1}\left[c_2(L+2M)\Lambda_1^2 + c_1M\Lambda_2^2\right]F \tag{6.4-12d}$$

$$\sigma_{xx} = 2c_0c_2\frac{\partial^2}{\partial y^2}\Lambda_1\nabla^2F \tag{6.4-12e}$$

$$\sigma_{yy} = 2c_0c_2\frac{\partial^2}{\partial x^2}\Lambda_1\nabla^2F \tag{6.4-12f}$$

$$\sigma_{xy} = \sigma_{yx} = -2c_0c_2\frac{\partial^2}{\partial x\partial y}\Lambda_1\nabla^2F \tag{6.4-12g}$$

$$
\begin{aligned}
H_{xx} = & -c_1 c_2 R \frac{\partial}{\partial y} \nabla^2 \nabla^2 \nabla^2 F \\
& + R^{-1} K_0 \frac{\partial}{\partial y} [c_2 (L + 2M) \Lambda_1^2 + c_1 M \Lambda_2^2] F
\end{aligned} \tag{6.4-12h}
$$

$$
\begin{aligned}
H_{xy} = & \, c_1 c_2 R \frac{\partial}{\partial x} \nabla^2 \nabla^2 \nabla^2 F \\
& - R^{-1} K_0 \frac{\partial}{\partial x} [c_2 (L + 2M) \Lambda_1^2 + c_1 M \Lambda_2^2] F
\end{aligned} \tag{6.4-12i}
$$

$$
H_{yx} = -c_1 c_2 R \frac{\partial}{\partial x} \nabla^2 \nabla^2 \nabla^2 F - c_0 K_0 \frac{\partial}{\partial y} \Lambda_1 \Lambda_2 F \tag{6.4-12j}
$$

$$
H_{yy} = -c_1 c_2 R \frac{\partial}{\partial y} \nabla^2 \nabla^2 \nabla^2 F + c_0 K_0 \frac{\partial}{\partial x} \Lambda_1 \Lambda_2 F \tag{6.4-12k}
$$

with

$$
c_0 = R(L + M), \quad K_0 = K_1 - K_2 \tag{6.4-13}
$$

The results were given by Li and Fan [15] and Li *et al.* [16]. Recently, Li and Fan [14] have derived the final governing equation of elasticity of the same point groups through the stress potential method, and the resulting equation is also quadruple harmonic equation, and of course the unknown function is the stress potential. Due to the limitation of space, the derivation about this is omitted, but the application will be shown in Chapter 8 for solving the notch problem of point group $5, \overline{5}$ and point group $10, \overline{10}$ two-dimensional quasicrystals.

The frequent appearance of quadruple harmonic equations in (6.2-7), (6.3-8), (6.4-11), etc. shows that this kind of equations is very important in theory and practice.

6.5 Plane Elasticity of Point Group 12*mm* of Dodecagonal Quasicrystals

Point group 12*mm* dodecagonal quasicrystals belong to the two-dimensional, and the Penrose tiling and the diffraction pattern are shown in Figs. 6.3 and 6.4, respectively. If taking the z-axis as the periodic arrangement direction, and supposing that the field variables

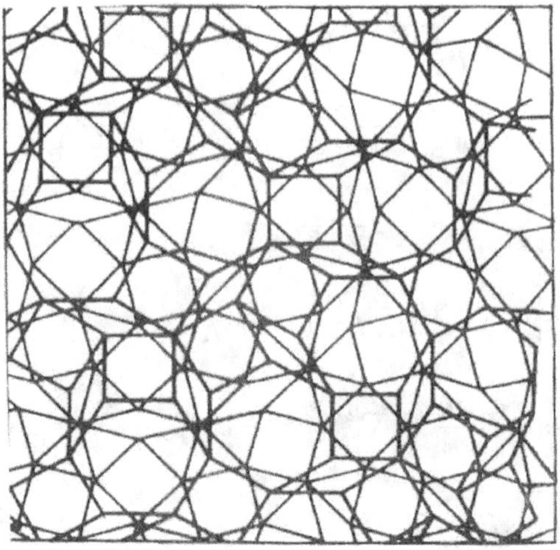

Fig. 6.3. Penrose tiling of dodecagonal quasicrystal.

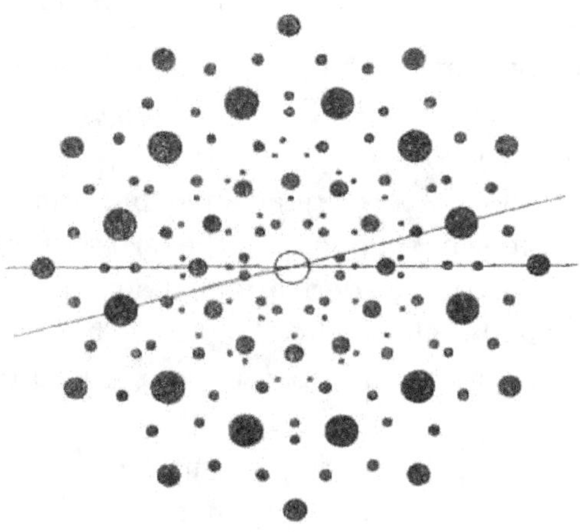

Fig. 6.4. Diffraction pattern of quasicrystal of 12 fold symmetry, in the condition of the phason strain field being uniformly.

are independent from the z-coordinate, then the elasticity problem can be decomposed into a superposition of a plane elasticity and anti-plane elasticity.

As mentioned in previous sections, the quasiperiodic plane is taken as the x–y plane, so the z-axis represents the 12-fold symmetry axis. Like that in other two-dimensional quasicrystals, the quasiperiodic plane is an elastic isotropic plane. There are two non-zero independent elastic constants in the plane, i.e., L and M

$$L = C_{12}, \quad M = (C_{11} - C_{12})/2 = C_{66} \tag{6.5-1}$$

Here, we have non-zero K_{ijkl} as

$$\left.\begin{array}{ll} K_{1111} = K_{2222} = K_1, & K_{1122} = K_{2211} = K_2 \\ K_{1221} = K_{2112} = K_3, & K_{2121} = K_{1212} = K_1 + K_2 + K_3 \end{array}\right\} \tag{6.5-2}$$

and others are zero. The results can also be expressed as

$$\begin{aligned} K_{ijkl} = & (K_1 - K_2 - K_3)(\delta_{ik} - \delta_{il}) + K_2\delta_{ij}\delta_{kl} \\ & + K_3\delta_{il}\delta_{jk} + 2(K_2 + K_3) \\ & \times (\delta_{i1}\delta_{j2}\delta_{k1}\delta_{l2} + \delta_{i2}\delta_{j1}\delta_{k2}\delta_{l1}) \quad (i, j, k, l = 1, 2) \end{aligned} \tag{6.5-3}$$

In addition, the phonon and phason are decoupled, i.e.,

$$R_{ijkl} = 0 \tag{6.5-4}$$

The elastic constant matrix is

$$[CKR] = \begin{bmatrix} L+2M & L & 0 & 0 & 0 & 0 & 0 & 0 \\ L & L+2M & 0 & 0 & 0 & 0 & 0 & 0 \\ 0 & 0 & M & M & 0 & 0 & 0 & 0 \\ 0 & 0 & M & M & 0 & 0 & 0 & 0 \\ 0 & 0 & 0 & 0 & K_1 & K_2 & 0 & 0 \\ 0 & 0 & 0 & 0 & K_2 & K_1 & 0 & 0 \\ 0 & 0 & 0 & 0 & 0 & 0 & K_1+K_2+K_3 & K_3 \\ 0 & 0 & 0 & 0 & 0 & 0 & K_3 & K_1+K_2+K_3 \end{bmatrix}$$

$$= \begin{bmatrix} L+2M & L & 0 & 0 & 0 & 0 & 0 & 0 \\ & L+2M & 0 & 0 & 0 & 0 & 0 & 0 \\ & & M & M & 0 & 0 & 0 & 0 \\ & & & M & 0 & 0 & 0 & 0 \\ & & & & K_1 & K_2 & 0 & 0 \\ & symmetry & & & & K_1 & 0 & 0 \\ & & & & & & K_1+K_2+K_3 & K_3 \\ & & & & & & & K_1+K_2+K_3 \end{bmatrix} \tag{6.5-5}$$

The relevant free energy density (or strain energy density) is

$$
\begin{aligned}
F = {} & \frac{1}{2}L(\nabla \cdot u)^2 + M\varepsilon_{ij}\varepsilon_{ij} + \frac{1}{2}K_1 w_{ij} w_{ij} \\
& + \frac{1}{2}K_2(w_{yx}^2 + w_{xy}^2 + 2w_{xx}w_{yy}) + \frac{1}{2}K_3(w_{yx} + w_{xy})^2
\end{aligned}
\tag{6.5-6}
$$

where ε_{ij} and w_{ij} denote the strain tensors

$$
\varepsilon_{ij} = \frac{1}{2}\left(\frac{\partial u_i}{\partial x_j} + \frac{\partial u_j}{\partial x_i}\right), \quad w_{ij} = \frac{\partial w_i}{\partial x_j} \tag{6.5-7}
$$

From (4.5-1) of Chapter 4 and (6.5-5) or (6.5-6), one can find the generalized Hooke's law for plane elasticity of dodecagonal quasicrystals as follows:

$$
\left.
\begin{aligned}
\sigma_{xx} &= L(\varepsilon_{xx} + \varepsilon_{yy}) + 2M\varepsilon_{xx} \\
\sigma_{yy} &= L(\varepsilon_{xx} + \varepsilon_{yy}) + 2M\varepsilon_{yy} \\
\sigma_{xy} &= \sigma_{yx} = 2M\varepsilon_{xy} \\
H_{xx} &= K_1 w_{xx} + K_2 w_{yy} \\
H_{yy} &= K_1 w_{yy} + K_2 w_{xx} \\
H_{xy} &= (K_1 + K_2 + K_3)w_{xy} + K_3 w_{yx} \\
H_{yx} &= (K_1 + K_2 + K_3)w_{yx} + K_3 w_{xy}
\end{aligned}
\right\} \tag{6.5-8}
$$

and there are the equilibrium equations when the body force is ignored

$$
\left.
\begin{aligned}
&\frac{\partial \sigma_{xx}}{\partial x} + \frac{\partial \sigma_{xy}}{\partial y} = 0, \quad \frac{\partial \sigma_{yx}}{\partial x} + \frac{\partial \sigma_{yy}}{\partial y} = 0 \\
&\frac{\partial H_{xx}}{\partial x} + \frac{\partial H_{xy}}{\partial y} = 0, \quad \frac{\partial H_{yx}}{\partial x} + \frac{\partial H_{yy}}{\partial y} = 0
\end{aligned}
\right\} \tag{6.5-9}
$$

Eliminating stress and strain components from (6.5-7), (6.5-8) and (6.5-9), we obtain the equilibrium equations expressed by displacement components as follows:

$$\left.\begin{aligned}&M\nabla^2 u_x+(L+M)\frac{\partial}{\partial x}\nabla\cdot\mathbf{u}=0\\&M\nabla^2 u_y+(L+M)\frac{\partial}{\partial y}\nabla\cdot\mathbf{u}=0\\&K_1\nabla^2 w_x+(K_2+K_3)\frac{\partial}{\partial y}\left(\frac{\partial w_x}{\partial y}+\frac{\partial w_y}{\partial x}\right)=0\\&K_1\nabla^2 w_y+(K_2+K_3)\frac{\partial}{\partial x}\left(\frac{\partial w_x}{\partial y}+\frac{\partial w_y}{\partial x}\right)=0\end{aligned}\right\}\tag{6.5-10}$$

If defining two displacement potentials $F(x,y)$ and $G(x,y)$ such as

$$\left.\begin{aligned}&u_x=(L+M)\frac{\partial^2 F}{\partial x\partial y}\\&u_y=-(L+2M)\frac{\partial^2 F}{\partial x^2}-M\frac{\partial^2 F}{\partial y^2}\\&w_x=(K_2+K_3)\frac{\partial^2 G}{\partial x\partial y}\\&w_y=-K_1\frac{\partial^2 G}{\partial x^2}-(K_1+K_2+K_3)\frac{\partial^2 G}{\partial y^2}\end{aligned}\right\}\tag{6.5-11}$$

then Eq. (6.5-10) will be reduced to

$$\nabla^2\nabla^2 F=0,\quad \nabla^2\nabla^2 G=0\tag{6.5-12}$$

which is found in Ref. [12] at first.

One can see that the problem is concluded to solve two biharmonic equations, and the theory and method studying this kind of equations are well developed in the theory of classical elasticity, which can be used in studying elasticity of quasicrystals. In this respect, the most systematic method is the complex analysis method. If we assume $\varphi_1(z),\psi_1(z),\pi_1(z)$ and $\chi_1(z)$are analytic functions of complex variable $z=x+iy(i=\sqrt{-1})$, then

$$\left.\begin{aligned}F(x,y)&=\mathrm{Re}[\bar{z}\varphi_1(z)+\textstyle\int\psi_1(z)dz]\\G(x,y)&=\mathrm{Re}[\bar{z}\pi_1(z)+\textstyle\int\chi_1(z)dz]\end{aligned}\right\}\tag{6.5-13}$$

where $\bar{z} = x - iy$ and Re denotes the real part of a complex number. The theory of analytic functions is a powerful tool to solve the boundary value problems of harmonic, biharmonic and multiple harmonic equations. The work on quadruple and sextuple harmonic equations is developed by the study of elasticity of quasicrystals, and the details will be given in the successive chapters. As a complete description of the new development of the method, some detailed summarization will be displayed in Chapter 11.

6.6 Plane Elasticity of Point Group $8mm$ of Octagonal Quasicrystals, Displacement Potential

Octagonal quasicrystals belong to two-dimensional quasicrystals, with the Penrose tiling of its quasiperiodic plane shown in Fig. 6.5 The stowing of planes along the third direction perpendicular to them will result in the quasicrystals.

In the following, we consider only a simple case of elasticity of the material, i.e., all of the field variables are independent from the axis along which the atom arrangement is periodic. In this case, the elasticity can be decomposed into a superposition of plane elasticity

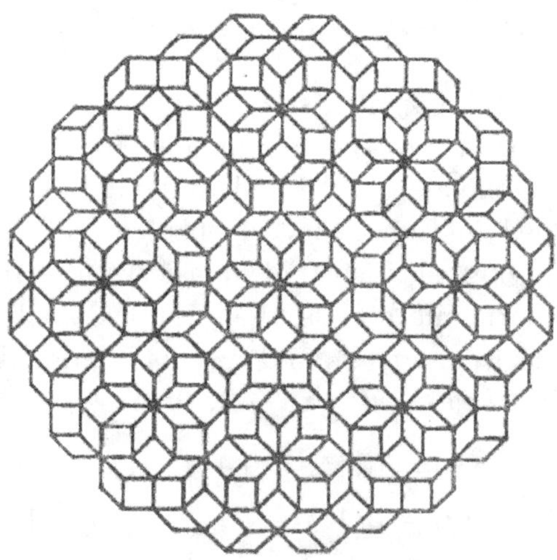

Fig. 6.5. Penrose tiling of 8 fold symmetry quasicrystal.

and anti-plane elasticity. We focus on the solution for the plane elasticity. Within the quasiperiodic plane, the phonon elasticity is isotropic, so C_{ijkl} are the same as those discussed in the previous sections. The phason elasticity here is anisotropic, but K_{ijkl} are the same with those of point group $12mm$ quasicrystals. Between phonon and phason fields, there is coupling, and the corresponding elastic constants R_{ijkl} are the same as given in (6.1-6), i.e.,

$$R_{ijkl} = R(\delta_{i1} - \delta_{i2})(\delta_{ij}\delta_{kl} - \delta_{ik}\delta_{jl} + \delta_{il}\delta_{jk}) \quad (i, j, k, l = 1, 2) \quad (6.6\text{-}1)$$

such that we have the elastic constants matrix

$$[CKR] = \begin{bmatrix} L+2M & L & 0 & 0 & R & R & 0 & 0 \\ L & L+2M & 0 & 0 & -R & -R & 0 & 0 \\ 0 & 0 & M & M & 0 & 0 & -R & R \\ 0 & 0 & M & M & 0 & 0 & -R & R \\ R & -R & 0 & 0 & K_1 & K_2 & 0 & 0 \\ R & -R & 0 & 0 & K_2 & K_1 & 0 & 0 \\ 0 & 0 & -R & -R & 0 & 0 & K_1+K_2+K_3 & K_3 \\ 0 & 0 & R & R & 0 & 0 & K_3 & K_1+K_2+K_3 \end{bmatrix}$$

$$= \begin{bmatrix} L+2M & L & 0 & 0 & R & R & 0 & 0 \\ & L+2M & 0 & 0 & -R & -R & 0 & 0 \\ & & M & M & 0 & 0 & -R & R \\ & & & M & 0 & 0 & -R & R \\ & & & & K_1 & K_2 & 0 & 0 \\ & \mathit{symmetry} & & & & K_1 & 0 & 0 \\ & & & & & & K_1+K_2+K_3 & K_3 \\ & & & & & & & K_1+K_2+K_3 \end{bmatrix} \quad (6.6\text{-}2)$$

The corresponding free energy is

$$\begin{aligned} F = {} & \frac{1}{2}L(\nabla \cdot \mathbf{u})^2 + M\varepsilon_{ij}\varepsilon_{ij} + \frac{1}{2}K_1 w_{ij}w_{ij} \\ & + \frac{1}{2}K_2(w_{xy}^2 + w_{yx}^2 + 2w_{xx}w_{yy}) + \frac{1}{2}K_3(w_{xy} + w_{yx})^2 \\ & + R[(\varepsilon_{xx} - \varepsilon_{yy})(w_{xx} + w_{yy}) + 2\varepsilon_{xy}(w_{yx} - w_{xy})] \end{aligned} \quad (6.6\text{-}3)$$

where ε_{ij} and w_{ij} are defined as before, and it is not listed again.

From (6.6-3) and (4.5-1), the generalized Hooke's law can be expressed as

$$\left.\begin{aligned}
\sigma_{xx} &= L(\varepsilon_{xx}+\varepsilon_{yy})+2M\varepsilon_{xx}+R(w_{xx}+w_{yy})\\
\sigma_{yy} &= L(\varepsilon_{xx}+\varepsilon_{yy})+2M\varepsilon_{yy}-R(w_{xx}+w_{yy})\\
\sigma_{xy} &= \sigma_{yx}=2M\varepsilon_{xy}+R(w_{yx}-w_{xy})\\
H_{xx} &= K_1w_{xx}+K_2w_{yy}+R(\varepsilon_{xx}-\varepsilon_{yy})\\
H_{yy} &= K_1w_{yy}+K_2w_{xx}+R(\varepsilon_{xx}-\varepsilon_{yy})\\
H_{xy} &= (K_1+K_2+K_3)w_{xy}+K_3w_{yx}-2R\varepsilon_{xy}\\
H_{yx} &= (K_1+K_2+K_3)w_{yx}+K_3w_{xy}+2R\varepsilon_{xy}
\end{aligned}\right\} \tag{6.6-4}$$

For saving the space, similar discussion on stress equilibrium equations is omitted.

Through a similar procedure, equilibrium equations given by displacement components are as follows:

$$\left.\begin{aligned}
&M\nabla^2u_x+(L+M)\frac{\partial}{\partial x}\nabla\cdot\mathbf{u}+R\left(\frac{\partial^2w_x}{\partial x^2}+2\frac{\partial^2w_y}{\partial x\partial y}-\frac{\partial^2w_x}{\partial y^2}\right)=0\\
&M\nabla^2u_y+(L+M)\frac{\partial}{\partial y}\nabla\cdot\mathbf{u}+R\left(\frac{\partial^2w_y}{\partial x^2}-2\frac{\partial^2w_x}{\partial x\partial y}-\frac{\partial^2w_y}{\partial y^2}\right)=0\\
&K_1\nabla^2w_x+(K_2+K_3)\left(\frac{\partial^2w_x}{\partial y^2}+\frac{\partial^2w_y}{\partial x\partial y}\right)\\
&\qquad+R\left(\frac{\partial^2u_x}{\partial x^2}-2\frac{\partial^2u_y}{\partial x\partial y}-\frac{\partial^2u_x}{\partial y^2}\right)=0\\
&K_1\nabla^2w_y+(K_2+K_3)\left(\frac{\partial^2w_x}{\partial x\partial y}+\frac{\partial^2w_y}{\partial x^2}\right)\\
&\qquad+R\left(\frac{\partial^2u_y}{\partial x^2}+2\frac{\partial^2u_x}{\partial x\partial y}-\frac{\partial^2u_y}{\partial y^2}\right)=0
\end{aligned}\right\} \tag{6.6-5}$$

It is evident that, if $K_2+K_3=0$, the equations will be reduced to (6.2-1). In fact, $K_2+K_3\neq 0$, so the equations are more complex than those given in the previous sections. But the final governing equation for this present case presents more interesting in mathematical physics, which we can see immediately.

At first, we introduce two auxiliary functions φ and ψ in such a way that

$$\left.\begin{aligned}
u_x &= (L+M)\frac{\partial^2\varphi}{\partial x\partial y} + M\frac{\partial^2\psi}{\partial x^2} + (L+2M)\frac{\partial^2\psi}{\partial y^2}\\
u_y &= -\left[(L+2M)\frac{\partial^2\varphi}{\partial x^2} + M\frac{\partial^2\varphi}{\partial y^2} + (L+M)\frac{\partial^2\psi}{\partial x\partial y}\right]\\
w_x &= -\omega\left(2\frac{\partial^2\varphi}{\partial x\partial y} + \frac{\partial^2\psi}{\partial x^2} - \frac{\partial^2\psi}{\partial y^2}\right)\\
w_y &= \omega\left(\frac{\partial^2\varphi}{\partial x^2} - \frac{\partial^2\varphi}{\partial y^2} - 2\frac{\partial^2\psi}{\partial x\partial y}\right)
\end{aligned}\right\} \tag{6.6-6}$$

where $\omega = M(L+M)/R$ so that (6.6-5) is simplified as

$$\left.\begin{aligned}
(\gamma\Pi_1 + \delta\Pi_2)\frac{\partial^2\varphi}{\partial x\partial y} + \left(\alpha\Pi_1\frac{\partial^2}{\partial y^2} - \beta\Pi_2\frac{\partial^2}{\partial x^2}\right)\psi = 0\\
(\gamma\Pi_2 + \delta\Pi_1)\frac{\partial^2\psi}{\partial x\partial y} + \left(\alpha\Pi_2\frac{\partial^2}{\partial y^2} - \beta\Pi_1\frac{\partial^2}{\partial x^2}\right)\varphi = 0
\end{aligned}\right\} \tag{6.6-7}$$

in which

$$\left.\begin{aligned}
&\Pi_1 = 3\frac{\partial^2}{\partial x^2} - \frac{\partial^2}{\partial y^2}, \quad \Pi_2 = 3\frac{\partial^2}{\partial y^2} - \frac{\partial^2}{\partial x^2}\\
&\alpha = R(L+2M) - \omega(K_1+K_2+K_3)\\
&\beta = RM - \omega K_1, \quad \delta = RM - \omega(K_1+K_2+K_3)\\
&\gamma = R(L+2M) - \omega K_1
\end{aligned}\right\} \tag{6.6-8}$$

and ω is given above. At last, the displacement potential $F(x,y)$ is introduced

$$\left.\begin{aligned}
\varphi &= \left(\beta\Pi_2\frac{\partial^2}{\partial x^2} - \alpha\Pi_1\frac{\partial^2}{\partial y^2}\right)F\\
\psi &= (\gamma\Pi_1 + \delta\Pi_2)\frac{\partial^2 F}{\partial x\partial y}
\end{aligned}\right\} \tag{6.6-9}$$

and (6.6-7) reduces to a single equation as follows:

$$(\nabla^2\nabla^2\nabla^2\nabla^2 - 4\varepsilon\nabla^2\nabla^2\Lambda^2\Lambda^2 + 4\varepsilon\Lambda^2\Lambda^2\Lambda^2\Lambda^2)F = 0 \tag{6.6-10}$$

where

$$\left.\begin{aligned}\nabla^2 &= \frac{\partial^2}{\partial x^2}+\frac{\partial^2}{\partial y^2}, \quad \Lambda^2=\frac{\partial^2}{\partial x^2}-\frac{\partial^2}{\partial y^2}\\ \varepsilon &= \frac{R^2(L+M)(K_2+K_3)}{[M(K_1+K_2+K_3)-R^2][(L+2M)K_1-R^2]}\end{aligned}\right\} \tag{6.6-11}$$

It is obvious that if $K_2+K_3=0$, then $\varepsilon=0$, Eq. (6.6-10) will be reduced to (6.2-7).

If $F(x,y)$ is the solution of (6.6-10), substituting it into (6.6-9) and then into (6.6-6), one can obtain the displacement field

$$\left.\begin{aligned}u_x &= \frac{\partial^2}{\partial x\partial y}\{[M\alpha\Pi_1+(L+2M)\beta\Pi_2]+4\omega(K_2+K_3)\\ &\quad\times\left[M\frac{\partial^2}{\partial x^2}+(L+2M)\frac{\partial^2}{\partial y^2}\right]\Lambda^2\right\}F\\ u_y &= \left[M\alpha\Pi_1\frac{\partial^2}{\partial y^2}-(L+2M)\beta\Pi_2\frac{\partial^2}{\partial x^2}\right]\nabla^2F\\ &\quad-4\omega(K_2+K_3)(L+M)\frac{\partial^4}{\partial x^2\partial y^2}\Lambda^2F\\ w_x &= \frac{\partial^2}{\partial x\partial y}\left[\omega(\alpha-\beta)\Pi_1\Pi_2-4\omega^2(K_2+K_3)\Lambda^2\Lambda^2\right]F\\ w_y &= -\omega\left[\alpha\Pi_1^2\frac{\partial^2}{\partial y^2}+\beta\Pi_2^2\frac{\partial^2}{\partial x^2}\right]F-8\omega^2(K_2+K_3)\frac{\partial^4}{\partial x^2\partial y^2}\Lambda^2F\end{aligned}\right\} \tag{6.6-12}$$

Similarly, the stress field has the following expression:

$$\begin{aligned}\sigma_{xx} &= 2M(L+M)\alpha\Pi_1\frac{\partial^3}{\partial y^3}\nabla^2F\\ &\quad+8M\omega(L+M)(K_2+K_3)\frac{\partial^5}{\partial x^2\partial y^3}\Lambda^2F\\ \sigma_{yy} &= 2M(L+M)\alpha\Pi_1\frac{\partial^3}{\partial x^2\partial y}\nabla^2F\\ &\quad+8M\omega(L+M)(K_2+K_3)\frac{\partial^5}{\partial x^4\partial y}\Lambda^2F\end{aligned}$$

$$
\begin{aligned}
\sigma_{xy} = \sigma_{yx} &= -2M(L+M)\alpha\Pi_1 \frac{\partial^3}{\partial x \partial y^2}\nabla^2 F \\
&\quad - 8M\omega(L+M)(K_2+K_3)\frac{\partial^5}{\partial x^3 \partial y^2}\Lambda^2 F \\
H_{xx} &= \left\{ R\frac{\partial}{\partial y}\left[M\alpha\Pi_1\Lambda^2 + 2(L+2M)\beta\Pi_2\frac{\partial^2}{\partial x^2}\right]\nabla^2 \right. \\
&\quad + 4R\omega(K_2+K_3)\frac{\partial^3}{\partial x^2 \partial y}\left[M\frac{\partial^2}{\partial x^2} + (2L+5M)\frac{\partial^2}{\partial y^2}\right]\Lambda^2 \\
&\quad + K_1\omega\frac{\partial^3}{\partial x^2 \partial y}\left[(\alpha-\beta)\Pi_1\Pi_2 - 4\omega(K_2+K_3)\Lambda^2\Lambda^2\right] \\
&\quad \left. - K_2\omega\left[\alpha\frac{\partial^3}{\partial y^3}\Pi_1^2 + \beta\frac{\partial^3}{\partial x^2 \partial y}\Pi_2^2 + 8\omega(K_2+K_3)\frac{\partial^5}{\partial x^2 \partial y^3}\Lambda^2\right]\right\} F \\
H_{yy} &= \left\{ R\frac{\partial}{\partial y}\left[M\alpha\Pi_1\Lambda^2 + 2(L+2M)\beta\Pi_2\frac{\partial^2}{\partial x^2}\right]\nabla^2 \right. \\
&\quad + 4R\omega(K_2+K_3)\frac{\partial^3}{\partial x^2 \partial y}\left[M\frac{\partial^2}{\partial x^2} + (2L+5M)\frac{\partial^2}{\partial y^2}\right]\Lambda^2 \\
&\quad + K_2\omega\frac{\partial^3}{\partial x^2 \partial y}\left[(\alpha-\beta)\Pi_1\Pi_2 - 4\omega(K_2+K_3)\Lambda^2\Lambda^2\right] \\
&\quad \left. - K_1\omega\left[\alpha\frac{\partial^3}{\partial y^3}\Pi_1^2 + \beta\frac{\partial^3}{\partial x^2 \partial y}\Pi_2^2 + 8\omega(K_2+K_3)\frac{\partial^5}{\partial x^2 \partial y^3}\Lambda^2\right]\right\} F \\
H_{xy} &= \left\{ -R\frac{\partial}{\partial x}\left[2M\alpha\frac{\partial^2}{\partial y^2}\Pi_1 - (L+2M)\beta\Pi_2\Lambda^2\right]\nabla^2 \right. \\
&\quad + 4R\omega(L+2M)(K_2+K_3)\frac{\partial^3}{\partial x \partial y^2}\Lambda^2\Lambda^2 \\
&\quad + (K_1+K_2+K_3)\omega\frac{\partial^3}{\partial x \partial y^2} \\
&\quad \times \left[(\alpha-\beta)\Pi_1\Pi_2 - 4\omega(K_2+K_3)\Lambda^2\Lambda^2\right] \\
&\quad \left. - K_3\omega\left[\alpha\frac{\partial^3}{\partial x \partial y^2}\Pi_1^2 + \beta\frac{\partial^3}{\partial x^3}\Pi_2^2 + 8\omega(K_2+K_3)\frac{\partial^5}{\partial x^3 \partial y^2}\Lambda^2\right]\right\} F
\end{aligned}
$$

$$
\begin{aligned}
H_{yx} = \bigg\{ & R\frac{\partial}{\partial x}\left[2M\alpha\frac{\partial^2}{\partial y^2}\Pi_1 - (L+2M)\beta\Pi_2\Lambda^2\right]\nabla^2 \\
& - 4R\omega(L+2M)(K_2+K_3)\frac{\partial^3}{\partial x\partial y^2}\Lambda^2\Lambda^2 \\
& + K_3\omega\frac{\partial^3}{\partial x\partial y^2}\left[(\alpha-\beta)\Pi_1\Pi_2 - 4\omega(K_2+K_3)\Lambda^2\Lambda^2\right] \\
& - (K_1+K_2+K_3)\omega \\
& \times \left[\alpha\frac{\partial^3}{\partial x\partial y^2}\Pi_1^2 + \beta\frac{\partial^3}{\partial x^3}\Pi_2^2 + 8\omega(K_2+K_3)\frac{\partial^5}{\partial x^3\partial y^2}\Lambda^2\right]\bigg\} F
\end{aligned}
\tag{6.6-13}
$$

A part of the above results was reported by Refs. [12, 15, 17].

6.7 Stress Potential of the Plane Field of Point Group $5, \overline{5}$ Pentagonal and Point Group $10, \overline{10}$ Decagonal Quasicrystals

In Section 4, we discussed the displacement potential of plane elasticity of point group $5, \overline{5}$ pentagonal and point group $10, \overline{10}$ decagonal quasicrystals. The stress potential for the quasicrystals is also beneficial, which will be introduced in this section.

From the basic formulas listed in Section 6.4, if we exclude the displacements, then there are the deformation compatibility equations

$$
\frac{\partial^2\varepsilon_{xx}}{\partial y^2} + \frac{\partial^2\varepsilon_{yy}}{\partial x^2} = 2\frac{\partial^2\varepsilon_{xy}}{\partial x\partial y}, \quad \frac{\partial w_{xy}}{\partial x} = \frac{\partial w_{xx}}{\partial y}, \quad \frac{\partial w_{yx}}{\partial y} = \frac{\partial w_{yy}}{\partial x}
\tag{6.7-1}
$$

If introducing stress potential functions $\varphi(x,y), \psi_1(x,y)$ and $\psi_2(x,y)$ such as

$$
\begin{aligned}
& \sigma_{xx} = \frac{\partial^2\varphi}{\partial y^2}, \quad \sigma_{yy} = \frac{\partial^2\varphi}{\partial x^2}, \quad \sigma_{xy} = \sigma_{yx} = -\frac{\partial^2\varphi}{\partial x\partial y} \\
& H_{xx} = \frac{\partial\psi_1}{\partial y}, \quad H_{xy} = -\frac{\partial\psi_1}{\partial x}, \quad H_{yx} = -\frac{\partial\psi_2}{\partial y}, \quad H_{yy} = \frac{\partial\psi_2}{\partial x}
\end{aligned}
\tag{6.7-2}
$$

then equilibrium equations $\partial\sigma_{ij}/\partial x_j = 0$ and $\partial H_{ij}/\partial x_j = 0$ will be automatically satisfied.

Based on the generalized Hooke's law (6.4-3), all strain components can be expressed by relevant stress components

$$
\begin{aligned}
\varepsilon_{xx} &= \frac{1}{4(L+M)}(\sigma_{xx}+\sigma_{yy}) + \frac{1}{4c}[(K_1+K_2)(\sigma_{xx}-\sigma_{yy}) \\
&\quad - 2R_1(H_{xx}+H_{yy}) - 2R_2(H_{xy}-H_{yx})] \\
\varepsilon_{yy} &= \frac{1}{4(L+M)}(\sigma_{xx}+\sigma_{yy}) - \frac{1}{4c}[(K_1+K_2)(\sigma_{xx}-\sigma_{yy}) \\
&\quad - 2R_1(H_{xx}+H_{yy}) - 2R_2(H_{xy}-H)] \\
\varepsilon_{xy} &= \varepsilon_{yx} = \frac{1}{2c}[(K_1+K_2)\sigma_{xy} - R_2(H_{xx}+H_{yy}) + R_1(H_{xy}-H_{yx})] \\
w_{xx} &= \frac{1}{2(K_1-K_2)}(H_{xx}-H_{yy}) \\
&\quad + \frac{1}{2c}[M(H_{xx}+H_{yy}) - R_1(\sigma_{xx}-\sigma_{yy}) - 2R_2\sigma_{xy}] \\
w_{yy} &= -\frac{1}{2(K_1-K_2)}(H_{xx}-H_{yy}) \\
&\quad + \frac{1}{2c}[M(H_{xx}+H_{yy}) - R_1(\sigma_{xx}-\sigma_{yy}) - 2R_2\sigma_{xy}] \\
w_{xy} &= \frac{1}{2c}[-R_2(\sigma_{xx}-\sigma_{yy}) + 2R_1\sigma_{xy}] \\
&\quad + \frac{1}{2(K_1-K_2)}(H_{xy}+H_{yx}) + \frac{M}{2c}(H_{xy}-H_{yx}) \\
w_{yx} &= \frac{1}{2c}[R_2(\sigma_{xx}-\sigma_{yy}) - 2R_1\sigma_{xy}] \\
&\quad + \frac{1}{2(K_1-K_2)}(H_{xy}+H_{yx}) - \frac{M}{2c}(H_{xy}-H_{yx})
\end{aligned}
\tag{6.7-3}
$$

in which

$$c = M(K_1 + K_2) - 2(R_1^2 + R_2^2) \tag{6.7-4}$$

So, the deformation compatibility equations (6.7-1) can be rewritten by the stresses σ_{ij}, H_{ij}, then by employing (6.4-3), one has

$$\begin{aligned}
&\left(\frac{1}{2(L+M)} + \frac{K_1+K_2}{2c}\right)\nabla^2\nabla^2\varphi + \frac{R_1}{c}\left(\frac{\partial}{\partial y}\Pi_1\psi_1 - \frac{\partial}{\partial x}\Pi_2\psi_2\right)\\
&\quad + \frac{R_2}{c}\left(\frac{\partial}{\partial x}\Pi_2\psi_1 + \frac{\partial}{\partial y}\Pi_1\psi_2\right) = 0\\
&\left(\frac{c}{K_1-K_2} + M\right)\nabla^2\psi_1 + R_1\frac{\partial}{\partial y}\Pi_1\varphi + R_2\frac{\partial}{\partial x}\Pi_2\varphi = 0\\
&\left(\frac{c}{K_1-K_2} + M\right)\nabla^2\psi_2 - R_1\frac{\partial}{\partial x}\Pi_2\varphi + R_2\frac{\partial}{\partial y}\Pi_1\varphi = 0
\end{aligned} \tag{6.7-5}$$

where

$$\nabla^2 = \frac{\partial^2}{\partial x^2} + \frac{\partial^2}{\partial y^2}, \quad \Pi_1 = 3\frac{\partial^2}{\partial x^2} - \frac{\partial^2}{\partial y^2}, \quad \Pi_2 = 3\frac{\partial^2}{\partial y^2} - \frac{\partial^2}{\partial x^2}$$

Equation (6.7-5) will be satisfied when we choose a new function G, which is called the stress function, such that

$$\begin{aligned}
&\varphi = c_1\nabla^2\nabla^2 G, \quad \psi_1 = -\left(R_1\frac{\partial}{\partial y}\Pi_1 + R_2\frac{\partial}{\partial x}\Pi_2\right)\nabla^2 G,\\
&\psi_2 = \left(R_1\frac{\partial}{\partial x}\Pi_2 - R_2\frac{\partial}{\partial y}\Pi_1\right)\nabla^2 G
\end{aligned} \tag{6.7-6}$$

in which

$$c_1 = \frac{c}{K_1 - K_2} + M \tag{6.7-7}$$

and

$$\nabla^2\nabla^2\nabla^2\nabla^2 G = 0 \tag{6.7-8}$$

The final governing equation based on the stress potential is the same as that based on the displacement potential.

6.8 Stress Potential of Point Group $8mm$ Octagonal Quasicrystals

The final governing equation of plane elasticity of point group $8mm$ octagonal quasicrystals was given in Section 6 by displacement potential; similarly we can also give a derivation by stress potential.

The strain compatibility equations and the definition of stress potentials are the same as those given by (6.7-1) and (6.7-2), and the strain–stress relations are as follows:

$$\varepsilon_{xx} = \frac{1}{(L+M)c}\left\{\frac{1}{4}\left[(K_1+K_2)(L+2M)-2R^2\right]\sigma_{xx} -\frac{1}{4}\left[(K_1+K_2)L+2R^2\right]\sigma_{yy}-\frac{1}{2}R(L+M)(H_{xx}+H_{yy})\right\} \tag{6.8-1a}$$

$$\varepsilon_{yy} = \frac{1}{(L+M)c}\left\{-\frac{1}{4}\left[(K_1+K_2)L+2R^2\right]\sigma_{xx} +\frac{1}{4}\left[(K_1+K_2)(L+2M)-2R^2\right]\sigma_{yy} +\frac{1}{2}R(L+M)(H_{xx}+H_{yy})\right\} \tag{6.8-1b}$$

$$\varepsilon_{xy} = \varepsilon_{yx} = \frac{1}{2c}[(K_1+K_2)\sigma_{xy}+R(H_{xy}-H_{yx})] \tag{6.8-1c}$$

$$w_{xx} = \frac{1}{(K_1-K_2)c}\left[\frac{1}{2}R(K_1-K_2)(\sigma_{yy}-\sigma_{xx}) + (K_1M-R^2)H_{xx}-(K_2M-R^2)H_{yy}\right] \tag{6.8-1d}$$

$$w_{yy} = \frac{1}{(K_1-K_2)c}\left[\frac{1}{2}R(K_1-K_2)(\sigma_{yy}-\sigma_{xx}) - (K_2M-R^2)H_{xx}+(K_1M-R^2)H_{yy}\right] \tag{6.8-1e}$$

$$w_{xy} = \frac{1}{(K_1+K_2+2K_3)c}\{R(K_1+K_2+2K_3)\sigma_{xy} + \left[(K_1+K_2+K_3)M-R^2\right]H_{xy}-(K_3M+R^2)H_{yx}\} \tag{6.8-1f}$$

$$w_{yx} = \frac{1}{(K_1 + K_2 + 2K_3)c} \{-R(K_1 + K_2 + 2K_3)\sigma_{xy} \\ -(K_3 M + R^2)H_{xy} + \left[(K_1 + K_2 + K_3)M - R^2\right] H_{yx}\} \tag{6.8-1g}$$

in which

$$c = M(K_1 + K_2) - 2R^2$$

If introducing stress potential functions $\varphi(x, y), \psi_1(x, y)$ and $\psi_2(x, y)$ such as

$$\sigma_{xx} = \frac{\partial^2 \varphi}{\partial y^2}, \quad \sigma_{yy} = \frac{\partial^2 \varphi}{\partial x^2}, \quad \sigma_{xy} = \sigma_{yx} = -\frac{\partial^2 \varphi}{\partial x \partial y}$$

$$H_{xx} = \frac{\partial \psi_1}{\partial y}, \quad H_{xy} = -\frac{\partial \psi_1}{\partial x}, \quad H_{yx} = -\frac{\partial \psi_2}{\partial y}, \quad H_{yy} = \frac{\partial \psi_2}{\partial x} \tag{6.8-2}$$

then equilibrium equations (6.5-9) (or $\partial \sigma_{ij}/\partial x_j = 0, \partial H_{ij}/\partial x_j = 0$) will be automatically satisfied.

Substituting the stress formulas (6.8-2) into the strain–stress relation (6.8-1) and then into the deformation compatibility equations (6.7-1), we have

$$c_1 \nabla^2 \nabla^2 \varphi + 2R(L + M)\frac{\partial}{\partial y}\Pi_1 \psi_1 - 2R(L + M)\frac{\partial}{\partial x}\Pi_2 \psi_2 = 0$$

$$-\frac{1}{2}R\frac{\partial}{\partial y}\Pi_1 \varphi + c_2 \frac{\partial^2 \psi_2}{\partial x \partial y} - c_3 \frac{\partial^2 \psi_1}{\partial x^2} - c_4 \frac{\partial^2 \psi_1}{\partial y^2} = 0$$

$$\frac{1}{2}R\frac{\partial}{\partial x}\Pi_2 \varphi + c_2 \frac{\partial^2 \psi_1}{\partial x \partial y} - c_4 \frac{\partial^2 \psi_2}{\partial x^2} - c_3 \frac{\partial^2 \psi_2}{\partial y^2} = 0 \tag{6.8-3}$$

where

$$\nabla^2 = \frac{\partial^2}{\partial x^2} + \frac{\partial^2}{\partial y^2}, \quad \Pi_1 = 3\frac{\partial^2}{\partial x^2} - \frac{\partial^2}{\partial y^2}, \quad \Pi_2 = 3\frac{\partial^2}{\partial y^2} - \frac{\partial^2}{\partial x^2}$$

$$c_1 = (K_1 + K_2)(L + 2M) - 2R^2,$$

$$c_2 = \frac{K_3 M + R^2}{K_1 + K_2 + 2K_3} + \frac{K_2 M + R^2}{K_1 - K_2}$$

$$c_3 = \frac{(K_1 + K_2 + K_3)M - R^2}{K_1 + K_2 + 2K_3}, \quad c_4 = \frac{K_1 M + R^2}{K_1 - K_2} \tag{6.8-4}$$

The manipulation of the second equation of (6.8-3) $\times\frac{\partial}{\partial x}\Pi_2+$ the third equation of (6.8-3) $\times\frac{\partial}{\partial y}\Pi_1$ leads to

$$\begin{aligned}&\left(c_2\frac{\partial^3}{\partial x\partial y^2}\Pi_1-c_3\frac{\partial^3}{\partial x^3}\Pi_2-c_4\frac{\partial^3}{\partial x\partial y^2}\Pi_2\right)\psi_1\\&\quad=\left(c_4\frac{\partial^3}{\partial x^2\partial y}\Pi_1+c_3\frac{\partial^3}{\partial y^3}\Pi_1-c_2\frac{\partial^3}{\partial x^2\partial y}\Pi_2\right)\psi_2\end{aligned}\tag{6.8-5}$$

There must be a function $A(x,y)$ so that

$$\begin{aligned}\psi_1&=\left(c_4\frac{\partial^3}{\partial x^2\partial y}\Pi_1+c_3\frac{\partial^3}{\partial y^3}\Pi_1-c_2\frac{\partial^3}{\partial x^2\partial y}\Pi_2\right)A\\\psi_2&=\left(c_2\frac{\partial^3}{\partial x\partial y^2}\Pi_1-c_3\frac{\partial^3}{\partial x^3}\Pi_2-c_4\frac{\partial^3}{\partial x\partial y^2}\Pi_2\right)A\end{aligned}\tag{6.8-6}$$

Substituting (6.8-6) into the third equation of (6.8-3), one finds that there exists a function $G(x,y)$ satisfying the relations such as

$$\begin{aligned}\varphi&=-c_3c_4\nabla^2\nabla^2G\\A&=\frac{1}{2}RG\end{aligned}\tag{6.8-7}$$

In the derivation, the relation $c_2=c_4-c_3$ has been used. The stress potentials φ,ψ_1,ψ_2 can be expressed by the new function $G(x,y)$, i.e.,

$$\begin{aligned}\varphi&=-c_3c_4\nabla^2\nabla^2G\\\psi_1&=\frac{1}{2}R\left(c_4\frac{\partial^3}{\partial x^2\partial y}\Pi_1+c_3\frac{\partial^3}{\partial y^3}\Pi_1-c_2\frac{\partial^3}{\partial x^2\partial y}\Pi_2\right)G\\\psi_2&=\frac{1}{2}R\left(c_2\frac{\partial^3}{\partial x\partial y^2}\Pi_1-c_3\frac{\partial^3}{\partial x^3}\Pi_2-c_4\frac{\partial^3}{\partial x\partial y^2}\Pi_2\right)G\end{aligned}\tag{6.8-8}$$

Substituting (6.8-8) into the first equation of (6.8-3) yields

$$\begin{aligned}&-c_1c_3c_4\nabla^2\nabla^2\nabla^2\nabla^2G+R^2(L+M)\\&\quad\times\left(c_4\frac{\partial^4}{\partial x^2\partial y^2}\Pi_1^2+c_3\frac{\partial^4}{\partial y^4}\Pi_1^2-2c_2\frac{\partial^4}{\partial x^2\partial y^2}\Pi_1\Pi_2\right.\\&\quad\left.+c_3\frac{\partial^4}{\partial y^4}\Pi_2^2+c_4\frac{\partial^4}{\partial x^2\partial y^2}\Pi_2^2\right)G=0\end{aligned}\tag{6.8-9}$$

Considering the following relations:

$$\frac{\partial^4}{\partial x^2 \partial y^2} = \frac{1}{4}(\nabla^2\nabla^2 - \Lambda^2\Lambda^2), \quad \Pi_1\Pi_2 = \nabla^2\nabla^2 - 4\Lambda^2\Lambda^2,$$

$$c_2 = c_4 - c_3$$

$$\nabla^2 = \frac{\partial^2}{\partial x^2} + \frac{\partial^2}{\partial y^2}, \quad \Lambda^2 = \frac{\partial^2}{\partial x^2} - \frac{\partial^2}{\partial y^2}$$

Eq. (6.8-9) can be simplified as

$$\nabla^2\nabla^2\nabla^2\nabla^2 G - 4\varepsilon\nabla^2\nabla^2\Lambda^2\Lambda^2 G + 4\varepsilon\Lambda^2\Lambda^2\Lambda^2\Lambda^2 G = 0 \qquad (6.8\text{-}10)$$

with

$$\begin{aligned}\varepsilon &= \frac{R^2(L+M)(c_3 - c_4)}{-c_1c_3 + R^2(L+M)c_3} \\ &= \frac{R^2(L+M)(K_2+K_3)}{[(K_1+K_2+K_3)M - R^2][K_1(L+2M) - R^2]}\end{aligned} \qquad (6.8\text{-}11)$$

The final governing equation in this case is exactly in agreement to that given by the displacement potential formulation discussed in Section 6.

6.9 The Three-dimensional Elasticity and the Field Equations of the First Kind of Two-dimensional Sold Quasicrystals

In the previous sections we list the basic equations of elasticity of the first kind of two-dimensional quasicrystals for plane field, while the basic equations of elasticity of theirs in three-dimensional case can also be obtained which is omitted to save the space.

6.10 Other Discussions

A part on plastic problems of this kind of two-dimensional quasicrystals and mathematical methods are discussed in Refs. [18–20], but the details are omitted.

References

[1] Chernikov M A, Ott H R, Bianchi A *et al.*, 1998, Elastic moduli of a single quasicrystal of decagonal Al–Ni–Co: evidence for transverse elastic isotropy, *Phys. Rev. Lett.*, **80**(2), 321–324.

[2] Abe H, Taruma N, Le Bolloc'h D *et al.*, 2000, Anomalous-X-ray scattering associated with short-range order in an $Al_{70}Ni_{15}Co_{15}$ decagonal quasicrystal, *Mater. Sci. and Eng. A*, **294–296**(12), 299–302.

[3] Jeong H C and Steinhardt P J, 1993, Finite-temperature elasticity phase transition in decagonal quasicrystals, *Phys. Rev. B*, **48**(13), 9394–9403.

[4] Edagawa K, 2007, Phonon-phason coupling in decagonal quasicrystals, *Phil. Mag.*, **87**(18–21), 2789–2798.

[5] Fan T Y and Mai Y W, 2004, Elasticity theory, fracture mechanics and relevant thermal properties of quasicrystalline materials, *Appl. Mech. Rev.*, **57**(5), 325–344.

[6] Hu C Z, Yang W G, Wang R H and Ding D H, 1997, Symmetry and physical properties of quasicrystals, *Prog. Phys.*, **17**(4), 345–374 (in Chinese).

[7] Bohsung J and Trebin H-R, 1989, *Introduction to the Quasicrystal Mathematics*, ed by Jaric M.V., Academic Press, New York.

[8] De P and Pelcovis R A, 1987, Disclinations in pentagonal quasicrystals, *Phys. Rev. B*, **35**(17), 9604–9607.

[9] Yang S H and Ding D H, 1998, *Foundation of Dislocation Theory of Crystals*, Vol. 2, Science Press, Beijing (in Chinese).

[10] De P and Pelcovits R A, 1987, Linear elasticity of pentagonal quasicrystals, *Phys. Rev. B*, **36**(13), 8609–8620.

[11] Ding D H, Wang R H, Yang W G and Hu C Z, 1995, General expressions for the elastic displacement fields induced by dislocations in quasicrystals, *J. Phys. Condens. Matter*, **7**(28), 5423–5426.

[12] Li X F and Fan T Y, 1998, New method for solving elasticity problems of some planar quasicrystals and solutions, *Chin. Phys. Lett.*, **15**(4), 278–280.

[13] Guo Y C and Fan T Y, 2001, Mode? Griffith crack in decagonal quasicrystals, *Appl. Math. Mech.*, English Edition, **22**(10), 1311–1317.

[14] Li L H and Fan T Y, 2006, Complex function method for solving notch problem of point 10 two-dimensional quasicrystal based on the stress potential function, *J. Phys.: Condens. Matter*, **18**(47), 10631–10641.

[15] Li X F, Fan T Y and Sun Y F, 1999, A decagonal quasicrystal with a Griffith crack, *Phil. Mag. A*, **79**(7), 1943–1952.

[16] Li X F, Duan X Y, Fan T.Y., *et al.*, 1999, Elastic field for a straight dislocation in a decagonal quasicryatal, *J. Phys.: Condens. Matter*, **11**(3), 703–711.

[17] Zhou W.M. and Fan T.Y., 2001, Plane elasticity of octagonal quasicrystals and solutions, *Chin. Phys.*, **10**(8), 743–747.
[18] Fan T Y, Trebin H-R, Messerschmidt U and Mai Y W, 2004, Plastic flow coupled with a Griffith crack in some one- and two-dimensional quasicrystals, *J. Phys.: Condens. Matter.*, **16**(37), 5419–5429.
[19] Fan T Y and Fan L, 2008, Plastic fracture of quasicrystals, *Phil. Mag.*, **88**(4), 523–535.
[20] Muskhelishvili N I, 1956, *Some Basic Problems in the Mathematical Theory of Elasticity*, English translation by Radok J R M, P. Noordhoff Ltd, Groringen.
[21] Sneddon I N, 1951, *Fourier Transforms*, McGraw-Hill, New York.

Chapter 7

Application I — Some Dislocation and Interface Problems and Solutions in One-dimensional and First Kind of Two-dimensional Quasicrystals

In Chapters 5 and 6, with the physical basis of quasicrystal elasticity based on the density wave model, we have performed some mathematical operations, e.g., by proper simplification, to reduce the original problems into the boundary value problems of high-order partial differential equations and to establish the standard solving procedure and the fundamental solution formulas. This work is the development of the boundary value problems in classical elasticity. Here, we need to pose a question: Do these mathematical operations contribute to solving the quasicrystal elasticity problems? This could only be answered by practice. The following two chapters will provide the applications of these theories, including the solutions to some dislocation, crack and interface problems in one- and two-dimensional quasicrystals. Our results indicate that the mathematical formulations created in Chapters 5 and 6 are powerful in solving these problems.

As we know, almost all monographs of classical elasticity do not place their main focus on dislocation and crack problems. These problems have been investigated in monographs relating to dislocation theory and fracture mechanics, respectively. As a monograph on theories of elasticity of quasicrystals, this chapter is not going to focus entirely on dislocations and cracks in quasicrystals. Therefore, in an attempt to examine our theories developed in Chapters 5 and 6 and to show their applications in the elasticity of quasicrystals, we present some calculation examples of certain realistic dislocation and crack problems in the elasticity of quasicrystals. The methods developed in this chapter can be used for studying other problems in quasicrystals.

Historically, soon after the discovery of quasicrystals, scientists proposed the possibility of existing dislocations in quasicrystals. De and Pelcovits [1, 2] first studied the elastic field around dislocations and disclinations in quasicrystals. Furthermore, Ding *et al.* [3, 4] conducted systematic investigations of this topic by using the Green function method.

Now, people have recognized that quasicrystal is a kind of ordered phase of quasiperiodic long range. Similar to crystals with the ordered phase of long range, the breaking of long-range regularity takes place usually through topological defects, i.e., dislocations lead to the long-range symmetry broken. As mentioned in Chapter 3, quasicrystals also possess orientational symmetries incompatible with those permitted in crystal theory. In quasicrystals, another defect, i.e., disclinations, also exists simultaneously, which leads to the breaking of the orientational symmetry in quasicrystals. In some cases, the crystalline phases often coexist with the quasicrystalline phases, resulting in another kind of defect: the interface between quasicrystal and crystal. The existence of these defects including the stacking fault dramatically affects the mechanical properties of quasicrystals. Therefore, it is important to study the elastic properties of quasicrystals with dislocations, disclinations, interface and stacking faults.

As aforementioned, during the study of dislocations in quasicrystals, physicists have developed some mathematical methods, such as the Green function method. These important methods are available in relevant literature. Hereafter, by using the elementary methods developed in Chapters 5 and 6, complex analysis and Fourier analysis, we study the displacement and stress fields around dislocations or interfaces in one-dimensional and two-dimensional quasicrystals quantitatively. The three-dimensional dislocation and dynamic dislocation problems will be discussed in Chapters 9 and 10, respectively.

7.1 Dislocations in One-dimensional Hexagonal Quasicrystals

Following the sequence from simplicity to complexity, we first study the dislocations in one-dimensional quasicrystals. A dislocation in an n-dimensional quasicrystal bears the Burgers vector $\mathrm{b}^{||}\oplus\mathrm{b}^{\perp}$, where $\mathrm{b}^{||}=(b_1^{||},b_2^{||},\ldots,b_n^{||})$ is its Burgers vector of phonon field and $\mathrm{b}^{\perp}=(b_1^{\perp},b_2^{\perp},\ldots,b_n^{\perp})$ is the Burgers vector of phase field. And $\mathrm{b}^{||}$ is located in the physical space $\mathrm{E}_{||}$, while $\mathrm{b}^{\perp}$ is in the supplementary space or the vertical space $\mathrm{E}_{\perp}$. For one-dimensional quasicrystals, $\mathrm{b}^{||}=(b_1^{||},b_2^{||},b_3^{||})$ and $\mathrm{b}^{\perp}=(0,0,b_3^{\perp})$ since $b_1^{\perp}=b_2^{\perp}=0$. Therefore, $\mathrm{b}^{||}\oplus\mathrm{b}^{\perp}=(b_1^{||},b_2^{||},b_3^{||},b_3^{\perp})$ here, which can be dealt with the superposition of $(b_1^{||},b_2^{||},0,0)$ and $(0,0,b_3^{||},b_3^{\perp})$. The $(b_1^{||},b_2^{||},0,0)$ corresponds to the blade dislocation in regular hexagonal crystals, whose elastic solution is available in common metal physics or dislocation monographs (e.g., Refs. [7,8]). The one correspondence to the component $(0,0,b_3^{||},b_3^{\perp})$ is the screw dislocation in one-dimensional quasicrystals. We solve the elastic field induced by this dislocation in the following.

In Section 5.2, we have obtained the governing equations for anti-plane strain problems in one-dimensional quasicrystal elasticity such as

$$\nabla^2 u_z=0,\quad \nabla^2 w_z=0 \tag{7.1-1}$$

The boundary conditions corresponding to the screw dislocation with the Burgers vector $(0,0,b_3^{||},b_3^{\perp})$ are

$$\left.\begin{aligned} u_z\Big|_{y=0^+}-u_z\Big|_{y=0^-}&=b_3^{||}\\ w_z\Big|_{y=0^+}-w_z\Big|_{y=0^-}&=b_3^{\perp}\end{aligned}\right\} \tag{7.1-2a}$$

or

$$\int_\Gamma du_z=b_3^{||},\quad \int_\Gamma dw_z=b_3^{\perp} \tag{7.1-2b}$$

where Γ indicates an arbitrary contour surrounding the dislocation core. The solution to the boundary-value problem (7.1.-1) and (7.1-2) is

$$u_z = \frac{b_3^{\|}}{2\pi}\arctan\frac{y}{x}, \quad w_z = \frac{b_3^{\perp}}{2\pi}\arctan\frac{y}{x} \tag{7.1-3}$$

The stress components can be extracted according to the stress–stain relation (5.2-8) in Chapter 5 as

$$\left.\begin{aligned}
\sigma_{xz} &= \sigma_{zx} = -\frac{C_{44}b_3^{\|}}{2\pi}\left(\frac{y}{x^2+y^2}\right) - \frac{R_3 b_3^{\perp}}{2\pi}\left(\frac{y}{x^2+y^2}\right)\\
\sigma_{yz} &= \sigma_{zy} = -\frac{C_{44}b_3^{\|}}{2\pi}\left(\frac{x}{x^2+y^2}\right) + \frac{R_3 b_3^{\perp}}{2\pi}\left(\frac{x}{x^2+y^2}\right)\\
H_{zx} &= -\frac{K_2 b_3^{\perp}}{2\pi}\left(\frac{y}{x^2+y^2}\right) - \frac{R_3 b_3^{\|}}{2\pi}\left(\frac{y}{x^2+y^2}\right)\\
H_{zy} &= -\frac{K_2 b_3^{\perp}}{2\pi}\left(\frac{x}{x^2+y^2}\right) + \frac{R_3 b_3^{\|}}{2\pi}\left(\frac{x}{x^2+y^2}\right)
\end{aligned}\right\} \tag{7.1-4}$$

The displacement and stress field corresponding to the Burgers vector $\mathrm{b}^{\|}\oplus\mathrm{b}^{\perp} = (b_1^{\|}, b_2^{\|}, b_3^{\|}, b_3^{\perp})$ can be obtained by superposing the above elastic field on that of the regular hexagonal crystals.

The elastic strain energy induced by the screw dislocation is

$$\begin{aligned}
W &= \frac{1}{2}\iint_{\Omega}\left(\sigma_{zj}\frac{\partial u_z}{\partial x_j} + H_{zj}\frac{\partial w_z}{\partial x_j}\right)dx_1dx_2\\
&= \frac{1}{2}\int_{r_0}^{R_0}\int_0^{2\pi} r\left(\sigma_{zj}\frac{\partial u_z}{\partial x_j} + H_{zj}\frac{\partial w_z}{\partial x_j}\right)drd\theta\\
&= \left(C_{44}b_3^{\|2} + K_2 b_3^{\perp 2} + 2R_3 b_3^{\|}b_3^{\perp}\right)\frac{1}{4\pi}\ln\frac{R_0}{r_0}
\end{aligned} \tag{7.1-5}$$

where r_0 is the size of the dislocation core and R_0 is the size of the dislocation network or inclusion available in theory of dislocations in regular hexagonal crystals; here, we use the same concept.

7.2 Dislocations in Quasicrystals with Point Groups $5m$ and $10mm$ Symmetries

Consider a dislocation in pentagonal point group $5m$ or decagonal point group $10mm$ quasicrystals with the Burgers vector $\mathrm{b}^{||}\oplus\mathrm{b}^{\perp} = (b_1^{||},b_2^{||},b_3^{||},b_1^{\perp},b_2^{\perp})$, where $\mathrm{b}^{||}$ is its Burgers vector of phonon field and $\mathrm{b}^{\perp}$ is its Burgers vector of phase field. Here, $b_3^{\perp} = 0$ due to $w_z = 0$.

This dislocation problem can be dealt with by the superposition of two dislocation problems of Burgers vectors $(b_1^{||},b_2^{||},0,b_1^{\perp},b_2^{\perp})$ and $(0,0,b_3^{||})$, respectively. On the dislocation line parallel to the direction of periodic arrangement (z-direction), the elastic field induced by the dislocation of the Burgers vector $(b_1^{||},b_2^{||},0,b_1^{\perp},b_2^{\perp})$ can be simplified as an in-plane elastic field in quasicrystal elasticity, which can be analyzed by the methods discussed in Sections 6.1 and 6.2. The elastic field induced by $(0,0,b_3^{||})$ is governed by the equation $\nabla^2 u_z = 0$, and its solution has been obtained in Section 7.1. Therefore, in the following two sections, we always assume the dislocation line is parallel to the direction of atom periodic arrangement in quasicrystals, and we only study the elastic field induced by the dislocation with the Burgers vector $(b_1^{||}, b_2^{||}, 0, b_1^{\perp}, b_2^{\perp})$.

To demonstrate the methodology concisely, we first consider a special case of $(b_1^{||}, 0, 0, b_1^{\perp}, 0)$. During this procedure, in order to further simplify the mathematical process, we first analyze the case of $b_1^{||} \neq 0$ and $b_2^{||} = b_1^{\perp} = b_2^{\perp} = 0$. In this case, we consider the problem in the upper half-space of $y \geq 0$, and the problem has the following boundary condition:

$$\left.\begin{array}{ll} \sigma_{ij}(x,y) \to 0, H_{ij}(x,y) \to 0, & \left(\sqrt{x^2+y^2} \to \infty\right) \\ \sigma_{yy}(x,0) = 0 & \\ w_x(x,0) = w_y(x,0) = 0 & \\ u_x(x,0^+) - u_y(x,0^-) = b_1^{||}\,, x < 0, \text{ or } & \displaystyle\int_\Gamma du_x = b_1^{||} \end{array}\right\} \tag{7.2-1}$$

where Γ indicates an arbitrary contour surrounding the dislocation core.

To solve this dislocation problem under the boundary condition (7.2-1), we need to solve the following equation (6.2-7), i.e.,

$$\nabla^2\nabla^2\nabla^2\nabla^2 F = 0$$

By introducing the Fourier transform

$$\hat{F}(\xi, y) = \int_{-\infty}^{+\infty} F(x, y)e^{i\xi x}dx \tag{7.2-2}$$

where ξ is the Fourier transform parameter, the above equation can be reduced into

$$\left(\frac{d^2}{dy^2} - \xi^2\right)^4 \hat{F}(\xi, y) = 0 \tag{7.2-3}$$

This is an ordinary partial differential equation with constant coefficients, and its general solution is

$$\begin{aligned}\hat{F}(\xi, y) = {} & [A_1 + B_1 y + C_1 y^2 + D_1 y^3]e^{-|\xi|y} \\ & + [A_2 + B_2 y + C_2 y^2 + D_2 y^3]e^{|\xi|y}\end{aligned} \tag{7.2-4}$$

where $A_1, B_1, \ldots, D_2$ are functions with respect to ξ to be determined by the boundary conditions.

We are going to use Fourier transform to search for problem solutions in Chapters 7–10. For convenience, we only consider the problem in the upper half-space (or the lower half-space) based on the symmetry or anti-symmetry of the problems. We should be cautious that here the symmetry or anti-symmetry is in the view of macroscopic continuum media, namely, the even and odd characteristics of the displacement function $F(x, y)$ or stress function $G(x, y)$ with respect to x but not in the sense of crystalline structures. Furthermore, during the use of Fourier transform, we attempt to make the boundary condition homogeneous at infinity (zero boundary condition). For example, the boundary condition (7.2-1) is homogeneous at infinity. Therefore, we only consider the case in the upper half-space, and the formal solution (7.2-4) can be simplified as

$$\hat{F}(\xi, y) = [A_1 + B_1 y + C_1 y^2 + D_1 y^3]e^{-|\xi|y} \tag{7.2-5}$$

In the evident reason, the suffix of A_1, B_1, C_1 and D_1 can be removed in the following. For conciseness in writing, introduce the following indicator:

$$X = (A, B, C, D), \quad Y = (1, y, y^2, y^3)^T \tag{7.2-6}$$

where the symbol T indicates the matrix transpose. Therefore, (7.2-5) can be rewritten as

$$\hat{F}(\xi, y) = XYe^{-|\xi|y} \tag{7.2-5'}$$

Since A, B, C and D are arbitrary functions with respect to ξ, without losing generality, (7.2-5) can be rewritten as

$$\hat{F}(\xi, y) = (4\xi^4)^{-1}XYe^{-|\xi|y} \tag{7.2-7}$$

By performing Fourier transform on the displacement expressions (6.2-8) and (6.2-9) and stress expressions (6.2-10) and (6.2-11), and using the notation (7.2-7), we have

$$\hat{u}_x(\xi, y) = i\xi^{-1}X[2n\xi^2Y' + (m - 5n)|\xi|Y'' - (2m - 5n)Y''']e^{-|\xi|y} \tag{7.2-8a}$$

$$\hat{u}_y(\xi, y) = |\xi|^{-1}X[2n\xi^2Y' - (m + 5n)|\xi|Y'' + (2m + 5n)Y''']e^{-|\xi|y} \tag{7.2-8b}$$

$$\begin{aligned}\hat{w}_x(\xi, y) &= -i\omega(\alpha - \beta)\xi^{-1}X \\ &\quad \times [2n|\xi|^3 Y - 12\xi^2Y' + 15|\xi|Y'' - 10Y''']e^{-|\xi|y}\end{aligned} \tag{7.2-9a}$$

$$\begin{aligned}\hat{w}_y(\xi, y) &= -\omega(\alpha - \beta)|\xi|^{-1}X[4|\xi|^3 Y \\ &\quad - 12\xi^2Y' + 15|\xi|Y'' - (10 + e_0 - e_1)Y''']e^{-|\xi|y}\end{aligned} \tag{7.2-9b}$$

$$\hat{\sigma}_{xx}(\xi, y) = 2M\alpha(L + M)X(-2\xi^2Y' + 8|\xi|Y'' - 13Y''']e^{-|\xi|y} \tag{7.2-10a}$$

$$\hat{\sigma}_{yy}(\xi, y) = 2M\alpha(L + M)X(2\xi^2Y' - 4|\xi|Y'' + 3Y''']e^{-|\xi|y} \tag{7.2-10b}$$

$$\begin{aligned}\hat{\sigma}_{xy}(\xi, y) &= \hat{\sigma}_{yx}(\xi, y) \\ &= i2M\alpha(L + M)\xi^{-1}|\xi|X(2\xi^2Y' - 6|\xi|Y'' + \delta Y''']e^{-|\xi|y}\end{aligned} \tag{7.2-10c}$$

$$\hat{H}_{xx}(\xi, y) = -\omega(\alpha-\beta)(K_1-K_2)\,|\xi|^{-1}\,X \times [4\,|\xi|^3\,Y - 16\xi^2 Y' + 27\,|\xi|\,Y'' - (25+e_2)Y''']e^{-|\xi|y} \tag{7.2-11a}$$

$$\hat{H}_{yy}(\xi, y) = -\omega(\alpha-\beta)(K_1-K_2)X \times [-4\,|\xi|^3\,Y + 12\xi^2 Y' - 15\,|\xi|\,Y'' + (10-e_1)Y''']e^{-|\xi|y} \tag{7.2-11b}$$

$$\hat{H}_{xy}(\xi, y) = i\omega(\alpha-\beta)(K_1-K_2)\xi^{-1}\,|\xi|\,X \times [-4\,|\xi|^3\,Y + 12\xi^2 Y' - 15\,|\xi|\,Y'' + (10+e_2)Y''']e^{-|\xi|y} \tag{7.2-11c}$$

$$\hat{H}_{yx}(\xi, y) = -i\omega(\alpha-\beta)(K_1-K_2)\xi^{-1}\,|\xi|\,X \times [-4\,|\xi|^3\,Y + 16\xi^2 Y' - 27\,|\xi|\,Y'' + (25-e_1)Y''']e^{-|\xi|y} \tag{7.2-11d}$$

where

$$\left.\begin{array}{l} m = M\alpha + (L+2M)\beta, \quad n = M\alpha - (L+2M)\beta \\ e_1 = \dfrac{2\alpha\beta}{\omega(\alpha-\beta)(K_1-K_2)}, \quad e_2 = \dfrac{2\alpha\beta}{\omega(\alpha-\beta)(K_1-K_2)} + \dfrac{(\alpha+\beta)}{(\alpha-\beta)} \end{array}\right\} \tag{7.2-12}$$

and α, β, δ and ω are given in (6.2-5).

According to the inverse of Fourier transform, it has

$$F(x, y) = \frac{1}{2\pi}\int_{-\infty}^{+\infty} \hat{F}(\xi, y)e^{-i\xi x}d\xi \tag{7.2-13}$$

Similarly, the inverses of displacements and stresses are obtained as follows:

$$u_j(x, y) = \frac{1}{2\pi}\int_{-\infty}^{+\infty} \hat{u}_j(\xi, y)e^{-i\xi x}d\xi \tag{7.2-14}$$

$$w_j(x, y) = \frac{1}{2\pi}\int_{-\infty}^{+\infty} \hat{w}_j(\xi, y)e^{-i\xi x}d\xi \tag{7.2-15}$$

$$\sigma_{jk}(x, y) = \frac{1}{2\pi}\int_{-\infty}^{+\infty} \hat{\sigma}_{jk}(\xi, y)e^{-i\xi x}d\xi \tag{7.2-16}$$

$$H_{jk}(x, y) = \frac{1}{2\pi}\int_{-\infty}^{+\infty} \hat{H}_{jk}(\xi, y)e^{-i\xi x}d\xi \tag{7.2-17}$$

In the above dislocation problem, $u_x(x, y)$ is an odd function with respect to x, therefore $F(x, y)$ must be an odd function with respect to x according to the first expression in (6.2-8). For an odd function, the Fourier transform (7.2-2) is rewritten as

$$\hat{F}(\xi, y) = \int_0^\infty F(x, y) \sin(\xi x) dx \tag{7.2-18}$$

and its inversion is

$$F(x, y) = \frac{2}{\pi} \int_0^\infty \hat{F}(\xi, y) \sin(\xi x) d\xi \tag{7.2-19}$$

If $F(x, y)$ is an even function with respect to x, it has

$$\hat{F}(\xi, y) = \int_0^\infty F(x, y) \cos(\xi x) dx \tag{7.2-20}$$

and its inversion is

$$F(x, y) = \frac{2}{\pi} \int_0^\infty \hat{F}(\xi, y) \cos(\xi x) d\xi \tag{7.2-21}$$

Therefore, the corresponding integrals relating to u_j, w_j, σ_{jk} and H_{jk} can be simplified.

So far, the displacement and stress components in forms of integrals have been obtained, which all include the unknown functions $A(\xi)$, $B(\xi)$, $C(\xi)$ and $D(\xi)$ to be determined by the second expressions in the boundary condition (7.2-1) such that

$$A = (9J \operatorname{sgn} \xi)/4\xi^2, \quad B = 2J\xi^2, \quad C = (J \operatorname{sgn} \xi)/2\xi^2, \quad D = 0 \tag{7.2-22}$$

where J is a constant to be determined. By using the last expression in (7.2-1), i.e., the dislocation condition, J can be expressed as

$$J = \frac{b_1^{\parallel}}{8(n - m)} \tag{7.2-23}$$

Substitution of above expressions into (7.2-8)–(7.2-11) and then into the inversion formulas (7.2-14)–(7.2-17) leads to the displacement and stress fields induced by dislocation with the Burgers vector

$(b_1^{\parallel}, 0, 0, 0)$ in quasicrystals of point group $5m$ five fold symmetry and $10mm$ 10 fold symmetry as follows:

$$u_x = \frac{b_1^{\parallel}}{2\pi}\left[\arctan\left(\frac{y}{x}\right) + \frac{(L+M)K_1}{(L+M)K_1 + (MK_1 - R^2)}\left(\frac{xy}{r^2}\right)\right] \tag{7.2-24a}$$

$$u_y = \frac{b_1^{\parallel}}{2\pi}\left[-\frac{(MK_1 - R^2)}{(L+M)K_1 + (MK_1 - R^2)}\left(\ln\frac{r}{a}\right) + \frac{(L+M)K_1}{(L+M)K_1 + (MK_1 - R^2)}\left(\frac{y^2}{r^2}\right)\right] \tag{7.2-24b}$$

$$w_x = \left(\frac{b_1^{\parallel}}{2\pi}\right)\frac{(L+M)K_1}{(L+M)K_1 + (MK_1 - R^2)}\left(\frac{2x^3y}{r^4}\right) \tag{7.2-25a}$$

$$w_y = \left(\frac{b_1^{\parallel}}{2\pi}\right)\frac{(L+M)K_1}{(L+M)K_1 + (MK_1 - R^2)}\left(\frac{2x^2y^2}{r^4}\right) \tag{7.2-25b}$$

$$\sigma_{xx} = -A\frac{y(3x^2 + y^2)}{r^4} \tag{7.2-26a}$$

$$\sigma_{yy} = A\frac{y(x^2 - y^2)}{r^4} \tag{7.2-26b}$$

$$\sigma_{xy} = \sigma_{yx} = A\frac{x(x^2 - y^2)}{r^4} \tag{7.2-26c}$$

$$H_{xx} = -A\frac{R(K_1 - K_2)}{MK_1 - R^2}\left[\frac{x^2y(3x^2 - y^2)}{r^6}\right] \tag{7.2-27a}$$

$$H_{yy} = -A\frac{R(K_1 - K_2)}{MK_1 - R^2}\left[\frac{x^2y(3y^2 - x^2)}{r^6}\right] \tag{7.2-27b}$$

$$H_{xy} = A\frac{R(K_1 - K_2)}{MK_1 - R^2}\left[\frac{xy^2(3x^2 - y^2)}{r^6}\right] \tag{7.2-27c}$$

$$H_{yx} = -A\frac{R(K_1 - K_2)}{MK_1 - R^2}\left[\frac{x^3(3y^2 - x^2)}{r^6}\right] \tag{7.2-27d}$$

where $r = \sqrt{x^2 + y^2}$, a is the size of the dislocation core, and

$$A = \left(\frac{b_1^{\parallel}}{\pi}\right)\frac{(L-M)(MK_1 - R^2)}{(L+M)K_1 + (MK_1 - R^2)} \tag{7.2-28}$$

Considering $L = C_{12}$ and $M = (C_{11}-C_{12})/2 = C_{66}$ in (6.1-1), and substituting them into (7.2-24a) and (7.2-25), we find that our solutions are exactly consistent with those given by Ding *et al.* [4,7]. They used the Green function method. This examines the correction of our derivation.

In the case of $R = 0$, the above solution recovers the solution of dislocation in hexagonal crystals [8]. If the material is isotropic, $L = \lambda$ and $M = \mu$. Substitution of L and M into (7.2-28) yields

$$A = \frac{\mu b_1^{\|}}{2\pi(1-\nu)} \tag{7.2-29}$$

where λ and μ are the Lamé constants and ν is Poisson's ratio of phonon field. In this case, u_i and σ_{ij} recover like those in a regular isotropic crystal with an edge dislocation, while $w_i = 0$ and $H_{ij} = 0$.

Let us consider another special case of $b_1^{\perp} \neq 0$ and $b_1^{\|} = b_2^{\|} = b_2^{\perp} = 0$. In this case, the boundary condition can be described as

$$\left.\begin{aligned}
&\sigma_{ij}(x,y) \to 0, H_{ij}(x,y) \to 0, \sqrt{x^2+y^2} \to \infty \\
&H_{yy}(x,0) = 0 \\
&u_x(x,0) = u_y(x,0) = 0 \\
&w_x(x,0^+) - w_x(x,0^-) = b_1^{\perp}, x < 0, \text{ or } \int_{\Gamma} dw_x = b_1^{\perp}
\end{aligned}\right\} \tag{7.2-30}$$

where the meaning of Γ is aforementioned.

By using a similar procedure, the displacement and stress field can be determined as

$$u_x = \frac{b_1^{\perp} k_0}{2\pi c_2}\left[\frac{xy}{r^2} - \left(\frac{c_1 - c_2}{2c_1}\right)\frac{2xy^3}{r^4}\right] \tag{7.2-31a}$$

$$u_y = \frac{b_1^{\perp} k_0}{2\pi c_2}\left[-\frac{xy}{r^2} + \left(\frac{c_1 - c_2}{2c_1}\right)\frac{y^2(x^2-y^2)}{r^4}\right] \tag{7.2-31b}$$

$$w_x = \frac{b_1^{\perp}}{2\pi}\left[\arctan\left(\frac{y}{x}\right) + \left(\frac{c_0 k_0}{2c_1 c_2}\right)\frac{xy(3x^2-y^2)(3y^2-x^2)}{3r^6}\right] \tag{7.2-31c}$$

$$w_y = \frac{b_1^{\perp}}{2\pi}\left[\left(1 - \frac{L+2M}{2c_1} - \frac{M}{2c_2}\right)\ln\frac{r}{a} + \left(\frac{c_0 k_0}{2c_1 c_2}\right)\frac{y^2(3x^2-y^2)^2}{3r^6}\right] \tag{7.2-31d}$$

$$\sigma_{xx} = -\left(\frac{c_0 b_1^{\perp} k_0}{\pi c_1 R}\right)\frac{x^2 y(3x^2-y^2)}{r^6} \tag{7.2-32a}$$

$$\sigma_{yy} = -\left(\frac{c_0 b_1^{\perp} k_0}{\pi c_1 R}\right)\frac{y^3(3x^2-y^2)}{r^6} \tag{7.2-32b}$$

$$\sigma_{xy} = \sigma_{yx} = \left(\frac{c_0 b_1^{\perp} k_0}{\pi c_1 R}\right)\frac{2xy^2(y^2-x^2)}{r^6} \tag{7.2-32c}$$

$$H_{xx} = \frac{k_0 b_1^{\perp}}{2\pi e_1}\left[(e_1+e_2)\frac{y}{r^2} - \frac{2x^2y(3x^2-y^2)(3y^2-x^2)}{r^8}\right] \tag{7.2-33a}$$

$$H_{yy} = -\left(\frac{k_0 b_1^{\perp} y}{2\pi e_1}\right)\left[\frac{2(x^2-y^2)}{r^4} + \frac{(x^2-y^2)(3x^2-y^2)(3y^2-x^2)}{r^8}\right] \tag{7.2-33b}$$

$$H_{xy} = \frac{k_0 b_1^{\perp}}{2\pi e_1}\left[(e_1+e_2)\frac{x}{r^2} + \frac{2xy^2(3x^2-y^2)(3y^2-x^2)}{r^8}\right] \tag{7.2-33c}$$

$$H_{yx} = -\left(\frac{k_0 b_1^{\perp} x}{2\pi e_1}\right)\frac{2(x^2-y^2)}{r^4} + \frac{(x^2-y^2)(3x^2-y^2)(3y^2-x^2)}{r^8} \tag{7.2-33d}$$

where

$$e_1 = \frac{2c_1c_2}{c_0k_0}, \quad e_2 = \frac{c_1c_2}{c_0k_0}\left(\frac{c_1'}{c_1} + \frac{c_2'}{c_2}\right)$$

$$c_1' = (L+2M)K_2 - R^2, \quad c_2' = MK_2 - R^2$$

and c_0, c_1, c_2 and k_0 are given in Chapter 6, i.e.,

$$c_0 = (L+2M)R, \quad c_1 = (L+2M)K_1 - R^2, \quad c_2 = MK_1 - R^2$$

$$k_0 = R(K_1 - K_2)$$

Superposition of the above two solutions yields the solution of dislocation $(b_1^{||}, 0, b_1^{\perp}, 0)$. The solution of $(0, b_2^{||}, 0, b_2^{\perp})$ can be determined similarly. Their superposition yields the solution of $(b_1^{||}, b_2^{||}, b_1^{\perp}, b_2^{\perp})$.

Part of this work can be found in the paper by Li and Fan [5]. Readers may examine that the present solutions are identical to those given by Ding *et al.* [4,7] using the method of Green's functions (note that $L = C_{12}$ and $M = (C_{11}\text{-}C_{12})/2 = C_{66}$).

7.3 Dislocations in Quasicrystals with Point Groups $5, \bar{5}$ fold and $10, \overline{10}$ fold Symmetries

Similar to those discussed in the proceeding section, as two-dimensional quasicrystals, the Burgers vector of the dislocation is $\mathrm{b}^{||} \oplus \mathrm{b}^{\perp} = (b_1^{||}, b_2^{||}, b_3^{||}, b_1^{\perp}, b_2^{\perp})$. Due to $w_3 = 0$ and our assumption that dislocation line is parallel to the direction of periodic arrangement (z-direction), the above Burgers vector can be considered as the superposition of $(b_1^{||}, b_2^{||}, b_1^{\perp}, b_2^{\perp})$ and $(0,0,b_3^{\perp})$. Under this assumption, the field variables do not vary with the change of the variable z. Therefore, the elastic field corresponding to $(b_1^{||},b_2^{||},b_1^{\perp},b_2^{\perp})$ can be determined by using the theory of plane elasticity of two-dimensional quasicrystals discussed in Section 6.4. The anti-plane elastic field induced by $(0,0,b_3^{\perp})$ is very simple, and can be solved by using the method discussed in Section 7.1.

From Section 6.4, we have obtained the governing equation of plane elastic field of quasicrystals with point groups $5, \bar{5}$ 5 fold and 10 fold $10, \overline{10}$ 10 fold symmetries:

$$\nabla^2\nabla^2\nabla^2\nabla^2 F = 0 \tag{7.3-1}$$

where $F(x, y)$ is the displacement potential function, defined in (6.4-10). By performing Fourier transform of (7.2-2) and considering the case of upper half-space ($y \geq 0$), its Fourier transform $\hat{F}(\xi, y)$ has the form

$$\hat{F}(\xi, y) = (4\xi^4 R^2)^{-1} XY e^{-|\xi|y} \tag{7.3-2}$$

where X and Y have the same meaning as in Section 7.2, ξ is the parameter of Fourier transform and

$$R^2 = R_1^2 + R_2^2 \tag{7.3-3}$$

After some algebraic operations with (6.2-8)–(6.2-11), the displacement components in Fourier transform domain are

$$\hat{u}_x(\xi, y) = i\xi^{-1}\bar{R}_0 X[2n\xi^2 Y' + (m - 5n)\,|\xi|\,Y'' - (2m - 5n)Y''']e^{-|\xi|y} \tag{7.3-4a}$$

$$\hat{u}_y(\xi, y) = |\xi|^{-1}\,\bar{R}_0 X \times[2n\xi^2 Y' - (m + 5n)\,|\xi|\,Y'' + (2m + 5n)Y''']e^{-|\xi|y} \tag{7.3-4b}$$

$$\hat{w}_x(\xi, y) = ic_0\xi^{-1}\bar{R}_0^2 X[4\,|\xi|^3\,Y - 12\xi^2 Y' + 15\,|\xi|\,Y'' - 10Y''']e^{-|\xi|y} \tag{7.3-4c}$$

$$\hat{w}_y(\xi, y) = c_0\,|\xi|^{-1}\,\bar{R}_0^2 X \times[4\,|\xi|^3\,Y - 12\xi^2 Y' + 15\,|\xi|\,Y'' - (10 + e_0 R_0^2)Y''']e^{-|\xi|y} \tag{7.3-4}$$

where

$$\left.\begin{array}{l} m = c_2 + c_1, \quad n = c_2 - c_1, \quad e_0 = -[C_{66}c_1 + C_{11}c_2]/(Rc_0) \\ R_0 = (R_1 + iR_2\ \mathrm{sgn}\,\xi)/R, \quad \bar{R}_0 = (R_1 - iR_2\ \mathrm{sgn}\,\xi)/R \end{array}\right\} \tag{7.3-5d}$$

Similarly, the stress components in Fourier transform domain are

$$\hat{\sigma}_{xx}(\xi, y) = 2c_0c_2R^{-1}\bar{R}_0 X(-2\xi^2 Y' + 8\,|\xi|\,Y'' - 13Y''')e^{-|\xi|y} \tag{7.3-6a}$$

$$\hat{\sigma}_{yy}(\xi, y) = 2c_0c_2R^{-1}\bar{R}_0 X(2\xi^2 Y' - 4\,|\xi|\,Y'' + 3Y''')e^{-|\xi|y} \tag{7.3-6b}$$

$$\hat{\sigma}_{xy}(\xi, y) = \hat{\sigma}_{yx}(\xi, y) = i2c_0c_2R^{-1}\bar{R}_0(\mathrm{sgn}\xi)X(2\xi^2 Y' - 6\,|\xi|\,Y'' + 7Y''']e^{-|\xi|y} \tag{7.3-6c}$$

$$\hat{H}_{xx}(\xi, y) = c_0k_0\bar{R}_0^2 X[4\,|\xi|^3\,Y - 16\xi^2 Y' + 27\,|\xi|\,Y'' - (25 + e_2R_0^2)Y''']e^{-|\xi|y} \tag{7.3-6d}$$

$$\hat{H}_{yy}(\xi, y) = c_0k_0\bar{R}_0^2 X[-4\,|\xi|^3\,Y + 12\xi^2 Y' - 15\,|\xi|\,Y'' + (10 - e_1R_0^2)Y''']e^{-|\xi|y} \tag{7.3-6e}$$

$$\hat{H}_{xy}(\xi, y) = ic_0k_0\bar{R}_0^2(\text{sgn}\xi)X[-4|\xi|^3 Y + 12\xi^2 Y' - 15|\xi| Y'' + (10 + e_2R_0^2)Y''']e^{-|\xi|y} \tag{7.3-6f}$$

$$\hat{H}_{yx}(\xi, y) = ic_0k_0\bar{R}_0^2(\text{sgn}\xi)X[-4|\xi|^3 Y + 16\xi^2 Y' - 27|\xi| Y'' + (25 - e_1R_0^2)Y''']e^{-|\xi|y} \tag{7.3-6g}$$

$$\left.\begin{aligned} e_1 &= \frac{2c_1c_2}{c_0k_0R}, \quad e_2 = \frac{c_1c_2}{c_0k_0R}\left(\frac{c_1'}{c_1} + \frac{c_2'}{c_2}\right) \\ c_1' &= C_{11}K_2 - R^2, \quad c_2' = C_{66}K_2 - R^2 \end{aligned}\right\} \tag{7.3-7}$$

where c_0 and K_0 are given in (6.4-13).

As introduced in the proceeding section, X includes four unknowns $A(\xi)$, $B(\xi)$, $C(\xi)$ and $D(\xi)$, which can be determined by boundary conditions. Once $A(\xi)$, $B(\xi)$, $C(\xi)$ and $D(\xi)$ are determined, u_i, w_i, σ_{jk} and H_{jk} can be determined by the Fourier inverse.

In the case of dislocation problem $\mathrm{b}^{||}\oplus\mathrm{b}^{\perp} = (b_1^{||}, 0, b_1^{\perp}, 0)$, we try to determine these functions and their Fourier inverses in the following.

In order to simplify the calculation, we first consider the case of $b_1^{||} \neq 0$ and $b_1^{\perp} = 0$. In this case, the boundary condition is

$$\left.\begin{aligned} &\sigma_{ij}(x, y) \to 0, H_{ij}(x, y) \to 0, \quad \left(\sqrt{x^2 + y^2} \to \infty\right) \\ &\sigma_{yy}(x, 0) = 0 \\ &w_x(x, 0) = w_y(x, 0) = 0 \\ &\int_\Gamma du_x = b_1^{||}, \quad \int_\Gamma du_y = 0 \end{aligned}\right\} \tag{7.3-8}$$

In Fourier domain, solutions (7.3-4d) and (7.3-6) and the first three expressions in boundary condition (7.3-8) yield

$$A(\xi) = 9C(\xi)/4, \quad B(\xi) = 2C(\xi), \quad D(\xi) = 0 \tag{7.3-9}$$

and the dislocation condition leads to

$$\left.\begin{aligned} \int_\Gamma du_x &= -4c_1R^{-1}(R_1\text{Re}C + R_2 \text{ sgn } \xi\text{Im}C) \\ \int_\Gamma du_y &= 4c_1R^{-1}(R_2\text{Re}C - R_1 \text{ sgn } \xi\text{Im}C) \end{aligned}\right\} \tag{7.3-10}$$

where the indicators Re and Im stand for the real and imaginary parts of a complex variable, respectively. Finally, expressions (7.3-10) and (7.3-8) define

$$C(\xi) = ReC + i\text{Im}C = -\frac{(R_1 + iR_2 \text{ sgn } \xi)b_1^{||}}{4\pi R c_1} = -\frac{R_0 b_1^{||}}{4\pi c_1} \quad (7.3\text{-}11)$$

where R_0 is given by (7.3-5d) and c_1 is given by (6.4-9), i.e.,

$$c_1 = (L + 2M)K_1 - R^2, \quad R^2 = R_1^2 + R_2^2 \quad (7.3\text{-}12)$$

Therefore,

$$u_x = \frac{b_1^{||}}{2\pi}\left[\arctan\left(\frac{y}{x}\right) + \frac{c_1 - c_2}{c_1}\left(\frac{xy}{r^2}\right)\right] \quad (7.3\text{-}13a)$$

$$u_y = \frac{b_1^{||}}{2\pi}\left[-\ln\frac{r}{a} + \frac{c_1 - c_2}{c_1}\left(\ln\frac{r}{a} + \frac{y^2}{r^2}\right)\right] \quad (7.3\text{-}13b)$$

$$w_x = \frac{c_0 b_1^{||}}{2\pi c_1}\left[\left(\frac{R_1}{R}\right)\frac{2x^3 y}{r^4} + \left(\frac{R_2}{R}\right)\frac{y^2(3x^2 + y^2)}{r^4}\right] \quad (7.3\text{-}13c)$$

$$w_y = \frac{c_0 b_1^{||}}{2\pi c_1}\left[\left(\frac{R_1}{R}\right)\frac{y^2(3x^2 + y^2)}{r^4} + \left(\frac{R_2}{R}\right)\frac{2x^3 y}{r^4}\right] \quad (7.3\text{-}13d)$$

where c_0 is given by (6.4-13) and a indicates the radius of the dislocation core.

The corresponding stress components can be determined by (7.3-13) and generalized Hooke's law (6.4-3):

$$\sigma_{xx} = -\left(\frac{c_0 c_2 b_1^{||}}{\pi c_1 R}\right)\frac{y(3x^2 + y^2)}{r^4} \quad (7.3\text{-}14a)$$

$$\sigma_{yy} = \left(\frac{c_0 c_2 b_1^{||}}{\pi c_1 R}\right)\frac{y(x^2 - y^2)}{r^4} \quad (7.3\text{-}14b)$$

$$\sigma_{xy} = \sigma_{yx} = \left(\frac{c_0 c_2 b_1^{||}}{\pi c_1 R}\right)\frac{x(x^2 - y^2)}{r^4} \quad (7.3\text{-}14c)$$

$$H_{xx} = -\frac{c_0 k_0 b_1^{||}}{\pi c_1}\left[\left(\frac{R_1}{R}\right)\frac{x^2 y(3x^2 - y^2)}{r^6} + \left(\frac{R_2}{R}\right)\frac{x^3(3y^2 - x^2)}{r^6}\right] \quad (7.3\text{-}14d)$$

$$H_{yy} = -\frac{c_0 k_0 b_1^{||}}{\pi c_1}\left[\left(\frac{R_1}{R}\right)\frac{x^2 y(3x^2 - y^2)}{r^6} - \left(\frac{R_2}{R}\right)\frac{xy^2(3x^2 - y^2)}{r^6}\right] \tag{7.3-14e}$$

$$H_{xy} = -\frac{c_0 k_0 b_1^{||}}{\pi c_1}\left[\left(\frac{R_1}{R}\right)\frac{xy^2(3x^2 - y^2)}{r^6} + \left(\frac{R_2}{R}\right)\frac{x^2 y(3y^2 - x^2)}{r^6}\right] \tag{7.3-14f}$$

$$H_{yx} = -\frac{c_0 k_0 b_1^{||}}{\pi c_1}\left[-\left(\frac{R_1}{R}\right)\frac{x^3(3y^2 - x^2)}{r^6} + \left(\frac{R_2}{R}\right)\frac{x^2 y(3x^2 - y^2)}{r^6}\right] \tag{7.3-14g}$$

Now, let us consider the case of $b_1^{||} = 0$ and $b_1^{\perp} \neq 0$. In this case, the corresponding boundary condition is

$$\left.\begin{array}{l}
\sigma_{ij}(x,y) \to 0, \quad H_{ij}(x,y) \to 0, \quad \left(\sqrt{x^2+y^2} \to \infty\right) \\
H_{yy}(x,0) = 0 \\
u_x(x,0) = u_y(x,0) = 0 \\
\int_\Gamma dw_x = b_1^{\perp}, \quad \int_\Gamma dw_y = 0
\end{array}\right\} \tag{7.3-15}$$

By using a similar analysis, the corresponding displacement and stress fields can be determined as (7.3-16) and (7.3-17)

$$\begin{aligned}
u_x = \frac{c_1 b_1^{\perp}}{\pi c_0 e_1}\Bigg\{&\frac{R_1}{R}\left[\frac{xy}{r^2} - \left(\frac{c_1 - c_2}{c_1}\right)\frac{2xy^3}{r^4}\right] \\
&+ \frac{R_2}{R}\left[\frac{y^2}{r^2} + \left(\frac{c_1 - c_2}{c_1}\right)\frac{y^2(x^2 - y^2)}{r^4}\right]\Bigg\}
\end{aligned} \tag{7.3-16a}$$

$$\begin{aligned}
u_y = \frac{c_1 b_1^{\perp}}{\pi c_0 e_1}\Bigg\{&-\frac{R_1}{R}\left[\frac{y^2}{r^2} - \left(\frac{c_1 - c_2}{c_1}\right)\frac{y^2(x^2 - y^2)}{r^4}\right] \\
&+ \frac{R_2}{R}\left[\frac{xy}{r^2} + \left(\frac{c_1 - c_2}{c_1}\right)\frac{2xy^3}{r^4}\right]\Bigg\}
\end{aligned} \tag{7.3-16b}$$

$$\begin{aligned}
w_x = \frac{b_1^{\perp}}{2\pi}\Bigg[&\arctan\left(\frac{y}{x}\right) + \left(\frac{R_1^2 - R_2^2}{e_1 R^2}\right)\frac{xy(3x^2 - y^2)(3y^2 - x^2)}{3r^6} \\
&+ \left(\frac{2R_1 R_2}{e_1 R^2}\right)\frac{y^2(3x^2 - y^2)^2}{3r^6}\Bigg]
\end{aligned} \tag{7.3-16c}$$

$$
\begin{aligned}
w_y = \frac{b_1^\perp}{2\pi e_1}\bigg[& e_2 \ln\frac{r}{a} + \left(\frac{R_1^2 - R_2^2}{R^2}\right)\frac{y^2(3x^2-y^2)^2}{3r^6} \\
& - \left(\frac{2R_1R_2}{e_1R^2}\right)\frac{xy(3x^2-y^2)(3y^2-x^2)}{3r^6}\bigg]
\end{aligned}
\tag{7.3-16d}
$$

$$
\sigma_{xx} = -\frac{2c_2b_1^\perp}{\pi e_1 R}\left[\frac{R_1}{R}\frac{x^2y(3x^2-y^2)}{r^6} + \frac{R_2}{R}\frac{x^3(3y^2-x^2)}{r^6}\right] \tag{7.3-17a}
$$

$$
\sigma_{yy} = -\frac{2c_2b_1^\perp}{\pi e_1 R}\left[\frac{R_1}{R}\frac{y^3(3x^2-y^2)}{r^6} + \frac{R_2}{R}\frac{xy^2(3y^2-x^2)}{r^6}\right] \tag{7.3-17b}
$$

$$
\sigma_{xy} = \sigma_{yx} = -\frac{2c_2b_1^\perp}{\pi e_1 R}\left[\frac{R_1}{R}\frac{xy^2(3x^2-y^2)}{r^6} + \frac{R_2}{R}\frac{x^2(3y^2-x^2)}{r^6}\right] \tag{7.3-17c}
$$

$$
\begin{aligned}
H_{xx} = \frac{K_0b_1^\perp}{2\pi e_1}\bigg[& -(e_1+e_2)\frac{y}{r^2} \\
& + x\left(\frac{R_1^2-R_2^2}{R^2}h_{21}(x,y) - \frac{2R_1R_2}{R^2}h_{22}(x,y)\right)\bigg]
\end{aligned}
\tag{7.3-17d}
$$

$$
H_{yy} = -\frac{K_0b_1^\perp y}{2\pi e_1}\left[\frac{R_1^2-R_2^2}{R^2}h_{22}(x,y) + \frac{2R_1R_2}{R^2}h_{21}(x,y)\right] \tag{7.3-17e}
$$

$$
\begin{aligned}
H_{xy} = \frac{K_0b_1^\perp}{2\pi e_1}\bigg[& (e_1+e_2)\frac{x}{r^2} \\
& + y\left(\frac{R_1^2-R_2^2}{R^2}h_{21}(x,y) - \frac{2R_1R_2}{R^2}h_{22}(x,y)\right)\bigg]
\end{aligned}
\tag{7.3-17f}
$$

$$
H_{yx} = -\frac{K_0b_1^\perp x}{2\pi e_1}\left[\frac{R_1^2-R_2^2}{R^2}h_{22}(x,y) + \frac{2R_1R_2}{R^2}h_{21}(x,y)\right] \tag{7.3-17g}
$$

in which

$$
\begin{aligned}
h_{21}(x,y) &= \frac{2xy(3x^2-y^2)(3y^2-x^2)}{r^8} \\
h_{22}(x,y) &= \frac{2(x^2-y^2)}{r^4} + \frac{(x^2-y^2)(3x^2-y^2)(3y^2-x^2)}{r^8}
\end{aligned}
\tag{7.3-18}
$$

The solution to the dislocation problem $(b_1^{\|},0,b_1^\perp,0)$ can be determined by superposing (7.3-13) and (7.3-14) onto (7.3-16) and

(7.3-17). The solution to the dislocation problem $(0,b_2^{||},0,b_2^{\perp})$ can be determined in a similar way. As a result, the solution to the dislocation problem with the Burgers vector $(b_1^{||},b_2^{||},b_1^{\perp},b_2^{\perp})$ can be determined completely.

The above work can be found in the literature [6].

7.4 Dislocations in Quasicrystals with Eight fold Symmetry

The final governing equation of the elasticity of two-dimensional quasicrystals with eight fold symmetry can be expressed as

$$(\nabla^2\nabla^2\nabla^2\nabla^2 - 4\varepsilon\nabla^2\nabla^2\Lambda^2\Lambda^2 + 4\varepsilon\Lambda^2\Lambda^2\Lambda^2\Lambda^2)F = 0 \tag{7.4-1}$$

in which

$$\left.\begin{aligned} &\nabla^2 = \frac{\partial^2}{\partial x^2} + \frac{\partial^2}{\partial y^2}, \quad \Lambda^2 = \frac{\partial^2}{\partial x^2} - \frac{\partial^2}{\partial y^2} \\ &\varepsilon = \frac{R^2(L+M)(K_2+K_3)}{[M(K_1+K_2+K_3)-R^2][(L+2M)K_1-R^2]} \end{aligned}\right\} \tag{7.4-2}$$

(see (6.6-10) and (6.6-11) for details). Equation (7.4-1) is more complicated than those of (6.2-7) and (7.3-1), the solution of which is also more complex than those discussed in the previous sections. Due to the limitation of space, we cannot list the whole procedure of the solution but only give some main results of which in the following, in which both Fourier analysis and complex analysis methods are respectively used.

7.4.1 *Fourier transform method*

Consider the dislocation problem $\mathrm{b}^{||}\oplus\mathrm{b}^{\perp} = (b_1^{||},0,b_1^{\perp},0,0)$, we try to determine the displacement field under the action of the boundary conditions

$$\left.\begin{aligned} &\sigma_{ij}(x,y) \to 0, H_{ij}(x,y) \to 0 \quad \left(\sqrt{x^2+y^2} \to \infty\right) \\ &\sigma_{yy}(x,0) = 0, \ \ H_{yy}(x,0) = 0 \\ &\int_\Gamma du_x = b_1^{||}, \ \int_\Gamma dw_x = b_1^{\perp} \end{aligned}\right\} \tag{7.4-3}$$

By performing Fourier transform of (7.4-1), it can be reduced to

$$\left[\left(\frac{d^2}{dy^2}-\xi^2\right)^4-4\varepsilon\left(\frac{d^2}{dy^2}-\xi^2\right)^2+4\varepsilon\left(\frac{d^2}{dy^2}+\xi^2\right)^4\right]\hat{F}=0 \tag{7.4-4}$$

The eigen roots of equation (7.4-4) depend on the value of the parameter ε; Zhou [9] gave a detailed discussion for the solutions corresponding to case (1): $0<\varepsilon<1$ and case (2): $\varepsilon<0$, but the calculation is tremendously complex and lengthy which cannot be included here. For case (1), the solution is

$$\begin{aligned}
u_x(x,y)=&\frac{1}{2\pi}\left\{\frac{b_1^{\|}}{2}\left[\operatorname{arc\,tan}\left(\frac{\lambda_1^2+\lambda_2^2}{\lambda_1}\frac{y}{x}+\frac{\lambda_2}{\lambda_1}\right)\right.\right.\\
&\left.+\arctan\left(\frac{\lambda_1^2+\lambda_2^2}{\lambda_1}\frac{y}{x}-\frac{\lambda_2}{\lambda_1}\right)\right]\\
&+(F_3C+F_4D)\left[\arctan\frac{2\lambda_3xy}{x^2-(\lambda_3^2+\lambda_4^2)y^2}\right.\\
&\left.\left.-\arctan\frac{2\lambda_1xy}{x^2-(\lambda_1^2+\lambda_2^2)y^2}\right]\right\}\\
&+\frac{1}{4\pi}\left[F_5\ln\frac{x^2+2\lambda_2xy+(\lambda_1^2+\lambda_2^2)y^2}{x^2-2\lambda_2xy+(\lambda_1^2+\lambda_2^2)y^2}\right.\\
&\left.+F_6\ln\frac{x^2+2\lambda_4xy+(\lambda_3^2+\lambda_4^2)y^2}{x^2-2\lambda_4xy+(\lambda_3^2+\lambda_4^2)y^2}\right]\\
u_y=&\frac{1}{2\pi}\left\{H_1\left[\arctan\frac{2\lambda_1\lambda_2y^2}{x^2+(\lambda_1^2-\lambda_2^2)y^2}-2\arctan\frac{\lambda_2}{\lambda_1}\right]\right.\\
&\left.+H_2\left[\arctan\frac{2\lambda_3\lambda_4y^2}{x^2+(\lambda_3^2-\lambda_4^2)y^2}-2\arctan\frac{\lambda_4}{\lambda_3}\right]\right\}\\
&+\frac{1}{4\pi}\left\{H_3\ln\left[1+\frac{x^4+2(\lambda_1^2-\lambda_2^2)x^2y^2}{(\lambda_1^2+\lambda_2^2)^2y^4}\right]\right.\\
&\left.+H_4\ln\left[1+\frac{x^4+2(\lambda_3^2-\lambda_4^2)x^2y^2}{(\lambda_3^2+\lambda_4^2)^2y^4}\right]\right\}
\end{aligned}$$

$$
\begin{aligned}
w_x(x,y) = \frac{1}{2\pi}\Bigg\{ & \frac{b_1^{\perp}}{2}\left[\arctan\left(\frac{\lambda_1^2+\lambda_2^2}{\lambda_1}\frac{y}{x}+\frac{\lambda_2}{\lambda_1}\right)\right. \\
& \left.+\arctan\left(\frac{\lambda_1^2+\lambda_2^2}{\lambda_1}\frac{y}{x}-\frac{\lambda_2}{\lambda_1}\right)\right] \\
& +(G_3C+G_4D)\times\left[\arctan\frac{2\lambda_3 xy}{x^2-(\lambda_3^2+\lambda_4^2)y^2}\right. \\
& \left.-\arctan\frac{2\lambda_1 xy}{x^2-(\lambda_1^2+\lambda_2^2)y^2}\right]\Bigg\} \\
& +\frac{1}{4\pi}\left[G_5\ln\frac{x^2+2\lambda_2 xy+(\lambda_1^2+\lambda_2^2)y^2}{x^2-2\lambda_2 xy+(\lambda_1^2+\lambda_2^2)y^2}\right. \\
& \left.+G_6\ln\frac{x^2+2\lambda_4 xy+(\lambda_3^2+\lambda_4^2)y^2}{x^2-2\lambda_4 xy+(\lambda_3^2+\lambda_4^2)y^2}\right] \\
w_y = \frac{1}{2\pi}\Bigg\{ & I_1\left[\arctan\frac{2\lambda_1\lambda_2 y^2}{x^2+(\lambda_1^2-\lambda_2^2)y^2}-2\arctan\frac{\lambda_2}{\lambda_1}\right] \\
& +I_2\left[\arctan\frac{2\lambda_3\lambda_4 y^2}{x^2+(\lambda_3^2-\lambda_4^2)y^2}-2\arctan\frac{\lambda_4}{\lambda_3}\right]\Bigg\} \\
& +\frac{1}{4\pi}\Bigg\{I_3\ln\left[1+\frac{x^4+2(\lambda_1^2-\lambda_2^2)x^2y^2}{(\lambda_1^2+\lambda_2^2)^2y^4}\right] \\
& +I_4\ln\left[1+\frac{x^4+2(\lambda_3^2-\lambda_4^2)x^2y^2}{(\lambda_3^2+\lambda_4^2)^2y^4}\right]\Bigg\} \qquad (7.4\text{-}5)
\end{aligned}
$$

in which $F_1,\ldots,F_6, G_1,\ldots,G_6, H_1,\ldots,H_4$ and $I_1,\ldots,I_4$ are some functions of $\lambda_1,\lambda_2,\lambda_3$ and λ_4 which are constants constituted from the original material constants M, L, K_1, K_2, K_3 and R, and the expressions are very complicated and lengthy, so we have to omit them.

By using a similar procedure, the solution for case (2) can also be obtained. But the solving procedure is very tedious due to the complexity of the final governing equation (7.4-1). We omit them for simplicity.

7.4.2 Complex analysis method

Equation (7.4-1) can also be solved by the complex variable function method. For this purpose, the equation can be rewritten as

$$\left[\frac{\partial^8}{\partial x^8} + 4(1-4\varepsilon)\frac{\partial^8}{\partial x^6 \partial y^2} + 2(3+16\varepsilon)\frac{\partial^8}{\partial x^4 \partial y^4} + 4(1-4\varepsilon)\frac{\partial^8}{\partial x^2 \partial y^6} + \frac{\partial^8}{\partial y^8}\right] F = 0 \tag{7.4-6}$$

The solution of Eq. (7.4-6) can be expressed in terms of four analytic functions $F_k(z_k)$ of complex variable z_k $(k = 1, 2, 3, 4)$, i.e.,

$$F(x, y) = 2\mathrm{Re}\sum_{k=1}^{4} F_k(z_k), z_k = x + \mu_k y \tag{7.4-7}$$

and $\mu_k = \alpha_k + i\beta_k$ $(i = 1, 2, 3, 4)$ are complex parameters and are determined by the roots of the following eigenvalue equation:

$$\mu^8 + 4(1-4\varepsilon)\mu^6 + 2(3+16\varepsilon)\mu^4 + 4(1-4\varepsilon)\mu^2 + 1 = 0 \tag{7.4-8}$$

Based on the displacement expressions (6.6-12) and the dislocation condition

$$\left.\int_\Gamma du_x = b_1^{\|}, \quad \int_\Gamma du_y = b_2^{\|}, \quad \int_\Gamma dw_x = b_1^{\perp}, \quad \int_\Gamma dw_y = b_2^{\perp}\right\} \tag{7.4-9}$$

we can obtain the solution as follows:

$$u_x = 2\ \mathrm{Re}\sum_{k=1}^{4} a_{1k} f_k(z_k), \quad u_y = 2\ \mathrm{Re}\sum_{k=1}^{4} a_{2k} f_k(z_k)$$
$$w_x = 2\ \mathrm{Re}\sum_{k=1}^{4} a_{3k} f_k(z_k), \quad w_y = 2\ \mathrm{Re}\sum_{k=1}^{4} a_{4k} f_k(z_k)$$

where

$$f_k(z_k) = \frac{\partial^6 F_k(z_k)}{\partial z_k^6} \tag{7.4-10}$$

Some details of this work can be found in Ref. [10].

7.5 Dislocations in Dodecagon Quasicrystals

In monograph [7], Ding *et al.* give the solution in terms of the Green function method which has not been introduced in this book, so the detail is omitted and we only list the results, such as the phonon displacement field

$$
\begin{aligned}
u_1 &= \frac{b_1^{\|}}{2\pi}\left(\arctan(x_2/x_1) + \frac{L+M}{L+2M}\frac{x_1x_2}{r^2}\right)\\
&\quad + \frac{b_2^{\|}}{2\pi}\left(\frac{M}{L+2M}\ln(r/r_0) + \frac{L+M}{L+2M}\frac{x_2^2}{r^2}\right)\\
u_2 &= -\frac{b_1^{\|}}{2\pi}\left(\frac{M}{L+2M}\ln(r/r_0) + \frac{L+M}{L+2M}\frac{x_1^2}{r^2}\right)\\
&\quad + \frac{b_2^{\|}}{2\pi}\left(\arctan(x_2/x_1) - \frac{L+M}{L+2M}\frac{x_1x_2}{r^2}\right)\\
u_3 &= 0
\end{aligned}
\tag{7.5-1}
$$

and the phason displacement field

$$
\begin{aligned}
w_1 &= \frac{b_1^{\perp}}{2\pi}\left(\arctan(x_2/x_1) - \frac{(K_1+K_2)(K_2+K_3)}{2K_1(K_1+K_2+K_3)}\frac{x_1x_2}{r^2}\right)\\
&\quad + \frac{b_2^{\perp}}{4\pi}\left(-\frac{K_2(K_1+K_2+K_3)-K_1K_3}{K_1(K_1+K_2+K_3)}\ln(r/r_0)\right.\\
&\quad \left. + \frac{(K_1+K_2)(K_2+K_3)}{K_1(K_1+K_2+K_3)}\frac{x_2^2}{r^2}\right)\\
w_2 &= \frac{b_1^{\perp}}{4\pi}\left(\frac{K_2(K_1+K_2+K_3)-K_1K_3}{K_1(K_1+K_2+K_3)}\ln(r/r_0)\right.\\
&\quad \left. - \frac{(K_1+K_2)(K_2+K_3)}{K_1(K_1+K_2+K_3)}\frac{x_1^2}{r^2}\right)\\
&\quad + \frac{b_2^{\perp}}{2\pi}\left(\arctan(x_2/x_1) + \frac{(K_1+K_2)(K_2+K_3)}{2K_1(K_1+K_2+K_3)}\frac{x_1x_2}{r^2}\right)
\end{aligned}
\tag{7.5-2}
$$

and the signs L, M are phonon elastic constants and the others are the same meaning as before. Because the phonons and phasons are decoupled for dodecagonal quasicrystals, the coupling constant $R = 0$.

7.6 Interface between Quasicrystal and Crystal

In the previous sections, we discussed dislocation problems in one- and two-dimensional quasicrystals, and a series of analytic solutions are obtained. In Chapter 8, we discuss crack problems in the material. Apart from dislocations and cracks, interface is a kind of defect in quasicrystals too, which is of particular significance for some physical processes.

We know that all quasicrystals observed to date are alloys. This kind of alloy can be in crystalline phase, or in quasicrystalline phase, or in crystal–quasicrystal coexisting phase. Li *et al.* [16, 17] observed the crystal–quasicrystal phase transition, which is a continuous transition. During the process, there must be interface between the crystal and the quasicrystal, so the interface is an important problem. In this section, we give a phenomenological study on the elastic behaviour of the interface for one-dimensional quasicrystal-isotropic crystal. References [15–17] pointed out that the phase transition is induced by the phason strains. However, this is a very complicated problem far unsolved. We focus on only the determination of the strains in this section, and further studies will be given in Chapter 9 for the icosahedral quasicrystal–cubic crystal interface.

Consider an orthorhombic quasicrystal laying at the upper half-plane (i.e., $y > 0$) whose phonon–phason coupling problem is governed by (5.4-3), i.e.,

$$\left(a_1\frac{\partial^4}{\partial x^4} + a_3\frac{\partial^4}{\partial x^2\partial y^2} + a_5\frac{\partial^4}{\partial y^4}\right)F = 0 \tag{7.6-1}$$

in which $F(x, y)$ denotes the displacement potential, a_1, a_3, a_5 the material constants composed from C_{ij}, K_i and R_i defined by (5.3-6). We assume that the crystal coexisting with the quasicrystal is laying at the lower half-plane with thickness h (i.e., $-h < y < 0$), then the plane $y = 0$ is the interface between the quasicrystal and crystal, see Fig. 7.1. For simplicity, suppose the crystal is an isotropic material characterized by elastic constants $C_{ij}^{(c)}(E^{(c)}, \mu^{(c)})$. At the interface,

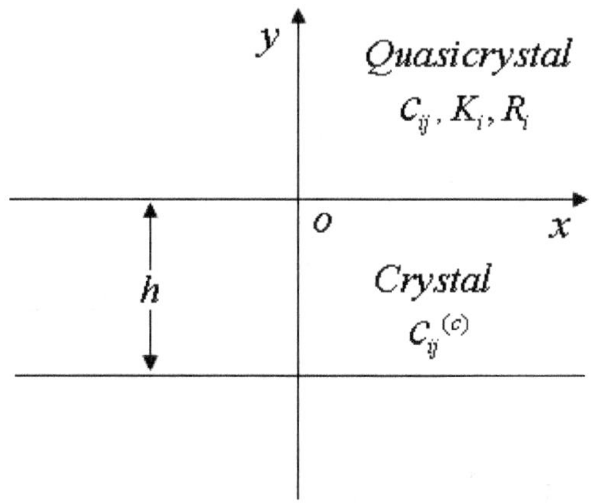

Fig. 7.1. Coexisting phase of quasicrystal-crystal.

there are the following boundary conditions:

$$y = 0, -\infty < x < \infty: \quad \sigma_{zy} = \tau f(x) + ku(x), \quad H_{zy} = 0 \tag{7.6-2}$$

in which $f(x)$ is the distribution function of applied stress at the interface, $u(x) = u_z(x, 0)$the value of displacement component of phonon field at the interface, τ a constant shear stress and k a material constant to be taken as

$$k = \frac{\mu^{(c)}}{h} \tag{7.6-3}$$

where $\mu^{(c)}$ and h are the shear modulus and thickness of the crystal. We further assume that the outer boundaries are stress-free.

Taking the Fourier transform

$$\hat{F}(\xi, y) = \int_{-\infty}^{\infty} F(x, y) e^{i\xi x} dx \tag{7.6-4}$$

to Eq. (7.6-1) yields

$$\left(a_5 \frac{d^4}{dy^4} + a_3 \xi^2 \frac{d^2}{dy^2} + a_1 \xi^4 \right) \hat{F} = 0 \tag{7.6-5}$$

If put solution of Eq. (7.6-5) is

$$\hat{F}(\xi, y) = e^{-\lambda |\xi| y}$$

where $y > 0$ and λ is a parameter, and substituting it into (7.6-5) leads to

$$a_5\lambda^4 - a_3\lambda^2 + a_1 = 0 \tag{7.6-6}$$

with roots of which

$$\lambda_1, \lambda_2, \lambda_3, \lambda_4 = \sqrt{\frac{2a_3 \pm \sqrt{a_3^2 - 4a_1a_5}}{2a_5}}, -\sqrt{\frac{2a_3 \pm \sqrt{a_3^2 - 4a_1a_5}}{2a_5}} \tag{7.6-7}$$

so that

$$\hat{F}(\xi, y) = Ae^{-\lambda_1|\xi|y} + Be^{-\lambda_2|\xi|y} + Be^{-\lambda_3|\xi|y} + De^{-\lambda_4|\xi|y}$$

By considering the condition of stress-free at $y = \infty$, $C = D = 0$, i.e.,

$$\hat{F}(\xi, y) = Ae^{-\lambda_1|\xi|y} + Be^{-\lambda_2|\xi|y} \tag{7.6-8}$$

According to Section 5.4, we have

$$\begin{aligned}
\sigma_{zy} &= (R_6C_{44} - R_5C_{55})\frac{\partial^3 F}{\partial x^2 \partial y} \\
H_{zy} &= -\left(C_{55}\frac{\partial^2}{\partial x^2} + C_{44}\frac{\partial^2}{\partial y^2}\right)\left(K_1\frac{\partial}{\partial x} + K_2\frac{\partial}{\partial y}\right)F \\
u_z &= \left(R_6\frac{\partial^2}{\partial x^2} + R_5\frac{\partial^2}{\partial y^2}\right)F
\end{aligned} \tag{7.6-9}$$

The Fourier transforms of the stresses and displacements are, e.g.,

$$\begin{aligned}
\hat{\sigma}_{zy} &= -(R_6C_{44} - R_5C_{55})\xi^2\frac{d\hat{F}}{dy} \\
\hat{H}_{zy} &= -iC_{55}K_1|\xi|\xi^2\hat{F} + C_{55}K_2\xi^2\frac{d\hat{F}}{dy} \\
&\quad + iC_{44}K_1|\xi|\frac{d^2\hat{F}}{dy^2} - C_{44}K_2\frac{d^3\hat{F}}{dy^3} \\
\hat{\sigma}_{zy} &= -(R_6C_{44} - R_5C_{55})\xi^2\frac{d\hat{F}}{dy}
\end{aligned}$$

$$\hat{H}_{zy} = -iC_{55}K_1 |\xi| \xi^2 \hat{F} + C_{55}K_2\xi^2 \frac{d\hat{F}}{dy}$$
$$+ iC_{44}K_1 |\xi| \frac{d^2\hat{F}}{dy^2} - C_{44}K_2 \frac{d^3\hat{F}}{dy^3}$$
$$\hat{u}_z = \left(-\xi^2 R_6 + R_5 \frac{d^2}{dy^2}\right) \hat{F} \tag{7.6-10}$$

Substituting (7.6-8) into the second one of (7.6-10) and then into the second one of (7.6-2) yields

$$B = \alpha A \tag{7.6-11}$$

where

$$\alpha = \frac{-\lambda_1 c_2 - \lambda_1^3 c_4 + i(c_1 + \lambda_1^2 c_3)}{\lambda_2 c_2 + \lambda_2^3 c_4 + i(-c_1 - \lambda_2^2 c_3)}$$
$$c_1 = C_{55}K_1, c_2 = C_{55}K_{2,} c_3 = C_{44}K_1, c_4 = C_{44}K_2 \tag{7.6-12}$$

The Fourier transform of the stress and displacement components is

$$\hat{\sigma}_{zy}(\xi, 0) = A(\xi) |\xi| (\lambda_1 e^{-\lambda_1 |\xi| y} + \alpha\lambda_2 e^{-\lambda_2 |\xi| y})$$
$$\hat{u}_z(\xi, 0) = A(\xi)\xi^2 [(-R_6 + \lambda_1^2 R_5) e^{-\lambda_1 |\xi| y} + \alpha(R_6 - \lambda_2^2 R_5) e^{-\lambda_2 |\xi| y}] \tag{7.6-13}$$

The Fourier transform of the first one of (7.6-2) is

$$\hat{\sigma}_{zy}(\xi, 0) = \tau \hat{f}(\xi) + k\hat{u}(\xi, 0) \tag{7.6-14}$$

From (7.6-13) and (7.6-14), one determines the unknown function

$$A(\xi) = \frac{\tau \hat{f}(\xi)}{|\xi| (\lambda_1 + \alpha\lambda_2) - k\xi^2[-R_6 + \lambda_1^2 R_5 + \alpha(R_6 - \lambda_1^2 R_5)]} \tag{7.6-15}$$

So, all stress and displacement components for phonon and phason fields can be evaluated, e.g.,

$$\sigma_{zy} = \frac{1}{2\pi} \int_{-\infty}^{\infty} \hat{\sigma}_{zy} e^{-i\xi x} d\xi$$
$$= \frac{1}{2\pi} \int_{-\infty}^{\infty} A(\xi) |\xi| (\lambda_1 e^{-\lambda_1 |\xi| y} + \alpha\lambda_2 e^{-\lambda_2 |\xi| y}) e^{-i\xi x} d\xi$$

$$
\begin{aligned}
u_z &= \frac{1}{2\pi}\int_{-\infty}^{\infty} \hat{u}_z e^{-i\xi x} d\xi \\
&= \frac{1}{2\pi}\int_{-\infty}^{\infty} A(\xi)\xi^2[(-R_6+\lambda_1^2 R_5)e^{-\lambda_1|\xi|y} \\
&\quad + \alpha(R_6-\lambda_2^2 R_5)e^{-\lambda_2|\xi|y}]e^{-i\xi x} d\xi \\
w_z &= \frac{1}{2\pi}\int_{-\infty}^{\infty} \hat{w}_z e^{-i\xi x} d\xi \\
&= -\frac{1}{2\pi}\int_{-\infty}^{\infty} A(\xi)\xi^2[(-C_{55}+\lambda_1^2 C_{66})e^{-\lambda_1|\xi|y} \\
&\quad + \alpha(C_{66}-\lambda_2^2 C_{55})e^{-\lambda_2|\xi|y}]e^{-i\xi x} d\xi
\end{aligned}
\tag{7.6-16}
$$

The phason strain field presents an important effect in the phase transition of crystal–quasicrystal, which is determined from the above solution such as

$$
\begin{aligned}
w_{zx} &= \frac{\partial w_z}{\partial x} \\
&= -i\frac{1}{2\pi}\int_{-\infty}^{\infty} A(\xi)\xi^3[(-C_{55}+\lambda_1^2 C_{66})e^{-\lambda_1|\xi|y} \\
&\quad + \alpha(C_{66}-\lambda_2^2 C_{55})e^{-\lambda_2|\xi|y}]e^{-i\xi x} d\xi \\
w_{zy} &= \frac{\partial w_z}{\partial y} \\
&= \frac{1}{2\pi}\int_{-\infty}^{\infty} A(\xi)\xi^2[\lambda_1|\xi|(-C_{55}+\lambda_1^2 C_{66})e^{-\lambda_1|\xi|y} \\
&\quad + \lambda_2|\xi|\alpha(C_{66}-\lambda_2^2 C_{55})e^{-\lambda_2|\xi|y}]e^{-i\xi x} d\xi
\end{aligned}
\tag{7.6-17}
$$

The calculation of integrals (7.6-17) depends on the form of function $\hat{f}(\xi) = \int_{-\infty}^{\infty} f(x)e^{i\xi x}dx$, in which $f(x)$ represents the stress distribution subjected at the interface. We can obtain the elementary form of the integrals by residual theorem for some cases, e.g., $f(x) = \delta(x)$, or

$$
f(x) = \begin{cases} 1 & |x| \le a/2 \\ 0 & |x| \ge a/2 \end{cases}
$$

The computed results vary from material constants C_{ij}, K_i, R_i of quasicrystals, and the material constant $\mu^{(c)}$ of crystals, applied stress τ and the size h of the crystals, so the results are very complicated.

Further discussion on interface and integrals calculation is given in Section 9.2 of Chapter 9.

7.7 Dislocation Pile Up, Dislocation Group and Plastic Zone

The previous discussion gives some solutions of single dislocation in some one- and two-dimensional quasicrystals. The dislocations can pile up when they meet some obstacles and lead to a dislocation group; this is a plastic zone in the material. Dislocations and plastic zone influence the behaviour of materials evidently. There is lack of theory of plasticity of quasicrystals so far. Observation on the plasticity of quasicrystals from the point of view of dislocation is a basic way at present. It provides not only the mechanism of plasticity of quasicrystals but also the tools for evaluating exactly some physical quantities concerning plastic deformation. This greatly helps the discussion in Chapters 8 and 14, respectively.

The slow movement of dislocations includes the forms of slip and clime. They relate with the plasticity of quasicrystals, refer to Messerschmidt [20], which is discussed in Chapters 8 and 14.

7.8 Discussions and Conclusions

The main object of this chapter lies in demonstration on the effect of our formulation in Chapters 5 and 6. It does not completely intend to explore the whole nature of dislocations and interfaces in quasicrystals. There are other works, e.g., [11–15, 18, 19], which can be referenced. Readers can find from the results listed above that dislocation leads to the appearance of singularity around the dislocation core, e.g., stresses $\sigma_{ij}, H_{ij} \sim 1/r \quad (r \to 0)$, which is the direct result of symmetry breaking due to the appearance of the topological defect, where r is the distance measured from the dislocation core. This is similar to that of crystals, i.e., the symmetry breaking also leads to the appearance of singularity in crystals. Physically, the high stress grade is called the stress concentration around the dislocation core which results in plastic flow in crystals as well as in quasicrystals. Therefore, dislocation and other defects in quasicrystals affect evidently the mechanical and other physical properties of the material. The dislocation solutions present important applications in studying plastic deformation and plastic fracture, which we can refer to [14] or Chapter 14 of this book. The discussion on the interface here is in the primary version, but it may be helpful for understanding the crystal–quasicrystal phase transition. However, this is a very complicated problem, and the data in this respect are very limited so far. The dislocations and interfaces and the solutions in three-dimensional quasicrystals are introduced in Chapter 9, and the dynamic dislocation problem can be referred to in Chapter 10.

References

[1] De P and Pelcovits R A, 1987, Linear elasticity theory of pentagonal quasicrystals, *Phys. Rev. B*, **35**(16), 8609–8620.

[2] De P and Pelcovits R A, 1987, Disclination in pentagonal quasicrystals, *Phys. Rev. B*, **36**(17), 9304–9307.

[3] Ding D H, Wang R H, Yang W G and Hu C Z, 1995, General expressions for the elastic displacement fields induced by dislocation in quasicrystals, *J. Phys. Condens. Matter.*, **7**(28), 5423–5436.

[4] Ding D H, Wang R H, Yang W G, Hu C Z and Qin Y L, 1995, Elasticity theory of straight dislocation in quasicrystals, *Phil. Mag. Lett.*, **72**(5), 353–359.

[5] Li X F and Fan T Y, 1998, New method for solving elasticity problems of some planar quasicrystals and solutions, *Chin. Phys. Lett.*, **15**(4), 278–280.

[6] Li X F, Duan X Y, Fan T Y *et al.*, 1999, Elastic field for a straight dislocation in a decagonal quasicrystal, *J. Phys.:Condens. Matter*, **11**(3), 703–711.

[7] Yang S H and Ding D H, 1998, *Fundamentals to Theory of Crystal Dislocations*, Vol II, Science Press, Beijing, (in Chinese).

[8] Firth J P and Lothe J, 1982, *Theory of Dislocations*, John Wiley and Sons, New York.

[9] Zhou W M, 2000, Dislocation, crack and contact problems in two- and three-dimensional quasicrystals, Dissertation, Beijing Institute of Technology, (in Chinese).

[10] Li L H, 2008, Study on complex variable function method and exact analytic solutions of elasticity of quasicrystals, Dissertation, Beijing Institute of Technology, (in Chinese).

[11] Fan T Y, Li X F and Sun Y F, 1999, A moving screw dislocation in a one-dimensional hexagonal quasicrystals, *Acta. Physica. Sinica.* (Oversea Edition), **8**(3), 288–295.

[12] Li X F and Fan T Y, 1999, A straight dislocation in one-dimensional hexagonal quasicrystals, *Phys. Stat. Sol.* (b), **212**(1), 19–26.

[13] Edagawa K, 2001, Dislocations in quasicrystals, *Mater. Sci. Eng. A* **309–310**(2), 528–538.

[14] Fan T Y, Trebin H-R, Messeschmidt U and Mai Y W, 2004, Plastic flow coupled with a crack in some one- and two-dimensional quasicrystals, *J. Phys.: Condens. Matter*, **16**(37), 5229–5240.

[15] Hu C Z, Wang R H and Ding D H, 2000, Symmetry groups, physical property tensors, elasticity and dislocations in quasicrystals, *Rep. Prog. Phys.*, **63**(1), 1–39.

[16] Li F H, 1993, *Crystal-Quasicrystal Transitions*, ed. by Jacaman M J and Torres M, Elsevier, pp. 13–47.

[17] Li F H, Teng C M, Huang Z R *et al.*, 1988, In between crystalline and quasicrystalline states, *Phil. Mag. Lett.*, **57** (1), 113–118.

[18] Fan T Y, Fan L, Wang Q Z and Xie L Y, 2009, Study on interface of quasicrystal-crystal, unpublished work.

[19] Kordak M, Fluckider T, Kortan A R *et al.*, 2004, Crystal-quasicrystal interface in Al-Pd-Mn, *Prog. Surface Sci*, **75**(3-8), 161–175.

[20] Messerschmidt U, 2010, *Dislocation Dynamics during Plastic Deformation*, Chapter 10, Springer-Verlag, Heidelberg.

Chapter 8

Application II — Solutions of Notch and Crack Problems of One- and Two-dimensional Quasicrystals

Quasicrystals are potential materials to be developed for structural use, and their strength and toughness attract the attention of researchers. Experimental observations [1, 2] have shown that quasicrystals are brittle under low and middle temperatures. With the common experience of conventional structural materials, we know that failure of brittle materials is related to the existence and growth of cracks. Chapter 7 indicates that dislocations have been observed in quasicrystals, and the accumulation of dislocations will eventually lead to cracking of the material. Now, let us study crack problems in quasicrystals that have both theoretical and practical values in the view of the application in future.

Chapters 5–7 have discussed some elasticity and dislocation problems in one- and two-dimensional quasicrystals. It has shown that when the quasicrystal configuration is independent of one coordinate, e.g., variable z, its elasticity problem can be decoupled into a plane problem and an anti-plane problem. In the case of one-dimensional quasicrystals, if the z-direction accords to the quasiperiodic axis, the above plane problem belongs to the classic elasticity problem, and the anti-plane problem is a coupling problem of phonon and phason fields. In the case of two-dimensional quasicrystals, if the z-direction represents the periodic axis, the above plane problem is a coupling problem of phonon and phason fields, and the anti-plane problem belongs to a classical elasticity problem. Due to the use of decomposition procedure, the resulting problem can be dramatically simplified. Chapters 5 and 6 have given their corresponding fundamental solutions, and Chapter 7 conducts the solutions of dislocations in detail. This chapter is going to focus on crack problems, and we continue using the above schemes, such as the fundamental solutions developed in Chapters 5 and 6 and the Fourier transform and complex analysis used in Chapter 7. Note that it emphasizes the complex analysis method, which will be developed in Sections 8.1, 8.2 and 8.4 and succeeded sections. Problems displayed in Sections 8.1 and 8.2 are relatively simple; the detailed introduction may help the reader to understand and further handle the principle and technique of the complex analysis method. Though the representation is beyond the classical Muskhelishvili [17] method, it is helpful to understand the solutions for more complicated problems displayed in Sections 8.4–8.8 and in the following chapters. A further summary of the method will be introduced in Chapter 11, considering that the contents in Sections 8.4–8.8 and Chapter 9 bring some new insights into the study and are beyond the classical Muskhelishvili method.

Based on the common nature of exact solutions of different static and dynamic cracks in different quasicrystal systems in linear and nonlinear deformation (which are discussed in this chapter and Chapters 9, 10 and 14), the fracture theory of quasicrystalline material is suggested in Chapter 15, which can be seen as a development of fracture mechanics of conventional structural materials.

8.1 Crack Problem and Solution of One-dimensional Quasicrystals

8.1.1 *Griffith crack*

As shown in Fig. 8.1, we assume a Griffith crack along the quasiperiodic axis (z-direction) of a one-dimensional hexangular quasicrystal, and under the action of external traction $\sigma_{yz}^{(\infty)} = \tau_1$ and/or $H_{zy}^{(\infty)} = \tau_2$. The deformation induced is often termed longitudinal shearing.

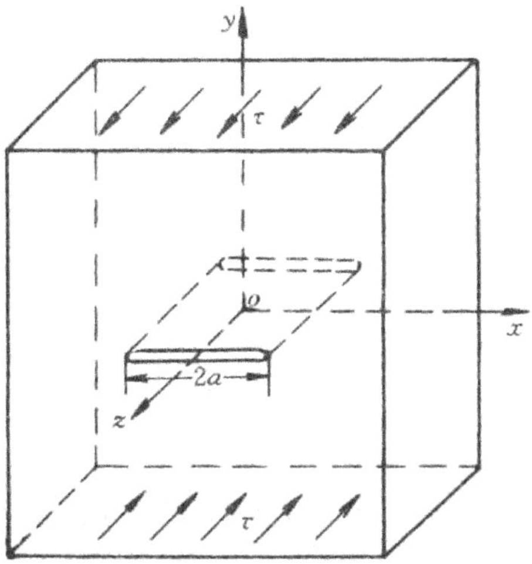

Fig. 8.1. A Griffith crack subjected to a longitudinal shear where $\tau = \tau_1$ corresponding to phonon field and $\tau = \tau_2$ corresponding to phason field.

Obviously, the geometry of the crack is independent of variable z. In this case, all field variables are independent of variable z. Therefore, based on the analysis in Chapter 5, this one-dimensional quasicrystal elasticity can be decomposed into a plane elasticity problem of regular crystal and an anti-plane elasticity problem of phonon–phason coupling field. Plane elasticity problems of a regular crystal have been studied extensively in classical elasticity, and its crack problems have also been studied in the classical fracture theory, such as Ref. [3]. Therefore, we skip the discussion of crack problems in regular crystal. The anti-plane elasticity problem of the phason–phonon coupling field is described by using the following basic equations:

$$\left.\begin{aligned} \sigma_{yz} &= \sigma_{zy} = 2C_{44}\varepsilon_{yz} + R_3 w_{zy} \\ \sigma_{zx} &= \sigma_{xz} = 2C_{44}\varepsilon_{zx} + R_3 w_{zx} \\ H_{zy} &= K_2 w_{zy} + 2R_3\varepsilon_{zy} \\ H_{zx} &= K_2 w_{zx} + 2R_3\varepsilon_{zx} \end{aligned}\right\} \tag{8.1-1}$$

$$\left.\begin{aligned} &\varepsilon_{yz}=\varepsilon_{zy}=\frac{1}{2}\frac{\partial u_z}{\partial y}, \quad \varepsilon_{zx}=\varepsilon_{xz}=\frac{1}{2}\frac{\partial u_z}{\partial x} \\ &w_{zy}=\frac{\partial w_z}{\partial y}, \quad w_{zx}=\frac{\partial w_z}{\partial x} \end{aligned}\right\} \tag{8.1-2}$$

$$\frac{\partial \sigma_{zx}}{\partial x}+\frac{\partial \sigma_{zy}}{\partial y}=0, \quad \frac{\partial H_{zx}}{\partial x}+\frac{\partial H_{zy}}{\partial y}=0 \tag{8.1-3}$$

The derivation in Chapter 5 shows that the above equations can be reduced to

$$\nabla^2 u_z=0, \quad \nabla^2 w_z=0 \tag{8.1-4}$$

where $\nabla^2=\partial^2/\partial x^2+\partial^2/\partial y^2$.

Figure 8.1 shows that the anti-plane deformation of a Griffith crack has the following boundary conditions:

$$\left.\begin{aligned} &\sqrt{x^2+y^2}\to\infty: \quad \sigma_{yz}=\tau_1, \quad H_{zy}=\tau_2, \quad \sigma_{zx}=H_{zx}=0 \\ &y=0, |x|<a: \quad \sigma_{yz}=0, \quad H_{zy}=0 \end{aligned}\right\} \tag{8.1-5}$$

where a is the half-length of the crack.

Based on the results of linear elasticity analysis, if the quasicrystal is traction-free at infinity instead there are a traction $\sigma_{yz}=-\tau_1$ and a generalized traction $H_{zy}=-\tau_2$ at the crack surface, then the boundary conditions stand for

$$\left.\begin{aligned} &\sqrt{x^2+y^2}\to\infty: \quad \sigma_{yz}=\sigma_{zx}=H_{zx}=H_{zy}=0 \\ &y=0, \ |x|<a: \quad \sigma_{yz}=-\tau_1, \quad H_{zy}=-\tau_2 \end{aligned}\right\} \tag{8.1-6}$$

Here, we give phason stress τ_2 at the crack surface from the physical point of view, though its measurement result has not been reported yet. For simplicity, we can assume $\tau_2=0$, sometimes.

In the following, we are going to solve the boundary value problem of (8.1-4) and (8.1-5) first. The complex analysis method will be used. To do so, we introduce the complex variable

$$t=x+iy=re^{i\theta}, \quad i=\sqrt{-1} \tag{8.1-7}$$

From Eq. (8.1-4), it is known that both $u_z(x,y)$ and $w_z(x,y)$ are harmonic functions that can be expressed in terms of the real part or the imaginary part of two arbitrary analytic functions $\phi_1(t)$ and $\psi_1(t)$ of complex variable t in a region occupied by the quasicrystal.

For simplicity, we can call $\phi_1(t)$ and $\psi_1(t)$ as complex potentials. Here, assume

$$\left.\begin{aligned} u_z(x,y) &= \operatorname{Re}\phi_1(t) \\ w_z(x,y) &= \operatorname{Re}\psi_1(t) \end{aligned}\right\} \tag{8.1-8}$$

in which the symbol Re indicates the real part of a complex number.

It is well known that if a function $F(t)$ is analytic, then

$$\frac{\partial F}{\partial x} = \frac{dF}{dt}, \quad \frac{\partial F}{\partial y} = \frac{idF}{dt} \tag{a}$$

Furthermore, assume

$$\begin{aligned} F(t) &= P(x,y) + iQ(x,y) \\ &= \operatorname{Re}F(t) + i\operatorname{Im}F(t) \end{aligned} \tag{b}$$

where the symbol Im denotes the imaginary part of a complex and $P(x,y)$ and $Q(x,y)$ represent the real and imaginary parts of $F(t)$, respectively. Therefore, the Cauchy–Riemann relation leads to

$$\frac{\partial P}{\partial x} = \frac{\partial Q}{\partial y}, \quad \frac{\partial P}{\partial y} = -\frac{\partial Q}{\partial x} \tag{c}$$

With the aid of relation (a), formula (8.1-8) and Eqs. (8.1-1) and (8.1-2) lead to

$$\left.\begin{aligned} \sigma_{yz} &= \sigma_{zy} = C_{44}\frac{\partial}{\partial y}\operatorname{Re}\phi_1 + R_3\frac{\partial}{\partial y}\operatorname{Re}\psi_1 \\ \sigma_{zx} &= \sigma_{xz} = C_{44}\frac{\partial}{\partial x}\operatorname{Re}\phi_1 + R_3\frac{\partial}{\partial x}\operatorname{Re}\psi_1 \\ H_{zx} &= K_2\frac{\partial}{\partial x}\operatorname{Re}\psi_1 + R_3\tfrac{\partial}{\partial x}\operatorname{Re}\phi_1 \\ H_{zx} &= K_2\frac{\partial}{\partial y}\operatorname{Re}\psi_1 + R_3\frac{\partial}{\partial y}\operatorname{Re}\phi_1 \end{aligned}\right\} \tag{8.1-9}$$

Based on the Cauchy–Riemann relation (c), the above equation can be rewritten as

$$\left.\begin{aligned} \sigma_{zx} - i\sigma_{zy} &= C_{44}\phi_1' + R_3\psi_1' \\ H_{zx} - iH_{zy} &= K_2\psi_1' + R_3\phi_1' \end{aligned}\right\} \tag{8.1-10}$$

where $\phi_1' = d\phi_1/dt$, $\psi_1' = d\psi_1/dt$.

According to formula (8.1-10), we have

$$\left.\begin{aligned} \sigma_{yz} &= \sigma_{zy} = -\mathrm{Im}(C_{44}\phi_1' + R_3\psi_1') \\ H_{zy} &= -\mathrm{Im}(K_2\psi_1' + R_3\phi_1') \end{aligned}\right\} \tag{d}$$

Furthermore, for an arbitrary complex function $F(t)$, its imaginary part is

$$\mathrm{Im}\, F(t) = \frac{1}{2i}(F - \overline{F})$$

where $\overline{F}$ indicates the conjugate of F. Then, the above formula (d) can be expressed as

$$\left.\begin{aligned} \sigma_{yz} = \sigma_{zy} &= -\frac{1}{2i}[C_{44}(\phi_1' - \overline{\phi_1'}) + R_3(\psi_1' - \overline{\psi_1'})] + \tau_1 \\ H_{zy} &= -\frac{1}{2i}[K_2(\psi_1' - \overline{\psi_1'}) + R_3(\phi_1' - \overline{\phi_1'})] + \tau_2 \end{aligned}\right\} \tag{8.1-11}$$

The expression given by (8.1-11) ensures the solution constructed in the following to automatically satisfy the boundary condition at infinity; the solution is

$$\left.\begin{aligned} \phi_1(t) &= \frac{ia(K_2\tau_1 - R_3\tau_2)}{C_{44}K_2 - R_3^2}\left(\frac{t}{a} - \sqrt{\left(\frac{t}{a}\right)^2 - 1}\right) \\ \psi_1(t) &= \frac{ia(C_{44}\tau_2 - R_3\tau_1)}{C_{44}K_2 - R_3^2}\left(\frac{t}{a} - \sqrt{\left(\frac{t}{a}\right)^2 - 1}\right) \end{aligned}\right\} \tag{8.1-12}$$

whose derivation can be obtained in terms of the strict complex analysis method as well as the Fourier method given in Section 8.12 — Appendix A.

From (8.1-12)

$$\begin{aligned} \phi_1'(t) &= \frac{i(K_2\tau_1 - R_3\tau_2)}{C_{44}K_2 - R_3^2}\left(1 - \frac{t}{\sqrt{t^2 - a^2}}\right) \\ \psi_1'(t) &= \frac{i(C_{44}\tau_2 - R_3\tau_1)}{C_{44}K_2 - R_3^2}\left(1 - \frac{t}{\sqrt{t^2 - a^2}}\right) \end{aligned} \tag{8.1-13}$$

and substitution of the above expressions into the first expression in (8.1-11) yields

$$\sigma_{zx} - i\sigma_{zy} = i\tau_1 \left(-\frac{t}{\sqrt{t^2 - a^2}}\right). \tag{8.1-14}$$

Separation of the real and imaginary parts of (8.1-14) leads to

$$\left.\begin{aligned} \sigma_{xz} = \sigma_{zx} &= \frac{\tau_1 r}{(r_1 r_2)^{1/2}} \sin\left(\theta - \frac{1}{2}\theta_1 - \frac{1}{2}\theta_2\right) \\ \sigma_{yz} = \sigma_{zy} &= \frac{\tau_1 r}{(r_1 r_2)^{1/2}} \cos\left(\theta - \frac{1}{2}\theta_1 - \frac{1}{2}\theta_2\right) \end{aligned}\right\} \tag{8.1-15}$$

where

$$t = re^{i\theta}, \quad t - a = r_1 2e^{i\theta_1}, \quad t + a = r_2 2e^{i\theta_2} \tag{8.1-16}$$

$$\left.\begin{aligned} &r = \sqrt{x^2 + y^2}, \quad r_1 = \sqrt{(x-a)^2 + y^2}, \quad r_2 = \sqrt{(x+a)^2 + y^2} \\ &\text{or} \\ &\theta = \arctan\left(\frac{y}{x}\right), \quad \theta_1 = \arctan\left(\frac{y}{x-a}\right), \quad \theta_2 = \arctan\left(\frac{y}{x+a}\right) \end{aligned}\right\} \tag{8.1-16$'$}$$

which can be shown in Fig. 8.2.

Similarly,

$$H_{zx} - iH_{zy} = i\tau_2 \left(-\frac{t}{\sqrt{t^2 - a^2}}\right) \tag{8.1-17}$$

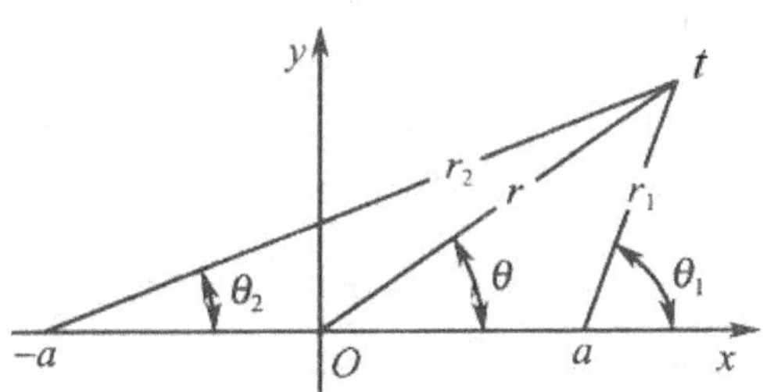

Fig. 8.2. The coordinate system of crack tip.

For these stress components, there are similar expressions like (8.1-15). As a result,

$$\sigma_{zy}(x,0)=\begin{cases}\dfrac{\tau_1 x}{\sqrt{x^2-a^2}}, & |x|>a\\ 0, & |x|<a\end{cases}\tag{8.1-18}$$

$$H_{zy}(x,0)=\begin{cases}\dfrac{\tau_2 x}{\sqrt{x^2-a^2}}-\tau_2, & |x|>a\\ 0, & |x|<a\end{cases}\tag{8.1-19}$$

The above two formulas indicate that at $y=0$ and $|x|<1:\sigma_{zy}=0,\ H_{zy}=0$. Therefore, the solution given above also satisfies the boundary condition at the crack surfaces.

Formulas (8.1-14) and (8.1-17) also show that when $\sqrt{x^2+y^2}\to\infty$, $\sigma_{yz}=\tau_1$, $\sigma_{xz}=0$ and $H_{zy}=\tau_2$, $H_{zx}=0$, namely, the solution given above satisfies the boundary condition at infinity.

8.1.2 *Brittle fracture theory*

The above formulas show that stresses have singular characteristics near crack tips, for example,

$$\begin{cases}\sigma_{zy}(x,0)=\dfrac{\tau_1 x}{\sqrt{x^2-a^2}}\to\infty, & x\to a^+\\ H_{zy}(x,0)=\dfrac{\tau_2 x}{\sqrt{x^2-a^2}}\to\infty, & x\to a^+\end{cases}\tag{8.1-20}$$

If defining the mode III stress intensity factors of the phonon and phason fields such that

$$K_{III}^{\parallel}=\lim_{x\to a^+}\sqrt{2\pi(x-a)}\,\sigma_{zy}(x,0),$$
$$K_{III}^{\perp}=\lim_{x\to a^+}\sqrt{2\pi(x-a)}\,H_{zy}(x,0),$$

we have

$$K_{III}^{\parallel}=\sqrt{\pi a}\,\tau_1,\quad K_{III}^{\perp}=\sqrt{\pi a}\,\tau_2,\tag{8.1-21}$$

where subscript "III" stands for model III (longitudinal shearing mode) [3].

Now, let us calculate the crack strain energy:

$$
\begin{aligned}
W_{III} &= 2\int_0^a (\sigma_{zy} \oplus H_{zy})(u_z \oplus w_z)dx \\
&= 2\int_0^a [\sigma_{zy}(x,0)u_z(x,0) + H_{zy}(x,0)w_z(x,0)]dx
\end{aligned} \tag{8.1-22}
$$

From (8.1-8) and (8.1-12), we have

$$
\left.
\begin{aligned}
u_z(x,0) &= \mathrm{Re}\,(\phi_1(t))_{t=x} = a\frac{K_2\tau_1 - R_3\tau_2}{C_{44}K_2 - R_3^2}\sqrt{1-\left(\frac{x}{a}\right)^2} \quad |x| < a \\
w_z(x,0) &= \mathrm{Re}\,(\psi_1(t))_{t=x} = a\frac{C_{44}\tau_2 - R_3\tau_1}{C_{44}K_2 - R_3^2}\sqrt{1-\left(\frac{x}{a}\right)^2} \quad |x| < a
\end{aligned}
\right\} \tag{8.1-23}
$$

In addition, considering the equivalency of problems (8.1-5) and (8.1-6), we can take $\sigma_{yz}(x,0) = -\tau_1$ and $H_{zy} = -\tau_2$ as $|x| < a$, then substitution of the above results into (8.1-22) yields

$$
W_{III} = \frac{K_2\tau_1^2 + C_{44}\tau_2^2 - 2R_3\tau_1\tau_2}{C_{44}K_2 - R_3^2}\pi a^2 \tag{8.1-24}
$$

From (8.1-24), we can determine the crack strain energy release rate (crack growth force) such that

$$
\begin{aligned}
G_{III} &= \frac{1}{2}\frac{\partial W_{III}}{\partial a} = \frac{K_2\tau_1^2 + C_{44}\tau_2^2 - 2R_3\tau_1\tau_2}{C_{44}K_2 - R_3^2}\pi a \\
&= \frac{K_2 K_{III}^{\parallel\,2} + C_{44}K_{III}^{\perp\,2} - 2R_3 K_{III}^{\parallel}K_{III}^{\perp}}{C_{44}K_2 - R_3^2}.
\end{aligned} \tag{8.1-25}
$$

Clearly, the crack energy and energy release rate are related not only to the phonon but also to phason and phonon–phason coupling fields.

If $\tau_2 = 0$, then

$$
G_{III} = \frac{K_2 K_{III}^{\parallel\,2}}{C_{44}K_2 - R_3^2} \tag{8.1-26}
$$

Furthermore, if $R_3 = 0$, we have

$$G_{III} = \frac{\pi a \tau_1^2}{C_{44}}, \quad \text{or} \quad G_{III} = \frac{(K_{III}^{\parallel})^2}{C_{44}}. \tag{8.1-27}$$

Since G_{III} comprehensively describes the coupling effect of the phonon and phason fields and the stress status near the crack tip, we recommend

$$G_{III} = G_{IIIC} \tag{8.1-28}$$

as the fracture criterion of quasicrystals under mode III deformation, where G_{IIIC} is the critical (threshold) value of G_{III}, called mode III fracture toughness to be determined by testing.

The stress intensity factor and strain energy release rate are fundamental physical parameters and constitute the basis of brittle fracture theory for both conventional crystalline as well as quasicrystalline materials.

8.2 Crack Problem in Finite-sized One-dimensional Quasicrystals

The preceding section discussed Griffith crack problems in one-dimensional quasicrystals and obtained their exact solutions. In this section, we are going to discuss the solution of another kind of important crack problem, which goes beyond the scope of the Griffith crack. In the previous section, we assumed that the size of a quasicrystal is much larger than that of the defect, while in this case we consider the quasicrystal as an infinite body. Therefore, this section is aimed at the study of crack problems in quasicrystals of finite size.

8.2.1 *Cracked quasicrystal strip with finite height*

As shown in Fig. 8.3, a one-dimensional hexagonal quasicrystal strip of height $2H$ has a semi-infinite crack embedded at the mid-plane with a crack tip corresponding to coordinate origin. Crack surfaces near the crack tip, i.e., $y = \pm 0$ and $-a < x < 0$, are assumed under the action of uniform shear traction $\sigma_{zy} = -\tau_1$, $H_{zy} = 0$, where a

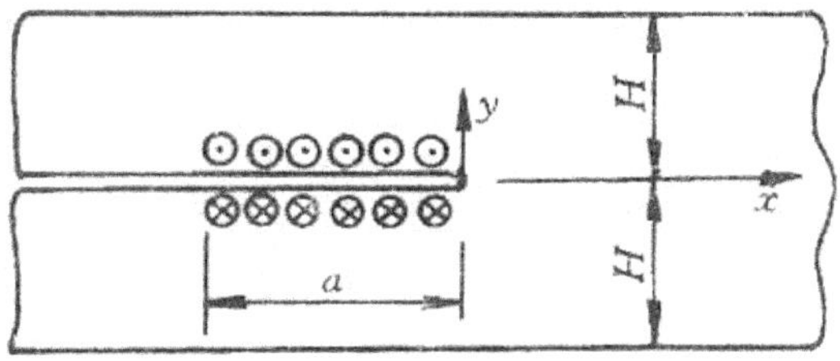

Fig. 8.3. Finite strip of hexagonal quasicrystal with a crack.

is a length used to simulate a finite-size crack. The upper and lower strip surfaces are assumed to be traction-free, therefore the boundary conditions are

$$\left.\begin{array}{r} y = \pm H, -\infty < x < \infty : \sigma_{zy} = 0, H_{zy} = 0 \\ x = \pm\infty, -H < y < H : \sigma_{zx} = 0, H_{zx} = 0 \\ y = \pm 0, -\infty < x < -a : \sigma_{zy} = 0, H_{zy} = 0; \\ -a < x < 0 : \sigma_{zy} = -\tau_1, H_{zy} = -\tau_2 \end{array}\right\} \tag{8.2-1}$$

For simplicity, we assume that $\tau_2 = 0$ in (8.2-1).

Formulas (8.1-1)–(8.1-11) in the preceding section still hold in this section, and the other notations and symbols are similar. Therefore, formula (8.1-12) should be modified as

$$\left.\begin{array}{l} C_{44}(\phi_1' - \overline{\phi_1'}) + R_3(\psi_1' - \overline{\psi_1'}) = 2i\tau_1 f(t) \\ K_2(\psi_1' - \overline{\psi_1'}) + R_3(\phi_1' - \overline{\phi_1'}) = 0 \end{array}\right\} \tag{8.2-2}$$

where

$$f(x) = \begin{cases} 0, & x < -a, \\ 1, & -a < x < 0 \end{cases} \tag{8.2-3}$$

By using the conformal mapping

$$t = \omega(\zeta) = \frac{H}{\pi} \ln\left[1 + \left(\frac{1+\zeta}{1-\zeta}\right)^2\right] \tag{8.2-4}$$

maps the domain in the t-plane onto the interior of the unit circle γ in the ζ-plane, $\zeta = \xi + i\eta$. Therefore, crack tip $t = 0$ corresponds to

$\zeta = -1$, and $t = -a$ accords to two points on the unit circle in the ζ-plane such that

$$\left.\begin{aligned} \sigma_{-a} &= \frac{-e^{-\pi a/H} + 2i\sqrt{1 - e^{-\pi a/H}}}{2 - e^{-\pi a/H}} \\ \overline{\sigma_{-a}} &= \frac{-e^{-\pi a/H} - 2i\sqrt{1 - e^{-\pi a/H}}}{2 - e^{-\pi a/H}} \end{aligned}\right\} \tag{8.2-5}$$

where $\sigma = e^{i\theta} = \zeta|_{|\zeta|=1}$ denotes the value of ζ at the unit circle γ.

Equations in Sections 8.1 and Appendix A of this chapter (i.e., Section 8.7) are useful. Nevertheless, the first $2i\omega'(\sigma)\tau_1$ in (8.7-4) should be changed to $2if\omega'(\sigma)\tau$, and the first term in (8.7-5)

$$\frac{2i\tau_1}{C_{44}}\frac{1}{2\pi i}\int_\gamma \frac{\omega'(\sigma)}{\sigma - \zeta}d\sigma$$

should be modified as

$$\frac{2i\tau_1}{C_{44}}\frac{1}{2\pi i}\int_\gamma f\frac{\omega'(\sigma)}{\sigma - \zeta}d\sigma$$

Thus, Eq. (8.7-5) for the present problem becomes

$$\left.\begin{aligned} &\phi'(\zeta) + \frac{R_3}{C_{44}}\psi'(\zeta) = \frac{2i\tau_1}{C_{44}}\frac{1}{2\pi i}\int_\gamma f\frac{\omega'(\sigma)}{\sigma - \zeta}d\sigma \\ &\psi'(\zeta) + \frac{R_3}{K_2}\phi'(\zeta) = 0 \end{aligned}\right\} \tag{8.2-6}$$

where f is the function given by formula (8.2-3), which takes values between $\overline{\sigma_{-a}}$ and σ_{-a} in the ζ-plane.

Integration of the right-hand side of Eq. (8.2-6) leads to

$$\begin{aligned} &\frac{1}{2\pi i}\int_\gamma f\frac{\omega'(\sigma)}{\sigma - \zeta}d\sigma \\ &= \frac{1}{2\pi\tau}\left[\frac{1}{1-\zeta}\ln(\sigma - 1) - \frac{1+\zeta}{(1-\zeta)(1+\zeta^2)}\ln(\sigma - \zeta)\right. \\ &\quad \left.-\frac{\zeta}{2(1+\zeta^2)}\ln(1+\sigma^2) + \frac{1}{2(1-\zeta^2)}\ln\frac{\sigma - i}{\sigma + i}\right]_{\sigma=\sigma_a}^{\sigma=\overline{\sigma_{-a}}} \equiv F(\zeta) \end{aligned} \tag{8.2-7}$$

where $\overline{\sigma_{-a}}$ and σ_{-a} are given by formula (8.2-5). Based on the above relation, (8.2-6) is further reduced to

$$\phi'(\zeta) = \frac{K_2\tau_1}{C_{44}K_2 - R_3^2}2iF(\zeta), \quad \psi'(\zeta) = \frac{-R_3\tau_1}{C_{44}K_2 - R_3^2}2iF(\zeta). \quad (8.2\text{-}8)$$

Now, let us calculate the stresses. During this process, the following relations will be used:

$$\phi_1'(t) = \frac{\phi'(\zeta)}{\omega'(\zeta)} = \frac{2iK_2\tau_1}{C_{44}K_2 - R_3^2}\cdot\frac{F'(\zeta)}{\omega'(\zeta)},$$

$$\psi_1'(t) = \frac{\psi'(\zeta)}{\omega'(\zeta)} = -\frac{2iR_3\tau_1}{C_{44}K_2 - R_3^2}\cdot\frac{F'(\zeta)}{\omega'(\zeta)}.$$

Substitution of the above expressions into (8.1-10) leads to

$$\left.\begin{aligned} &\sigma_{zx} - i\sigma_{zy} = i\tau_1\frac{F'(\zeta)}{\omega'(\zeta)} \\ &H_{zx} - iH_{zy} = 0 \end{aligned}\right\} \quad (8.2\text{-}9)$$

which shows that the stress distribution is independent of material constants.

Substitution of (8.2-7) into (8.2-9) leads to the explicit expressions of stresses. The forms of $F(\zeta)$ and $F'(\zeta)$ are relatively complex, while the expressions of stress in terms of variable $\zeta(=\xi + i\eta)$ are concise. In an attempt to invert them to the t-plane, the inverse of transform (8.2-4) must be used:

$$\zeta = \omega^{-1}(t) = \frac{-(e^{\pi t/H} - 2) \pm 2i\sqrt{e^{\pi t/H} - 2}}{e^{\pi t/H} - 2} \quad (8.2\text{-}10)$$

Substitution of (8.2-10) into (8.2-9) yields the final expressions of σ_{zx}, σ_{zy}, H_{zx} and H_{zx} in the t-plane, which are very complex. Here, we skip this procedure.

Now, let us calculate the stress intensity factors. According to (8.1-15) and (8.1-16), it is known that at the region such that $r_1/a \ll 1$,

$$\sigma_{zx} = -\frac{K_{III}^{\|}}{\sqrt{2\pi r_1}}\sin\frac{\theta_1}{2}, \quad \sigma_{zy} = -\frac{K_{III}^{\|}}{\sqrt{2\pi r_1}}\cos\frac{\theta_1}{2}$$

therefore

$$\sigma_{zx} - i\sigma_{zy} = -\frac{K_{III}^{\parallel}}{\sqrt{2\pi r_1}}\left(\sin\frac{\theta_1}{2} - i\cos\frac{\theta_1}{2}\right) = -\frac{K_{III}^{\parallel}}{\sqrt{2\pi t_1}} \tag{8.2-11}$$

where $t_1 = r_1 e^{i\theta_1} = x_1 + iy_1$.

By using conformal mapping $t = \omega(\zeta)$, we have

$$t_1 = \omega(\zeta_1) \tag{8.2-12}$$

where ζ_1 is the point in the ζ-plane corresponding to t_1. With relations (8.2-11) and (8.2-9), we define

$$K_{III}^{\parallel} = \lim_{\zeta \to -1} \sqrt{\pi}\tau_1 \frac{F'(\zeta)}{\sqrt{\omega''(\zeta)}}, \quad K_{III}^{\perp} = 0 \tag{8.2-13}$$

where $\zeta = -1$ is the point corresponding to the crack tip. Substitution of (8.2-4) and (8.2-7) into (8.2-13) yields

$$K_{III}^{\parallel} = \frac{\sqrt{2H}\tau_1}{2\pi} \ln \frac{2e^{\pi a/H} - 1 + 2e^{\pi a/H}\sqrt{1 - e^{-\pi a/H}}}{2e^{\pi a/H} - 1 - 2e^{\pi a/H}\sqrt{1 - e^{-\pi a/H}}} \tag{8.2-14}$$

If we do not assume $\tau_2 = 0$ in (8.2-1), then the stress intensity factor $K_{III}^{\perp}$ can be similarly evaluated, and the expression is similar to (8.2-14); this is the extension of work given by Ref. [4] for classical elasticity to quasicrystal elasticity.

8.2.2 *Finite strip with two cracks*

The configuration is shown in Fig. 8.4, and the following are the boundary conditions:

$$\left.\begin{array}{c} y = \pm H, -\infty < x < \infty : \sigma_{zy} = 0, H_{zy} = 0 \\ x = \pm\infty, -H < y < H : \sigma_{zx} = 0, H_{zx} = 0 \\ y = \pm 0, -\infty < x < -a : \sigma_{zy} = 0, H_{zy} = 0; \\ -a < x < 0 : \sigma_{zy} = -\tau_1, H_{zy} = -\tau_2; \\ L < x < \infty : \sigma_{zy} = 0, H_{zy} = 0 \end{array}\right\} \tag{8.2-15}$$

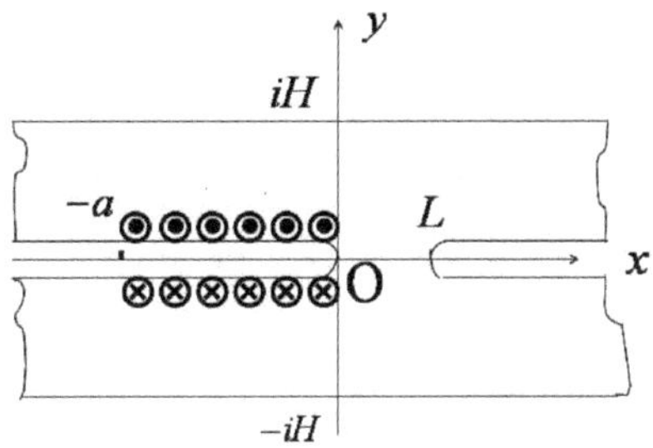

Fig. 8.4. Two cracks in a strip.

$$z = \omega(\zeta) = \frac{H}{\pi} \ln \frac{1 + \alpha \left(\frac{1-\zeta}{1+\zeta}\right)^2}{1 + \beta\alpha \left(\frac{1-\zeta}{1+\zeta}\right)^2} \tag{8.2-16}$$

For simplicity, we assume that $\tau_2 = 0$ in (8.2-15). The following conformal mapping transforms the region at z-plane onto the interior of the unit circle γ at ζ-plane, in which

$$\alpha = \frac{1 - e^{-\pi a/H}}{1 - e^{-\pi(a+L)/H}}, \quad \beta = e^{-\pi L/H} \tag{8.2-17}$$

Then, substituting (8.2-16) into (8.2-6), we can find the solution $\phi'(\varsigma)$, so the stress intensity factors are

$$\begin{aligned} K_{III}^{\| \, (0,0)} &= \lim_{\zeta \to 1+0} \left\{ -2\sqrt{2\pi\omega(\zeta)}\tau_1 \frac{F(\zeta)}{\omega'(\zeta)} \right\} \\ &= \frac{\sqrt{2H}\tau_1}{\pi\sqrt{1-\beta}} \left[\ln \frac{1+\sqrt{\alpha}}{1-\sqrt{\alpha}} - \sqrt{\beta} \ln \frac{1+\sqrt{\alpha\beta}}{1-\sqrt{\alpha\beta}} \right] \\ K_{III}^{\| \, (L,0)} &= \lim_{\zeta \to -1-0} \left\{ -2\sqrt{2\pi(L-\omega(\zeta))}\tau_1 \frac{F(\zeta)}{\omega'(\zeta)} \right\} \\ &= \frac{\sqrt{2H}\tau_1}{\pi\sqrt{1-\beta}} \left[\sqrt{\beta} \ln \frac{1+\sqrt{\alpha}}{1-\sqrt{\alpha}} - \ln \frac{1+\sqrt{\alpha\beta}}{1-\sqrt{\alpha\beta}} \right] \end{aligned} \tag{8.2-18}$$

in which

$$\phi'(\zeta) = \frac{K_2\tau_1}{C_{44}K_2 - R_3^2} 2iF(\zeta), \psi'(\zeta) = -\frac{R_3\tau_1}{C_{44}K_2 - R_3^2} 2iF(\zeta)$$

$$F(\zeta) = \frac{2H}{\pi^2}\left[\frac{\sqrt{\alpha}A}{(1+\zeta)^2+\alpha(1-\zeta)^2} - \frac{\sqrt{\alpha\beta}M}{(1+\zeta)^2+\alpha\beta(1-\zeta)^2}\right]$$
$$+ \frac{2H}{\pi^2}\frac{i\alpha(1-\beta)(1-\zeta^2)}{[(1+\zeta)^2+\alpha(1-\zeta)^2][(1+\zeta)^2+\alpha\beta(1-\zeta)^2]}$$
$$\times \ln\frac{i-\zeta}{1-i\zeta} \tag{8.2-19}$$
$$A = \ln((1+\sqrt{\alpha})/(1-\sqrt{\alpha})), M = \ln((1+\sqrt{\alpha\beta})/(1-\sqrt{\alpha\beta}))$$

If we do not assume $\tau_2 = 0$ in (8.2-15), then the stress intensity factor $K_{III}^{\perp}$ can be similarly evaluated, and the expression is similar to (8.2-18); this extends the study for the classical elasticity. The detail can be found in Refs. [5, 6], and some calculations on the function $F(\zeta)$ refer to Section A.1 of Major Appendix of this book.

8.3 Griffith Crack Problems in Point Groups 5m and 10mm Quasicrystal Based on Displacement Potential Function Method

Literature [2] reported that Chinese material scientists had started to characterize the fracture toughness of quasicrystals. Due to the lack of theoretical solutions to crack problems in quasicrystals, their experimental work was performed largely by indirectly measuring the fracture toughness. If the crack solution had been known then, the fracture toughness could have been determined by direct measurement, a much simpler and more accurate method.

This section aims to solve the mode I Griffith crack in quasicrystal with 10 fold symmetry by using the method of displacement functions. Therefore, the fundamental formulas in Section 6.2 and formulas using Fourier transform in Section 7.2 are the basis for this section. To save space, we do not plan to list those formulas in detail here, and interested readers may refer to the previous two chapters. In the following section, we are going to solve this problem using the method of stress functions. On the one hand, this demonstrates the problem-solving procedure based on the method of stress functions; on the other hand, this is targeted to examine the results obtained

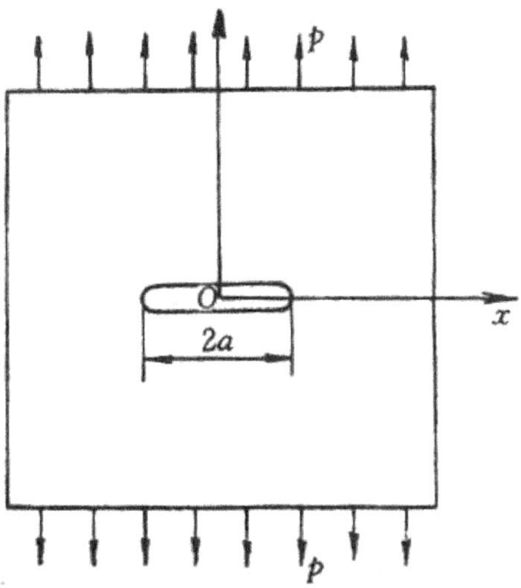

Fig. 8.5. Griffith crack along the periodic axis of quasicrystal and subjected to a tension.

by the method of displacement functions. For a correct solution, it can be examined using any available method.

Let us consider a Griffith crack under the action of external traction, i.e., $\sigma_{yy}^{(\infty)} = p$, and the crack is assumed to penetrate the periodic axis (z-direction) of the quasicrystals, as shown in Fig. 8.5. Similar to the analysis in the preceding section, within the framework of Griffith's theory, this problem can be replaced by an equivalent crack problem shown in Fig. 8.6. Furthermore, assume the external traction being independent of z, therefore the deformation of the quasicrystal is also independent of z, namely,

$$\frac{\partial u_i}{\partial z} = 0, \quad \frac{\partial w_i}{\partial z} = 0 \quad (i = 1, 2, 3) \tag{8.3-1}$$

According to the analysis performed in Chapter 6, under this case, the two-dimensional quasicrystal elasticity problem can be decoupled into a plane elasticity problem of phonon–phason coupling and an anti-plane pure elasticity problem. In this case, the latter only has a trivial solution under mode I external traction, which can be neglected. The plane elasticity problem of phonon–phason coupling with point groups $5m$ and $10mm$ has been studied in Section 6.2,

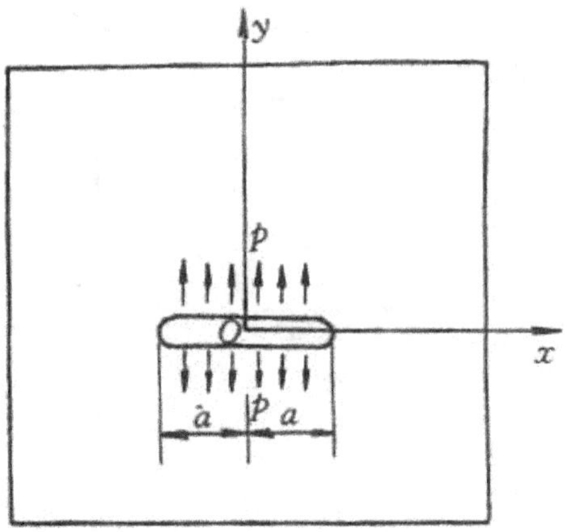

Fig. 8.6. The same Griffith crack as Fig. 8.5 with external traction acting on crack surfaces.

and its final governing equation is

$$\nabla^2\nabla^2\nabla^2\nabla^2 F = 0 \tag{8.3-2}$$

Here, $F(x, y)$ is the displacement potential function introduced in Section 6.2.

As shown in Fig. 8.6, the Griffith crack is under the action of uniform traction at crack surfaces and without far-field traction, i.e., $\sigma_{yy}(x, 0) = -p$, $|x| < a$. We modify this problem into the semi-plane problem, i.e., only study the case in the upper half-plane or the lower half-plane under the following conditions:

$$\left.\begin{array}{c}
\sqrt{x^2+y^2} \to \infty : \sigma_{ij} = H_{ij} = 0 \\
y = 0, |x| < a : \sigma_{yy} = -p, \sigma_{yx} = 0 \\
H_{yy} = 0, \quad H_{yx} = 0 \\
y = 0, |x| > a : \sigma_{yx} = 0, H_{yx} = 0 \\
u_y = 0, w_y = 0
\end{array}\right\} \tag{8.3-3}$$

By performing Fourier transform on Eq. (8.3-2),

$$\hat{F}(\xi, y) = \int_{-\infty}^{\infty} F(x, y) e^{i\xi x} dx \tag{8.3-4}$$

Equation (8.3-2) is therefore reduced to an ordinary differential equation such that

$$\left(\frac{d^2}{dy^2} - \xi^2\right)^4 \hat{F}(\xi, y) = 0 \tag{8.3-5}$$

If we choose the upper half-plane $y > 0$ for our study, the solution of the above equation is

$$\hat{F}(\xi, y) = (4\xi^4)^{-1} XY e^{-|\xi| y} \tag{8.3-6}$$

where $X = (A, B, C, D)$, $Y = (1, y, y^2, y^3)^T$ and A, B, C and D are arbitrary functions with respect to ξ to be determined according to boundary condition, and "T" stands for the transpose of a matrix. The Fourier transforms of the displacement and stress components can be expressed in terms of $\hat{F}(\xi, y)$, i.e., X and Y as discussed in Section 7.2.

Solution (8.2-6) has satisfied the boundary condition (8.2-3) at infinity, and the left boundary condition in (8.2-3) results in

$$\left.\begin{aligned} A(\xi) &= [21C(\xi)|\xi| - 3(32 - e_2)D(\xi)]/2|\xi|^3 \\ B(\xi) &= [6C(\xi)|\xi| - 21D(\xi)]/\xi^2 \end{aligned}\right\} \tag{8.3-7}$$

and the following set of dual integral equations:

$$\left.\begin{aligned} &\frac{2}{d_{11}} \int_0^\infty [C(\xi)\xi - 6D(\xi)] \cos(\xi x) d\xi = -p, \quad 0 < x < a \\ &\int_0^\infty \xi^{-1}[C(\xi)\xi - 6D(\xi)] \cos(\xi x) d\xi = 0, \quad x > a \\ &\frac{2}{d_{12}} \int_0^\infty D(\xi) \cos(\xi x) d\xi = 0, \quad 0 < x < a \\ &\int_0^\infty \xi^{-1} D(\xi) \cos(\xi x) d\xi = 0, \quad x > a \end{aligned}\right\}. \tag{8.3-8}$$

Here, e_2 is given by the second relation in (7.2-12), i.e.,

$$e_2 = \frac{2\alpha\beta}{\omega(\alpha - \beta)(K_1 - K_2)} + \frac{\alpha - \beta}{\alpha + \beta},$$

α and β are given by the second relation in (6.2-5), i.e.,

$$\alpha = R(L + 2M) - \omega K_1, \quad \beta = RM - \omega K_1, \quad \omega = M(L + 2M)/R$$

d_{11} and d_{22} are given as

$$\left.\begin{aligned} d_{11} &= nR/[(4M/L+M)(L+2M)(MK_1-R^2)] \\ d_{12} &= nR^2/d_0M(L+2M) \\ d_0 &= -\{(MK_1-R^2)[(L+2M)(K_1+K_2)-2R^2] \\ &\quad -[(L+2M)K_1-R^2][M(K_1+K_2)-2R^2]\} \end{aligned}\right\} \tag{8.3-9}$$

and n is determined by (7.2-12), namely,

$$n = M\alpha - (L+2M)\beta$$

The theory of dual integral equations is listed in the Major Appendix of this monograph. Accordingly, the solution to the set of Eq. (8.3-8) is

$$2C(\xi)\xi = d_{11}pa J_1(a\xi), \; D(\xi) = 0 \tag{8.3-10}$$

where $J_1(a\xi)$ is the first-order Bessel function of the first kind.

So far, the unknown functions $A(\xi)$, $B(\xi)$, $C(\xi)$ and $D(\xi)$ have been determined completely. In the view of mathematics, this problem has been solved. However, in the view of physics, we need to perform the Fourier inverse:

$$F(x,y) = \frac{1}{2\pi}\int_{-\infty}^{\infty} \hat{F}(\xi,y)e^{-i\xi x}d\xi \tag{8.3-11}$$

in order to express the field variables in the physical space.

Obviously, once $F(x,y)$ is determined from the integral (8.3-11), u_j, σ_{jk}, and H_{jk} can be determined by substituting $F(x,y)$ into (8.3-8)–(8.3-11). Alternatively, $\hat{u}_j(\xi,y)$, $\hat{w}_j(\xi,y)$, $\hat{\sigma}_{jk}(\xi,y)$ and $\hat{H}_{jk}(\xi,y)$ can be determined by substituting $X(\xi)$ and $Y(\xi)$ into (8.3-8)–(8.3-11), and then their Fourier inverses finally lead to u_j, σ_{jk} and H_{jk}. Luckily, the above integrals with the Bessel function can be expressed explicitly using elemental functions. Nevertheless, their final expressions in terms of variables x and y are extremely complex. However, the final expressions appear more concise if using (r,θ), (r_1,θ_1) and (r_2,θ_2) to stand for the three polar coordinate systems with the crack centre, left crack tip and right crack tip as origins, respectively, as shown in Fig. 8.2, similar to (8.1-16), i.e.,

$$\left.\begin{aligned} x &= r\cos\theta = a + r_1\cos\theta_1 = -a + r_2\cos\theta_2 \\ y &= r\sin\theta = r_1\sin\theta_1 = r_2\sin\theta_2 \end{aligned}\right\} \tag{8.3-12}$$

The following infinite integrals involving Bessel functions are used for stress calculation:

$$\left.\begin{aligned} \int_0^\infty J_1(a\xi)e^{-\xi z}d\xi &= \frac{1}{a}\left[1 - \frac{z}{(a^2-z^2)^{1/2}}\right] \\ \int_0^\infty \xi J_1(a\xi)e^{-\xi z}d\xi &= \frac{a}{(a^2-z^2)^{3/2}} \\ \int_0^\infty \xi^2 J_1(a\xi)e^{-\xi z}d\xi &= \frac{3az}{(a^2-z^2)^{5/2}} \\ \int_0^\infty \xi^3 J_1(a\xi)e^{-\xi z}d\xi &= \frac{3a(4z^2-a^2)}{(a^2-z^2)^{7/2}} \end{aligned}\right\} \tag{8.3-13}$$

where $z = x + iy$.

After proper calculation, we have

$$\left\{\begin{aligned} \sigma_{xx} &= -p[1 + \bar{r}(\overline{r_1}\,\overline{r_2})^{-3/2}\cos(\theta - \bar{\theta})] - p\bar{r}(\overline{r_1}\,\overline{r_2})^{-3/2}\sin\theta\sin 3\bar{\theta} \\ \sigma_{yy} &= -p[1 - \bar{r}(\overline{r_1}\,\overline{r_2})^{-3/2}\cos(\theta - \bar{\theta})] + p\bar{r}(\overline{r_1}\,\overline{r_2})^{-3/2}\sin\theta\sin 3\bar{\theta} \\ \sigma_{xy} &= \sigma_{yx} = p\bar{r}(\overline{r_1}\,\overline{r_2})^{-3/2}\sin\theta\cos 3\bar{\theta} \\ H_{xx} &= -4d_{21}p\bar{r}(\overline{r_1}\,\overline{r_2})^{-3/2}\sin\theta\cos 3\bar{\theta} \\ &\quad - 6d_{21}p\bar{r}^3(\overline{r_1}\,\overline{r_2})^{-5/2}\sin^2\theta\cos(\theta - 5\bar{\theta}) \\ H_{yy} &= -6d_{21}p\bar{r}^3(\overline{r_1}\,\overline{r_2})^{-5/2}\sin^2\theta\cos(\theta - 5\bar{\theta}) \\ H_{xy} &= 6d_{21}p\bar{r}^3(\overline{r_1}\,\overline{r_2})^{-5/2}\sin^2\theta\sin(\theta - 5\bar{\theta}) \\ H_{yx} &= 4d_{21}p\bar{r}(\overline{r_1}\,\overline{r_2})^{-3/2}\sin\theta\cos 3\bar{\theta} \\ &\quad + 6d_{21}p\bar{r}^3(\overline{r_1}\,\overline{r_2})^{-5/2}\sin^2\theta\sin(\theta - 5\bar{\theta}) \end{aligned}\right. \tag{8.3-14a}$$

in which $\bar{r} = r/a, \overline{r_1} = r_1/a, \overline{r_2} = r_2/a, \bar{\theta} = (\theta_1 + \theta_2)/2$

$$d_{21} = R(K_1 - K_2)/4(MK_1 - R^2) \tag{8.3-14b}$$

Similarly, displacement components u_j and w_j can be also expressed in terms of elemental functions. We list here only the component

$$u_y(x,0) = \begin{cases} 0 & |x| > a \\ \dfrac{p}{2}\left(\dfrac{K_1}{MK_1 - R^2} + \dfrac{1}{L+M}\right)\sqrt{a^2 - x^2} & |x| < a \end{cases} \tag{8.3-15}$$

From these results of (8.3-14a), we can find the stress intensity factor

$$K_I^{\|} = \lim_{x \to a^+} \sqrt{2\pi(x-a)}\sigma_{yy}(x,0) = \sqrt{\pi a}p \tag{8.3-16}$$

and the crack strain energy as

$$\begin{aligned} W_I &= 2\int_0^a (\sigma_{yy}(x,0) \oplus H_{yy}(x,0))(u_y(x,0) \oplus w_y(x,0))dx \\ &= \frac{\pi a^2 p^2}{4}\left(\frac{1}{L+M} + \frac{K_1}{MK_1 - R^2}\right) \end{aligned} \tag{8.3-17}$$

So, there is the crack energy release rate

$$G_I = \frac{1}{2}\frac{\partial W_1}{\partial a} = \frac{1}{4}\left(\frac{1}{L+M} + \frac{K_1}{MK_1 - R^2}\right)(K_I^{\|})^2 \tag{8.3-18}$$

It is evident that the crack energy release rate depends not only on the phonon elastic constants $L(= C_{12}), M(= (C_{11} - C_{12})/2)$ but also on the phason elastic constant K_1 and phonon–phason coupling elastic constant R.

Further meaning and applications of these quantities will be discussed in detail in Chapter 15. This provides the basis of fracture theory of quasicrystalline materials.

8.4 Stress Potential Function Formulation and Complex Analysis Method for Solving Notch/Crack Problem of Quasicrystals of Point Groups $5, \bar{5}$ and $10, \overline{10}$

The Fourier method cannot solve notch problems, while the complex analysis method with conformal mapping can solve them. In this section, we develop the complex potential method for notch/crack problems for plane elasticity of point groups $5, \bar{5}$ and $10, \overline{10}$ quasicrystals. We will use the stress potential formulation [9]. Of course, we can also use the displacement potential formulation for the plane elasticity as introduced in Section 6.4.

8.4.1 *Complex analysis method*

From Section 6.7, we find that, based on the stress potential method, the final governing equation of plane elasticity of point groups $5, \bar{5}$ and $10, \overline{10}$ decagonal quasicrystals is

$$\nabla^2\nabla^2\nabla^2\nabla^2 G = 0 \tag{8.4-1}$$

The general solution of Eq. (8.4-1) is

$$G = 2\mathrm{Re}\left[g_1(z) + \bar{z}g_2(z) + \frac{1}{2}\bar{z}^2 g_3(z) + \frac{1}{6}\bar{z}^3 g_4(z)\right] \tag{8.4-2}$$

where $g_j(z)$ $(j = 1, \ldots, 4)$ are the four analytic functions of a single complex variable $z = x+iy = re^{i\theta}$. The bar over the quantity denotes the complex conjugate hereinafter, i.e., $\bar{z} = x - iy = re^{-i\theta}$. We can see that the complex analysis here will be more complicated than that of Muskhelishvili method for plane elasticity of classical elasticity.

8.4.2 *The complex representation of stresses and displacements*

Substituting expression (8.4-2) into equations (6.7-6) and then into Eq. (6.7-2) leads to

$$\begin{aligned}
\sigma_{xx} &= -32c_1\,\mathrm{Re}[\Omega(z) - 2g_4'''(z)] \\
\sigma_{yy} &= 32c_1\,\mathrm{Re}[\Omega(z) + 2g_4'''(z)] \\
\sigma_{xy} &= \sigma_{yx} = 32c_1\,\mathrm{Im}\Omega(z) \\
H_{xx} &= 32R_1\,\mathrm{Re}[\Theta'(z) - \Omega(z)) - 32R_2\,\mathrm{Im}(\Theta'(z) - \Omega(z)] \\
H_{xy} &= -32R_1\mathrm{Im}[\Theta'(z) + \Omega(z)) - 32R_2\,\mathrm{Re}(\Theta'(z) + \Omega(z)] \\
H_{yx} &= -32R_1\mathrm{Im}[\Theta'(z) - \Omega(z)) - 32R_2\,\mathrm{Re}(\Theta'(z) - \Omega(z)] \\
H_{yy} &= -32R_1\mathrm{Re}[\Theta'(z) + \Omega(z)) + 32R_2\,\mathrm{Im}(\Theta'(z) + \Omega]
\end{aligned} \tag{8.4-3}$$

where

$$\begin{aligned}\Theta(z) &= g_2^{(\mathrm{IV})}(z) + \bar{z}g_3^{(\mathrm{IV})}(z) + \frac{1}{2}\bar{z}^2 g_4^{(\mathrm{IV})}(z)\\ \Omega(z) &= g_3^{(\mathrm{IV})}(z) + \bar{z}g_4^{(\mathrm{IV})}(z)\end{aligned} \tag{8.4-4}$$

in which the prime, two prime, three prime and superscript (IV) denote the first- to fourth-order differentiation of $g_i(z)$ to variable z, in addition $\Theta'(z) = d\Theta(z)/dz$.

We further derive the complex representations of displacement components of phonon and phason fields. The first two equations of Eq. (6.7-3) can be rewritten as

$$\begin{aligned}\varepsilon_{xx} &= c_2(\sigma_{xx} + \sigma_{yy}) - \frac{K_1 + K_2}{2c}\sigma_{yy}\\ &\quad - \frac{1}{2c}[R_1(H_{xx} + H_{yy}) + R_2(H_{xy} - H_{yx})]\\ \varepsilon_{yy} &= c_2(\sigma_{xx} + \sigma_{yy}) - \frac{K_1 + K_2}{2c}\sigma_{xx}\\ &\quad + \frac{1}{2c}[R_1(H_{xx} + H_{yy}) + R_2(H_{xy} - H_{yx})]\end{aligned} \tag{8.4-5}$$

where

$$c_2 = \frac{c + (L + M)(K_1 + K_2)}{4(L + M)c} \tag{8.4-6}$$

and c refers to (6.7-4). Substituting Eq. (8.4-3) into (8.4-5) and by integration yield

$$\begin{aligned}u_x &= 128c_1c_2\mathrm{Re}\, g_4''(z) - \frac{K_1 + K_2}{2c}\frac{\partial}{\partial x}\phi\\ &\quad + \frac{32(R_1^2 + R_2^2)}{c}\mathrm{Re}[g_3'''(z) + \bar{z}g_4'''(z) - g_4''(z)] + f_1(y)\\ u_y &= 128c_1c_2\mathrm{Im}\, g_4''(z) - \frac{K_1 + K_2}{2c}\frac{\partial}{\partial y}\phi\\ &\quad - \frac{32(R_1^2 + R_2^2)}{c}\mathrm{Im}[g_3'''(z) + \bar{z}g_4'''(z) + g_4''(z)] + f_2(x)\end{aligned}$$

With these results and other equations of Eq. (6.7-3), one finds that

$$-\frac{df_1(y)}{dy} = \frac{df_2(x)}{dx}$$

This means these two functions must be constants which only give rigid body displacements. Omitting the trial functions $f_1(y), f_2(x)$, one obtains

$$\begin{aligned} u_x + iu_y &= 32(4c_1c_2 - c_3 - c_1c_4)g_4''(z) \\ &\quad - 32(c_1c_4 - c_3)[\overline{g_3'''(z)} + z\overline{g_4'''(z)}] \end{aligned} \tag{8.4-7}$$

where c_1 refers to (6.7-7) and

$$c_3 = \frac{R_1^2 + R_2^2}{c}, \quad c_4 = \frac{K_1 + K_2}{c} \tag{8.4-8}$$

Similarly, the complex representations of displacement components of phason fields can be expressed as follows:

$$w_x + iw_y = \frac{32(R_1 - iR_2)}{K_1 - K_2}\overline{\Theta(z)} \tag{8.4-9}$$

8.4.3 *Elliptic notch problem*

To illustrate the effect of the stress potential and complex analysis method to the complicated stress boundary value problems of eight-order partial differential equations given above, we here calculate the stress and displacement fields induced by an elliptic notch L : $\frac{x^2}{a^2} + \frac{y^2}{b^2} = 1$; see Fig. 8.7(a), the edge of which is subjected to a uniform pressure p. This problem is equivalent to the case that the body is subjected to a tension at infinity and the surface of the notch is stress-free shown in Fig. 8.7(a) if $\alpha = \pi/2$. The equivalency is proved in Section 11.3.9 in Chapter 11.

For the problem shown in Fig. 8.7(a), the boundary conditions can be expressed as follows:

$$\begin{aligned} \sigma_{xx}\cos(\mathbf{n}, x) + \sigma_{xy}\cos(\mathbf{n}, y) &= T_x,\ \sigma_{xy}\cos(\mathbf{n}, x) + \sigma_{yy}\cos(\mathbf{n}, y) \\ &= T_y, (x, y) \in L \end{aligned} \tag{8.4-10}$$

$$\begin{aligned} H_{xx}\cos(\mathbf{n}, x) + H_{xy}\cos(\mathbf{n}, y) &= h_x, H_{yx}\cos(\mathbf{n}, x) + H_{yy}\cos(\mathbf{n}, y) \\ &= h_y, (x, y) \in L \end{aligned} \tag{8.4-11}$$

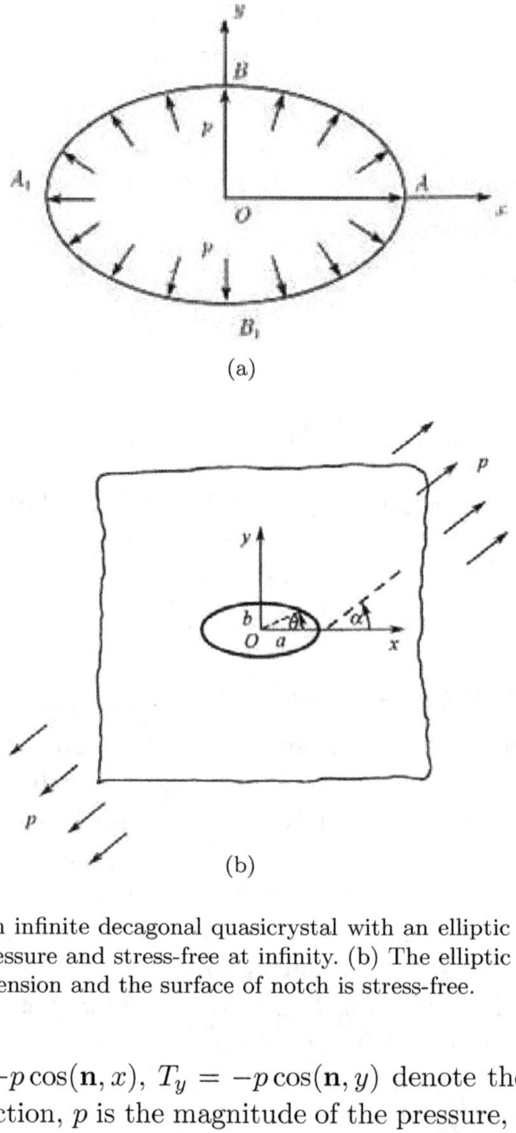

Fig. 8.7. (a) An infinite decagonal quasicrystal with an elliptic notch subjected to a uniform pressure and stress-free at infinity. (b) The elliptic notch subjected to an external tension and the surface of notch is stress-free.

where $T_x = -p\cos(\mathbf{n}, x)$, $T_y = -p\cos(\mathbf{n}, y)$ denote the components of surface traction, p is the magnitude of the pressure, h_x and h_y are the generalized surface tractions and $\mathbf{n}$ represents the outward unit normal vector of any point of the boundary. But the measurement of generalized tractions has not been reported so far; for simplicity, we assume that $h_x = 0$, $h_y = 0$.

From equations (8.4-3), (8.4-4) and (8.4-10), one has

$$g_4''(z) + \overline{g_3'''(z)} + z\overline{g_4'''(z)} = \frac{i}{32c_1}\int (T_x + iT_y)ds = -\frac{1}{32c_1}pz \quad z \in L \tag{8.4-12}$$

Taking conjugate on both sides of Eq. (8.4-12) yields

$$\overline{g_4''(z)} + g_3'''(z) + \bar{z}g_4'''(z) = -\frac{1}{32c_1}p\bar{z} \quad z \in L \tag{8.4-13}$$

From Eqs. (8.4-3), (8.4-4) and (8.4-11), we have

$$\begin{aligned} &R_1 \mathrm{Im}\,\Theta(z) + R_2 \mathrm{Re}\,\Theta(z) = 0 \\ &-R_1 \mathrm{Re}\,\Theta(z) + R_2 \mathrm{Im}\,\Theta(z) = 0 \end{aligned} \quad z \in L \tag{8.4-14}$$

Multiplying the second formula of (8.4-14) by i and adding it to the first, one obtains

$$\Theta(z) = 0 \quad z \in L \tag{8.4-15}$$

Because the function $g_1(z)$ does not appear in the displacement and stress formulas, boundary equations (8.4-12), (8.4-13) and (8.4-15) are enough for determining the unknown functions $g_2(z)$, $g_3(z)$ and $g_4(z)$. However, the calculation cannot be completed at the z-plane due to the complexity of the evaluation, and we must use the conformal mapping

$$z = \omega(\zeta) = R_0\left(\frac{1}{\zeta} + m\zeta\right) \tag{8.4-16}$$

to transform the region with the ellipse at the z-plane onto the interior of the unit circle γ at the ζ-plane, refer to Fig. 8.8, where $\zeta = \xi + i\eta = \rho e^{i\varphi}$ and $R_0 = \frac{a+b}{2}$, $m = \frac{a-b}{a+b}$. For simplicity, we introduce the following new symbols:

$$g_2^{(\mathrm{IV})}(z) = F_2(z), g_3'''(z) = F_3(z), g_4''(z) = F_4(z) \tag{8.4-17}$$

and we have

$$F_j(z) = F_j(\omega(\zeta)) = \Phi_j(\zeta),\ F_j'(z) = \frac{\Phi_j'(\zeta)}{\omega'(\zeta)} \quad (j = 1, \ldots, 4) \tag{8.4-18}$$

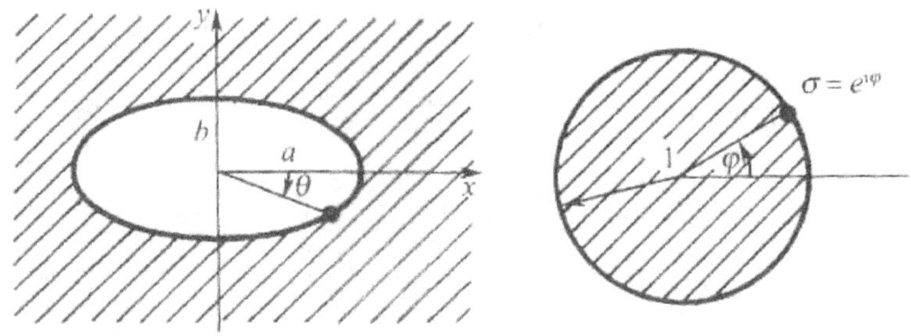

Fig. 8.8. Conformal mapping and boundary correspondence.

Substituting (8.4-17) into (8.4-12), (8.4-13) and (8.4-15), then multiplying both sides of equations by $\frac{1}{2\pi i}\frac{d\sigma}{\sigma-\zeta}$, and integrating along the unit circle, we have

$$
\begin{aligned}
&\frac{1}{2\pi i}\int_\gamma \frac{\Phi_4(\sigma)d\sigma}{\sigma-\zeta} + \frac{1}{2\pi i}\int_\gamma \frac{\overline{\Phi_3(\sigma)}d\sigma}{\sigma-\zeta} + \frac{1}{2\pi i}\int_\gamma \frac{\omega(\sigma)}{\overline{\omega(\sigma)}}\frac{\overline{\Phi_4'(\sigma)}d\sigma}{\sigma-\zeta} \\
&\quad = -\frac{p}{32c_1}\frac{1}{2\pi i}\int_\gamma \frac{\omega(\sigma)d\sigma}{\sigma-\zeta} \\
&\frac{1}{2\pi i}\int_\gamma \frac{\overline{\Phi_4(\sigma)}d\sigma}{\sigma-\zeta} + \frac{1}{2\pi i}\int_\gamma \frac{\Phi_3(\sigma)d\sigma}{\sigma-\zeta} + \frac{1}{2\pi i}\int_\gamma \frac{\overline{\omega(\sigma)}}{\omega(\sigma)}\frac{\Phi_4'(\sigma)d\sigma}{\sigma-\zeta} \\
&\quad = -\frac{p}{32c_1}\frac{1}{2\pi i}\int_\gamma \frac{\overline{\omega(\sigma)}d\sigma}{\sigma-\zeta} \\
&\frac{1}{2\pi i}\int_\gamma \frac{\Phi_2(\sigma)d\sigma}{\sigma-\zeta} + \frac{1}{2\pi i}\int_\gamma \frac{\overline{\omega(\sigma)}}{\omega'(\sigma)}\frac{\Phi_3'(\sigma)d\sigma}{\sigma-\zeta} \\
&\quad + \frac{1}{2\pi i}\left[\int_\gamma \frac{\overline{\omega(\sigma)}^2}{[\omega'(\sigma)]^2}\frac{\Phi_4''(\sigma)d\sigma}{\sigma-\zeta} - \int_\gamma \frac{\overline{\omega(\sigma)}^2\omega''(\sigma)}{[\omega'(\sigma)]^3}\frac{\Phi_4'(\sigma)d\sigma}{\sigma-\zeta}\right] \\
&\quad = 0
\end{aligned}
\tag{8.4-19}
$$

where $\sigma = e^{i\varphi}$ $(\rho = 1)$ represents the value of ζ at the unit circle.

According to the Cauchy integral formula and analytic extension of the complex analysis theory, similar to the calculations in Sections 8.1 and 8.2, from the first and the second equations of (8.4-19), one obtains

$$\Phi_3(\zeta) = \frac{p\,R_0}{32c_1}\frac{(1+m^2)\zeta}{m\zeta^2-1}$$

$$\Phi_4(\zeta) = -\frac{p\,R_0}{32c_1}m\zeta \tag{8.4-20}$$

Substitution of

$$\frac{\overline{\omega(\sigma)}}{\omega'(\sigma)} = \sigma\frac{\sigma^2+m}{m\sigma^2-1}, \quad \frac{\overline{\omega(\sigma)}^2\omega''(\sigma)}{\omega'(\sigma)^3} = -\frac{2\sigma(\sigma^2+m)^2}{(m\sigma^2-1)^3}$$

and (8.4-20) into the third equation of (8.4-19) yields

$$\frac{1}{2\pi i}\int_\gamma \frac{\Phi_2(\sigma)d\sigma}{\sigma-\zeta}$$
$$+\frac{1}{2\pi i}\int_\gamma \sigma\frac{\sigma^2+m}{m\sigma^2-1}\frac{\Phi_3'(\sigma)d\sigma}{\sigma-\zeta} + \frac{1}{2\pi i}\int_\gamma \frac{\sigma(\sigma^2+m)^2}{(m\sigma^2-1)^3}\frac{\Phi_4'(\sigma)d\sigma}{\sigma-\zeta} = 0$$

By Cauchy's integral formula, we have

$$\frac{1}{2\pi i}\int_\gamma \frac{\Phi_2(\sigma)d\sigma}{\sigma-\zeta} = \Phi_2(\zeta)$$

$$\frac{1}{2\pi i}\int_\gamma \sigma\frac{\sigma^2+m}{m\sigma^2-1}\frac{\Phi_3'(\sigma)d\sigma}{\sigma-\zeta} = \zeta\frac{\zeta^2+m}{m\zeta^2-1}\Phi_3'(\zeta)$$

$$\frac{1}{2\pi i}\int_\gamma \frac{\sigma(\sigma^2+m)^2}{(m\sigma^2-1)^3}\frac{\Phi_4'(\sigma)d\sigma}{\sigma-\zeta} = \frac{\zeta(\zeta^2+m)^2}{(m\zeta^2-1)^3}\Phi_4'(\zeta)$$

Substituting these equations into (8.4-19), one finds that at last

$$\Phi_2(\zeta) = \frac{p\,R_0}{32c_1}\frac{\zeta(\zeta^2+m)[(1+m^2)(1+m\zeta^2)-(\zeta^2+m)]}{(m\zeta^2-1)^3} \tag{8.4-21}$$

Utilizing the above-mentioned results, the phonon and phason stresses can be determined at the ζ-plane. We here only give a simple example, i.e., along the edge of notch ($\rho = 1$), there are the phonon stress components such as

$$\sigma_{\varphi\varphi} = p\frac{1 - 3m^2 + 2m\cos 2\varphi}{1 + m^2 - 2m\cos 2\varphi}, \quad \sigma_{\rho\rho} = -p, \quad \sigma_{\rho\varphi} = \sigma_{\varphi\rho} = 0$$

which are identical to the well-known results of the classical elasticity theory.

8.4.4 *Elastic field caused by a Griffith crack*

The solution of the Griffith crack subjected to uniform pressure has been observed by Li *et al.* [7] in terms of the Fourier transform method, which can also be obtained by the notch solution in corresponding to the case $m = 1$, $R_0 = \frac{a}{2}$ of this work. For explicitness, we express the solution in z-plane. The inversion of transformation (8.4-16) is as $m = 1$

$$\zeta = \frac{1}{a}(z - \sqrt{z^2 - a^2}) \tag{8.4-22}$$

From Eqs. (8.4-20), (8.4-25) and (8.4-22), we have

$$\begin{aligned} g_2^{(\mathrm{IV})}(z) &= -\frac{pa^2}{128c_1}\frac{z^2}{\sqrt{(z^2 - a^2)^3}} \\ g_3'''(z) &= -\frac{p}{64c_1}\frac{a^2}{\sqrt{z^2 - a^2}} \\ g_4''(z) &= \frac{p}{64c_1}(\sqrt{z^2 - a^2} - z) \end{aligned} \tag{8.4-23}$$

So, the stresses and the displacements can be expressed with complex variable z.

Similar to (8.1-16), we introduce three pairs of the polar coordinates (r, θ), (r_1, θ_1) and (r_2, θ_2) with the origin at the crack centre, at the right crack tip and at the left crack tip, i.e., $z = re^{i\theta}$, $z - a = r_1 e^{i\theta_1}$, and $z + a = r_2 e^{i\theta_2}$, respectively, the analytic expressions for the stress and displacement fields can be obtained.

Moreover, the stress intensity factor and free energy of the crack and so on can be evaluated as the direct results of the solution. Here we only list the stress intensity factor and energy release rate as follows:

$$K_I^{\|} = \lim_{x \to a^+} \sqrt{2\pi(x-a)}\sigma_{yy}(x,0) = \sqrt{\pi a}\, p$$

$$\begin{aligned} G_I &= \frac{1}{2}\frac{\partial}{\partial a}\left\{2\int_0^a [(\sigma_{yy}(x,0) \oplus H_{yy}(x,0))((u_y(x,0) \oplus w_y(x,0))]dx\right\} \\ &= \frac{L(K_1+K_2)+2(R_1^2+R_2^2)}{8(L+M)c}(K_I^{\|})^2 \end{aligned} \quad (8.4\text{-}24)$$

where $c = M(K_1+K_2) - 2(R_1^2+R_2^2)$ and $L = C_{12}$, $M{=}(C_{11}-C_{12})/2 = C_{66}$. The crack energy release rate G_I is dependent upon not only the phonon elastic constants $L(=C_{12})$, $M(=(C_{11}-C_{12})/2)$, but also the phason elastic constants K_1, K_2 and phonon–phason coupling elastic constants R_1, R_2, though we have assumed the generalized tractions $h_x = h_y = 0$.

It is evident that the present solution covers the solution for point groups $5m$ and $10mm$ quasicrystals, or say the solution of the latter is a special case of the present problem.

The details for some further principle of the complex analysis method will be discussed in-depth in Chapter 11. For further implications and applications of the results to the fracture theory of quasicrystals, refer to Chapter 15.

8.5 Solutions of Crack/Notch Problems in Two-dimensional Octagonal Quasicrystals

Zhou and Fan [11] and Zhou [12] obtained the solution of a Griffith crack in octagonal quasicrystals in terms of the Fourier transform and dual integral equations, and the calculation is very complex and lengthy, which cannot be listed here.

Li [13] gave solutions for a notch/Griffith crack problem in terms of complex analysis method based on the stress potential formulation; an outline of the algorithm is listed in Section 6.8.

The final governing equation of plane elasticity of octagonal quasicrystals of point group $8mm$ based on the stress potential formulation is the same as that of (6.6-12), but here we rewrite it as

$$\left[\frac{\partial^8}{\partial x^8} + 4(1-4\varepsilon)\frac{\partial^8}{\partial x^6 \partial y^2} + 2(3+16\varepsilon)\frac{\partial^8}{\partial x^4 \partial y^4} + 4(1-4\varepsilon)\frac{\partial^8}{\partial x^2 \partial y^6} + \frac{\partial^8}{\partial y^8}\right] G = 0 \tag{8.5-1}$$

in which $G(x, y)$ is the stress potential function and the material constant ε is the same as that given in Chapters 6 and 7. The complex representation of Eq. (8.5-1) is

$$G(x, y) = 2\mathrm{Re}\sum_{k=1}^{4} G_k(z_k),\ z_k = x + \mu_k y \tag{8.5-2}$$

in which unknown functions $G_k(z_k)$ are analytic functions of complex variable z_k $(k = 1, 2, 3, 4)$, to be determined, and $\mu_k = \alpha_k + i\beta_k$ $(k = 1, 2, 3, 4)$ are complex parameters and are determined by the roots of the following eigenvalue equation:

$$\mu^8 + 4(1-4\varepsilon)\mu^6 + 2(3+16\varepsilon)\mu^4 + 4(1-4\varepsilon)\mu^2 + 1 = 0 \tag{8.5-3}$$

The stresses can be expressed by functions $G_k(z_k)$ such as

$$\sigma_{xx} = -2c_3c_4\mathrm{Re}\sum_{k=1}^{4}(\mu_k^2 + 2\mu_k^4 + \mu_k^6)g_k'(z_k) \tag{8.5-4a}$$

$$\sigma_{yy} = -2c_3c_4\mathrm{Re}\sum_{k=1}^{4}(1 + 2\mu_k^2 + \mu_k^4)g_k'(z_k) \tag{8.5-4b}$$

$$\sigma_{xy} = \sigma_{yx} = 2c_3c_4\mathrm{Re}\sum_{k=1}^{4}(\mu_k + 2\mu_k^3 + \mu_k^5)g_k'(z_k) \tag{8.5-4c}$$

$$H_{xx} = R\,\mathrm{Re}\sum_{k=1}^{4}[(4c_4 - c_3)\mu_k^2 + 2(3c_3 - 2c_4)\mu_k^4 - c_3\mu_k^6)]g_k'(z_k) \tag{8.5-4d}$$

$$H_{xy} = -R\,\mathrm{Re}\sum_{k=1}^{4}[(4c_4 - c_3)\mu_k + 2(3c_3 - 2c_4)\mu_k^3 - c_3\mu_k^5)]g_k'(z_k) \tag{8.5-4e}$$

$$H_{yx} = -R\,\mathrm{Re}\sum_{k=1}^{4}[c_3\mu_k + 2(c_4 - 2c_3)\mu_k^3 - (4c_4 - c_3)\mu_k^5)]g'_k(z_k) \tag{8.5-4f}$$

$$H_{yy} = R\mathrm{Re}\sum_{k=1}^{4}[c_3 + 2(c_4 - 2c_3)\mu_k^2 - (4c_4 - c_3)\mu_k^4)]g'_k(z_k) \tag{8.5-4g}$$

in which

$$g_k(z_k) = \frac{\partial^6 G_k(z_k)}{\partial z_k^6}, \quad g'_k(z_k) = \frac{dg_k(z_k)}{dz_k}$$

$$c_3 = \frac{(K_1 + K_2 + K_3)M - R^2}{K_1 + K_2 + 2K_3}, \quad c_4 = \frac{K_1 M - R^2}{K_1 - K_2}$$

We now consider an elliptic hole $L : x^2/a^2 + y^2/b^2 = 1$, at which there are the boundary conditions:

$$\begin{aligned}\sigma_{xx}\cos(\mathbf{n}, x) + \sigma_{xy}\cos(\mathbf{n}, y) &= T_x,\ \sigma_{xy}\cos(\mathbf{n}, x) + \sigma_{yy}\cos(\mathbf{n}, y)\\ &= T_y,\ (x, y) \in L\\ H_{xx}\cos(\mathbf{n}, x) + H_{xy}\cos(\mathbf{n}, y) &= h_x,\ H_{yx}\cos(\mathbf{n}, x) + H_{yy}\cos(\mathbf{n}, y)\\ &= h_y,\ (x, y) \in L\end{aligned} \tag{8.5-5}$$

The complex variable z_k can be rewritten as

$$\begin{aligned}z_k &= x_k + iy_k\\ x_k &= x + \alpha_k y,\ y_k = \beta_k y\end{aligned} \tag{8.5-6}$$

The second formula of Eq. (8.5-6) represents a coordinate transformation.

8.6 Approximate Analytic Solutions of Notch/Crack of Two-dimensional Quasicrystals with Five- and 10-fold Symmetries

Fan and Tang [14] solved finite bending specimens with elliptic notch/crack of two-dimensional quasicrystal (Fig. 8.9), in which (a) is the model of Muskhelishvili in calculating conventional structural materials, and our model is shown in (b).

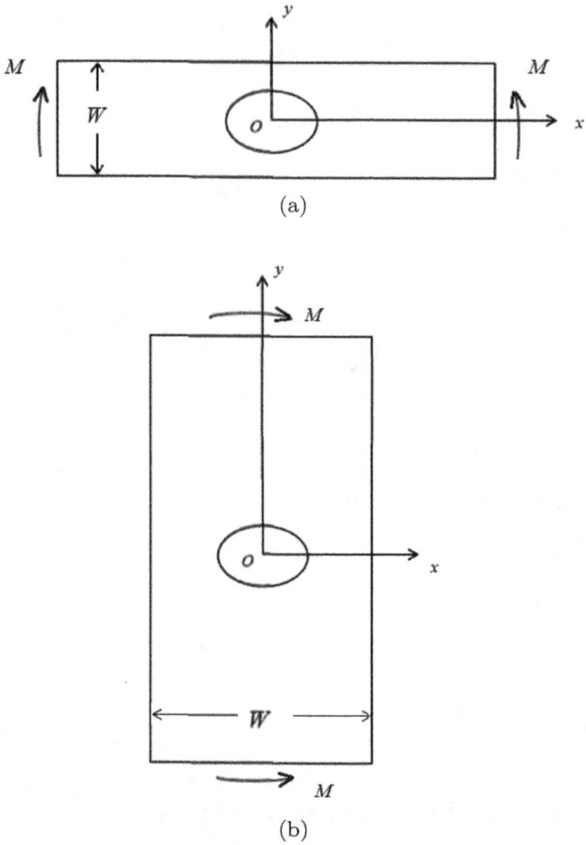

Fig. 8.9. Specimen with elliptical notch and finite width of two-dimensional quasicrystals. (a) Model of Muskhelishvili (for conventional structural materials), (b) Fan and Tang's model [14] (for quasicrystals).

According to the assumption of Muskhelishvili, the width is larger than the of size of the elliptic notch, so conformal mapping (8.4-16) can still be used. The problem has the following boundary conditions:

$$\begin{cases} \boldsymbol{y} \to \pm\infty, |\boldsymbol{x}| < \boldsymbol{W}/2 : \boldsymbol{B} \displaystyle\int_{-\boldsymbol{W}/2}^{\boldsymbol{W}/2} \sigma_{yy} x dx = M, \sigma_{yx} = 0, \\ \qquad \boldsymbol{H}_{yy} = 0, \boldsymbol{H}_{yx} = 0 \\ y = 0, |\boldsymbol{x}| < \boldsymbol{a} : \sigma_{yy} = 0, \sigma_{yx} = 0, \boldsymbol{H}_{yy} = 0, \boldsymbol{H}_{yx} = 0 \\ x = \pm\boldsymbol{W}/2, -\infty < \boldsymbol{y} < \infty : \sigma_{xx} = 0, \sigma_{xy} = 0, \boldsymbol{H}_{xx} = 0, \boldsymbol{H}_{xy} = 0 \end{cases} \tag{8.6-1}$$

The solution of Muskhelishvili [15, p. 344] for conventional structural materials is given for model (a) in Fig. 8.9; he used conformal mapping

$$z = \omega(\zeta) = R_0 \left(\frac{1}{\zeta} + m\zeta \right) \tag{8.6-2}$$

whose results are not suitable for model (b) in Fig. 8.9. The configuration of the solution of Lokchine [16] for conventional structural materials is identical to that of ours, but he used an elliptic coordinate in deriving the solution, which is quite different from complex analysis. Our interest aims to develop the complex analysis here. From the discussion of Chapter 6 referring to solution (6.9-7) of bending specimen without a notch

$$\sigma_{yy} = \frac{M}{I} x, \quad I = \frac{BW^3}{12} \tag{8.6-3}$$

In the following, we take $B = 1$. In further calculation, for simplicity, we remove the boundary conditions at the upper and lower surfaces of the specimen listed in (8.6-1) and instead add an equivalent boundary condition at the notch surface. Then, at the surface of the notch/crack, there is the applied stress (8.6-3), and other stress components are zero. If we consider only the crack problem, then $m = 1$, $R_0 = a/2$ in (8.6-2).

As we have known from 8.4, the key equation in system (8.4-19) concerning the crack stress intensity factor is

$$\begin{aligned} &\frac{1}{2\pi i} \int_\gamma \frac{\Phi_4(\sigma) d\sigma}{\sigma - \zeta} + \frac{1}{2\pi i} \int_\gamma \frac{\overline{\Phi_3(\sigma)} d\sigma}{\sigma - \zeta} + \frac{1}{2\pi i} \int_\gamma \frac{\omega(\sigma)}{\overline{\omega'(\sigma)}} \frac{\overline{\Phi_4'(\sigma)} d\sigma}{\sigma - \zeta} \\ &\quad = \frac{1}{32 c_1} \frac{1}{2\pi i} \int_\gamma \frac{\mathbf{t} d\sigma}{\sigma - \zeta} \end{aligned} \tag{8.6-4}$$

where $\mathbf{t} = i \int (T_x + iT_y) ds$. Substituting conformal mapping (8.6-2) and the boundary condition of crack surface (8.6-3) into Eq. (8.6-4) yields the solution

$$\Phi_4(\zeta) = \frac{Aa^2}{8} \frac{1}{\zeta^2} + \frac{Aa^2}{4} \tag{8.6-5}$$

where

$$A = 12 \frac{M}{BW^3} \tag{8.6-6}$$

Bearing in mind, the well-known results around the crack tip

$$\sigma_{xx}+\sigma_{yy}=(2/\pi r_1)^{1/2}K_I^{\parallel}\cos(\theta_1/2)$$

and

$$\sigma_{xx}+\sigma_{yy}=-(2/\pi r_1)^{1/2}K_{II}^{\parallel}\sin(\theta_1/2)$$

and define the complex stress intensity factor

$$K=K_I^{\parallel}-iK_{II}^{\parallel}$$

Note that $z-z_1=r_1e^{i\theta_1}$, where z_1 represents the location of the crack tip, then we have

$$\sigma_{xx}+\sigma_{yy}=2\mathrm{Re}(K/\sqrt{2\pi(z-z_1)})$$

and find again

$$\sigma_{xx}+\sigma_{yy}=128c_1\,\mathrm{Re}\,g_4'''(z)=128c_1\,\mathrm{Re}\,\Phi_4(\zeta)$$

and the expression of complex stress intensity factor

$$\mathbf{K}=K_I^{\parallel}-iK_{II}^{\parallel}=32c_1\left(2\sqrt{\pi}\lim_{\zeta\to-1}\frac{\Phi_4'(\zeta)}{\sqrt{\omega''(\zeta)}}\right)$$

Substituting (8.6-2) into the above formula yields

$$K_I^{(1)}=\sqrt{\pi a}\sigma_N \tag{8.6-7}$$

in which

$$\sigma_N=6\frac{Ma}{BW^3}$$

here $B=1$, so further the energy release rate

$$\begin{aligned}G_I&=\frac{1}{4}\left(\frac{1}{L+M}+\frac{K_1}{MK_1-R^2}\right)\left(K_I^{\parallel}\right)^2\\&=\frac{1}{4}\left(\frac{1}{L+M}+\frac{K_1}{MK_1-R^2}\right)\pi a(\sigma_N)^2\end{aligned} \tag{8.6-8}$$

The solution obtained here is approximate, and it presents enough accuracy provided

$$2a/W\le 1/3 \tag{8.6-9}$$

8.7 Cracked Strip with Finite Height of Two-dimensional Quasicrystals with 5- and 10-fold Symmetries and Exact Analytic Solution

The power of complex analysis lies in the application of conformal mapping to some extent. To demonstrate further the effect of conformal mapping, we give another example for quasicrystals of point group $5m$ and $10mm$, which is given by Fan and Tang [17], refer to Fig. 8.10.

The boundary conditions are as follows:

$$\left.\begin{array}{l} y = \pm H, -\infty < x < \infty : \sigma_{yy} = 0, \sigma_{yx} = 0, H_{yy} = 0, H_{yx} = 0 \\ x = \pm\infty, -H < y < H : \sigma_{xx} = 0, \sigma_{xy} = 0, H_{xx} = 0, H_{xy} = 0 \\ y = \pm 0, -\infty < x < -a : \sigma_{yy} = 0, \sigma_{yx} = 0, H_{yx} = 0, H_{yx} = 0; \\ -a < x < 0 : \sigma_{yy} = -p, \sigma_{yx} = -\tau, H_{yx} = 0, H_{yy} = 0 \end{array}\right\} \tag{8.7-1}$$

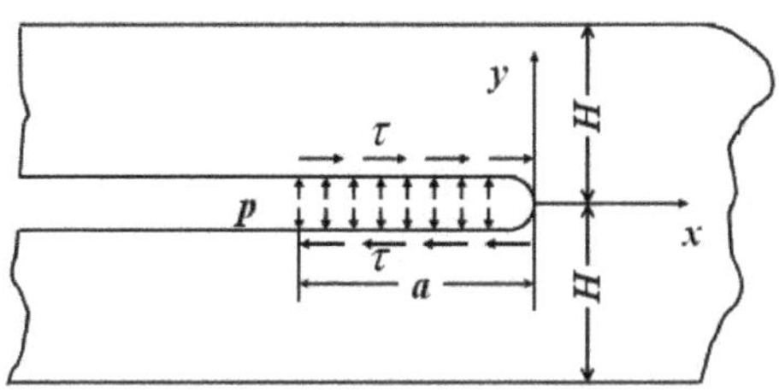

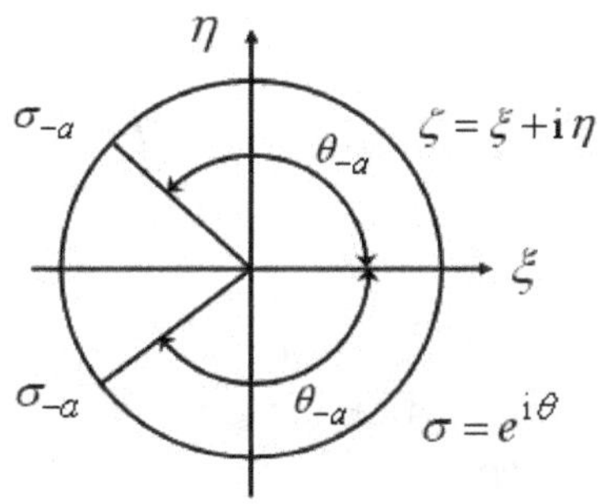

Fig. 8.10. Finite height specimen of quasicrystals (a) and conformal mapping onto the unit circle at ζ-plane (b).

So, the problem is quite complicated, and solving it is very difficult if we taking other approachs. The complex analysis is effective for the solution. The conformal mapping here used is

$$z = \omega(\zeta) = \frac{H}{\pi} \ln\left[1 + \left(\frac{1+\zeta}{1-\zeta}\right)^2\right] \tag{8.7-2}$$

which maps the region at the physical plane shown in Fig. 8.10(a) onto the interior of the unit circle γ. This conformal mapping was used in solving the problem of one-dimensional quasicrystals; refer to Section 8.2. Fan and Tang [17] developed the approach given by [4]. The application of the mapping yields the function equations along the unit circle γ due to the boundary conditions

$$\left\{\begin{aligned}
&\frac{1}{2\pi i}\int_\gamma \frac{\Phi_4(\sigma)d\sigma}{\sigma-\zeta} + \frac{1}{2\pi i}\int_\gamma \frac{\overline{\Phi_3(\sigma)}d\sigma}{\sigma-\zeta} + \frac{1}{2\pi i}\int_\gamma \frac{\omega(\sigma)}{\overline{\omega'(\sigma)}}\frac{\overline{\Phi_4'(\sigma)}d\sigma}{\sigma-\zeta}\\
&\quad = \frac{1}{32c_1}\frac{1}{2\pi i}\int_\gamma \frac{\mathbf{t}d\sigma}{\sigma-\zeta}\\
&\frac{1}{2\pi i}\int_\gamma \frac{\overline{\Phi_4(\sigma)}d\sigma}{\sigma-\zeta} + \frac{1}{2\pi i}\int_\gamma \frac{\Phi_3(\sigma)d\sigma}{\sigma-\zeta} + \frac{1}{2\pi i}\int_\gamma \frac{\overline{\omega(\sigma)}}{\omega'(\sigma)}\frac{\Phi_4'(\sigma)d\sigma}{\sigma-\zeta}\\
&\quad = \frac{1}{32c_1}\frac{1}{2\pi i}\int_\gamma \frac{\bar{\mathbf{t}}d\sigma}{\sigma-\zeta}\\
&\frac{1}{2\pi i}\int_\gamma \frac{\Phi_2(\sigma)d\sigma}{\sigma-\zeta} + \frac{1}{2\pi i}\int_\gamma \frac{\overline{\omega(\sigma)}}{\omega'(\sigma)}\frac{\Phi_3'(\sigma)d\sigma}{\sigma-\zeta}\\
&\qquad + \frac{1}{2\pi i}\left[\int_\gamma \frac{\overline{\omega(\sigma)}^2}{[\omega'(\sigma)]^2}\frac{\Phi_4''(\sigma)d\sigma}{\sigma-\zeta} - \int_\gamma \frac{\overline{\omega(\sigma)}^2\omega''(\sigma)}{[\omega'(\sigma)]^3}\frac{\Phi_4'(\sigma)d\sigma}{\sigma-\zeta}\right]\\
&\quad = \frac{1}{R_1 - iR_2}\frac{1}{2\pi i}\int_\gamma \frac{\mathbf{h}d\sigma}{\sigma-\zeta}
\end{aligned}\right. \tag{8.7-3}$$

The solution of these function equations will determine the complex stress potentials, in which

$$\mathbf{t} = i\int (T_x + iT_y)ds, \bar{\mathbf{t}} = -i\int (T_x - iT_y)ds, \mathbf{h} = i\int (h_1 + ih_2)ds$$

and

$$\begin{cases} g_2^{(\mathrm{IV})}(z) = h_2(z), g_3'''(z) = h_3(z), g_4''(z) = h_4(z) \\ h_2(z) = h_2(\omega(z)) = \Phi_2(\zeta), h_3(z) = h_3(\omega(z)) = \Phi_3(\zeta) \\ h_4(z) = h_4(\omega(z)) = \Phi_4(\zeta), h_1(z) = h_1(\omega(z)) = \Phi_1(\zeta) = 0 \end{cases} \tag{8.7-4}$$

The inversion of conformal mapping (8.7-1) is

$$\zeta = \omega^{-1}(z) = \frac{-e^{-\pi z/H} \pm 2i\sqrt{1 - e^{-\pi z/H}}}{2 - e^{-\pi z/H}} \tag{8.7-5}$$

The points at γ in the mapping plane corresponding to $z = (-a, 0^+)$ and $z = (-a, 0^-)$ in physical plane are

$$\begin{cases} \sigma_{-a} = \dfrac{-e^{-\pi a/H} + 2i\sqrt{1 - e^{-\pi a/H}}}{2 - e^{-\pi a/H}} \\ \overline{\sigma_{-a}} = \dfrac{-e^{-\pi a/H} - 2i\sqrt{1 - e^{-\pi a/H}}}{2 - e^{-\pi a/H}} \end{cases} \tag{8.7-6}$$

In addition, the crack tip is mapped to $\zeta = -1$. Substituting the mapping into the equation

$$\begin{aligned} &\Phi_4(\zeta) + \frac{1}{2\pi i}\int_\gamma \overline{G(\sigma)}\frac{\overline{\Phi_4'(\sigma)}}{\sigma - \zeta}d\sigma + \overline{\Phi_3(0)} \\ &= \frac{1}{32c_1}\int_{\sigma_{-a}}^{\overline{\sigma_{-a}}} \frac{i\int(T_x + iT_y)ds}{\sigma - \zeta}d\sigma \end{aligned} \tag{8.7-7}$$

in which

$$G(\zeta) = \frac{\overline{\omega(1/\overline{\zeta})}}{\omega'(\zeta)} \tag{8.7-8}$$

leads to

$$\Phi_4(\zeta) + \overline{G(0)\Phi_4'(0)} + \overline{\Phi_3(0)} = \frac{1}{32c_1}\int_{\sigma_{-a}}^{\overline{\sigma_{-a}}} \left\{\frac{i\int(T_x + iT_y)ds}{\sigma - \zeta}\right\}d\sigma \tag{8.7-9}$$

and the solution at last

$$\Phi_4'(\zeta) = \frac{1}{32c_1} \cdot \frac{1}{2\pi i} \left[\frac{1}{1-\zeta} \ln(\sigma - 1) - \frac{1+\zeta}{(1-\zeta)(1+\zeta^2)} \ln(\sigma - \zeta) \right.$$
$$\left. - \frac{\zeta}{2(1+\zeta^2)} \ln(1+\sigma^2) + \frac{1}{2(1-\zeta^2)} \ln \frac{\sigma - i}{\sigma + i} \right]_{\sigma=\sigma_a}^{\sigma=\overline{\sigma_{-a}}} \tag{8.7-10}$$

Similarly, $\Phi_2(\zeta)$ and $\Phi_3(\zeta)$ can also be determined. If it is calculating stress intensity factor, the information on $\Phi_4'(\zeta)$ is enough because

$$\mathbf{K} = K_I^{\parallel} - iK_{\text{II}}^{\parallel} = 32c_1 \left(2\sqrt{\pi} \lim_{\zeta \to -1} \frac{\Phi_4'(\zeta)}{\sqrt{\omega''(\zeta)}} \right) \tag{8.7-11}$$

Substituting the above results into this formula (note that $T_x = 0$, $T_y = -p$, $h_x = h_y = 0$) yields

$$K_I^{\parallel} = \frac{\sqrt{2}p\sqrt{H}}{2\pi} F(a/H), \quad K_{II}^{\parallel} = 0 \tag{8.7-12}$$

in which the configuration factor is

$$F(a/H) = \ln \frac{2e^{\pi a/H} - 1 + 2e^{\pi a/H}\sqrt{1 - e^{-\pi a/H}}}{2e^{\pi a/H} - 1 - 2e^{\pi a/H}\sqrt{1 - e^{-\pi a/H}}} \tag{8.7-13}$$

Further, we have the crack energy release rate

$$G_I = \frac{1}{4}\left(\frac{1}{L+M} + \frac{K_1}{MK_1 - R^2} \right) \left(K_I^{\parallel} \right)^2 \tag{8.7-14}$$

If the applied stress at the crack surface is pure shear stress, then

$$K_{II}^{\parallel} = \frac{\sqrt{2}\tau\sqrt{H}}{2\pi} F(a/H), \quad K_I^{\parallel} = 0 \tag{8.7-15}$$

and

$$G_{II} = \frac{1}{4}\left(\frac{1}{L+M} + \frac{K_1}{MK_1 - R^2} \right) \left(K_{II}^{\parallel} \right)^2 \tag{8.7-16}$$

8.8 Exact Analytic Solution of Single Edge Crack in a Finite Width Specimen of a Two-dimensional Quasicrystal of 10-fold Symmetry

We have mentioned that the power of a complex analysis lies in the application of conformal mapping to some extent. To further demonstrate the powerful approach, let us consider another example of a cracked specimen of quasicrystal of point group $5m$ and $10mm$, shown in Fig. 8.11, the exact solution of which is difficult due to the finite width of the specimen.

The applied stresses are tensile at the top and bottom surfaces of the specimen (which is equivalent to inner pressure along the crack surface) or shear at external surface of the sample (which is equivalent to shearing along the crack surface). For simplicity, we consider the case of action of inner pressure along the crack surface or of shear along the crack surface only (this ensures stress-free at the region far away from the interior of the specimen), so we have the boundary conditions for the case of action of inner pressure

$$\begin{cases} \sigma_{yy} = \sigma_{xy} = 0, \ H_{yy} = H_{yx} = 0, & y \pm \infty, -a < x < l - a; \\ \sigma_{xx} = \sigma_{xy} = 0, \ H_{xx} = H_{xy} = 0, & \begin{array}{c} -\infty < y < +\infty, \ x = -a, \\ x = l - a; \end{array} \\ \sigma_{yy} = -p, \ \sigma_{xy} = 0, \ H_{yy} = H_{yx} = 0, & y = \pm 0, \ -a < x < 0 \end{cases} \tag{8.8-1}$$

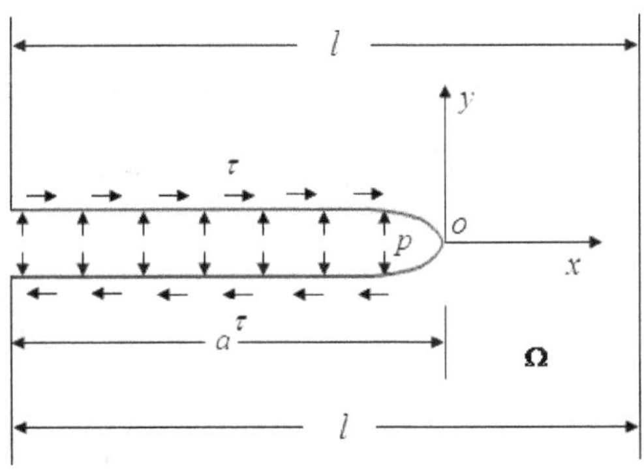

Fig. 8.11. Single-edge cracked specimen subjected to tension (or inner pressure along the crack surface) or shearing (or shearing along the crack surface).

The formulations are Eqs. (8.4-1)–(8.4-9), in which the key equation for the boundary value problem is

$$F_4(z) + \overline{F_3(z)} + z\overline{F_4'(z)} = f_0(z), \quad (x, y) \in S_t \text{ (or } z \in S_t) \tag{8.8-2}$$

where

$$\begin{aligned} f_0(z) &= \frac{i}{32c_1}\int_{-a}^{z}(T_x + iT_y)ds \\ &= -\frac{1}{32c_1}\int_{-a}^{z} pdz = \begin{cases} -\dfrac{1}{32c_1}p(z+a), & -a < x < 0, \\ 0, & x \notin (-a, 0) \end{cases} \end{aligned} \tag{8.8-3}$$

For solving the specimen shown in (8.8-1), we need to use the conformal mapping

$$z = \omega(\zeta) = \left(\frac{2l}{\pi}\right)\arctan\left[\sqrt{1-\zeta^2}\cdot\tan\left(\frac{\pi a}{2l}\right)\right] - a \tag{8.8-4}$$

This mapping function originated from Ref. [18], which is devoted to solving the crack problem in traditional materials (we should point out that the final results of Ref. [18] are wrong). The conformal mapping maps the region at the physical plane onto the upper half-plane (or lower half-plane) at ζ-plane. The complex potentials, e.g., $\Phi_3(\zeta)$ and $\Phi_4(\zeta)$ (refer to Section 8.4), transformed from $F_3(z)$ and $F_4(z)$ after the mapping, respectively, and satisfy the boundary equations (transformed from (8.8-2) after the mapping)

$$\Phi_4(\zeta) + \frac{1}{2\pi i}\int_\gamma \frac{\omega(\sigma)}{\overline{\omega'(\sigma)}}\frac{\overline{\Phi_4'(\sigma)}}{\sigma - \zeta}d\sigma + \overline{\Phi_3(0)} = \frac{1}{2\pi i}\int_\gamma \frac{f_0}{\sigma - \zeta}d\sigma \tag{8.8-5a}$$

$$\overline{\Phi_4(0)} + \frac{1}{2\pi i}\int_\gamma \frac{\overline{\omega(\sigma)}}{\omega'(\sigma)}\frac{\Phi_4'(\sigma)}{\sigma - \zeta}d\sigma + \Phi_3(\zeta) = \frac{1}{2\pi i}\int_\gamma \frac{\overline{f_0}}{\sigma - \zeta}d\sigma \tag{8.8-5b}$$

where σ represents the value of ζ along γ at the ζ-plane. We know that the function $\frac{\omega(\zeta)}{\overline{\omega'(\zeta)}}\overline{\Phi_4'(\zeta)}$ is analytic in the lower half-plane $\eta < 0$, and according to the condition of stress-free in the region far away from the interior of the specimen

$$\lim_{z\to\infty} z\overline{F_4'(z)} = 0 \tag{8.8-6}$$

so that $\lim_{\zeta\to\infty} \frac{\omega(\zeta)}{\omega'(\zeta)}\overline{\Phi_4'(\zeta)} = \lim_{z\to\infty} z\overline{F_4'(z)} = 0$. By applying Cauchy's integral theory, there is

$$\frac{1}{2\pi i}\int_\gamma \frac{\omega(\sigma)}{\overline{\omega'(\sigma)}}\frac{\overline{\Phi_4'(\sigma)}}{\sigma-\zeta}d\sigma = 0 \tag{8.8-7}$$

From (8.8-7), (8.8-5a) and (8.8-3),

$$\Phi_4(\zeta) = -\frac{1}{32c_1}\frac{1}{2\pi i}\int_{-1}^{1}\frac{p[\omega(\sigma)+a]}{\sigma-\zeta}d\sigma \tag{8.8-8}$$

Further, integrating by parts after derivation to (8.8-8) with respect to ζ, we have

$$\Phi_4'(\zeta) = -\frac{1}{32c_1}\frac{1}{2\pi i}\int_{-1}^{1}\frac{p\omega'(\sigma)}{\sigma-\zeta}d\sigma \tag{8.8-9}$$

By the way from (8.8-4), there is

$$\omega''(0) = -\frac{2l}{\pi}\tan\left(\frac{\pi a}{2l}\right)\cos^2\left(\frac{\pi a}{2l}\right) \tag{8.8-10}$$

According to the definition of complex stress intensity factor, for the crack under the action of inner pressure, one has

$$K_I^{||} = \frac{2}{\sqrt{\pi}}\sqrt{\frac{2l}{\pi a}\tan\frac{\pi a}{2l}}\sqrt{\pi a}p, \quad K_{II}^{||} = 0 \tag{8.8-11}$$

Similarly, for the case of action of inner shear, there is

$$K_I^{||} = 0, \quad K_{II}^{||} = \frac{2}{\sqrt{\pi}}\sqrt{\frac{2l}{\pi a}\tan\frac{\pi a}{2l}}\sqrt{\pi a}\tau \tag{8.8-12}$$

The work can be found in Ref. [19], and the detail is given in Appendix A.5 in Major Appendix of this book.

8.9 Perturbation Solution of Three-dimensional Elliptic Disk Crack in One-dimensional Hexagonal Quasicrystals [20]

Elasticity for all quasicrystals is three-dimensional in fact, even for one-dimensional quasicrystals; we have previously mentioned this point. To simplify solving the procedure in the previous and succeeded descriptions, we often decompose a complex three-dimensional problem into a plane and anti-plane elasticity to solve, which helps us to construct some solutions. When the so-called decomposition is not available, one must solve the three-dimensional problem instead. Here we consider an example in one-dimensional hexagonal quasicrystals, in which there is a three-dimensional elliptic disk crack with major and minor semi-axes a and b, respectively, shown in Fig. 8.12, and the crack is located at the centre, subjected to tension at the region far away from the crack.

For simplicity, assume that the externally applied stress is removed and at the crack surface is subjected to uniform pressure; this results in the following boundary conditions:

$$\begin{aligned}
&\sqrt{x^2+y^2+z^2}\to\infty: && \sigma_{ii}=0, H_{ii}=0\\
&z=0,(x,y)\in\Omega: && \sigma_{zz}=-p, H_{zz}=-q, \sigma_{xz}=\sigma_{yz}=0\\
&z=0,(x,y)\notin\Omega: && \sigma_{xz}=\sigma_{yz}=0, u_z=0, w_z=0
\end{aligned} \tag{8.9-1}$$

where Ω denotes the crack surface.

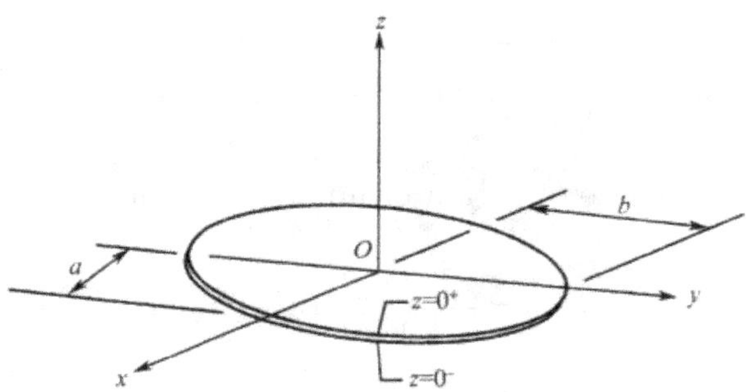

Fig. 8.12. Elliptic disc-shaped crack.

The governing equations of this problem are introduced in Chapter 5, which are not needed to list again.

Due to the coupling between phonon and phason fields, the problem is very complicated, and a strict analysis is difficult. In the following, an approximate solution based on perturbation is discussed.

Assume the coupling elastic constants are much smaller than those of phonon and phason, i.e.,

$$\frac{R_i}{C_{jk}}, \frac{R_i}{K_j} \sim \varepsilon \ll 1 \tag{8.9-2}$$

Take perturbation expansions such as

$$u_i = \sum_{n=0}^{\infty} \varepsilon^n u_i^{(n)} \quad (i = x, y, z), \quad w_z = \sum_{n=0}^{\infty} \varepsilon^n w_z^{(n)} \tag{8.9-3}$$

Substituting (8.9-3) into the governing equations and considering condition (8.9-2), the zero-order solutions expressed by displacements are as follows:

$$\begin{aligned} u_x^{(0)} &= \frac{\partial}{\partial x}(F_1 + F_2) - \frac{\partial}{\partial y}F_3, \quad u_y^{(0)} = \frac{\partial}{\partial y}(F_1 + F_2) + \frac{\partial}{\partial x}F_3 \\ u_z^{(0)} &= \frac{\partial}{\partial z}(m_1 F_1 + m_2 F_2), \qquad u_z^{(0)} = F_4 \end{aligned} \tag{8.9-4}$$

in which the displacement potentials satisfy the generalized harmonic equations

$$\nabla_i^2 F_i = 0 \quad (i = 1, 2, 3, 4) \tag{8.9-5}$$

where

$$\nabla_i^2 = \frac{\partial^2}{\partial x^2} + \frac{\partial^2}{\partial y^2} + \gamma_i^2 \frac{\partial^2}{\partial z^2} \tag{8.9-6}$$

$$\begin{aligned} \gamma_i^2 &= \frac{C_{44} + (C_{13} + C_{44})m_i}{C_{11}} = \frac{C_{33} m_i}{C_{13} + C_{44} + C_{44} m_i} (i = 1, 2), \\ \gamma_3^2 &= \frac{C_{44}}{C_{66}}, \qquad \gamma_4^2 = \frac{K_1}{K_2} \end{aligned} \tag{8.9-7}$$

Without losing generality, we can take

$$F_2 = F_3 = 0 \tag{8.9-8}$$

so that

$$u_z^{(0)} = m_1 \frac{\partial F_1}{\partial z}, \quad w_z^{(0)} = F_4 \tag{8.9-9}$$

From the basic formulas of Chapter 5, it stands for the zero-order perturbation solution

$$\begin{aligned}
\sigma_{zz}^{(0)} &= -C_{13}\gamma_1^2 \frac{\partial^2}{\partial z^2} F_1 + R_2 \frac{\partial}{\partial z} F_4 \\
\sigma_{zx}^{(0)} &= C_{44}(m_1+1) \frac{\partial^2}{\partial x \partial z} F_1 + R_3 \frac{\partial}{\partial x} F_4 \\
\sigma_{zy}^{(0)} &= C_{44}(m_1+1) \frac{\partial^2}{\partial y \partial z} F_1 + R_2 \frac{\partial}{\partial y} F_4 \\
H_{zz}^{(0)} &= -R_1\gamma_1^2 \frac{\partial^2}{\partial z^2} F_1 + K_1 \frac{\partial}{\partial z} F_4
\end{aligned} \tag{8.9-10}$$

the stresses $\sigma_{xx}^{(0)}, \sigma_{yy}^{(0)}, \sigma_{xy}^{(0)}$ are not listed because they have no direct connection with the following calculation. Thus, the zero-order perturbation solution on relevant stress components are expressed by

$$\begin{aligned}
\sigma_{zz}^{(0)} &\approx -C_{13}\gamma_1^2 \frac{\partial^2}{\partial z^2} F_1^{(0)} \\
\sigma_{zx}^{(0)} &\approx C_{44}(m_1+1) \frac{\partial^2 F_1^{(0)}}{\partial x \partial z} \\
\sigma_{zy}^{(0)} &\approx C_{44}(m_1+1) \frac{\partial^2 F_1^{(0)}}{\partial y \partial z} \\
H_{zz}^{(0)} &\approx K_1 \frac{\partial F_4^{(0)}}{\partial z}
\end{aligned} \tag{8.9-11}$$

in which the zero-order perturbations of $F_1^{(0)}$ and $F_4^{(0)}$satisfy

$$\nabla_1^2 F_1^{(0)} = 0, \quad \nabla_4^2 F_4^{(0)} = 0 \tag{8.9-12}$$

Due to the above treatment, the boundary conditions reduce to

$$\begin{aligned}
&\sqrt{x^2+y^2+z^2}\to\infty: \quad f=0, \quad g=0 \\
&z=0,(x,y)\in\Omega: \qquad 2C_{13}\gamma_1^2\frac{\partial^2 F_1^{(0)}}{\partial z^2}=-p, \quad K_1\frac{\partial F_4^{(0)}}{\partial z}=-q \\
&z=0,(x,y)\notin\Omega: \qquad \frac{\partial F_1^{(0)}}{\partial z}=0, \quad F_4^{(0)}=0
\end{aligned} \tag{8.9-13}$$

The boundary value problem of (8.9-12) and (8.9-13) is the Lamb problem appearing in fluid dynamics [21, 22], and the solution is known as

$$\begin{aligned}
F_1^{(0)}(x,y,z)&=\frac{A}{2}\int_\xi^\infty\left[\frac{x^2}{a^2+s}+\frac{y^2}{b^2+s}+\frac{z^2}{s}\right]\frac{ds}{\sqrt{Q(s)}} \\
F_4^{(0)}(x,y,z)&=\frac{B}{2}\int_\xi^\infty\left[\frac{x^2}{a^2+s}+\frac{y^2}{b^2+s}+\frac{z^2}{s}-1\right]\frac{ds}{\sqrt{Q(s)}}
\end{aligned} \tag{8.9-14}$$

where

$$Q(s)=s(a^2+s)(b^2+s), \tag{8.9-15}$$

A and B are unknown constants to be determined and ξ denotes the ellipsoid coordinate. After a lengthy calculation (see, e.g., [3] or Appendix B of this chapter), the unknown constants are determined as

$$A=\frac{-ab^2p}{4C_{13}\gamma_1^2E(k)}, \qquad B=\frac{-ab^2q}{2K_{13}E(k)} \tag{8.9-16}$$

where $E(k)$ is the complete elliptic integral of second kind, and $k^2=(a^2-b^2)/a^2$. So far, the problem in the sense of the zero-order perturbation is solved already. The stress intensity factors of zero-order approximation are

$$\begin{aligned}
K_I^{\parallel}&=\lim_{r_1\to 0}\sigma_{zz}=\frac{p\sqrt{\pi}}{E(k)}\left(\frac{b}{a}\right)^{1/2}(a^2\sin^2\phi+b^2\cos^2\phi)^{1/4} \\
K_I^{\perp}&=\lim_{r_1\to 0}H_{zz}=\frac{q\sqrt{\pi}}{E(k)}\left(\frac{b}{a}\right)^{1/2}(a^2\sin^2\phi+b^2\cos^2\phi)^{1/4}
\end{aligned} \tag{8.9-17}$$

and $\phi = \arctan(y/x)$. For the details of the derivation, please refer to Appendix 2 of this chapter.

A modification to the zero-order approximation can be done. Substituting the zero-order approximate solution into the original equations that have not been approximated yields the one-order approximate solution, e.g.,

$$\begin{aligned} \sigma_{zz}^{(1)} &= -C_{13}\gamma_1^2 \frac{\partial^2}{\partial z^2} F_1^{(0)} + R_2 \frac{\partial}{\partial z} F_4^{(0)}, \\ H_{zz}^{(1)} &= K_1 \frac{\partial}{\partial z} F_4^{(0)} - R_1 \gamma_1^2 \frac{\partial^2}{\partial z^2} F_1^{(0)} \end{aligned} \tag{8.9-18}$$

In this case, the stress components are in phonon–phason coupling already, furthermore, the stress intensity factors of one-order approximation and other quantities can be evaluated.

8.10 Other Crack Problems in One- and Two-dimensional Quasicrystals

Peng and Fan [23] reported the study on circular disk-shaped cracks, which is a special case of the work of Fan and Guo [20].

Liu and Liu *et al.* gave some solutions to crack problems in terms of complex analysis. In particular, the discussion on those of one-dimensional quasicrystals is comparatively comprehensive, which are summarized in thesis [24].

Li *et al.* [25] studied the crack solution of one-dimensional hexagonal quasicrystals under the action of thermal stress.

The exact analytic solutions of quasicrystals provide a basis for the fracture theory of the material, which is summarized in Chapter 15.

8.11 Plastic Zone Around Crack Tip

The previous discussion lets us understand the fact that there is high stress concentration near the crack tip; the maximum value of some stress components is beyond the yielding limit of the material and leads to plastic deformation in the material. That is, a plastic zone appears around the crack tip. How can we estimate the size of the plastic zone and its effects? In classical fracture theory concerning structural materials (or engineering materials), the relevant analysis can be carried out by the theory of plasticity. However, there is no theory for quasicrystalline material so far. In principle, the analysis of plastic deformation for quasicrystals is not available at present.

In Chapter 7, we studied the solutions of dislocations of quasicrystals. Formation of dislocation means the starting of plastic deformation [26]. The movement of dislocations meets an obstacle, and the dislocations will be piled up and the pile-up forms a dislocation group, constructing a plastic zone macroscopically. By using a continuous distribution model of dislocations, the size of the plastic zone can be evaluated. The size of the plastic zone around the crack tip may be big or small. If the size can be compared with the crack size, this case is called the large-scale plastic zone, and the fracture behaviour is dominated by plastic deformation. If the size is small in comparison with the crack size, this case is then called a small-scale plastic zone, and the fracture behaviour is still controlled by elastic deformation. In Chapter 14, we will give some analyses about this.

8.12 Appendix A: Some Derivations in Secition 8.1

Due to similarity to that of Section 8.4, some mathematical details of 8.1 are omitted. However, the complex analysis is very important, which is also the basis of 8.2; here we give some additional details of the derivation.

For simplicity, let us consider boundary value problem (8.1-5) as an example. From (8.1-11), it is known that

$$\left.\begin{aligned} \sigma_{yz} = \sigma_{zy} &= -\frac{1}{2i}[C_{44}(\phi_1' - \overline{\phi_1'}) + R_3(\psi_1' - \overline{\psi_1'})] + \tau_1 \\ H_{zy} &= -\frac{1}{2i}[K_2(\psi_1' - \overline{\psi_1'}) + R_3(\phi_1' - \overline{\phi_1'})] + \tau_2 \end{aligned}\right\} \tag{8.12-1}$$

In the following, denote L as the crack surface, then the second formula of boundary conditions (8.1-5) can be rewritten as

$$\left.\begin{aligned} &C_{44}(\phi_1' - \overline{\phi_1'}) + R_3(\psi_1' - \overline{\psi_1'}) - 2i\tau_1 = 0 \quad t \in L \\ &K_2(\psi_1' - \overline{\psi_1'}) + R_3(\phi_1' - \overline{\phi_1'}) - 2i\tau_2 = 0 \quad t \in L \end{aligned}\right\} \tag{8.12-2}$$

By applying conformal mapping

$$t = \omega(\zeta) = \frac{a}{2}\left(\zeta + \frac{1}{\zeta}\right) \tag{8.12-3}$$

to transform the domain with Griffith crack at the t-plane onto the interior of the unit circle γ at the ζ-plane ($\zeta = \xi + i\eta = \rho e^{i\varphi}$), L corresponds to the unit circle γ (similar to Fig. 8.8).

At the unit circle γ, $\zeta = \sigma \equiv e^{i\varphi}$, $\rho = 1$. Under the mapping (8.12-3) unknown functions $\phi_1(t)$ and $\psi_1(t)$, their derivatives are expressed as

$$\left.\begin{aligned} &\phi_1(t) = \phi_1[\omega(\zeta)] = \phi(\zeta), \quad \psi_1(t) = \psi_1[\omega(\zeta)] = \psi(\zeta) \\ &\phi_1'(t) = \phi'(\zeta)/\omega'(\zeta), \qquad\qquad \psi_1'(t) = \psi'(\zeta)/\omega'(\zeta) \end{aligned}\right\} \tag{8.12-4}$$

Similar to 8.4, the boundary conditions are reduced to

$$\left.\begin{aligned} &\frac{1}{2\pi i}\int_\gamma \frac{\phi'(\sigma)}{\sigma - \zeta}d\sigma - \frac{1}{2\pi i}\int_\gamma \frac{\omega'(\sigma)}{\overline{\omega'(\sigma)}}\overline{\phi'(\sigma)}\frac{d\sigma}{\sigma - \zeta} + \frac{R_3}{C_{44}}\frac{1}{2\pi i}\int_\gamma \frac{\psi'(\sigma)}{\sigma - \zeta}d\sigma \\ &\quad - \frac{R_3}{C_{44}}\frac{1}{2\pi i}\int_\gamma \frac{\omega'(\sigma)}{\overline{\omega'(\sigma)}}\overline{\psi'(\sigma)}\frac{d\sigma}{\sigma - \zeta} = \frac{2i\tau_1}{C_{44}}\frac{1}{2\pi i}\int_\gamma \frac{\omega'(\sigma)}{\sigma - \zeta}d\sigma \\ &\frac{1}{2\pi i}\int_\gamma \frac{\psi'(\sigma)}{\sigma - \zeta}d\sigma - \frac{1}{2\pi i}\int_\gamma \frac{\omega'(\sigma)}{\overline{\omega'(\sigma)}}\overline{\psi'(\sigma)}\frac{d\sigma}{\sigma - \zeta} + \frac{R_3}{K_2}\frac{1}{2\pi i}\int_\gamma \frac{\phi'(\sigma)}{\sigma - \zeta}d\sigma \\ &\quad - \frac{R_3}{K_2}\frac{1}{2\pi i}\int_\gamma \frac{\omega'(\sigma)}{\overline{\omega'(\sigma)}}\overline{\phi'(\sigma)}\frac{d\sigma}{\sigma - \zeta} = \frac{2i\tau_2}{K_2}\frac{1}{2\pi i}\int_\gamma \frac{\omega'(\sigma)}{\sigma - \zeta}d\sigma \end{aligned}\right\} \tag{8.12-5}$$

It is analogue to Section 8.4; there are

$$\left.\begin{array}{l}\dfrac{1}{2\pi i}\displaystyle\int_\gamma \dfrac{\phi'(\sigma)}{\sigma-\zeta}d\sigma=\phi'(\zeta),\ \dfrac{1}{2\pi i}\displaystyle\int_\gamma \dfrac{\omega'(\sigma)}{\overline{\omega'(\sigma)}}\overline{\phi'(\sigma)}\dfrac{d\sigma}{\sigma-\zeta}=0\\ \dfrac{1}{2\pi i}\displaystyle\int_\gamma \dfrac{\psi'(\sigma)}{\sigma-\zeta}d\sigma=\psi'(\zeta),\ -\dfrac{1}{2\pi i}\displaystyle\int_\gamma \dfrac{\omega'(\sigma)}{\overline{\omega'(\sigma)}}\overline{\psi'(\sigma)}\dfrac{d\sigma}{\sigma-\zeta}=0\\ \dfrac{1}{2\pi i}\displaystyle\int_\gamma \dfrac{\omega'(\sigma)}{\sigma-\zeta}d\sigma=\dfrac{a}{2}\end{array}\right\} \tag{a}$$

so the solution of Eq. (8.12-5) is

$$\phi'(\zeta)=ia\frac{K_2\tau_1-R_3\tau_2}{C_{44}K_2-R_3^2},\quad \psi'(\zeta)=ia\frac{C_{44}\tau_2-R_3\tau_1}{C_{44}K_2-R_3^2} \tag{8.12-6}$$

Integrating (8.12-6) yields

$$\phi(\zeta)=ia\frac{K_2\tau_1-R_3\tau_2}{C_{44}K_2-R_3^2}\zeta,\quad \psi(\zeta)=ia\frac{C_{44}\tau_2-R_3\tau_1}{C_{44}K_2-R_3^2}\zeta \tag{8.12-7}$$

The single-valued inversion of conformal mapping (8.12-3) is

$$\zeta=\omega^{-1}(t)=\frac{t}{a}-\sqrt{\left(\frac{t}{a}\right)^2-1} \tag{8.12-8}$$

due to $|t|=\infty$ corresponding to $\zeta=0$, and substituting this into (8.10-8), we obtain

$$\begin{aligned}\phi(\zeta)=\phi(\omega^{-1}(t))=\phi_1(t)&=ia\frac{K_2\tau_1-R_3\tau_2}{C_{44}K_2-R_3^2}\zeta\\ &=ia\frac{K_2\tau_1-R_3\tau_2}{C_{44}K_2-R_3^2}\left(\frac{t}{a}-\sqrt{\left(\frac{t}{a}\right)^2-1}\right)\end{aligned}$$

$$\psi(\zeta) = \psi(\omega^{-1}(t)) = \psi_1(t) = ia\frac{C_{44}\tau_2 - R_3\tau_1}{C_{44}K_2 - R_3^2}\zeta$$

$$= ia\frac{C_{44}\tau_2 - R_3\tau_1}{C_{44}K_2 - R_3^2}\left(\frac{t}{a} - \sqrt{\left(\frac{t}{a}\right)^2 - 1}\right)$$

These are the complex potentials of (8.1-12).

8.13 Appendix B: Some Further Derivation of Solution in Section 8.9

To determine the unknown constants A, B, we must do some complex calculations and need to introduce ellipsoid coordinates and elliptic functions.

The region Ω is defined by an ellipse, and outside it is the region marked by $(Z-\Omega)$. They can also be defined in terms of the ellipsoid coordinates ξ, η, ζ, which are of some roots s of the ellipsoid equation

$$\frac{x^2}{a^2+s} + \frac{y^2}{b^2+s} + \frac{z^2}{s} - 1 = 0$$

in which

$$-a^2 \le \zeta \le -b^2 \le \eta \le 0 \le \xi \le \infty$$

The relations between ellipsoid coordinates ξ, η, ζ and rectilinear coordinates x, y, z are as follows:

$$\left.\begin{aligned} a^2(a^2-b^2)x^2 &= (a^2+\xi)(a^2+\eta)(a^2+\zeta) \\ b^2(b^2-a^2)y^2 &= (b^2+\xi)(b^2+\eta)(b^2+\zeta) \\ a^2b^2z^2 &= \xi\eta\zeta \end{aligned}\right\} \tag{8.13-1}$$

As $\xi = 0$, corresponding to $z = 0$, $(x, y) \in \Omega$, while $\eta = 0$, corresponding to $z = 0$, $(x, y) \in (Z-\Omega)$. So, the boundary conditions (8.9-1) can be rewritten in a more explicit version if using the ellipsoid coordinates:

$$\xi = 0: \quad \sigma_{zz} = -p_0, \sigma_{xy} = \sigma_{yz} = 0 \tag{8.9-1'}$$

$$\eta = 0: \quad u_z = 0, \sigma_{xz} = \sigma_{yz} = 0 \tag{8.9-2'}$$

The derivatives of ellipsoid coordinate ξ with respect to rectilinear coordinates

$$\frac{\partial \xi}{\partial x} = \frac{x}{2h_1^2(a^2+\xi)}, \quad \frac{\partial \xi}{\partial y} = \frac{y}{2h_1^2(b^2+\xi)}, \quad \frac{\partial \xi}{\partial z} = \frac{z}{2\xi h_1^2} \tag{8.13-2}$$

where

$$4h_1^2 Q(\xi) = (\xi - \eta)(\xi - \zeta) \tag{8.13-3}$$

and Q (ξ) is defined by (8.9-15).

The derivative of the first formula of (8.9-14) with respect to z is

$$\frac{\partial F_1^{(0)}}{\partial z} = Az \int_\xi^\infty \frac{ds}{\sqrt{Q(s)}} \tag{8.13-4}$$

For convenience of calculations afterwards, the right-hand side of the above equation can be changed as

$$\frac{\partial F_1^{(0)}}{\partial z} = Az \left[\frac{2}{\sqrt{Q(s)}} - \int_\xi^\infty \frac{(2s + a^2 + b^2)ds}{(a^2+s)(b^2+s)\sqrt{Q(s)}} \right]$$

Putting a substitution

$$\xi = \frac{a^2 cn^2 u}{sn^2 u} = a^2(sn^{-2}u - 1) \tag{8.13-5}$$

where u, cnu, snu, $sn^{-2}u$ are elliptic functions, their definitions and manipulations refer to the introduction hereafter. Inserting (8.13-5) into the above equation, the partial derivative $\partial F_1^{(0)}/\partial z$ is expressed by elliptic functions:

$$\frac{\partial F_1^{(0)}}{\partial z} = \frac{2Az}{ab^2} \left[\frac{snudnu}{cnu} - E(u) \right] \tag{8.13-4'}$$

in which E(u) denotes the complete elliptic integral of the second kind (see the following introduction), which can also be expressed by the integral of the elliptic function $\mathrm{dn}^2 u$, i.e.,

$$E(u) = \int_0^u dn^2\beta d\beta \tag{8.13-6}$$

Taking the derivative of (8.13-4′) with respect to z and applying the formula (8.13-2), one finds that

$$\frac{\partial^2 F_1^{(0)}}{\partial z^2} = A\left\{\frac{2\xi^{1/2}[\xi(a^2b^2-\eta\zeta)-a^2b^2(\eta+\zeta)-(a^2+b^2)\eta\zeta]}{a^2b^2(\xi-\eta)(\xi-\zeta)(a^2+\xi)^{1/2}(b^2+\xi)^{1/2}}\right.$$
$$\left.-\frac{2}{ab^2}\left[E(u)-\frac{\mathrm{snucnu}}{dnu}\right]\right\} \tag{8.13-7}$$

From (8.13-5), as $\xi = 0$, then $u = \pi/2$, and from the succeeded formulas, there is $E(u) = E(k)$, $snucnu/dnu = 0$. Comparison between (8.13-7) and the second formula of (8.9-1) determines the unknown constant A. Similarly, constant B can be determined.

After the determination of constants A and B, function (8.9-14) is determined. Naturally, the stress and displacement fields are completely solved. Of course, some calculations are still complicated.

Substituting (8.9-14) into the expression of normal stress of phonon field, and for $\eta = 0$, there is

$$\begin{cases} \sigma_{zz}(x,y,0) = \dfrac{p_0}{E(k)}\left\{\dfrac{ab^2}{\sqrt{Q(\xi)}} - \left[E(u) - \dfrac{\mathrm{snucnu}}{dnu}\right]\right\} \\ (x,y) \in (\mathrm{Z}-\Omega) \end{cases} \tag{8.13-8}$$

This is the normal stress of phonon field at plane $z = 0$ and outside the crack surface.

We now calculate the stress intensity factor K_{I} according to the definition

$$K_I^{\parallel} = \lim_{r_1\to 0}\sqrt{2\pi r_1}\sigma_{zz}(x,y,0)\Big|_{(x,y)\in(Z-D)} \tag{8.13-9}$$

where r_1 characterizes a geometrical parameter of the crack tip, and $r_1 \ll a$, $r_1 \ll b$ (Fig. 8.13).

The limit process in (8.13-9) can also be expressed through the ellipsoid coordinate $\xi \to 0$. In Fig. 8.13, the location near the crack

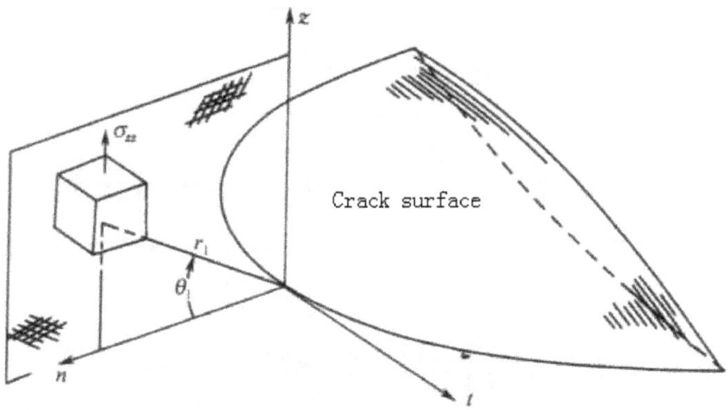

Fig. 8.13. The coordinate system of crack tip.

tip can be described by (8.13-10):

$$\left.\begin{aligned} z &= r_1 \sin\theta_1 \\ \xi &= \frac{2ab}{(\Pi_0)^{1/2}} r_1 \cos^2\frac{\theta_1}{2} \end{aligned}\right\} \tag{8.13-10}$$

and r_1, θ_1 are depicted in the figure; in addition,

$$\Pi_0 = a^2\sin^2\phi + b^2\cos^2\phi$$

where ϕ is the polar angle at any point of the contour of the ellipse (3.11.1) (Fig. 8.14).

As previous pointed out, as $\xi \to 0$, $E(u)$ in (8.13-8) reduces to the complete elliptic integral of the second kind, and $(snu/cnu)/dnu$ tends to zero. The first term presents singularity with order $(2r_1)^{-1/2}$, and the coefficient containing the singular term is

$$\frac{p_0}{E(k)}\Pi_0^{1/4}\left(\frac{b}{a}\right)^{1/2}$$

It follows that according to (8.13-9),

$$K_I^{\|} = \frac{p\sqrt{\pi}}{E(k)}\left(\frac{b}{a}\right)^{1/2}(a^2\sin^2\phi + b^2\cos^2\phi)^{1/4} \tag{8.13-11}$$

which is the first formula of (8.9-17), and the second one is similarly obtained.

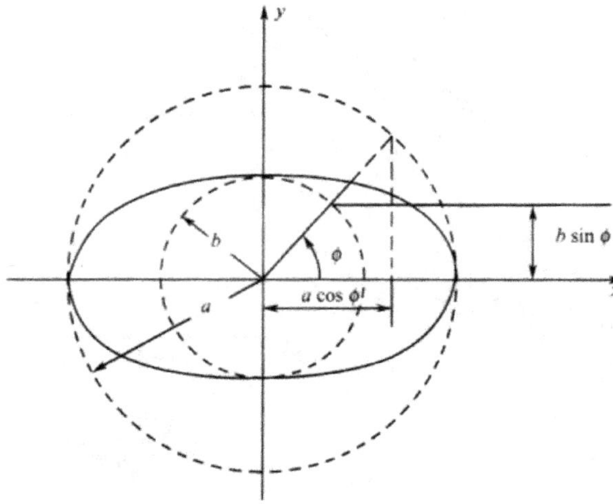

Fig. 8.14. The geometry relation concerning major and minor semi-axes of the elliptic crack.

In the above derivations, some elliptic functions are used and listed as follows:

Denoting

$$u = \int_0^{\Phi} (1 - k^2 \sin^2 t)^{-1/2} dt = F(\phi, k) \tag{a}$$

which is defined as u to be a function (multi-valued function) of $\mathrm{x} = \sin\phi$; in contrast, equation (a) defines ϕ or $\sin\phi$ as a function of u (maybe a multi-valued function). Putting notation

$$\phi = \mathrm{am}u = \mathrm{am}(u, k) \tag{b}$$

means it to be a function of modulus k and argument u. There are the following basic functions:

$$\mathrm{sn}u = \mathrm{sn}(u, k) = \sin(\mathrm{am}u) \tag{c}$$

$$\mathrm{cn}u = \mathrm{cn}(u, k) = \cos(\mathrm{am}u) \tag{d}$$

$$\mathrm{dn}u = \mathrm{dn}(u, k) = \Delta(\mathrm{am}u, k) = [1 - k^2 \sin^2(\mathrm{am}u)]^{1/2} \tag{e}$$

The ranges of variation of these functions are

$$-1 \leq snu \leq 1, -1 \leq cnu \leq 1, k' \leq dnu \leq 1 \tag{f}$$

Apart from that, there are the following functions:

$$\left.\begin{array}{l} \mathrm{ns}u = 1/\mathrm{sn}u, \quad \mathrm{nc}u = 1/\mathrm{cn}u \\ \mathrm{nd}u = 1/\mathrm{dn}u \\ \mathrm{cs}u = \mathrm{cn}u/\mathrm{sn}u, \quad \mathrm{sc}u = \mathrm{sn}u/\mathrm{cn}u \\ \mathrm{sd}u = \mathrm{sn}u/\mathrm{dn}u \\ \mathrm{ds}u = \mathrm{dn}u/\mathrm{sn}u, \quad \mathrm{dc}u = \mathrm{dn}u/\mathrm{cn}u \\ \mathrm{cd}u = \mathrm{cn}u/\mathrm{dn}u \end{array}\right\} \tag{g}$$

The above functions are named Jacobi elliptic functions.

At $u = 0$, put

$$\mathrm{sn}0 = 0, \mathrm{cn}0 = \mathrm{dn}0 = 1 \tag{h}$$

Besides, there is

$$\mathrm{cn}K = 0 \tag{i}$$

Periods of elliptic function $sn(u,k)$ are $4K$, $i2K'$; periods of $cn(u,k)$ are $4K$, $2K + i2K'$; periods of $dn(u,k)$ are $2K$, $i4K'$.

Other properties of elliptic functions are listed as a part which is concerned in the previous derivations, e.g.,

$$\mathrm{sn}^2u + cn^2u = 1 \tag{j}$$

$$k^2\mathrm{sn}^2u + \mathrm{dn}^2u = 1 \tag{k}$$

$$\mathrm{dn}^2u - k^2\mathrm{cn}^2u = k'^2 \tag{l}$$

$$k'^2\mathrm{sn}^2u + \mathrm{cn}^2u = \mathrm{dn}^2u \tag{m}$$

References

[1] Hu C Z, Yang W H, Wang R H and Ding D H, 1997, Symmetry and physical properties of quasicrystals, *Adv. Phys.*, **17**(4), 345–376 (in Chinese).

[2] Meng X M, Tong B Y and Wu Y Q, 1994, Mechanical properties of quasicrystals $Al_{65}Cu_{20}Co_{15}$, *Acta. Metal. Sinica*, **30**(2), 61–64 (in Chinese).

[3] Fan T Y 2014, *Foundation of Defect and Fracture Theory of Solid and Soft Matter*, Beijing, Science Press (in Chinese).

[4] Fan T Y 1991, Exact analytic solutions of stationary and fast propagating cracks in a strip. *Science in China*, A. **34**(5), 560–569; Fan T Y, 1999, *Mathematical Theory of Elasticity of Quasicrystals*, Beijing, Beijing Institute of Technology Press (in Chinese).

[5] Li L H and Fan T Y, 2008, Exact solutions of two semi-infinite collinear cracks in a strip of one-dimensional hexagonal quasicrystal, *Appl. Math. Comput.*, **196**(1), 1–5.
[6] Shen D W and Fan T Y, 2003, Exact solutions of two semi-infinite collinear cracks in a strip, *Eng. Fracture Mech.*, **70**(8), 813–822.
[7] Li X F, Fan T Y and Sun Y F, 1999, A decagonal quasicrystal with a Griffith crack *Phil. Mag.* A, **79**(8), 1943–1952.
[8] Li L H and Fan T Y, 2007, Complex function method for solving notch problem of point group 10, $\overline{10}$ two-dimensional quasicrystals based on the stress potential function, *J. Phys.: Condens. Matter*, **18**(47), 10631–10641.
[9] Liu G T and Fan T Y, 2003, The complex method of the plane elasticity in 2D quasicrystals point group $10mm$ ten-fold rotation symmetry notch problems, *Science in China*, E, **46**(3), 326–336.
[10] Li X F, 1999, Elastic fields of dislocations and cracks in one- and two-dimensional quasicrystals, Dissertation, Beijing Institute of Technology (in Chinese).
[11] Zhou W M and Fan T Y, 2001, Plane elasticity problem of two-dimensional octagonal quasicrystal and crack problem, *Chin. Phys.*, **10**(8), 743–747.
[12] Zhou W M, 2000, Mathematical analysis of elasticity and defects of two- and three-dimensional quasicrystals, Dissertation, Beijing Institute of Technology (in Chinese).
[13] Li L H, 2008, Study on complex function method and analytic solutions of elasticity of quasicrystals, Dissertation, Beijing Institute of Technology (in Chinese).
[14] Fan T Y and Tang Z Y, 2014, Bending problem for a two-dimensional quasicrystal with an elliptic notch and nonlinear analysis, unpublished work.
[15] Muskhelishvili N I, 1953, *Some Basic Problems of the Mathematical Theory of Elasticity*, Noordhoff Ltd, Groningen.
[16] Lokchine M A, 1930, Sur l'influence d'un trou elliptique dans la poutre qui eprouve une flexion, *C.R.Paris*, **190**, 1178–1179.
[17] Fan T Y and Tang Z Y, 2014, Crack solution of a strip in a two-dimensional quasicrystal, unpublished work.
[18] Fan T Y, Yang X C and Li H X, 1998, Exact analytic solution for a finite width strip with a single crack, *Chin. Phys. Lett.*, 18(1), 18–21.
[19] Li W, 2011, Analytic solutions of a finite width strip with a single edge crack of two-dimensional quasicrystals, *Chin. Phys. B*, 20, 116201.
[20] Fan T Y and Guo R P, 2013, Three-dimensional elliptic crack in one-dimensional hexagonal quasicrystals, unpublished work.
[21] Lamb H, Hydrodynamics, 1933, Dover, New York.

[22] Green A E and Sneddon I N, 1950, The distribution of stress in the neighbourhood of a at elliptic crack in an elastic solid, *Proc. Camb. Phil. Soc.*, 46(2), 159–163.

[23] Peng Y Z and Fan T Y, 2000, Elastic theory of 1D quasiperiodic stacking of 2D crystals, *J. Phys.: Condens. Matter*, **12**(45), 9381–9387.

[24] Liu G T, 2004, Elasticity and defects of quasicrystals and auxiliary function method of nonlinear evolution equations, Dissertation, Beijing Institute of Technology (in Chinese).

[25] Li X Y, Fundamental solutions of a penny shaped embedded crack and half-infinite plane crack in infinite space of one-dimensional hexagonal quasicrystals under thermal loading, *Proc. Roy. Soc. A*, **469**, 20130023, 2013.

[26] Messerschmidt U, 2010, *Dislocation Dynamics during Plastic Deformation*, Chapter 10, Springer-Verlag, Heidelberg.

Chapter 9

Theory of Elasticity of Three-dimensional Quasicrystals and Their Applications

In Chapters 5–8, we discussed the theories of elasticity of one- and two-dimensional quasicrystals and their applications. In this chapter, the theory and applications of elasticity of three-dimensional quasicrystals are dealt with. The three-dimensional quasicrystals include icosahedral quasicrystals and cubic quasicrystals. In all individual quasicrystals observed to date, there are almost half icosahedral quasicrystals, so they play the central role in this kind of solid. This suggests the major importance of elasticity of icosahedral quasicrystals in the study of the mechanical behaviour of quasicrystalline materials.

There are some polyhedrons with the icosahedral symmetry: one of them is shown in Fig. 9.1, which consists of 20 right triangles and contains 12 five fold symmetric axes $A5$, 20 three fold symmetric axes $A3$ and 30 two fold symmetric axes $A2$. One of the diffraction patterns is shown in Fig. 3.1, and the stereographic structure of one of the icosahedral point groups is also depicted in Fig. 3.1.

The elasticity of icosahedral quasicrystals was studied immediately after the discovery of the structure, which is the pioneering work of the field, and promoted the development of the discipline of research. The outlook about this was figured out in Chapter 4, in which the contribution of pioneers such as P. Bak *et al.* was introduced. Afterwards, Ding *et al.* [1] set up the physical framework

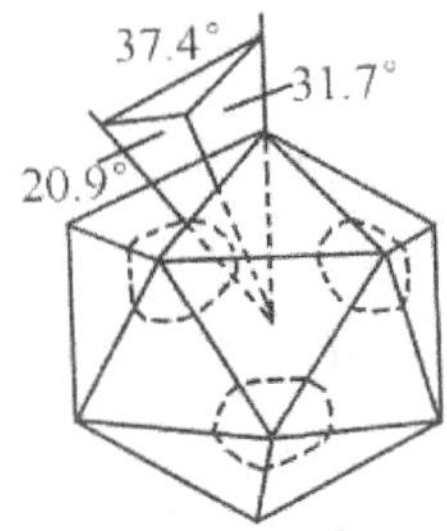

Fig. 9.1. Icosahedral.

of elasticity of icosahedral quasicrystals; they [2] also summarized the basic relationship of elasticity of cubic quasicrystals. In terms of the Green function method, Yang *et al.* [3] gave an approximate solution on dislocation for a special case, i.e., the phonon–phason decoupled plane elasticity of icosahedral quasicrystal. In this chapter, we mainly discuss the general theory of elasticity of icosahedral quasicrystals and its application; in addition, those for cubic quasicrystals are also concerned. We focus on the mathematical theory of the elasticity and the analytic solutions. Because of the large number of field variables and field equations involving the elasticity of these two kinds of three-dimensional quasicrystals, the solution presents tremendous difficulty. We continue to develop the decomposition and superposition procedure adopted in the previous chapters; this can reduce the number of field variables and field equations, and three-dimensional elasticity can be simplified to two-dimensional elasticity to solve cases with important practical applications. The introduction of displacement potentials or stress potentials [4, 5] can further simplify the problems. In this chapter, some systematic and direct methods of mathematical physics and function theory have been developed, and a series of analytic solutions are constructed, which will be included in this chapter. Because the calculations are very complex, we would like to introduce them in as much detail as possible in order to facilitate comprehension of the text.

9.1 Basic Equations of Elasticity of Icosahedral Quasicrystals

The equations of deformation geometry are

$$\varepsilon_{ij} = \frac{1}{2}\left(\frac{\partial u_i}{\partial x_j} + \frac{\partial u_j}{\partial x_i}\right), \quad w_{ij} = \frac{\partial w_i}{\partial x_j} \tag{9.1-1}$$

which are similar in form to those given in previous chapters, but here, u_i and w_i have six components and ε_{ij} and w_{ij} have 15 components in total.

The equilibrium equations are as follows:

$$\frac{\partial \sigma_{ij}}{\partial x_j} = 0, \quad \frac{\partial H_{ij}}{\partial x_j} = 0 \tag{9.1-2}$$

which are also similar in form to those listed in the previous chapters, however here adding σ_{ij} and H_{ij} gives 15 stress components.

Between the stresses and strains, there is the generalized Hooke's law such as

$$\sigma_{ij} = C_{ijkl}\varepsilon_{kl} + R_{ijkl}w_{kl} \quad H_{ij} = R_{klij}\varepsilon_{kl} + K_{ijkl}w_{kl} \tag{9.1-3}$$

in which the phonon elastic constants are described by

$$C_{ijkl} = \lambda\delta_{ij}\delta_{kl} + \mu(\delta_{ik}\delta_{jl} + \delta_{il}\delta_{jk}) \tag{9.1-4}$$

where λ and μ (= G in some references) are the Lamé constants.

If the strain components are arranged as a vector according to the order

$$[\varepsilon_{ij}, w_{ij}] = [\varepsilon_{11}\varepsilon_{22}\varepsilon_{33}\varepsilon_{23}\varepsilon_{31}\varepsilon_{12}w_{11}w_{22}w_{33}w_{23}w_{32}w_{12}w_{32}w_{13}w_{21}] \tag{9.1-5$'$}$$

and the stress components are also arranged according to the same order, i.e.,

$$[\sigma_{ij}, H_{ij}] = [\sigma_{11}\sigma_{22}\sigma_{33}\sigma_{23}\sigma_{31}\sigma_{12}H_{11}H_{22}H_{33}H_{23}H_{32}H_{12}H_{32}H_{13}H_{21}] \tag{9.1-5$''$}$$

then phason and phonon–phason coupling elastic constants can be expressed by the matrixes of [K] and [R]

$$[K] = \begin{bmatrix} K_1 & 0 & 0 & 0 & K_2 & 0 & 0 & K_2 & 0 \\ 0 & K_1 & 0 & 0 & -K_2 & 0 & 0 & K_2 & 0 \\ 0 & 0 & K_2 + K_1 & 0 & 0 & 0 & 0 & 0 & 0 \\ 0 & 0 & 0 & K_1 - K_2 & 0 & K_2 & 0 & 0 & -K_2 \\ K_2 & -K_2 & 0 & 0 & K_1 - K_2 & 0 & 0 & 0 & 0 \\ 0 & 0 & 0 & K_2 & 0 & K_1 & -K_2 & 0 & 0 \\ 0 & 0 & 0 & 0 & 0 & -K_2 & K_1 - K_2 & 0 & -K_2 \\ K_2 & K_2 & 0 & 0 & 0 & 0 & 0 & K_1 - K_2 & 0 \\ 0 & 0 & 0 & -K_2 & 0 & 0 & -K_2 & 0 & K_1 \end{bmatrix}$$

$$[R] = R \begin{bmatrix} 1 & 1 & 1 & 0 & 0 & 0 & 0 & 1 & 0 \\ -1 & -1 & 1 & 0 & 0 & 0 & 0 & -1 & 0 \\ 0 & 0 & -2 & 0 & 0 & 0 & 0 & 0 & 0 \\ 0 & 0 & 0 & 0 & 0 & -1 & 1 & 0 & -1 \\ 1 & -1 & 0 & 0 & 1 & 0 & 0 & 0 & 0 \\ 0 & 0 & 0 & -1 & 0 & -1 & 0 & 0 & 1 \\ 0 & 0 & 0 & 0 & 0 & -1 & 1 & 0 & -1 \\ 1 & -1 & 0 & 0 & 1 & 0 & 0 & 0 & 0 \\ 0 & 0 & 0 & -1 & 0 & -1 & 0 & 0 & 1 \end{bmatrix} \tag{9.1-6}$$

Equations (9.1-1)–(9.1-3) are the basic equations of elasticity of icosahedral quasicrystals; there are 36 field equations in total, and the number of field variables is also 36. It is consistent and solvable mathematically.

Due to the huge number of field variables and field equations, the mathematical solution is highly complex. One way to solve the elasticity problem is to reduce the number of field variables and field equations mentioned above. For this purpose, we can utilize the eliminating element method in classical mathematical physics.

Based on the matrix expression of the generalized Hooke's law of (4.5-3) in Chapter 4, i.e.,

$$\begin{bmatrix} \sigma_{ij} \\ H_{ij} \end{bmatrix} = \begin{bmatrix} [C] & [R] \\ [R]^T & [K] \end{bmatrix} \begin{bmatrix} \varepsilon_{ij} \\ w_{ij} \end{bmatrix}$$

where

$$\begin{bmatrix} \sigma_{ij} \\ H_{ij} \end{bmatrix} = [\sigma_{ij}, H_{ij}]^T$$

$$\begin{bmatrix} \varepsilon_{ij} \\ w_{ij} \end{bmatrix} = [\varepsilon_{ij}, w_{ij}]^T$$

one has the explicit relationship between stresses and strains as follows:

$$\begin{aligned}
\sigma_{xx} &= \lambda\theta + 2\mu\varepsilon_{xx} + R(w_{xx} + w_{yy} + w_{zz} + w_{xz}) \\
\sigma_{yy} &= \lambda\theta + 2\mu\varepsilon_{yy} - R(w_{xx} + w_{yy} - w_{zz} + w_{xz}) \\
\sigma_{zz} &= \lambda\theta + 2\mu\varepsilon_{zz} - 2Rw_{zz} \\
\sigma_{yz} &= 2\mu\varepsilon_{yz} + R(w_{zy} - w_{xy} - w_{yx}) = \sigma_{zy} \\
\sigma_{zx} &= 2\mu\varepsilon_{zx} + R(w_{xx} - w_{yy} - w_{zx}) = \sigma_{xz} \\
\sigma_{xy} &= 2\mu\varepsilon_{xy} + R(w_{yx} - w_{yz} - w_{xy}) = \sigma_{yx} \\
H_{xx} &= R(\varepsilon_{xx} - \varepsilon_{yy} + 2\varepsilon_{zx}) + K_1 w_{xx} + K_2(w_{zx} + w_{xz}) \\
H_{yy} &= R(\varepsilon_{xx} - \varepsilon_{yy} - 2\varepsilon_{zx}) + K_1 w_{yy} + K_2(w_{xz} - w_{zx}) \\
H_{zz} &= R(\varepsilon_{xx} + \varepsilon_{yy} - 2\varepsilon_{zz}) + (K_1 + K_2) w_{zz}
\end{aligned}$$

$$
\begin{aligned}
H_{yz} &= -2R\varepsilon_{xy} + (K_1 - K_2)w_{yz} + K_2(w_{xy} - w_{yx}) \\
H_{zx} &= 2R\varepsilon_{zx} + (K_1 - K_2)w_{zx} + K_2(w_{xx} - w_{yy}) \\
H_{xy} &= -2R(\varepsilon_{yz} + \varepsilon_{xy}) + K_1 w_{xy} + K_2(w_{yz} - w_{zy}) \\
H_{zy} &= 2R\varepsilon_{yz} + (K_1 - K_2)w_{zy} - K_2(w_{xy} + w_{yx}) \\
H_{xz} &= R(\varepsilon_{xx} - \varepsilon_{yy}) + K_2(w_{xx} + w_{yy}) + (K_1 - K_2)w_{xz} \\
H_{yx} &= 2R(\varepsilon_{xy} - \varepsilon_{yz}) + K_1 w_{yx} - K_2(w_{yz} + w_{zy})
\end{aligned}
\tag{9.1-7}
$$

where $\theta = \varepsilon_{xx} + \varepsilon_{yy} + \varepsilon_{zz}$ denotes the volume strain and ε_{ij} and w_{ij} are defined by (9.1-1). This explicit expression was first given by Ding *et al.* [1].

Substituting (9.1-7) into (9.1-2) yields one form of the final governing equations — the equilibrium equations in terms of displacements are as follows:

$$
\begin{aligned}
&\mu\nabla^2 u_x + (\lambda+\mu)\frac{\partial}{\partial x}\nabla\cdot\mathrm{u} \\
&\quad + R\left(\frac{\partial^2 w_x}{\partial x^2} + 2\frac{\partial^2 w_x}{\partial x\partial z} - \frac{\partial^2 w_x}{\partial y^2} + 2\frac{\partial^2 w_y}{\partial x\partial y} - 2\frac{\partial^2 w_y}{\partial y\partial z} + 2\frac{\partial^2 w_z}{\partial x\partial z}\right) = 0 \\
&\mu\nabla^2 u_y + (\lambda+\mu)\frac{\partial}{\partial y}\nabla\cdot\mathrm{u} \\
&\quad + R\left(-2\frac{\partial^2 w_x}{\partial x\partial y} - 2\frac{\partial^2 w_x}{\partial y\partial z} + \frac{\partial^2 w_y}{\partial x^2} - 2\frac{\partial^2 w_y}{\partial x\partial z} - \frac{\partial^2 w_y}{\partial y^2} + 2\frac{\partial^2 w_z}{\partial y\partial z}\right) = 0 \\
&\mu\nabla^2 u_z + (\lambda+\mu)\frac{\partial}{\partial z}\nabla\cdot\mathrm{u} \\
&\quad + R\left(\frac{\partial^2 w_x}{\partial x^2} - \frac{\partial^2 w_x}{\partial y^2} - 2\frac{\partial^2 w_y}{\partial x\partial y} + \frac{\partial^2 w_z}{\partial x^2} + \frac{\partial^2 w_z}{\partial y^2} - 2\frac{\partial^2 w_z}{\partial z^2}\right) = 0 \\
&K_1\nabla^2 w_x + K_2\left(2\frac{\partial^2 w_x}{\partial x\partial z} - \frac{\partial^2 w_x}{\partial z^2} + 2\frac{\partial^2 w_y}{\partial y\partial z} + \frac{\partial^2 w_z}{\partial x^2} - \frac{\partial^2 w_z}{\partial y^2}\right) \\
&\quad + R\left(\frac{\partial^2 u_x}{\partial x^2} - \frac{\partial^2 u_x}{\partial y^2} + 2\frac{\partial^2 u_x}{\partial x\partial z} - 2\frac{\partial^2 u_y}{\partial x\partial y} - 2\frac{\partial^2 u_y}{\partial y\partial z} + \frac{\partial^2 u_z}{\partial x^2} - \frac{\partial^2 u_z}{\partial y^2}\right) = 0 \\
&K_1\nabla^2 w_y + K_2\left(2\frac{\partial^2 w_x}{\partial y\partial z} - 2\frac{\partial^2 w_y}{\partial x\partial z} - 2\frac{\partial^2 w_z}{\partial x\partial y} - \frac{\partial^2 w_y}{\partial z^2}\right) \\
&\quad + R\left(2\frac{\partial^2 u_x}{\partial x\partial y} - 2\frac{\partial^2 u_x}{\partial y\partial z} + \frac{\partial^2 u_y}{\partial x^2} - \frac{\partial^2 u_y}{\partial y^2} - 2\frac{\partial^2 u_y}{\partial x\partial z} - 2\frac{\partial^2 u_z}{\partial x\partial y}\right) = 0
\end{aligned}
$$

$$(K_1 - K_2)\nabla^2 w_z + K_2\left(\frac{\partial^2 w_x}{\partial x^2} - \frac{\partial^2 w_x}{\partial y^2} - 2\frac{\partial^2 w_y}{\partial y \partial x} + 2\frac{\partial^2 w_z}{\partial z^2}\right)$$

$$+R\left(2\frac{\partial^2 u_x}{\partial x \partial z} + 2\frac{\partial^2 u_x}{\partial x \partial z} + \frac{\partial^2 u_z}{\partial x^2} + \frac{\partial^2 u_z}{\partial y^2} - 2\frac{\partial^2 u_z}{\partial z^2}\right) = 0 \qquad (9.1\text{-}8)$$

where $\nabla^2 = \frac{\partial^2}{\partial x^2} + \frac{\partial^2}{\partial y^2} + \frac{\partial^2}{\partial z^2}$, $\nabla \cdot u = \frac{\partial u_x}{\partial x} + \frac{\partial u_y}{\partial y} + \frac{\partial u_z}{\partial z}$.

Equations (9.1-8) are six partial differential equations of second order on displacements u_i and w_i. So, the number of field variables and the number of field equations are reduced already. But obtaining a solution is still very difficult; one of the reasons is that the boundary conditions for quasicrystals are much more complicated than those of the classical theory of elasticity. In the subsequent sections, we will make a great effort to solve some complex boundary value problems through different approaches.

It is obvious that the material constants of λ, μ, K_1, K_2 and R are very important for the stress analysis for different icosahedral quasicrystals, which are experimentally measured through various methods (e.g., X-ray diffraction and neutron scattering) and listed in Tables 9.1–9.3 respectively as follows.

It is needed to point out that Eq. (9.1-8) is not the only form of final governing equation of elasticity of icosahedral quasicrystals, there are other forms which are discussed in Section 9.5.

Table 9.1. Phonon elastic constants of various icosahedral quasicrystals.

Alloys	λ	$\mu(G)$	B	v	Refs.
Al–Li–Cu	30	35	53	0.23	[6]
Al–Li–Cu	30.4	40.9	57.7	0.213	[7]
Al–Cu–Fe	59.1	68.1	104	0.213	[8]
Al–Cu–Fe–Ru	48.4	57.9	87.0	0.228	[8]
Al–Pd–Mn	74.9	72.4	123	0.254	[8]
Al–Pd–Mn	74.2	70.4	121	0.256	[9]
Ti–Zr–Ni	85.5	38.3	111	0.345	[10]
Cu–Yh	35.28	25.28	52.13	0.2913	[11]
Zn–Mg–Y	33.0	46.5	64.0	0.208	[12]

Note: The measurement unit of λ, μ and B is GPa, $B = (3\lambda + 2\mu)/3$ represents the bulk modulus, and $v = \lambda/2(\lambda + \mu)$ Poisson's ratio, respectively.

Table 9.2. Phason elastic constants of various icosahedral quasicrystals.

Alloys	Source	Meas. temp.	K_1 (MPa)	K_2 (MPa)	Refs.
Al–Pd–Mn	X-ray	R.T.	43	−22	[13]
Al–Pd–Mn	Neutron	R.T.	72	−37	[13]
Al–Pd–Mn	Neutron	1043 K	125	−50	[13]
Zn–Mg–Sc	X-ray	R.T.	300	−45	[14]

Table 9.3. Phonon–phason coupling elastic constant of various icosahedral quasicrystals.

Alloys	Source	R	Refs.
Mg–Ga–Al–Zn	X-ray	-0.04μ	[15]
Al–Cu–Fe	X-ray	0.004μ	[15]

9.2 Anti-plane Elasticity of Icosahedral Quasicrystals and Problem of Interface of Quasicrystal–Crystal

People can find that Eq. (9.1-8) are very complex, but they can be simplified for some meaningful cases physically. One of them is the so-called anti-plane case where the non-zero displacements are only u_z and w_z, and the other displacements vanish. In particular, these two displacements and relevant strains and stresses are independent of the coordinate x_3 (or z). If there is a Griffith crack along the x-axis (see Fig. 9.2) or a straight dislocation line along the direction, etc., in addition, the applied external fields are independent from variable z, so the field variables and field equations are free from the coordinate z for the configuration:

$$\frac{\partial}{\partial x_3}\left(=\frac{\partial}{\partial z}\right)=0 \tag{9.2-1}$$

Because there are only two components u_z and w_z and the others have vanished, the corresponding strains are only

$$\varepsilon_{yz}=\varepsilon_{zy}=\frac{1}{2}\frac{\partial u_z}{\partial y},\quad \varepsilon_{xz}=\varepsilon_{zx}=\frac{1}{2}\frac{\partial u_z}{\partial x},\quad w_{zy}=\frac{\partial w_z}{\partial y},\quad w_{zx}=\frac{\partial w_z}{\partial x} \tag{9.2-2}$$

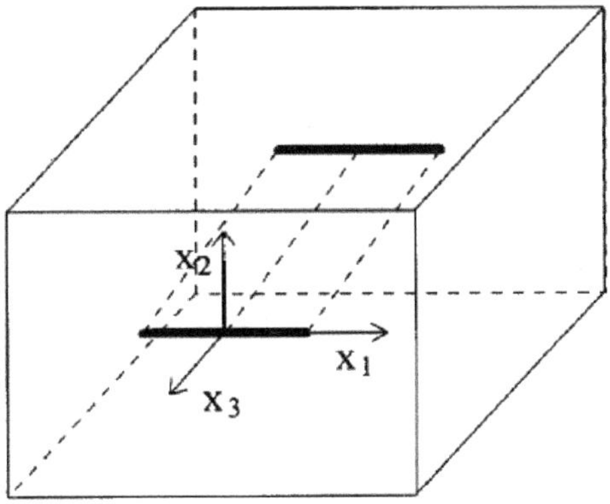

Fig. 9.2. One of the configurations of plane or anti-plane elasticity.

From the formulas listed in Section 9.1, the non-zero stress components are

$$\begin{aligned}
\sigma_{xz} &= \sigma_{zx} = 2\mu\varepsilon_{xz} + Rw_{zx} \\
\sigma_{yz} &= \sigma_{zy} = 2\mu\varepsilon_{yz} + Rw_{zy} \\
H_{zx} &= (K_1 - K_2)w_{zx} + 2R\varepsilon_{xz} \\
H_{zy} &= (K_1 - K_2)w_{zy} + 2R\varepsilon_{yz} \\
H_{xx} &= 2R\varepsilon_{xz} + K_2 w_{zx} \\
H_{yy} &= -2R\varepsilon_{xz} - K_2 w_{zx} \\
H_{xy} &= -2R\varepsilon_{yz} - K_2 w_{zy} \\
H_{yx} &= -2R\varepsilon_{yz} - K_2 w_{zy}
\end{aligned} \tag{9.2-3}$$

and the equilibrium equations stand for

$$\left.\begin{aligned}
\frac{\partial \sigma_{zx}}{\partial x} + \frac{\partial \sigma_{zy}}{\partial y} = 0, \quad \frac{\partial H_{zx}}{\partial x} + \frac{\partial H_{zy}}{\partial y} = 0 \\
\frac{\partial H_{xx}}{\partial x} + \frac{\partial H_{xy}}{\partial y} = 0, \quad \frac{\partial H_{yx}}{\partial x} + \frac{\partial H_{yy}}{\partial y} = 0
\end{aligned}\right\} \tag{9.2-4}$$

The problem described in Eqs. (9.2-2)–(9.2-4) is an anti-plane elasticity problem; we have the final governing equations

$$\nabla_1^2 u_z = 0, \quad \nabla_1^2 w_z = 0 \tag{9.2-5}$$

where $\nabla_1^2 = \frac{\partial^2}{\partial x^2} + \frac{\partial^2}{\partial y^2}$.

One can see that Eq. (9.2-5) is similar to (5.2-11), which can be solved in a procedure similar to that adopted in Chapters 5, 7 and 8.

As an example of the solution of an anti-plane elasticity of icosahedral quasicrystals, we discuss the interface problem between centre-body cubic crystals and icosahedral quasicrystals.

The physical model is similar to that proposed in Section 7.6, i.e., the icosahedral quasicrystal is located in the upper half-space $y > 0$, whose governing equations are listed above, while the centre-body cubic crystal lies in the lower space $y < 0$ with finite thickness h (refer to Fig. 7.1 in Chapter 7) and is governed by the following equation:

$$\nabla^2 u_z^{(c)} = 0 \tag{9.2-6}$$

with the following stress–strain relations:

$$\sigma_{zy}^{(c)} = \sigma_{yz}^{(c)} = 2\mu^{(c)}\varepsilon_{yz}^{(c)}, \quad \sigma_{zx}^{(c)} = \sigma_{xz}^{(c)} = 2\mu^{(c)}\varepsilon_{xz}^{(c)}$$

for crystals, in which $\varepsilon_{ij}^{(c)} = (\partial u_i^{(c)}/\partial x_j + \partial u_j^{(c)}/\partial x_i)/2, \mu^{(c)} = C_{44}^{(c)}$.

After Fourier transform, the solution of (9.2-6) is very easy to obtain such that

$$\begin{aligned} u_z(x,y) &= \frac{1}{2\pi}\int_{-\infty}^{\infty} A(\xi)e^{-|\xi|y-i\xi x}d\xi, \\ w_z(x,y) &= \frac{1}{2\pi}\int_{-\infty}^{\infty} B(\xi)e^{-|\xi|y-i\xi x}d\xi \end{aligned} \tag{9.2-7}$$

and the relevant stresses, e.g.,

$$\begin{aligned} \sigma_{zy} &= -2\mu\frac{1}{2\pi}\int_{-\infty}^{\infty} |\xi|A(\xi)e^{-|\xi|y-i\xi x}d\xi - R\frac{1}{2\pi}\int_{-\infty}^{\infty} |\xi|B(\xi)e^{-|\xi|y-i\xi x}d\xi \\ H_{zy} &= -R\frac{1}{2\pi}\int_{-\infty}^{\infty} |\xi|A(\xi)e^{-|\xi|y-i\xi x}d\xi \\ &\quad - (K_1 - K_2)\frac{1}{2\pi}\int_{-\infty}^{\infty} |\xi|B(\xi)e^{-|\xi|y-i\xi x}d\xi \end{aligned} \tag{9.2-8}$$

in which $y > 0$ and $A(\xi)$ and $B(\xi)$ are arbitrary functions of ξ to be determined.

According to the second boundary condition, at the interface,

$$y = 0, -\infty < x < \infty : \quad \sigma_{zy} = \tau f(x) + ku(x), \quad H_{zy} = 0 \quad (9.2\text{-}9)$$

where

$$k = \frac{\mu^{(c)}}{h} \quad (9.2\text{-}10)$$

and the relation between the two unknown functions is

$$B(\xi) = -\frac{R}{K_1 - K_2} A(\xi) \quad (9.2\text{-}11)$$

From (9.2-7) and (9.2-8), we have

$$\hat{\sigma}_{zy} = -\left(2\mu + \frac{R^2}{K_1 - K_2}\right) A(\xi)\,|\xi|$$

and from the first one of condition (9.2-9), one determines the unknown function

$$A(\xi) = -\frac{\tau \hat{f}(\xi)}{\left(\frac{R^2}{K_1 - K_2} + \mu\right)|\xi| + k} \quad (9.2\text{-}12)$$

So,

$$B(\xi) = \frac{R\tau \hat{f}(\xi)}{(K_1 - K_2)\left[\left(\frac{R^2}{K_1 - K_2} + \mu\right)|\xi| + k\right]} \quad (9.2\text{-}13)$$

in which τ and k are defined by (9.2-9) and (9.2-10), respectively. Thus, the problem is solved. The phason strain field can be determined as

$$w_{zy}(x, y) = -R\tau \frac{1}{2\pi} \int_{-\infty}^{\infty} \frac{|\xi|\, \hat{f}(\xi)}{(R^2 - \mu(K_1 - K_2))\,|\xi| + k(K_1 - K_2)} e^{-|\xi| y - i\xi x} d\xi$$

$$w_{zx}(x, y) = iR\tau \frac{1}{2\pi} \int_{-\infty}^{\infty} \frac{\xi \hat{f}(\xi)}{(R^2 - \mu(K_1 - K_2))\,|\xi| + k(K_1 - K_2)} e^{-|\xi| y - i\xi x} d\xi$$

$$(9.2\text{-}14)$$

Note that $y > 0$. The integrals in (9.2-14) for some cases can be evaluated through the residual theorem introduced in the Major Appendix of this book.

As the first example, assume that $f(x) = 1$, as $-a/2 < x < a/2$, and $f(x) = 0$, as $x < -a/2$ and $x > a/2$, so $\hat{f}(\xi) = \frac{2}{\xi}\sin\left(\frac{a}{2}\xi\right)$, then from solution (9.2-14), we obtain

$$\begin{aligned}
w_{zy}(x,y) &= \frac{a}{h}\frac{\mu^{(c)}R(K_1-K_2)\tau}{[\mu(K_1-K_2)-R^2]^2}\sin\left(\frac{1}{2}\frac{\mu^{(c)}(K_1-K_2)}{\mu(K_1-K_2)-R^2}\frac{a}{h}\right)\\
&\quad\times\exp\left(-\frac{\mu^{(c)}(K_1-K_2)}{\mu(K_1-K_2)-R^2}\frac{y}{h}\right)\cos\left[\frac{\mu^{(c)}(K_1-K_2)}{\mu(K_1-K_2)-R^2}\frac{x}{h}\right]\\
w_{zx}(x,y) &= \frac{a}{h}\frac{\mu^{(c)}R(K_1-K_2)\tau}{[\mu(K_1-K_2)-R^2]^2}\sin\left(\frac{1}{2}\frac{\mu^{(c)}(K_1-K_2)}{\mu(K_1-K_2)-R^2}\frac{a}{h}\right)\\
&\quad\times\exp\left(-\frac{\mu^{(c)}(K_1-K_2)}{\mu(K_1-K_2)-R^2}\frac{y}{h}\right)\sin\left[\frac{\mu^{(c)}(K_1-K_2)}{\mu(K_1-K_2)-R^2}\frac{x}{h}\right]
\end{aligned} \tag{9.2-15}$$

in which we have $k = \mu^{(c)}/h$ and have used the normalized expression, i.e., $x/h, y/h$. Then, considering the second example $f(x) = \delta(x)$, the integrals (9.2-14) will be

$$\begin{aligned}
w_{zy}(x,y) &= \frac{\mu^{(c)}R(K_1-K_2)\tau}{[\mu(K_1-K_2)-R^2]^2}\\
&\quad\times\exp\left(-\frac{\mu^{(c)}(K_1-K_2)}{\mu(K_1-K_2)-R^2}\frac{y}{h}\right)\sin\left[\frac{\mu^{(c)}(K_1-K_2)}{\mu(K_1-K_2)-R^2}\frac{x}{h}\right]\\
w_{zx}(x,y) &= \frac{\mu^{(c)}R(K_1-K_2)\tau}{[\mu(K_1-K_2)-R^2]^2}\\
&\quad\times\exp\left(-\frac{\mu^{(c)}(K_1-K_2)}{\mu(K_1-K_2)-R^2}\frac{y}{h}\right)\cos\left[\frac{\mu^{(c)}(K_1-K_2)}{\mu(K_1-K_2)-R^2}\frac{x}{h}\right]
\end{aligned} \tag{9.2-16}$$

The detail of the evaluation is given in the Major Appendix of this book.

The results are quite interesting. In the first example, the phason strain field is dominated by the elastic constants μ, K_1, K_2, R and $\mu^{(c)}$ of the quasicrystal and crystal, applied stress τ and geometry parameters a and h, while in the second example, the geometry parameter is only h. For different $\tau/\mu, \mu^{(c)}/\mu$, a/h and given values of μ, K_1, K_2 and R, one can find a rich set of numerical results. The computation used the measured values of these quantities for Al–Pd–Mn icosahedral quasicrystals which are provided in Tables 9.1, 9.2 and 9.3

$$\mu = 72.4\,\text{GPa}, \quad K_1 = 125\,\text{MPa}, \quad K_2 = -50\,\text{MPa}, \quad R = 0.04\mu$$

In the first example, the strain field of phason is dependent on elastic constants μ, K_1, K_2, R of the quasicrystal, the elastic constant $\mu^{(c)}$ of the crystal, the applied stress τ and the geometry parameters a and h; in the second example, the geometry parameter is only h. For different values of τ/μ, $\mu^{(c)}/\mu$, a/h under given values of μ, K_1, K_2 and R, some significant results are found and shown in Figs. 9.3–9.6, respectively. The quasicrystalline material constants are taken from Tables 9.1–9.3, i.e.,

$$\mu = 72.4\,\text{GPa}, \quad K_1 = 125\,\text{MPa}, \quad K_2 = -50\,\text{MPa}, \quad R = 0.04\mu$$

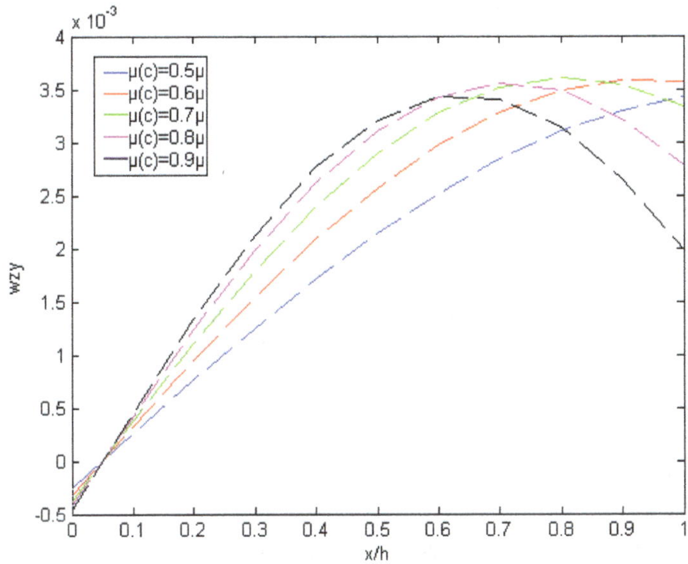

Fig. 9.3. Variation of w_{zy} versus x.

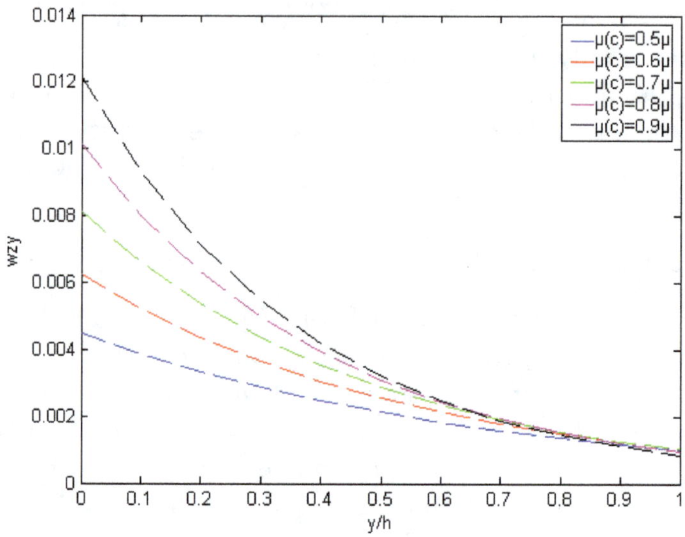

Fig. 9.4. Variation of w_{zy} versus y.

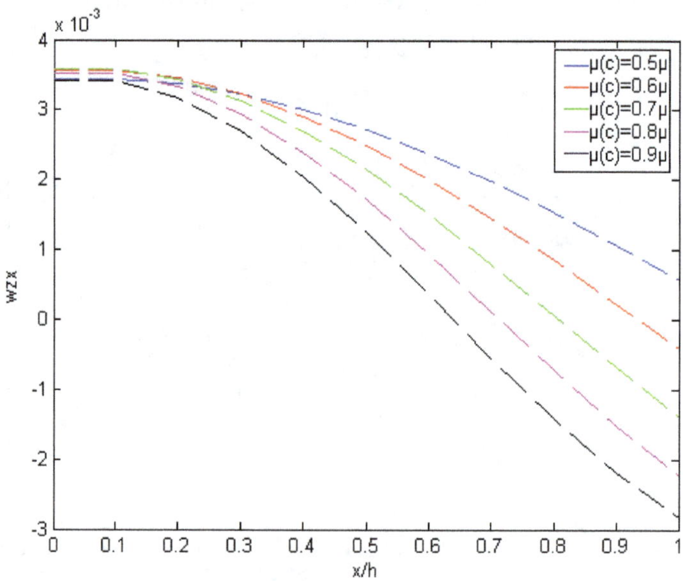

Fig. 9.5. Variation of w_{zx} versus x.

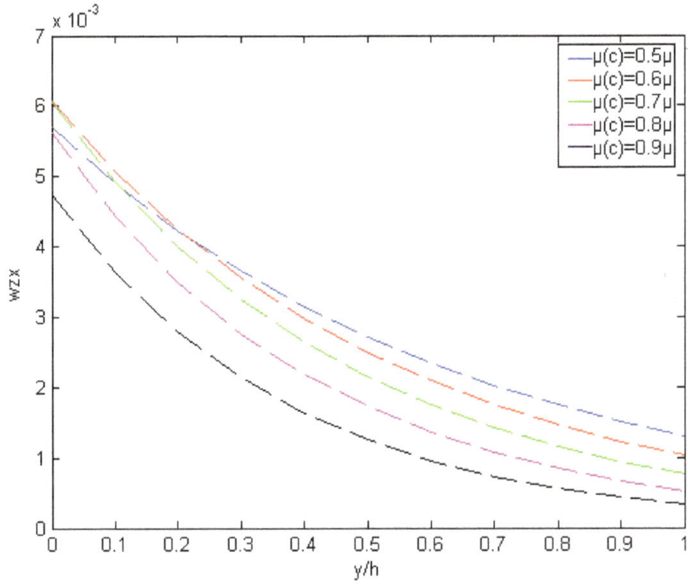

Fig. 9.6. Variation of w_{zx} versus y.

The numerical results show that the influence of the ratio $\mu^{(c)}/\mu$ of shear modulus of the crystal and quasicrystal is very evident. In addition, the influence of the applied stress τ/μ is also very important. Besides, the influence of a/h is not evident for the first example.

Another feature of the solution here is quite different from that in Section 7.6 due to the reason for the difference in quasicrystalline systems.

This work is reported in Ref. [24].

9.3 Phonon–Phason Decoupled Plane Elasticity of Icosahedral Quasicrystals

Yang *et al.* [3] presented an approximate solution of a straight dislocation in icosahedral quasicrystals under assumptions

$$\frac{\partial}{\partial z} = 0 \tag{9.3-1}$$

and

$$R = 0 \tag{9.3-2}$$

The conditions (9.3-1) and (9.3-2) result in a phonon–phason decoupled plane elasticity, which leads to

$$\varepsilon_{zz} = 0, \quad w_{zz} = w_{yz} = w_{xz} = 0 \tag{9.3-3}$$

Based on conditions (9.3-1) and (9.3-2), the final governing Eq. (9.1-8) reduce to

$$\begin{aligned}
&\mu\nabla_1^2 u_x + (\lambda+\mu)\frac{\partial}{\partial x}\nabla_1\cdot\mathbf{u_1} = 0\\
&\mu\nabla_1^2 u_y + (\lambda+\mu)\frac{\partial}{\partial y}\nabla_1\cdot\mathbf{u_1} = 0\\
&\mu\nabla_1^2 u_z = 0\\
&K_1\nabla_1^2 w_x + K_2\left(\frac{\partial^2 w_z}{\partial x^2} - \frac{\partial^2 w_z}{\partial y^2}\right) = 0\\
&K_1\nabla_1^2 w_y - 2K_2\frac{\partial^2 w_z}{\partial x\partial y} = 0\\
&(K_1-K_2)\nabla_1^2 w_z + K_2\left(\frac{\partial^2 w_x}{\partial x^2} - 2\frac{\partial^2 w_y}{\partial x\partial y} - \frac{\partial^2 w_y}{\partial y^2}\right) = 0
\end{aligned} \tag{9.3-4}$$

where

$$\nabla_1^2 = \frac{\partial^2}{\partial x^2} + \frac{\partial^2}{\partial y^2}, \quad \mathbf{u_1} = (u_x, u_y), \quad \nabla_1\cdot\mathbf{u_1} = \frac{\partial u_x}{\partial x} + \frac{\partial u_y}{\partial y}$$

Because the phonons and phasons are decoupled, the first three equations of (9.3-4) are pure phonon equilibrium equations; in addition, u_z is independent of u_x and u_y, and the second three equations in (9.3-4) are pure phason equilibrium equations.

Yang *et al.* [3] solved the equations under the dislocation conditions

$$\int_\Gamma du_i = b_i^{\parallel}, \quad \int_\Gamma dw_i = b_i^{\perp} \tag{9.3-5}$$

where Γ represents a path enclosing the dislocation core. The authors used the Green function method to calculate. The results are

$$
\begin{aligned}
u_x &= \frac{b_1^{\|}}{2\pi}\left(\arctan\frac{y}{x} + \frac{\lambda+\mu}{\lambda+2\mu}\frac{xy}{r^2}\right) + \frac{b_2^{\|}}{2\pi}\left(\frac{\mu}{\lambda+2\mu}\ln\frac{r}{r_0} + \frac{\lambda+\mu}{\lambda+2\mu}\frac{x^2}{r^2}\right)\\
u_y &= -\frac{b_1^{\|}}{2\pi}\left(\frac{\mu}{\lambda+2\mu}\ln\frac{r}{r_0} + \frac{\lambda+\mu}{\lambda+2\mu}\frac{y^2}{r^2}\right) + \frac{b_2^{\|}}{2\pi}\left(\arctan\frac{y}{x} - \frac{\lambda+\mu}{\lambda+2\mu}\frac{xy}{r^2}\right)\\
u_z &= \frac{b_3^{\|}}{2\pi}\arctan\frac{y}{x}\\
w_x &= \frac{b_1^{\perp}}{2\pi}\left[\arctan\frac{y}{x} + \frac{K_2^2}{2K_5}\left(\frac{2xy^3}{r^4} - \frac{xy}{r^2}\right)\right] - \frac{b_2^{\perp}}{4\pi}\frac{K_2^2}{K_5}\left(\ln\frac{r}{r_0} + \frac{2x^2y^2}{r^4}\right)\\
&\quad + \frac{b_3^{\perp}}{2\pi}\frac{K_2}{K_1}\frac{xy}{r^2}\\
w_y &= \frac{b_1^{\perp}}{4\pi}\frac{K_2^2}{K_5}\left(\ln\frac{r}{r_0} - \frac{2x^2y^2}{r^4}\right) + \frac{b_2^{\perp}}{2\pi}\left[\arctan\frac{y}{x} + \frac{K_2^2}{2K_5}\left(\frac{xy^3 - x^3y}{r^4}\right)\right]\\
&\quad - \frac{b_3^{\perp}}{2\pi}\frac{K_2}{K_1}\frac{y^2}{r^2}\\
w_z &= \frac{b_1^{\perp}}{2\pi}\frac{K_1K_2}{K_5}\frac{xy}{r^2} - \frac{b_2^{\perp}}{2\pi}\frac{K_1K_2}{K_5}\frac{y^2}{r^2} + \frac{b_3^{\perp}}{2\pi}\arctan\frac{y}{x}
\end{aligned}
\tag{9.3-6}
$$

in which

$$
r = \sqrt{x^2+y^2}, \quad K_5 = K_1^2 - K_1K_2 - K_2^2 \tag{9.3-7}
$$

The first three of (9.3-6) are the well-known solutions of a pure phonon field in the classical theory of dislocation, and the second three of (9.3-6) are the new results for a pure phason field. Due to the lack of coupling terms, the interaction between phonons and phasons could not be revealed.

9.4 Phonon–Phason Coupled Plane Elasticity of Icosahedral Quasicrystals — Displacement Potential Formulation

In the previous section, Yang *et al.* [3] introduced the assumption (9.3-2), which is not valid and leads to losing a lot of information coming from phonon–phason coupling. The studies of Fan and Guo [4], Zhu and Fan [16], Zhu *et al.* [17], considered the coupling effects, i.e.,

$$R \neq 0$$

and obtained the complete theory for the plane elasticity of quasicrystals. In this chapter, the assumption (9.2-1) or (9.3-1) still maintains, i.e.,

$$\frac{\partial}{\partial z} = 0 \tag{9.4-1}$$

In this case, the three-dimensional elasticity can be reduced into a plane elasticity. From condition (9.3-1) directly, we have

$$\varepsilon_{zz} = w_{zz} = w_{xz} = w_{yz} = 0 \tag{9.4-2}$$

Thus, the number of field variables and field equations from 36 is reduced to 32. Though the reduction of the total number is not so much, the resulting equation system has been greatly simplified with the following form:

$$\begin{aligned}
&\mu\nabla_1^2 u_x + (\lambda+\mu)\frac{\partial}{\partial x}\nabla_1\cdot\mathbf{u_1} + R\left(\frac{\partial^2 w_x}{\partial x^2} + 2\frac{\partial^2 w_y}{\partial x\partial y} - \frac{\partial^2 w_y}{\partial y^2}\right) = 0\\
&\mu\nabla_1^2 u_y + (\lambda+\mu)\frac{\partial}{\partial y}\nabla_1\cdot\mathbf{u_1} + R\left(\frac{\partial^2 w_y}{\partial x^2} - 2\frac{\partial^2 w_x}{\partial x\partial y} - \frac{\partial^2 w_y}{\partial y^2}\right) = 0\\
&\mu\nabla_1^2 u_z + R\left(\frac{\partial^2 w_x}{\partial x^2} - 2\frac{\partial^2 w_y}{\partial x\partial y} - \frac{\partial^2 w_x}{\partial y^2} + \nabla_1^2 w_z\right) = 0\\
&K_1\nabla_1^2 w_x + K_2\left(\frac{\partial^2 w_z}{\partial x^2} - \frac{\partial^2 w_z}{\partial y^2}\right)\\
&\quad + R\left(\frac{\partial^2 u_x}{\partial x^2} - 2\frac{\partial^2 u_y}{\partial x\partial y} - \frac{\partial^2 u_x}{\partial y^2} + \frac{\partial^2 u_z}{\partial x^2} - \frac{\partial^2 u_z}{\partial y^2}\right) = 0\\
&K_1\nabla_1^2 w_y - 2K_2\frac{\partial^2 w_z}{\partial x\partial y} + R\left(\frac{\partial^2 u_y}{\partial x^2} + 2\frac{\partial^2 u_x}{\partial x\partial y} - \frac{\partial^2 u_y}{\partial y^2} - 2\frac{\partial^2 u_z}{\partial x\partial y}\right) = 0\\
&(K_1 - K_2)\nabla_1^2 w_z + K_2\left(\frac{\partial^2 w_x}{\partial x^2} - 2\frac{\partial^2 w_y}{\partial x\partial y} - \frac{\partial^2 w_y}{\partial y^2}\right) + R\nabla_1^2 u_z = 0
\end{aligned} \tag{9.4-3}$$

where ∇_1^2 and $\nabla_1 \cdot \mathbf{u}_1$ are the same as those in Section 9.3, but the suffix 1 of the two-dimensional Laplace operator will be omitted in the following for simplicity.

The equation set is much simpler than that of (9.1-8) but still quite complicated, too. If we introduce a displacement potential $F(x, y)$ such as

$$\begin{aligned}
u_x &= R\frac{\partial^2}{\partial x \partial y}\nabla^2\nabla^2 \left[\mu\alpha\Pi_1 + \beta(\lambda + 2\mu)\Pi_2\right] F \\
&\quad + c_0 R \frac{\partial^2}{\partial x \partial y}\Lambda\left[(3\mu - \lambda)\frac{\partial^4}{\partial x^4} + 10(\lambda + \mu)\frac{\partial^4}{\partial x^2 \partial y^2} - (5\lambda + 9\mu)\frac{\partial^4}{\partial y^4}\right] F \\
u_y &= R\nabla^2\nabla^2\left[\mu\alpha\frac{\partial^2}{\partial y^2}\Pi_1 - \beta(\lambda + 2\mu)\frac{\partial^2}{\partial x^2}\Pi_2\right] F \\
&\quad + c_0 R\Lambda^2\left[(\lambda + 2\mu)\frac{\partial^6}{\partial x^6} - 5(2\lambda + 3\mu)\frac{\partial^6}{\partial x^4 \partial y^2} + 5\lambda\frac{\partial^6}{\partial x^2 \partial y^4} + \mu\frac{\partial^6}{\partial y^6}\right] F \\
u_z &= c_1\frac{\partial^2}{\partial x \partial y}\left[(\alpha - \beta)\Lambda^2\Pi_1\Pi_2 + \alpha\frac{\partial^2}{\partial y^2}\Pi_1^2 + \beta\frac{\partial^2}{\partial x^2}\Pi_2^2\right] F \\
w_x &= -\omega\frac{\partial^2}{\partial x \partial y}\nabla^2\left[2c_0\Lambda^2\nabla^2 - (\alpha - \beta)\Pi_1\Pi_2\right] F \\
w_y &= -\omega\nabla^2\left[c_0\Lambda^2\Lambda^2\nabla^2 + \alpha\frac{\partial^2}{\partial y^2}\Pi_1^2 + \beta\frac{\partial^2}{\partial x^2}\Pi_2^2\right] F \\
w_z &= c_2\frac{\partial^2}{\partial x \partial y}\left[(\alpha - \beta)\Lambda^2\Pi_1\Pi_2 + \alpha\frac{\partial^2}{\partial y^2}\Pi_1^2 + \beta\frac{\partial^2}{\partial x^2}\Pi_2^2\right] F
\end{aligned} \tag{9.4-4}$$

then the field equations mentioned above will be satisfied if

$$\nabla^2\nabla^2\nabla^2\nabla^2\nabla^2\nabla^2 F(x, y) + \nabla^2 L F(x, y) = 0 \tag{9.4-5}$$

where

$$\begin{aligned}
&\alpha = (\lambda + 2\mu)R^2 - \omega K_1 \quad \beta = \mu R^2 - \omega K_1 \quad \omega = \mu(\lambda + 2\mu) \\
&c_0 = \omega\frac{\mu K_2^2 + (K_1 - 3K_2)R^2}{\mu(K_1 - K_2) - R^2} \qquad c_1 = \frac{(K_1 - 2K_2)R\omega}{\mu(K_1 - K_2) - R^2} \\
&c_2 = \frac{(K_2\mu - R^2)\omega}{\mu(K_1 - K_2) - R^2}
\end{aligned}$$

$$\Pi_1 = 3\frac{\partial^2}{\partial x^2} - \frac{\partial^2}{\partial y^2}, \quad \Pi_2 = 3\frac{\partial^2}{\partial y^2} - \frac{\partial^2}{\partial x^2},$$
$$\nabla^2 = \frac{\partial^2}{\partial x^2} + \frac{\partial^2}{\partial y^2}, \quad \Lambda^2 = \frac{\partial^2}{\partial x^2} - \frac{\partial^2}{\partial y^2} \tag{9.4-6}$$

in which the suffix 1 of the two-dimensional Laplace operator is omitted and

$$L = \frac{c_0}{\beta}\left[-\frac{\partial^{10}}{\partial x^{10}} + 5\left(4 - 5\frac{\alpha}{\beta}\right)\frac{\partial^{10}}{\partial x^8 \partial y^2} - 10\left(11 - 10\frac{\alpha}{\beta}\right)\frac{\partial^{10}}{\partial x^6 \partial y^4} \right.$$
$$\left. + 10\left(10 - 11\frac{\alpha}{\beta}\right)\frac{\partial^{10}}{\partial x^4 \partial y^6} - 5\left(5 - 4\frac{\alpha}{\beta}\right)\frac{\partial^{10}}{\partial x^2 \partial y^8} - \frac{\alpha}{\beta}\frac{\partial^{10}}{\partial y^{10}}\right] \tag{9.4-7}$$

Assuming

$$R^2/(\mu K_1) \ll 1 \tag{9.4-8}$$

(this is understandable because the coupling effect is weaker than that of phonon), from Eqs. (9.4-6) and (9.4-7)

$$\beta/\alpha \to 1, \quad \nabla^2 L = \frac{c_0}{\beta}\nabla^2\nabla^2\nabla^2\nabla^2\nabla^2\nabla^2 \tag{9.4-9}$$

Substituting (9.4-9) into (9.4-5), we find that

$$\nabla^2\nabla^2\nabla^2\nabla^2\nabla^2\nabla^2 F(x, y) = 0 \tag{9.4-10}$$

This is our final governing equation for the plane elasticity of quasicrystals based on the displacement potential formulation. With the aid of the generalized Hooke's law, the phonon and phason stress components can also be expressed in terms of the potential function $F(x, y)$ and these expressions are omitted here due to the limitation of space.

In other words, equation set (9.4-4) gives a fundamental solution in terms of $F(x, y)$ for the plane elasticity problem of an icosahedral quasicrystal. Once the function $F(x, y)$ satisfying Eq. (9.4-10) is determined for prescribed boundary conditions, the entire elastic field of an icosahedral quasicrystal can be found from (9.4-4). This formulation is reported briefly by Fan and Guo [4]. In some extent which is a development of Li and Fan [18] for the elasticity of two-dimensional quasicrystals. The application of the formulation and relevant solution are given in Section 9.6.

9.5 Phonon–Phason Coupled Plane Elasticity of Icosahedral Quasicrystals — Stress Potential Formulation

In the previous section, the displacement formulation exhibits its effect, which reduces a very complicated partial differential equation set into a single partial differential equation with higher order (12th order); the latter will be easy to solve. In the meantime, the stress potential formulation is effective too. In this section, we will introduce the formulation.

To contrast the displacement potential formulation, we here maintain the stress components and exclude the displacement components. From the deformation geometry equations (9.1-1) and considering (9.3-1) and (9.3-2), we obtain the deformation compatibility equations as follows:

$$\begin{gathered}\frac{\partial^2 \varepsilon_{xx}}{\partial y^2}+\frac{\partial^2 \varepsilon_{yy}}{\partial x^2}=2\frac{\partial^2 \varepsilon_{xy}}{\partial x \partial y}, \quad \frac{\partial \varepsilon_{yz}}{\partial x}=\frac{\partial \varepsilon_{zx}}{\partial y}\\ \frac{\partial w_{xy}}{\partial x}=\frac{\partial w_{xx}}{\partial y}, \quad \frac{\partial w_{yy}}{\partial x}=\frac{\partial w_{yx}}{\partial y}, \quad \frac{\partial w_{zy}}{\partial x}=\frac{\partial w_{zx}}{\partial y}\end{gathered} \tag{9.5-1}$$

Thus, the displacements are excluded already.

By getting the expressions of strains through stresses from the inversion of equation set (9.1-7) and substituting them into (9.5-1), we obtain the deformation compatibility equations expressed by stress components so that the strain components have been excluded up to now (those equations are too lengthy and so here we do not list them).

So far, one has the deformation compatibility equations expressed by stresses and equilibrium equations only.

If we introduce stress potential functions

$$\varphi_1(x,y), \varphi_2(x,y), \psi_1(x,y), \psi_2(x,y), \psi_3(x,y)$$

such as

$$\begin{aligned}
&\sigma_{xx} = \frac{\partial^2 \varphi_1}{\partial y^2}, \quad \sigma_{xy} = -\frac{\partial^2 \varphi_1}{\partial x \partial y}, \quad \sigma_{yy} = \frac{\partial^2 \varphi_1}{\partial x^2} \\
&\sigma_{zx} = \frac{\partial \varphi_2}{\partial y}, \quad \sigma_{zy} = -\frac{\partial \varphi_2}{\partial x} \\
&H_{xx} = \frac{\partial \psi_1}{\partial y}, \quad H_{xy} = -\frac{\partial \psi_1}{\partial x}, \quad H_{yx} = \frac{\partial \psi_2}{\partial y} \\
&H_{yy} = -\frac{\partial \psi_2}{\partial x}, \quad H_{zx} = \frac{\partial \psi_3}{\partial y}, \quad H_{zy} = -\frac{\partial \psi_3}{\partial x}
\end{aligned} \tag{9.5-2}$$

with

$$\begin{aligned}
&\varphi_1 = c_2 c_3 R \frac{\partial}{\partial y}\left[2\frac{\partial^2}{\partial x^2}\Pi_2 - \Lambda^2 \Pi_1\right]\nabla^2\nabla^2 G \\
&\varphi_2 = -c_3 c_4 \nabla^2\nabla^2\nabla^2\nabla^2\nabla^2 G \\
&\psi_1 = c_1 c_2 R \frac{\partial^2}{\partial y^2}\left[2\frac{\partial^2}{\partial x^2}\Pi_1\Pi_2 - \Lambda^2 \Pi_1^2\right]\nabla^2 G \\
&\qquad + c_2 c_4 \Lambda^2 \nabla^2\nabla^2\nabla^2\nabla^2 G \\
&\psi_2 = c_1 c_2 R \frac{\partial^2}{\partial x \partial y}\left[2\frac{\partial^2}{\partial x^2}\Pi_2^2 - \Lambda^2 \Pi_1\Pi_2\right]\nabla^2 G \\
&\qquad - 2c_2 c_4 \frac{\partial^2}{\partial x \partial y}\nabla^2\nabla^2\nabla^2\nabla^2 G \\
&\psi_3 = -\frac{1}{R} K_2 c_3 c_4 \nabla^2\nabla^2\nabla^2\nabla^2\nabla^2 G
\end{aligned} \tag{9.5-3}$$

then the equilibrium equations and the deformation compatibility equations will be automatically satisfied if

$$\nabla^2\nabla^2\nabla^2\nabla^2\nabla^2\nabla^2 G = 0 \tag{9.5-4}$$

under the approximation $R^2/K_1\mu \ll 1$, which is the final governing equation of the plane elasticity of icosahedral quasicrystals, and the function $G(x, y)$ is named the stress potential in which

$$\begin{aligned}
c_1 &= \frac{R(2K_2 - K_1)(\mu K_1 + \mu K_2 - 3R^2)}{2(\mu K_1 - 2R^2)} \\
c_2 &= \frac{1}{R}K_2(\mu K_2 - R^2) - R(2K_2 - K_1) \\
c_3 &= \mu(K_1 - K_2) - R^2 - \frac{(\mu K_2 - R^2)^2}{\mu K_1 - 2R^2} \\
c_4 &= c_1 R + \frac{1}{2}c_3\left(K_1 + \frac{\mu K_1 - 2R^2}{\lambda + \mu}\right) \\
\Pi_1 &= 3\frac{\partial^2}{\partial x^2} - \frac{\partial^2}{\partial y^2}, \ \Pi_2 = 3\frac{\partial^2}{\partial y^2} - \frac{\partial^2}{\partial x^2} \\
\nabla^2 &= \frac{\partial^2}{\partial x^2} + \frac{\partial^2}{\partial y^2}, \ \Lambda^2 = \frac{\partial^2}{\partial x^2} - \frac{\partial^2}{\partial y^2}
\end{aligned} \tag{9.5-5}$$

In the derivation of (9.5-4), the approximation (9.4-8) is used in the last step.

This is given in Ref. [5], which may be seen as a development of the study for two-dimensional quasicrystals given by Guo and Fan (see, e.g., [19, 20]).

9.6 A Straight Dislocation in an Icosahedral Quasicrystal

The formulations exhibited in the previous sections are meaningful, which have greatly simplified the complicated equations involving elasticity. Their applications will be addressed in this and subsequent sections, in which the Fourier analysis and complex analysis method play important roles.

We introduced the dislocation solution of Yang *et al.* [3] in an icosahedral quasicrystal. The authors of Ref. [3] assumed that the coupling effect between phonons and phasons is omitted, i.e., $R = 0$. In this case, the phonon solution of the dislocation is the same as that of the classical isotropic elastic solution of an edge dislocation. And the phason solution of the problem is newly found, which is independent of the phonon field. In this section, we try to give a complete analysis of the problem in which the phonon–phason coupling effect is taken into account.

For a dislocation along the x_3-axis (or z-axis) in icosahedral quasicrystal with the core at the origin, the Burgers vector is denoted as $b = b^{\|} \oplus b^{\perp} = (b_1^{\|}, b_2^{\|}, b_3^{\|}, b_1^{\perp}, b_2^{\perp}, b_3^{\perp})$ where the dislocation conditions are

$$\int_{\Gamma} du_j = b_j^{\|} \quad \int_{\Gamma} dw_j = b_j^{\perp} \tag{9.6-1}$$

in which $x_1 = x$, $x_2 = y$, $x_3 = z$, and the integrals in (9.6-1) should be taken along the Burgers circuit surrounding the dislocation core in space $E_{\|}$. By using the superposition principle, we here calculate first the elastic field for a special case, i.e., which corresponds to $b_1^{\|} \neq 0$, $b_1^{\perp} \neq 0$ and $b_2^{\|} = b_3^{\|} = 0$, $b_2^{\perp} = b_3^{\perp} = 0$.

For simplicity, we can solve a half-plane problem by considering symmetry and anti-symmetry of relevant field variables, so the following boundary conditions include the dislocation condition:

$$\sigma_{yy}(x,0) = \sigma_{zy}(x,0) = 0 \tag{9.6-2a,b}$$

$$H_{yy}(x,0) = H_{zy}(x,0) = 0 \tag{9.6-2c,d}$$

$$\int_{\Gamma} du_x = b_1^{\|}, \quad \int_{\Gamma} dw_x = b_1^{\perp} \tag{9.6-2e,f}$$

In addition, there are boundary conditions at infinity:

$$\sigma_{ij}(x,y) \to 0, \quad H_{ij}(x,y) \to 0, \quad \sqrt{x^2+y^2} \to \infty \tag{9.6-3}$$

In the following, we use the formulation of Section 9.4 to solve the above boundary value problem. By performing the Fourier transform to Eq. (9.4-10) and the above boundary conditions, we obtain the solution at the transformed domain; then taking inversion of the Fourier transform, we obtain the solution as follows:

$$u_x = \frac{1}{2\pi}\left(b_1^{\|}\arctan\frac{y}{x} + c_{12}\frac{xy}{r^2} + c_{13}\frac{xy^3}{r^4}\right)$$

$$u_y = \frac{1}{2\pi}\left(-c_{21}\ln\frac{r}{r_0} + c_{22}\frac{y^2}{r^2} + c_{23}\frac{y^2(y^2-x^2)}{2r^4}\right)$$

$$u_z = \frac{1}{2\pi}\left(-c_{31}\arctan\frac{y}{x} + c_{32}\frac{xy}{r^2} + c_{33}\frac{xy^3}{r^4}\right)$$

$$w_x = \frac{1}{2\pi}\left(b_1^{\perp}\arctan\frac{y}{x} + c_{42}\frac{xy}{r^2} + c_{43}\frac{xy^3}{r^4}\right)$$

$$w_y = \frac{1}{2\pi}\left(-c_{51}\ln\frac{r}{r_0} + c_{52}\frac{y^2}{r^2} + c_{53}\frac{y^2(y^2-x^2)}{2r^4}\right)$$

$$w_z = \frac{1}{2\pi}\left(-c_{61}\arctan\frac{y}{x} + c_{62}\frac{xy}{r^2} + c_{63}\frac{xy^3}{r^4}\right) \tag{9.6-4}$$

in which $r^2 = x^2 + y^2$ and r_0, the radius of the dislocation core, and c_{ij} are constants shown as follows:

$$c_{12} = \frac{2c_0[\mu(2R^2 + c_0\mu)(\lambda^2 + 3\lambda\mu + 2\mu^2]b_1^{\parallel} + R[-e(\lambda+\mu) + 2\mu c_0(\lambda+2\mu)^2]b_1^{\perp}}{-e[2e + \mu c_0(\lambda+2\mu)] + \mu c_0(\lambda+2\mu)[e + 2\mu c_0(\lambda+2\mu)]}$$

$$c_{13} = \frac{2c_0 R(\lambda+\mu)[2R\mu(\lambda+\mu)b_1^{\parallel} + 2\mu c_0(\lambda+2\mu)b_1^{\perp}]}{-e[2e + \mu c_0(\lambda+2\mu)] + \mu c_0(\lambda+2\mu)[e + 2\mu c_0(\lambda+2\mu)]}$$

$$c_{21} = \frac{[2c_0^2\mu^3(\lambda+2\mu) - 2e^2]b_1^{\parallel} + 2c_0 R(\lambda+3\mu)eb_1^{\perp}}{-e[2e + \mu c_0(\lambda+2\mu)] + \mu c_0(\lambda+2\mu)[e + 2\mu c_0(\lambda+2\mu)]}$$

$$c_{22} = \frac{2c_0[-\mu^2(\lambda+\mu)(-2R^2 + c_0(\lambda+2\mu)]b_1^{\parallel} + R[-(\lambda+\mu)e + 2c_0\mu^2)b_1^{\perp}]}{-e[2e + \mu c_0(\lambda+2\mu)] + \mu c_0(\lambda+2\mu)[e + 2\mu c_0(\lambda+2\mu)]}$$

$$c_{23} = \frac{2c_0 R(\lambda+\mu)[2R\mu(\lambda+\mu)b_1^{\parallel} + 2c_0\mu^2 b_1^{\perp}]}{-e[2e + \mu c_0(\lambda+2\mu)] + \mu c_0(\lambda+2\mu)[e + 2\mu c_0(\lambda+2\mu)]}$$

$$c_{31} = \frac{-3c_1 e\left\{\begin{array}{c} 2(c_0\mu + 7e)\mu c_0(\lambda+2\mu)b_1^{\parallel} + R[54c_0^2(\lambda^2 + 3\lambda\mu + \mu^2) \\ -2(\alpha-\beta)(e + \mu c_0(\lambda+2\mu))]b_1^{\perp} \end{array}\right\}}{4c_0 R[-e(2e + \mu c_0(\lambda+2\mu)] + \mu c_0(\lambda+2\mu)[e + 2\mu c_0(\lambda+2\mu)]}$$

$$c_{32} = \frac{3c_1 e\left\{2\mu[-e + \mu c_0(\lambda+2\mu)]b_1^{\parallel} + R[-2e + 2\mu c_0(\lambda+2\mu)]b_1^{\perp}\right\}}{-e[2e + \mu c_0(\lambda+2\mu)] + \mu c_0(\lambda+2\mu)[e + 2\mu c_0(\lambda+2\mu)]}$$

$$c_{33} = \frac{-3ec_1[2R\mu(\lambda+\mu)b_1^{\parallel} + 2\mu c_0(\lambda+2\mu)b_1^{\perp}]}{-e[2e + \mu c_0(\lambda+2\mu)] + \mu c_0(\lambda+2\mu)[e + 2\mu c_0(\lambda+2\mu)]}$$

$$c_{42} = \frac{-2e[2R\mu(\lambda+\mu)b_1^{\parallel} + 2\mu c_0(\lambda+2\mu)b_1^{\perp}]}{-e[2e + \mu c_0(\lambda+2\mu)] + \mu c_0(\lambda+2\mu)[e + 2\mu c_0(\lambda+2\mu)]}$$

$$c_{43} = 0$$

$$c_{51} = -\frac{\left\{\begin{array}{l} -4e\mu^2 c_0(\lambda+2\mu)b_1^{\|} + R[2(\lambda+2\mu)(e+0.5\mu c_0) \\ \quad +\mu(2\beta^2\mu + 2c_0^2(\lambda+2\mu)^2 + c_0(\lambda+2\mu)(-\beta\mu + R^2(\lambda+\mu))]b_1^{\perp} \end{array}\right\}}{R(-e(2e+\mu c_0(\lambda+2\mu)) + \mu c_0(\lambda+2\mu)(e+2\mu c_0(\lambda+2\mu)))}$$

$$c_{52} = -\frac{2e[2R\mu(\lambda+\mu)b_1^{\|} + 2\mu c_0(\lambda+2\mu)b_1^{\perp}]}{-e[2e+\mu c_0(\lambda+2\mu)] + \mu c_0(\lambda+2\mu)[e+2\mu c_0(\lambda+2\mu)]}$$

$$c_{53} = 0$$

$$c_{61} = -\frac{3c_2 e\left\{\begin{array}{l} 2(c_0\mu + 7e)\mu c_0(\lambda+2\mu)b_1^{\|} \\ \quad + R[54c_0^2(\lambda^2+3\lambda\mu+\mu^2) - 2(\alpha-\beta)(e+\mu c_0(\lambda+2\mu))]b_1^{\perp} \end{array}\right\}}{4c_0 R[-e(2e+\mu c_0(\lambda+2\mu)] + \mu c_0(\lambda+2\mu)[e+2\mu c_0(\lambda+2\mu)]}$$

$$c_{62} = \frac{3ec_2[2\mu(-e+\mu c_0(\lambda+2\mu))b_1^{\|} + R(-2e+2\mu c_0(\lambda+2\mu))b_1^{\perp}]}{-e[2e+\mu c_0(\lambda+2\mu)] + \mu c_0(\lambda+2\mu)[e+2\mu c_0(\lambda+2\mu)]}$$

$$c_{63} = \frac{-3ec_2[2R\mu(\lambda+\mu)b_1^{\|} + 2\mu c_0(\lambda+2\mu)b_1^{\perp}]}{-e[2e+\mu c_0(\lambda+2\mu)] + \mu c_0(\lambda+2\mu)[e+2\mu c_0(\lambda+2\mu)]} \tag{9.6-5}$$

with $e = -(\lambda+\mu)R^2$.

For the other two typical problems, in which the Burgers vector of a dislocation is denoted by $(0, b_2^{\|}, 0, 0, b_2^{\perp}, 0)$ and $(0, 0, b_3^{\|}, 0, 0, b_3^{\perp})$, respectively, a completely similar consideration will yield similar results, which are omitted here. Alternatively, the expressions are denoted as $u_j^{(2)}$, $w_j^{(2)}$ and $u_j^{(3)}$, $w_j^{(3)}$.

Analytic expressions for the elastic field for a dislocation $(b_1^{\|}, b_2^{\|}, b_3^{\|}, b_1^{\perp}, b_2^{\perp}, b_3^{\perp})$ in an icosahedral quasicrystal can be obtained by the superposition of the corresponding expressions for the elastic fields for $(b_1^{\|}, 0, 0, b_1^{\perp}, 0, 0)$, $(0, b_2^{\|}, 0, 0, b_2^{\perp}, 0)$ and $(0, 0, b_3^{\|}, 0, 0, b_3^{\perp})$, namely,

$$u_j = u_j^{(1)} + u_j^{(2)} + u_j^{(3)} \quad w_j = w_j^{(1)} + w_j^{(2)} + w_j^{(3)} \quad i, j = 1, 2, 3 \tag{9.6-6}$$

We can see that the interaction between phonon–phonon, phason–phason and phonon–phason is very evident, so the solution (9.6-4) is quite different from the solution given by Yang *et al.* [3] (whose solution for the phonon displacement field is given by the first three formulas of Eq. (9.3-6) and will be quoted again in the following; see formula (9.6-7)), where they took $R = 0$, i.e., they assumed the phonon and phason are decoupled, so the solution for phonon is the same as the classical solution for crystals. It is obvious that our solution given in (9.6-4) explores that the realistic case for quasicrystals is quite different from that of a crystal. To illustrate the coupling effect, we give some numerical results in Figs. 9.7 and 9.8 for the normalized displacement $u_1/b_1^{\|}$ versus x and y, respectively, in which the results exhibit the influence of the coupling constant R which is significant. In the calculation, we take the data of the elastic moduli as

$$\lambda = 74.9, \quad \mu = 72.4(\text{GPa}), \quad K_1 = 72, \quad K_2 = -73(\text{MPa})$$

and the phonon–phason coupling elastic constant for three different cases, i.e., $R/\mu = 0, R/\mu = 0.004$ and $R/\mu = 0.006$, in which the first one corresponds to the decoupled case. The results are depicted in Fig. 9.7 for $u_1/b_1^{\|}$ versus x, and Fig. 9.8 for $u_1/b_1^{\|}$ versus y, respectively, as follows:

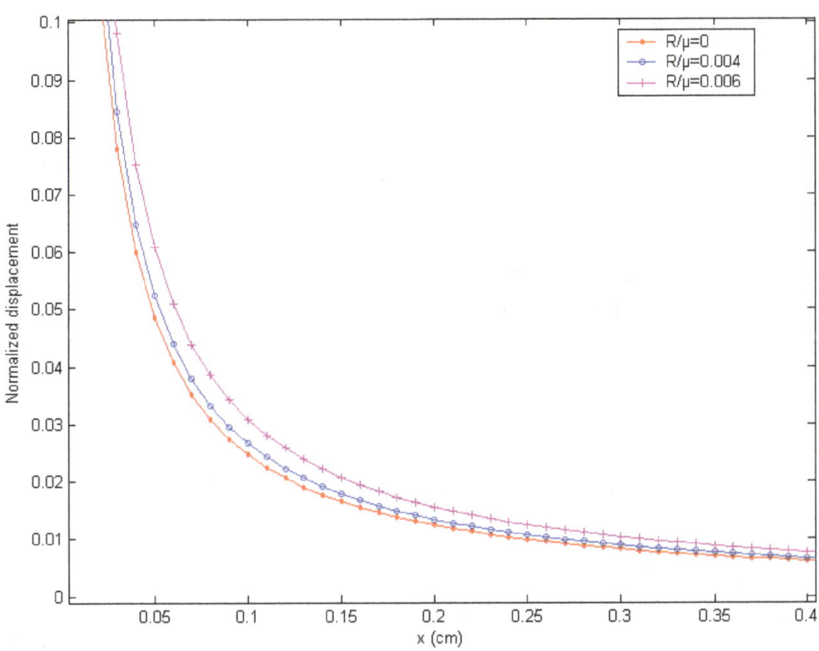

Fig. 9.7. The displacement $u_1/b_1^{\|}$ verses x for different coupling elastic constants.

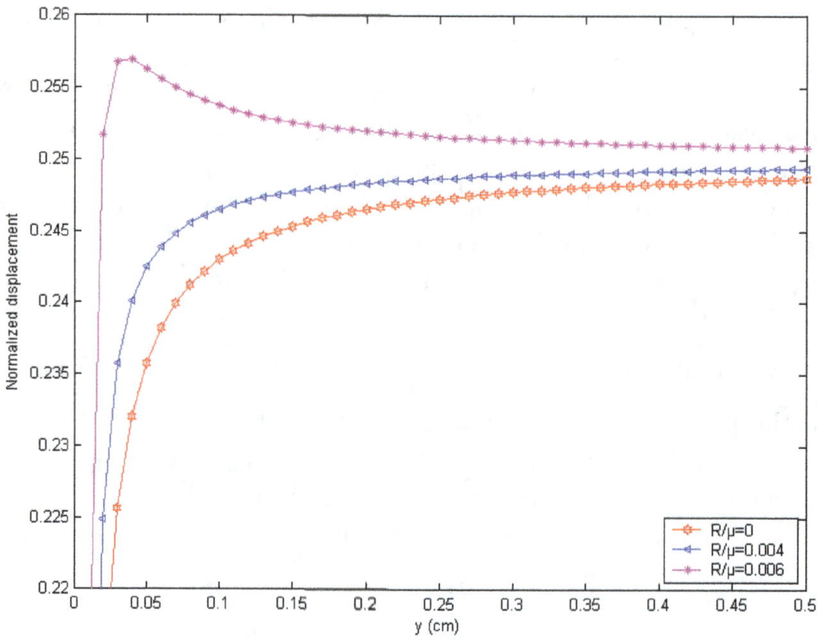

Fig. 9.8. The displacement $u_1/b_1^{\|}$ verses y for different coupling constants.

The figures show that the coupling effect is very important, and the values of displacements are increasing with the growth of the value of R.

For icosahedral quasicrystals with the presence of a dislocation, there are five independent elastic constants. If $R = 0$, $w_i = 0$, our solution is exactly reduced to the solution of the dislocation of crystals, i.e.,

$$u_x = \frac{b_1^{\|}}{2\pi}\left(\arctan\frac{y}{x} + \frac{\lambda+\mu}{\lambda+2\mu}\frac{xy}{r^2}\right) + \frac{b_2^{\|}}{2\pi}\left(\frac{\mu}{\lambda+2\mu}\ln\frac{r}{r_0} + \frac{\lambda+\mu}{\lambda+2\mu}\frac{x^2}{r^2}\right)$$

$$u_y = -\frac{b_1^{\|}}{2\pi}\left(\frac{\mu}{\lambda+2\mu}\ln\frac{r}{r_0} + \frac{\lambda+\mu}{\lambda+2\mu}\frac{x^2}{r^2}\right) + \frac{b_2^{\|}}{2\pi}\left(\arctan\frac{y}{x} - \frac{\lambda+\mu}{\lambda+2\mu}\frac{xy}{r^2}\right)$$

$$u_z = \frac{b_3^{\|}}{2\pi}\arctan\frac{y}{x} \tag{9.6-7}$$

The present solution reveals the interactions between phonon–phonon, phason–phason and phonon–phason for a dislocation in icosahedral quasicrystals.

The displacement potential function formulation establishes the basis for solving the defects problem in icosahedral quasicrystals. The formulation greatly simplifies the solution process. In the subsequent steps, a systematic Fourier analysis is developed, which provides a constructive procedure to find the analytic solution; it is effective not only for the dislocation problem but also for more complicated mixed boundary value problems (e.g., crack problems refer to the following section or Ref. [18]). The solution is explicit with closed form. As a complete solution of the dislocation of icosahedral quasicrystals, it is significant.

The present solution can be used as a fundamental solution for a dislocation in an icosahedral quasicrystal. Therefore, many elasticity problems in an icosahedral quasicrystal can be directly solved with the aid of this fundamental solution by superposition.

This work has been published in Ref. [17].

9.7 Application of Displacement Potential to Crack Problem of Icosahedral Quasicrystal

In the previous section, the application of displacement potential to dislocation problem of icosahedral quasicrystal is given; we now discuss an application of the potential method to crack the problem of the matter, with the help of the Fourier analysis and dual integral equation theory, an analytic solution is obtained, which are reported in [25] and presented as follows.

The cracked specimen is shown in Fig. 9.9; it is sufficient to consider a half of the sample, so it has the following boundary conditions:

$$\begin{array}{ll}\sigma_{yy}(x,0)=-p,\ H_{yy}(x,0)=0, & |x|\le a\\ u_y(x,0)=0,\ w_y(x,0)=0, & |x|>a\\ \sigma_{yx}(x,0)=\sigma_{yz}(x,0)=0,\ H_{yx}(x,0)=H_{yz}(x,0)=0, & -\infty<x<\infty\end{array} \tag{9.7-1}$$

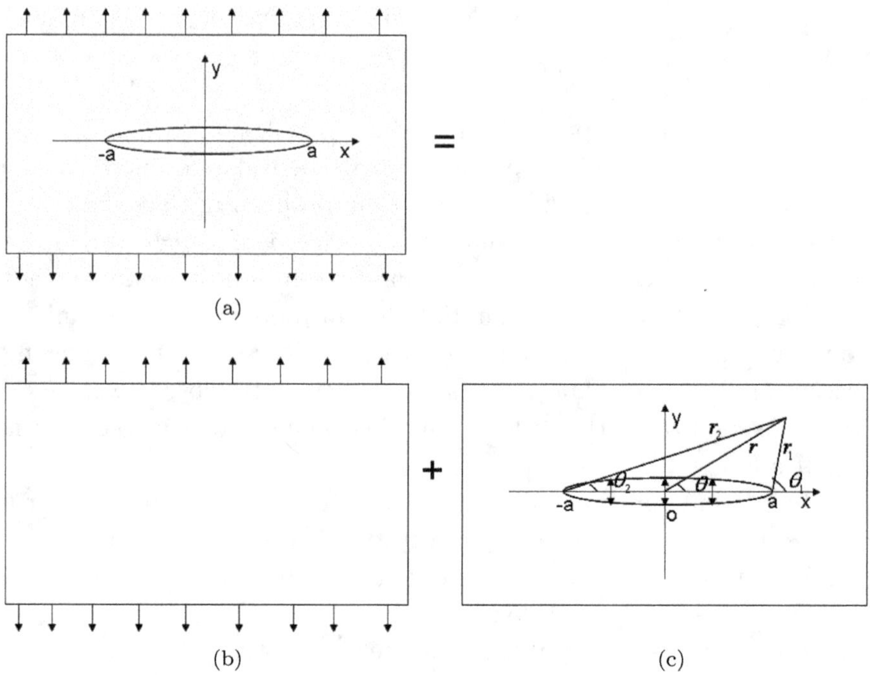

Fig. 9.9. A Griffith crack in an icosahedral quasicrystal.

and

$$\sigma_{ij} \to 0, H_{ij} \to 0, \quad \sqrt{x^2 + y^2} \to \infty \tag{9.7-2}$$

After Fourier transforming to Eq. (9.4-10), and by considering the boundary condition at infinity, the formal solution in Fourier transform domain is

$$\tilde{F} = XYe^{-|\xi|y} \tag{9.7-3}$$

where $X = (A, B, C, D, E, F)$, $Y = (1, y, y^2, y^3, y^4, y^5)^T$ and A, B, C, D, E and F are unknown functions of ξ to be determined, and the superscript T denotes the transpose operator of the matrix.

To determine the unknown functions in (9.7-3), with the Fourier transforms of displacements and stresses and boundary conditions, the problem is reduced to solve the following dual integral equations:

$$\begin{cases} \int_0^\infty \xi A_j \cos(\xi x) dx = p\pi b_j & 0 < x \le a \\ \int_0^\infty A_j \cos(\xi x) dx = 0 & x \ge a \end{cases} \tag{9.7-4}$$

where

$$A_1 = \xi^8 A, \quad A_2 = \left|\xi^7\right| B, \quad A_3 = 2\xi^6 C, \quad A_4 = 6\left|\xi^5\right| D,$$
$$A_5 = 24\xi^4 E, \quad A_6 = 120\left|\xi^3\right| F \tag{9.7-5}$$

$$b_j = (-1)^j \frac{\Delta_j}{\Delta} \quad j = 1, \ldots, 6 \tag{9.7-6}$$

$$\Delta = \begin{vmatrix} b_{11} & b_{12} & b_{13} & b_{14} & b_{15} & b_{16} \\ b_{21} & b_{22} & b_{23} & b_{24} & b_{25} & b_{26} \\ b_{31} & b_{32} & b_{33} & b_{34} & b_{35} & b_{36} \\ b_{41} & b_{42} & b_{43} & b_{44} & b_{45} & b_{46} \\ b_{51} & b_{52} & b_{53} & b_{54} & b_{55} & b_{56} \\ b_{61} & b_{62} & b_{63} & b_{64} & b_{65} & b_{66} \end{vmatrix}$$

$$\Delta_j = \begin{vmatrix} b_{21} & \cdots & b_{2,j-1} & b_{2,j+1} & \cdots & b_{26} \\ b_{31} & \cdots & b_{3,j-1} & b_{3,j+1} & \cdots & b_{36} \\ b_{41} & \cdots & b_{4,j-1} & b_{4,j+1} & \cdots & b_{46} \\ b_{51} & \cdots & b_{5,j-1} & b_{5,j+1} & \cdots & b_{56} \\ b_{61} & \cdots & b_{6,j-1} & b_{6,j+1} & \cdots & b_{66} \end{vmatrix} \tag{9.7-7}$$

b_{ij} are constants composed from the basic elastic constants, given subsequently.

Equations (9.7-4) belong to Titchmarsh–Busbridge dual integral equations; by using the standard solving procedure (refer to Major Appendix II of this book), we find the solution

$$A_j = ap\pi b_j J_1(\xi a)/\xi \tag{9.7-8}$$

where $J_1(x)$ is the first kind of Bessel function of first order (see the Major Appendix). After the determination of these functions, the problem is solved mathematically; then taking inversion of the Fourier transform, all field variables can be evaluated through a series of integrations. It is fortunate, and all integrals here can be expressed by elementary functions. We list only the displacements as follows:

$$\begin{aligned}
u_x/p = {} & c_{11}[(r_1r_2)^{1/2}\cos\bar{\theta} - r\cos\theta] + c_{12}r^2(r_1r_2)^{-1/2}\sin\theta\sin(\theta-\bar{\theta}) \\
& + 1/2c_{13}r^2(r_1r_2)^{-3/2}a^2\sin^2\theta\cos 3\bar{\theta} \\
& - 1/2c_{14}r^4(r_1r_2)^{-5/2}a^2\sin^3\theta\sin(\theta-5\bar{\theta}) \\
& - 1/8c_{15}r^4(r_1r_2)^{-7/2}a^2\sin^4\theta h_{21} \\
& + 1/8c_{16}r^6(r_1r_2)^{-9/2}a^2\sin^5\theta h_{22} \\
u_y/p = {} & c_{21}[(r_1r_2)^{1/2}\sin\bar{\theta} - r\sin\theta] \\
& + c_{22}r[1 - r(r_1r_2)^{-1/2}\cos(\theta-\bar{\theta})]\sin\theta \\
& - 1/2c_{23}r^2(r_1r_2)^{-3/2}a^2\sin^2\theta\sin 3\bar{\theta} \\
& + 1/2c_{24}r^4(r_1r_2)^{-5/2}a^2\sin^3\theta\cos(\theta-5\bar{\theta}) \\
& + 1/8c_{25}r^4(r_1r_2)^{-7/2}a^2\sin^4\theta h_{11} \\
& - 1/8c_{26}r^6(r_1r_2)^{-9/2}a^2\sin^5\theta h_{12} \\
u_z/p = {} & c_{31}[(r_1r_2)^{1/2}\cos\bar{\theta} - r\cos\theta] + c_{32}r^2(r_1r_2)^{-1/2}\sin\theta\sin(\theta-\bar{\theta}) \\
& + 1/2c_{33}r^2(r_1r_2)^{-3/2}a^2\sin^2\theta\cos 3\bar{\theta} \\
& - 1/2c_{34}r^4(r_1r_2)^{-5/2}a^2\sin^3\theta\sin(\theta-5\bar{\theta}) \\
& - 1/8c_{35}r^4(r_1r_2)^{-7/2}a^2\sin^4\theta h_{21} \\
& + 1/8c_{36}r^6(r_1r_2)^{-9/2}a^2\sin^5\theta h_{22} \\
w_x/p = {} & c_{41}[(r_1r_2)^{1/2}\cos\bar{\theta} - r\cos\theta] + c_{42}r^2(r_1r_2)^{-1/2}\sin\theta\sin(\theta-\bar{\theta}) \\
& + 1/2c_{43}r^2(r_1r_2)^{-3/2}a^2\sin^2\theta\cos 3\bar{\theta} \\
& - 1/2c_{44}r^4(r_1r_2)^{-5/2}a^2\sin^3\theta\sin(\theta-5\bar{\theta}) \\
& - 1/8c_{45}r^4(r_1r_2)^{-7/2}a^2\sin^4\theta h_{21} \\
& + 1/8c_{46}r^6(r_1r_2)^{-9/2}a^2\sin^5\theta h_{22} \\
w_y/p = {} & c_{51}[(r_1r_2)^{1/2}\sin\bar{\theta} - r\sin\theta] \\
& + c_{52}r[1 - r(r_1r_2)^{-1/2}\cos(\theta-\bar{\theta})]\sin\theta
\end{aligned}$$

$$
\begin{aligned}
&- 1/2c_{53}r^2(r_1r_2)^{-3/2}a^2\sin^2\theta\sin 3\bar{\theta}\\
&+ 1/2c_{54}r^4(r_1r_2)^{-5/2}a^2\sin^3\theta\cos(\theta-5\bar{\theta})\\
&+ 1/8c_{55}r^4(r_1r_2)^{-7/2}a^2\sin^4\theta h_{11}\\
&- 1/8c_{56}r^6(r_1r_2)^{-9/2}a^2\sin^5\theta h_{12}
\end{aligned}
$$

$$
\begin{aligned}
w_z/p = {}& c_{61}[(r_1r_2)^{1/2}\cos\bar{\theta} - r\cos\theta] + c_{62}r^2(r_1r_2)^{-1/2}\sin\theta\sin(\theta-\bar{\theta})\\
&+ 1/2c_{63}r^2(r_1r_2)^{-3/2}a^2\sin^2\theta\cos 3\bar{\theta}\\
&- 1/2c_{64}r^4(r_1r_2)^{-5/2}a^2\sin^3\theta\sin(\theta-5\bar{\theta})\\
&- 1/8c_{65}r^4(r_1r_2)^{-7/2}a^2\sin^4\theta h_{21}\\
&+ 1/8c_{66}r^6(r_1r_2)^{-9/2}a^2\sin^5\theta h_{22}
\end{aligned}
\tag{9.7-9}
$$

where

$$
\begin{aligned}
h_{11} &= a^2\sin 7\bar{\theta} - 4r^2\sin(2\theta-7\bar{\theta}),\\
h_{12} &= 3a^2\cos(\theta-9\bar{\theta}) + 4r^2\cos(3\theta-9\bar{\theta})\\
h_{21} &= a^2\cos 7\bar{\theta} + 4r^2\cos(2\theta-7\bar{\theta}),\\
h_{22} &= 3a^2\sin(\theta-9\bar{\theta}) + 4r^2\sin(3\theta-9\bar{\theta})
\end{aligned}
\tag{9.7-10}
$$

with constants c_{ij}

$$
\begin{aligned}
&c_{i1} = \sum_{j=1}^{6} a_{j1}b_j, \quad c_{i2} = \sum_{j=1}^{5} a_{j1}b_{j+1}, \quad c_{i3} = \sum_{j=1}^{4} a_{j1}b_{j+2},\\
&c_{i4} = \sum_{j=1}^{3} a_{j1}b_{j+3}, \quad c_{i5} = \sum_{j=1}^{2} a_{j1}b_{j+4}, \quad c_{i6} = \sum_{j=1}^{1} a_{j1}b_{j+5},\\
&\qquad\qquad\qquad\qquad\qquad\qquad i = 1, 2, \ldots, 6
\end{aligned}
\tag{9.7-11}
$$

$\theta = (\theta_1 + \theta_2)/2$, which are defined by Fig. 9.9(c), and constants a_{ij} are listed as follows:

$$
\begin{aligned}
a_{11} &= R(2c_0(5\lambda + 9\mu) - \alpha\mu)\\
a_{12} &= R(\alpha\mu - 2c_0(39\lambda + 67\alpha\mu))\\
a_{13} &= 2R(-6\alpha\mu + 8\beta(\lambda + 2\mu) + c_0(111\lambda + 179\mu))
\end{aligned}
$$

$$a_{14} = -2R(157c_0\lambda + 28\beta\lambda + 249c_0\mu - 16\alpha\mu + 56\beta\mu)$$

$$a_{15} = 5R(-7\alpha\mu + 16\beta(\lambda + 2\mu) + c_0(50\lambda + 82\mu))$$

$$a_{16} = R(21\alpha\mu - 58\beta(\lambda + 2\mu) - 8c_0(15\lambda + 26\mu))$$

$$a_{21} = c_0R(32\lambda + 27\mu)$$

$$a_{22} = -14c_0R(8\lambda + 5\mu)$$

$$a_{23} = R(c_0(176\lambda + 59\mu) + 16(\beta\lambda - \alpha\mu + 2\beta\mu))$$

$$a_{24} = -8(c_0(20\lambda - 2\mu) - 7\alpha\mu + 5\beta(\lambda + 2\mu))$$

$$a_{25} = 10R(c_0(9\lambda - 7\mu) + 4(-2\alpha\mu + \beta(\lambda + 2\mu)))$$

$$a_{26} = R(62\alpha\mu - 18\beta(\lambda + 2\mu) + c_0(-30\lambda + 62\mu))$$

$$a_{31} = 2c_1(24\alpha - 5\beta)$$

$$a_{32} = c_1(-192\alpha + 78\beta)$$

$$a_{33} = c_1(340\alpha - 226\beta)$$

$$a_{34} = c_1(-352\alpha + 306\beta)$$

$$a_{35} = 5c_1(47\alpha - 43\beta)$$

$$a_{36} = c_1(-103\alpha + 85\beta)$$

$$a_{41} = 0$$

$$a_{42} = -16\omega c_0$$

$$a_{43} = -16[10\mu(\lambda + 2\mu) - 7R^2(\lambda + \mu)]$$

$$a_{44} = -24\omega(2c_0 - \alpha + \beta)$$

$$a_{45} = 60\omega(c_0 - \alpha + \beta)$$

$$a_{46} = -2\omega(20c_0 - 27(\alpha - \beta))$$

$$a_{51} = 0$$

$$a_{52} = 32\omega(\alpha - \beta)$$

$$a_{53} = -16\omega(c_0 + 7\alpha - 7\beta)$$

$$a_{54} = 24\omega(2c_0 + 7\alpha - 7\beta)$$

$$a_{55} = -4\omega(17c_0 + 37\alpha - 33\beta)$$

$$a_{56} = 2\omega(28c_0 + 43\alpha - 27\beta)$$

$$a_{61} = 2c_2(24\alpha - 5\beta)$$

$$a_{62} = c_2(-192\alpha + 78\beta)$$

$$a_{63} = c_2(340\alpha - 226\beta)$$

$$a_{64} = c_2(-352\alpha + 306\beta)$$

$$a_{65} = 5c_2(47\alpha - 43\beta)$$

$$a_{66} = c_2(-103\alpha + 85\beta) \tag{9.7-12}$$

Besides, b_j are defined by matrix calculation:

$$b_j = (-1)^j \frac{\Delta_j}{\Delta} \quad j = 1, \ldots, 6 \tag{9.7-13}$$

$$\Delta = \begin{vmatrix} b_{11} & b_{12} & b_{13} & b_{14} & b_{15} & b_{16} \\ b_{21} & b_{22} & b_{23} & b_{24} & b_{25} & b_{26} \\ b_{31} & b_{32} & b_{33} & b_{34} & b_{35} & b_{36} \\ b_{41} & b_{42} & b_{43} & b_{44} & b_{45} & b_{46} \\ b_{51} & b_{52} & b_{53} & b_{54} & b_{55} & b_{56} \\ b_{61} & b_{62} & b_{63} & b_{64} & b_{65} & b_{66} \end{vmatrix}$$

$$\Delta_j = \begin{vmatrix} b_{21} & \cdots & b_{2,j-1} & b_{2,j+1} & \cdots & b_{26} \\ b_{31} & \cdots & b_{3,j-1} & b_{3,j+1} & \cdots & b_{36} \\ b_{41} & \cdots & b_{4,j-1} & b_{4,j+1} & \cdots & b_{46} \\ b_{51} & \cdots & b_{5,j-1} & b_{5,j+1} & \cdots & b_{56} \\ b_{61} & \cdots & b_{6,j-1} & b_{6,j+1} & \cdots & b_{66} \end{vmatrix} \tag{9.7-14}$$

with matrix elements b_{ij}

$$b_{11} = -R(\alpha\lambda\mu + c_0(22\lambda^2 + 73\lambda\mu + 54\mu^2))$$

$$b_{12} = R(32\omega(\alpha - \beta) + \alpha\lambda\mu + c_0(66\lambda^2 + 215\lambda\mu + 194\mu^2))$$

$$b_{13} = -R(c_0(32\omega + 66\lambda^2 + 347\lambda\mu + 258\mu^2) + 4(36\omega(\alpha - \beta) + \mu(8\beta(\lambda + 2\mu) - \alpha(\lambda + 8\mu))))$$

$$b_{14} = R(c_0(112\omega + 22\lambda^2 + 217\lambda\mu + 86\mu^2) + 8(32\omega(\alpha - \beta) + \mu(14\beta(\lambda + 2\mu) - \alpha(5\lambda + 18\mu))))$$

$$b_{15} = R(-4c_0(44\omega + (\lambda - 43\mu)\mu) + 16\omega(16\beta - 15\alpha) + \mu(-160\beta(\lambda + 2\mu) + \alpha(101\lambda + 272\mu)))$$

$$b_{16} = R(4c_0(41\omega - (25\lambda + 66\mu)\mu) - 12\omega(11\beta - 15\alpha) + \mu(116\beta(\lambda + 2\mu) - \alpha(121\lambda + 284\mu)))$$

$$b_{21} = 2(c_2K_2 + c_1R)(24\alpha - 5\beta) + R^2(-c_0(21\lambda + 8\mu) + \beta(\lambda + 2\mu) - 2\alpha\mu)$$

$$b_{22} = (c_2K_2 - c_1R)(-288\alpha + 98\beta) + R^2(c_0(109\lambda - 10\mu) - \beta(\lambda + 2\mu) + 14\alpha\mu)$$

$$b_{23} = 4(c_2K_2 + c_1R)(193\alpha - 98\beta) - 54\alpha\mu R^2 - c_0(16K_1\omega + R^2(220\lambda - 239\mu))$$

$$b_{24} = 2(-(c_2K_2 - c_1R)(612\alpha - 418\beta) - 12K_1\omega(\alpha - \beta) + 59\alpha\mu R^2) + 5c_0(16K_1\omega + R^2(44\lambda - 157\mu))$$

$$b_{25} = (c_2K_2 - c_1R)(1279\alpha - 1053\beta) + 108K_1\omega(\alpha - \beta) - 145\alpha\mu R^2 + c_0(-172K_1\omega - 5R^2(22\lambda - 216\mu))$$

$$b_{26} = -(c_2K_2 - c_1R)(925\alpha - 821\beta) - 198K_1\omega(\alpha - \beta) + 99\alpha\mu R^2 + c_0(208K_1\omega + R^2(22\lambda - 1325\mu))$$

$$b_{31} = -2(c_2(K_1 - K_2) + c_1R)(24\alpha - 5\beta) + RK_2(c_0(42\lambda + 45\mu) - \alpha\mu)$$

$$b_{32} = 2(c_2(K_1 - K_2) + c_1R)(144\alpha - 49\beta) - RK_2(c_0(242\lambda + 267\mu) - 3\alpha\mu)$$

$$b_{33} = -4(c_2(K_1 - K_2) + c_1R)(193\alpha - 98\beta) + RK_2(c_0(676\lambda + 773\mu) - 31\alpha\mu + 32\beta(\lambda + 2\mu)$$

$$b_{34} = 4(c_2(K_1 - K_2) + c_1R)(306\alpha - 209\beta) - RK_2(c_0(1172\lambda + 1391\mu) - 129\alpha\mu + 144\beta(\lambda + 2\mu)$$

$$b_{35} = -(c_2(K_1 - K_2) + c_1R)(1279\alpha - 1053\beta) + RK_2(2c_0(675\lambda + 839\mu) - 247\alpha\mu + 288\beta(\lambda + 2\mu)$$

$$\begin{aligned}
b_{36} &= (c_2(K_1 - K_2) + c_1 R)(925\alpha - 821\beta) \\
&\quad + RK_2(34c_0(31\lambda + 41\mu) - 265\alpha\mu + 332\beta(\lambda + 2\mu) \\
b_{41} &= -2(c_2 R + c_1\mu)(24\alpha - 5\beta) \\
b_{42} &= -4((c_2 R + c_1\mu)(96\alpha - 39\beta) + R\omega(20\alpha - 16\beta + 2c_0)) \\
b_{43} &= -4(19(c_2 R + c_1\mu)(7\alpha - 4\beta) + 28R\omega(\alpha - \beta)) \\
b_{44} &= 4((c_2 R + c_1\mu)(61\alpha - 49\beta) - R\omega(64\alpha - 36\beta - 24c_0)) \\
b_{45} &= (c_2 R + c_1\mu)(587\alpha - 521\beta) - R\omega(232\alpha - 216\beta - 40c_0) \\
b_{46} &= (c_2 R + c_1\mu)(338\alpha - 300\beta) + R\omega(200\alpha - 168\beta - 44c_0) \\
b_{51} &= R\mu(-c_0(42\lambda + 45\mu) + \alpha\mu) \\
b_{52} &= -2R(-4\omega(7\alpha - 5\beta) + \mu(11\alpha\mu - 4\beta(\lambda + 2\mu)) \\
&\quad + c_0(4\omega + \mu(76\lambda + 93\mu))) \\
b_{53} &= R(112\omega(\alpha - \beta) + \mu(29\alpha\mu - 32\beta(\lambda + 2\mu)) \\
&\quad + c_0(32\omega - \mu(476\lambda + 551\mu))) \\
b_{54} &= 4R(-\omega(92\alpha - 120\beta) + \mu(31\alpha\mu - 56\beta(\lambda + 2\mu)) \\
&\quad + 2c_0(28\omega + \mu(17\lambda - 7\mu))) \\
b_{55} &= R(16\omega(4\alpha - 3\beta) + \mu(147\alpha\mu - 176\beta(\lambda + 2\mu)) \\
&\quad + 2c_0(88\omega - \mu(327\lambda + 419\mu))) \\
b_{56} &= 2R(2\omega(7\alpha - 15\beta) - \mu(59\alpha\mu - 78\beta(\lambda + 2\mu)) \\
&\quad - c_0(78\omega - \mu(200\lambda + 278\mu))) \\
b_{61} &= -2(c_2 K_2 + c_1 R)(24\alpha - 5\beta) \\
&\quad - \omega(\alpha + c_0)(K_1 + R) + Rc_0(42\lambda + 46\mu) \\
b_{62} &= 6(c_2 K_2 + c_1 R)(32\alpha - 13\beta) \\
&\quad + \omega(9R\alpha - 23K_1\alpha + 32K_1\beta) - 8R^2\alpha\mu \\
&\quad + 3c_0(3K_1\omega + R(3\omega - 74R\lambda - 80R\mu))
\end{aligned}$$

$$
\begin{aligned}
b_{63} &= -2(c_2K_2 + c_1R)(170\alpha - 113\beta) \\
&\quad + \omega(-36R\alpha + 108K_1\alpha + 144K_1\beta) \\
&\quad + 8R^2(\alpha\mu + \beta\lambda) + c_0(-20K_1\omega + R(-36\omega + 510R\lambda + 523R\mu)) \\
b_{64} &= 2(c_2K_2 + c_1R)(176\alpha - 153\beta) + 2\omega(42R\alpha - 98K_1\alpha + 140K_1\beta) \\
&\quad + 4R^2(5\alpha\mu - 56\beta\lambda - 112\beta\mu) \\
&\quad + c_0(20K_1\omega + R(84\omega - 650R\lambda - 625R\mu)) \\
b_{65} &= -5(c_2K_2 + c_1R)(47\alpha - 43\beta) + 2\omega(-63R\alpha + 95K_1\alpha + 150K_1\beta) \\
&\quad + 5R^2(-9\alpha\mu + 32\beta\lambda + 64\beta\mu) \\
&\quad - 2c_0(5K_1\omega + R(63\omega - 25R(10\lambda + 9\mu))) \\
b_{66} &= (c_2K_2 + c_1R)(103\alpha - 85\beta) + 6\omega(21R\alpha - 18K_1\alpha + 31K_1\beta) \\
&\quad + R^2(37\alpha\mu - 116\beta\lambda - 232\beta\mu) \\
&\quad + 2c_0(K_1\omega + R(63\omega - 120R\lambda - 101R\mu))
\end{aligned}
\tag{9.7-15}
$$

With these data, any information concerning field variables in any point can be found.

In practical applications, people are interested in the stress field near the crack tip, i.e., the zone $r_1/a \ll 1$. Therefore, we consider the normal stress of a phonon field around the right crack tip, for example. In the obtained results, maintaining the term $(r_1/a)^{-1/2}$ in stress expressions is sufficient, and the others are higher-order small quantities. According to the definition of the phonon stress intensity factor of Mode I, we find that

$$
K_{\mathrm{I}}^{\|} = \lim_{x \to a^+} \{[2\pi(x-a)]^{1/2}\sigma_{yy}(x,0)\} = \sqrt{\pi a}p \tag{9.7-16}
$$

This is identical to that of conventional structural materials.

A more important quantity is the crack energy release rate, which correlates not only the stress field but also the displacement field. Near the crack tip,

$$
u_y(y,0) = \begin{cases} pM\sqrt{a^2 - x^2} & |x| \le a \\ 0 & |x| > a \end{cases} \tag{9.7-17}
$$

$$
\sigma_{yy}(x,0) = \begin{cases} -p & |x| \le a \\ p\left(\dfrac{|x|}{\sqrt{x^2-a^2}} - 1\right) & |x| > a \end{cases} \tag{9.7-18}
$$

where

$$
M = \sum_{j=1}^{6} a_{2j} b_j \tag{9.7-19}
$$

a_{2j} see (9.7-12), b_j refer to (9.7-13).

The crack strain energy and energy release rate are as follows, respectively:

$$
\begin{aligned}
W &= 2\int_0^a \sigma_{yy}(x,0)u_y(x,0)dx = M\pi a^2 p^2/2 \\
G_I &= \frac{1}{2}\frac{\partial W}{\partial a} = M(K_{\mathrm{I}}^{\parallel})^2/2
\end{aligned} \tag{9.7-20}
$$

which is shown in Fig. 9.10. In addition, the crack opening displacement is depicted in Fig. 9.11.

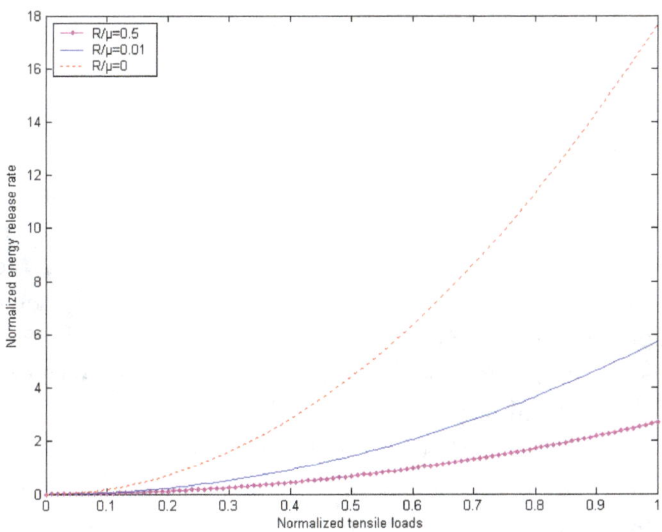

Fig. 9.10. Crack energy release rate versus applied stress, effect of coupling between phonons and phasons.

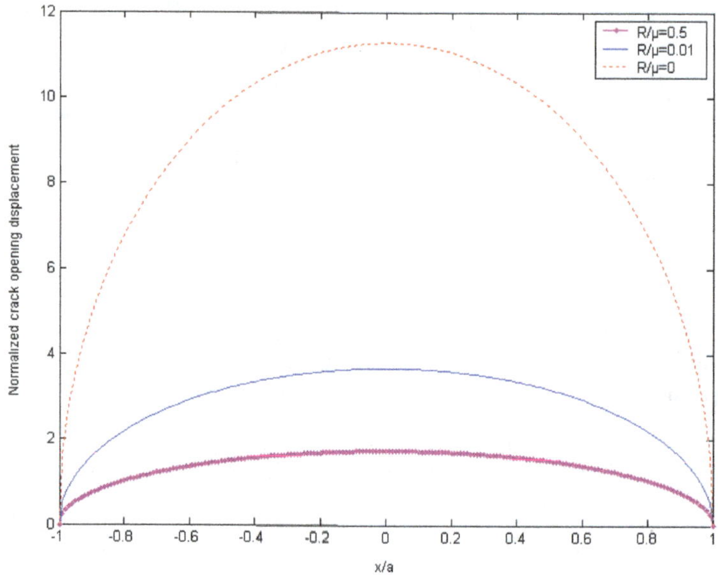

Fig. 9.11. Crack opening displacement and influence due to phonon–phason coupling.

9.8 An Elliptic Notch/Griffith Crack in an Icosahedral Quasicrystal

The notch problem for icosahedral quasicrystals has no solution before 2006; the difficulty lies in which cannot be solved by the Fourier transform method. We must develop other methods, in which the complex variable function-conformal mapping method is particular effect, and some analytic solutions are constructed by the method. The results are given in Ref. [21], which may be seen as a development of the same authors in Ref. [22].

In this section, we consider an icosahedral quasicrystal with an elliptic notch along the z-axis. On the basis of the general solution introduced in Section 9.5, explicit expressions of stress and displacement components of phonon and phason fields in the quasicrystals are given. With the help of conformal mapping, an analytic solution for elliptic notch problem of the quasicrystals is presented. The solution of the Griffith crack problem can be observed as a special case of the results, which will be reduced to the well-known results in a conventional material if the phason field is absent. The stress intensity factor and energy of the release rate are also obtained.

9.8.1 *The complex representation of stresses and displacements*

The solution of (9.5-4) can be expressed as

$$G(x,y) = \mathrm{Re}[g_1(z) + \bar{z}g_2(z) + \bar{z}^2 g_3(z) + \bar{z}^3 g_4(z) + \bar{z}^4 g_5(z) + \bar{z}^5 g_6(z)] \tag{9.8-1}$$

where $g_i(z)$ are arbitrary analytic functions of $z = x + iy$ and the bar over the complex variable or complex function denotes the complex conjugate.

By Eqs. (9.5-2)–(9.5-4) and (9.8-1), the stresses can be expressed as follows:

$$\begin{aligned}
&\sigma_{xx} + \sigma_{yy} = 48c_2c_3R\mathrm{Im}\Gamma'(z)\\
&\sigma_{yy} - \sigma_{xx} + 2i\sigma_{xy} = 8ic_2c_3R(12\overline{\Psi'(z)} - \Omega'(z))\\
&\sigma_{zy} - i\sigma_{zx} = -960c_3c_4f_6'(z) \quad \sigma_{zz} = \frac{24\lambda R}{(\mu+\lambda)}c_2c_3\mathrm{Im}\Gamma'(z)\\
&H_{xy} - H_{yx} - i(H_{xx} + H_{yy}) = -96c_2c_5\overline{\Psi'(z)} - 8c_1c_2R\Omega'(z)\\
&H_{yx} - H_{xy} + i(H_{xx} - H_{yy}) = -480c_2c_5\overline{f_6'(z)} - 4c_1c_2R\Theta'(z)\\
&H_{yz} + iH_{xz} = 48c_2c_6\Gamma'(z) - 4c_2R^2(2K_2 - K_1)\overline{\Omega'(z)}\\
&H_{zz} = \frac{24R^2}{(\mu+\lambda)}c_2c_3\mathrm{Im}\Gamma'(z)
\end{aligned} \tag{9.8-2}$$

where

$$c_1 = \frac{R(2K_2 - K_1)(\mu K_1 + \mu K_2 - 3R^2)}{2(\mu K_1 - 2R^2)}$$

$$c_2 = \frac{1}{R}K_2(\mu K_2 - R^2) - R(2K_2 - K_1)$$

$$c_3 = \mu(K_1 - K_2) - R^2 - \frac{(\mu K_2 - R^2)^2}{\mu K_1 - 2R^2}$$

$$c_4 = c_1 R + \frac{1}{2}c_3\left(K_1 + \frac{\mu K_1 - 2R^2}{\lambda + \mu}\right)$$

and

$$\Psi(z) = f_5(z) + 5\bar{z}f_6'(z)$$

$$\Gamma(z) = f_4(z) + 4\bar{z}f_5'(z) + 10\bar{z}^2 f_6''(z)$$

$$\Omega(z) = f_3(z) + 3\bar{z}f_4'(z) + 6\bar{z}^2 f_5''(z) + 10\bar{z}^3 f_6'''(z)$$

$$\Theta(z) = f_2(z) + 2\bar{z}f_3'(z) + 3\bar{z}^2 f_4''(z) + 4\bar{z}^3 f_5'''(z) + 5\bar{z}^4 f_6^{(IV)}(z)$$

$$c_5 = 2c_4 - c_1 R, \quad c_6 = (2K_2 - K_1)R^2 - 4c_4\frac{\mu K_2 - R^2}{\mu K_1 - 2R^2} \tag{9.8-3}$$

In the above expressions, the function $g_1(z)$ does not appear; this implies that for the stress boundary value problem in this formalism, only five complex potentials $g_2(z)$, $g_3(z)$, $g_4(z)$, $g_5(z)$ and $g_6(z)$ are needed and we can take $g_1(z) = 0$. For simplicity, we have introduced the following new symbols:

$$\begin{aligned} &g_2^{(9)}(z) = f_2(z), \quad g_3^{(8)}(z) = f_3(z), \quad g_4^{(7)}(z) = f_4(z) \\ &g_5^{(6)}(z) = f_5(z), \quad g_6^{(5)}(z) = f_6(z) \end{aligned} \tag{9.8-4}$$

where $g_i^{(n)}$ denotes the nth derivatives with the argument z; accordingly, $f_1(z) = 0$ (because $g_1(z) = 0$). Similar to the formalism in Chapter 8, the complex representations of displacement components can be written as follows (here, we have omitted the rigid body displacements):

$$u_y + iu_x = -6c_2 R\left(\frac{2c_3}{\mu + \lambda} + c_7\right) - 2c_2 c_7 R\Omega(z)$$

$$\begin{aligned} u_z = &\frac{4}{\mu(K_1 + K_2) - 3R^2}(240c_{10}\mathrm{Im} f_6(z)) \\ &+ c_1 c_2 R^2 \mathrm{Im}(\Theta(z) - 2\Omega(z) + 6\Gamma(z) - 24\Psi(z))) \end{aligned}$$

$$
\begin{aligned}
w_y + iw_x &= -\frac{R}{c_1(\mu K_1 - 2R^2)}(24c_9\overline{\Psi(z)} - c_8\Theta(z)) \\
w_z &= \frac{4(\mu K_2 - R^2)}{(K_1 - 2K_2)R(\mu(K_1 + K_2) - 3R^2)} \\
&\quad \times (240c_{10}\mathrm{Im} f_6(z)) + c_1c_2R^2\mathrm{Im}(\Theta(z) - 2\Omega(z) \\
&\quad + 6\Gamma(z) - 24\Psi(z))) \qquad (9.8\text{-}5)
\end{aligned}
$$

in which

$$
\begin{aligned}
c_7 &= \frac{c_3K_1 + 2c_1R}{\mu K_1 - 2R^2}, \quad c_8 = c_1c_2R(\mu(K_1 - K_2) - R^2) \\
c_9 &= c_8 + 2c_2c_4\left(c_3 - \frac{(\mu K_2 - R^2)^2}{\mu K_1 - 2R^2}\right), \qquad (9.8\text{-}6) \\
c_{10} &= c_1c_2R^2 - c_4(c_2R - c_3K_1)
\end{aligned}
$$

9.8.2 *Elliptic notch problem*

We consider an icosahedral quasicrystal solid with an elliptic notch, which penetrates through the medium along the z-axis direction, the edge of the elliptic notch subject to the uniform pressure p; see Fig. 9.12.

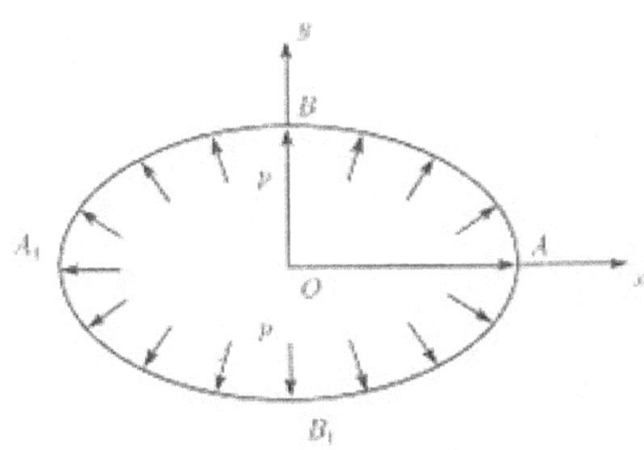

Fig. 9.12. An elliptic notch subject to an inner pressure in icosahedral quasicrystal.

The boundary conditions of this problem can be expressed as follows:

$$\sigma_{xx}\cos(\mathbf{n},x)+\sigma_{xy}\cos(\mathbf{n},y)=T_x,$$
$$\sigma_{xy}\cos(\mathbf{n},x)+\sigma_{yy}\cos(\mathbf{n},y)=T_y,\quad (x,y)\in L \tag{9.8-7}$$
$$H_{xx}\cos(\mathbf{n},x)+H_{xy}\cos(\mathbf{n},y)=h_x,$$
$$H_{yx}\cos(\mathbf{n},x)+H_{yy}\cos(\mathbf{n},y)=h_y,\quad (x,y)\in L \tag{9.8-8}$$
$$\sigma_{zx}\cos(\mathbf{n},x)+\sigma_{zy}\cos(\mathbf{n},y)=0,$$
$$H_{zx}\cos(\mathbf{n},x)+H_{zy}\cos(\mathbf{n},y)=0,\quad (x,y)\in L \tag{9.8-9}$$

where

$$\cos(\mathbf{n},x)=\frac{dy}{ds},\quad \cos(\mathbf{n},y)=-\frac{dx}{ds},$$
$$T_x=-p\cos(\mathbf{n},x),\quad T_y=-p\cos(\mathbf{n},y)$$

T_x, T_y denote the components of surface traction, p is the magnitude of the pressure, h_x, h_y are the components of the generalized surface traction, $\mathbf{n}$ is the outward unit normal vector of any point on the boundary and $L:\frac{x^2}{a^2}+\frac{y^2}{b^2}=1$ is the edge of the elliptic notch. Since the measurement of the generalized traction has not been reported so far, for simplicity, we assume that

$$h_x=0,\quad h_y=0$$

Utilizing Eqs. (9.8-2), (9.8-3) and (9.8-7), we obtain

$$\begin{aligned}&-4c_2c_3R[3(f_4(z)+4\bar{z}f_5'(z)+10\bar{z}^2f_6''(z))-(\overline{f_3(z)}\\&+3z\overline{f_4'(z)}+6z^2\overline{f_5''(z)}+10z^3\overline{f_6'''(z)})]\\&=\int(T_x+iT_y)ds=ipz\end{aligned} \tag{9.8-10}$$

Taking complex conjugate on both sides of Eq. (9.7-10) yields

$$\begin{aligned}&-4c_2c_3R[3(\overline{f_4(z)}+4z\overline{f_5'(z)}+10z^2\overline{f_6''(z)})-(f_3(z)+3\bar{z}f_4'(z)\\&+6z^2f_5'(z)+10\bar{z}^3f_6'''(z)=-ip\bar{z}\end{aligned} \tag{9.8-11}$$

From Eqs. (9.8-2), (9.8-3) and (9.8-8), we have

$$\begin{aligned} &48c_2(2c_4 - c_1 R)\mathrm{Re}\overline{\Psi(z)} + 2c_1c_2R\mathrm{Re}\Theta(z) = 0 \\ &-48c_2(2c_4 - c_1 R)\mathrm{Im}\overline{\Psi(z)} - 2c_1c_2R\mathrm{Im}\Theta(z) = 0 \end{aligned} \tag{9.8-12}$$

Multiplying the second formula of (9.8-12) by $-i$ and adding it to the first one, we can obtain

$$48c_2(2c_4 - c_1 R)\overline{\Psi(z)} + 2c_1c_2R\Theta(z) = 0 \tag{9.8-13}$$

By Eqs. (9.8-2), (9.8-3) and (9.8-9), one has

$$\begin{aligned} &f_6(z) + \overline{f_6(z)} = 0 \\ &4c_{11}\mathrm{Re}[f_5(z) + 5\bar{z}f_6'(z)] + (2K_2 - K_1)R\mathrm{Re}[f_4(z) + 4\bar{z}f_5'(z) \\ &\quad + 10\bar{z}^2 f_6''(z) + 20f_6(z)] = 0 \end{aligned} \tag{9.8-14}$$

in which

$$c_{11} = (2K_2 - K_1)R - \frac{4c_4(\mu K_2 - R^2)}{(\mu K_1 - 2R^2)R} \tag{9.8-15}$$

However, the further calculation will be very difficult at the z-plane owing to the complexity of the manipulation; we must employ the conformal mapping

$$z = \omega(\zeta) = R_0\left(\frac{1}{\zeta} + m\zeta\right) \tag{9.8-16}$$

to transform the region with the ellipse at the z-plane onto the interior of the unit circle γ at the ζ-plane, referring to Fig. 8.8, in which

$$R_0 = (a+b)/2, \quad m = (a-b)/(a+b)$$

Let

$$f_j(z) = f_j[\omega(\zeta)] = \Phi_j(\zeta) \quad (j = 2, \ldots, 6) \tag{9.8-17}$$

Substituting (9.8-16) into (9.8-10), (9.8-11), (9.8-13) and (9.8-14), then multiplying both sides of the equations by $d\sigma/[2\pi i(\sigma - \zeta)]$

(σ represents the value of ζ at the unit circle) and integrating along the unit circle γ, by means of the Cauchy integral formula and analytic extension of the complex analysis theory, we obtain (the details are given in Appendix of Chapter 11)

$$\Phi_2(\zeta) = \frac{R_0}{2c_2c_3R}\frac{ip\zeta(\zeta^2+m)(m^3\zeta^2+1)}{(m\zeta^2-1)^3} + \frac{(2K_2-K_1)R_0}{2c_2c_3C_{11}} \times \frac{pm\zeta^3(\zeta^2+m)[m^2\zeta^6-(m^3+4m)\zeta^4+(2m^4+4m^2+5)\zeta^2+m]}{(m\zeta^2-1)^5}$$

$$\Phi_3(\zeta) = \frac{R_0}{4c_2c_3R}\frac{ip\zeta(m^2+1)}{(m\zeta^2-1)} - \frac{(2K_2-K_1)R_0}{12c_2c_3C_{11}}\frac{pm\zeta^3(\zeta^2+m)(m\zeta^2-m^2-2)}{(m\zeta^2-1)^3}$$

$$\Phi_4(\zeta) = -\frac{R_0}{12c_2c_3R}ipm\zeta - \frac{(2K_2-K_1)R_0}{2c_2c_3C_{11}}\frac{pm\zeta(\zeta^2+m)}{(m\zeta^2-1)}$$

$$\Phi_5(\zeta) = -\frac{(2K_2-K_1)R_0}{48c_2c_3C_{11}}pm\zeta, \qquad \Phi_6(\zeta) = 0 \tag{9.8-18}$$

The elliptic notch problem has been solved.

The solution of the Griffith crack subjected to a uniform pressure can be obtained if we put $m = 1$, $R_0 = a/2$ in the above notch solution. The solution of the crack can be expressed explicitly on the z-plane, for example,

$$\begin{aligned}\sigma_{yy} = \mathrm{Im}\Bigg[& ip\left(\frac{z}{\sqrt{z^2-a^2}} + \frac{ia^2y}{(\sqrt{z^2-a^2})^3} - 1\right) \\ & + \frac{3(2K_2-K_1)R}{2c_{11}}\frac{ipa^2y}{\sqrt{(z^2-a^2)^3}} + \frac{(2K_2-K_1)R}{2c_{11}}\frac{ipy(2a^4-3z\bar{z})}{\sqrt{(z^2-a^2)^5}} \\ & - \frac{(2K_2-K_1)R}{4c_{11}}\frac{a^2pz(z\bar{z}-a^2)}{\sqrt{(z^2-a^2)^5}} + \frac{(2K_2-K_1)R}{4c_{11}}\frac{a^2p\bar{z}}{\sqrt{(z^2-a^2)^3}}\Bigg]\end{aligned} \tag{9.8-19}$$

$$\begin{aligned}u_y = -6c_2R\left(\frac{2c_3}{\mu+\lambda}+c_7\right)\mathrm{Re}\Bigg[& \frac{ip}{24c_2c_3R}(\overline{z-\sqrt{z^2-a^2}}) \\ & + \frac{2K_2-K_1}{24c_2c_3c_{11}}p\left(\frac{z\bar{z}}{\sqrt{z^2-a^2}} - \frac{a^2}{\sqrt{z^2-a^2}} - \sqrt{z^2-a^2}\right)\Bigg]\end{aligned}$$

$$
\begin{aligned}
&- 2c_2c_7\mathrm{Re}\left[\frac{ip}{8c_2c_3R}\left(\frac{z\bar{z}}{\sqrt{z^2-a^2}} - \frac{a^2}{\sqrt{z^2-a^2}} - \bar{z}\right)\right. \\
&- \frac{2K_2-K_1}{4c_2c_3c_{11}}ipy + \frac{2K_2-K_1}{16c_2c_3c_{11}}p\left(\frac{a^2[(z\bar{z}-a^2)+2iy\bar{z}]}{\sqrt{(z^2-a^2)^3}}\right) \\
&\left.+ \frac{2K_2-K_1}{16c_2c_3c_{11}}p\left(\frac{a^2}{\sqrt{z^2-a^2}} - \frac{2z\bar{z}}{\sqrt{z^2-a^2}} + 2\sqrt{z^2-a^2}\right)\right]
\end{aligned}
\tag{9.8-20}
$$

From Eqs. (9.7-19) and (9.7-20), the stress intensity factor and energy release rate can be evaluated as follows:

$$
\begin{aligned}
K_I^{\|} &= \sqrt{\pi a}p \\
G_I &= \frac{1}{2}\frac{\partial}{\partial a}\left[2\int_{-a}^{a}(\sigma_{yy}(x,0)\oplus H(x,0))(u_y(x,0)\oplus w_y(x,0))dx\right] \\
&= \frac{1}{2}\left(\frac{1}{\lambda+\mu}+\frac{c_7}{c_3}\right)\left(K_I^{\|}\right)^2
\end{aligned}
\tag{9.8-21}
$$

in which material constant c_3 is given by (9.5-5) and c_7 by (9.8-6); it is evident that the crack energy release rate depends not only on the phonon elastic constants λ, μ but also on the phason elastic constants K_1, K_2 and on the phonon–phason coupling elastic constant R though we assume phason tractions $h_x = h_y = 0$.

9.8.3 *Brief summary*

The notch problem can be solved only by the complex variable function method; the solution includes that of the Griffith crack problem naturally. Though the Fourier transform can solve the Griffith crack problem, referring to Zhu and Fan [25], it cannot solve the notch problem. Whatever the solution for the notch or crack here reveals the effects of not only the phonon but also the phason and phonon–phason coupling. The numerical examples on the crack opening displacement $\delta(x) = u_y(x,+0) - u_y(x,-0)$ and the energy release rate G_I for different values of R/μ are identical to those given in Figs. 9.10 and 9.11, respectively.

Both solutions given by the complex variable function method and the Fourier transform reduce to that of the classical theory when the phason is absent; this is helpful to examine the present result.

This study developed the previous work for the elasticity of two-dimensional quasicrystals given by Fan and co-workers. This work is helpful to understand quantitatively the influence of the elliptic notch and crack on the mechanical behaviour of icosahedral quasicrystals. The stress intensity factor and energy release rate are also obtained as the direct results of the solution, which are all important criteria of fracture mechanics.

The strict theory on the complex potential method is summarized in-depth in Chapter 11, which can also be referred to in [26].

9.9 Elasticity of Cubic Quasicrystals — The Anti-plane and Axisymmetric Deformation

Cubic quasicrystal is one of the important three-dimensional quasicrystals. Due to the complexity of the basic equations of the elasticity of the quasicrystal, there are few analytic solutions. Developing a systematic and direct method for solving the complicated boundary value problem of the elasticity of a quasicrystal is a fundamental task. Because the phasons in this case have the same irreducible representation as the phonons, the stress and strain tensors are symmetrical. With this feature, we can discuss two cases: one for the anti-plane and another for the axisymmetric elastic theory of cubic quasicrystal, and the latter can reveal the three-dimensional effect of the elasticity; this may be the only three-dimensional elastic analytic solution for quasicrystals so far. In addition, a penny-shaped crack problem under tensile loading in the material is investigated, and the exact analytic solution is obtained by using the Hankel transform and dual integral equations theory, and the stress intensity factor and the strain energy release rate are determined, which provide some useful information for studying the deformation and fracture of the quasicrystalline material.

From the physical basis provided by Hu *et al.* [2], we first discuss the anti-plane elasticity, such as, i.e., the stress–strain relation

$$\begin{aligned}
\sigma_{23} &= 2C_{44}\varepsilon_{23} + R_{44}w_{23}\\
\sigma_{31} &= 2C_{44}\varepsilon_{31} + R_{44}w_{31}\\
H_{23} &= 2R_{44}\varepsilon_{23} + K_{44}w_{23}\\
H_{31} &= 2R_{44}\varepsilon_{31} + K_{44}w_{31}
\end{aligned}$$

the deformation geometry

$$\varepsilon_{23} = \frac{1}{2}\frac{\partial u_3}{\partial x_2}, \quad \varepsilon_{31} = \frac{1}{2}\frac{\partial u_3}{\partial x_1}, \quad w_{23} = \frac{\partial w_3}{\partial x_2}, \quad w_{31} = \frac{\partial w_3}{\partial x_1}$$

and the equilibrium equations

$$\frac{\partial \sigma_{31}}{\partial x_1} + \frac{\partial \sigma_{32}}{\partial x_2} = 0, \quad \frac{\partial H_{31}}{\partial x_1} + \frac{\partial H_{32}}{\partial x_2} = 0$$

These equations are exactly similar to those of the anti-plane elasticity of one-dimensional and icosahedral quasicrystals, and so result in the final governing equations

$$\nabla^2 u_3 = 0, \quad \nabla^2 w_3 = 0$$

The solution of this can be derived from the relevant discussion in Chapters 5, 7 and 8 and Section 9.2 of this chapter, so it need not be mentioned again.

For the axisymmetric case, Zhou and Fan [23] developed a displacement potential theory to reduce the basic equations to a single partial differential equation with a high order in the circular cylindrical coordinate system (r, θ, z), i.e., assume

$$\frac{\partial}{\partial \theta} = 0 \tag{9.9-1}$$

and by the generalized Hooke's law,

$$\begin{aligned}
\sigma_{rr} &= C_{11}\varepsilon_{rr} + C_{12}(\varepsilon_{\theta\theta} + \varepsilon_{zz}) + R_{11}w_{rr} + R_{12}(w_{\theta\theta} + w_{zz})\\
\sigma_{\theta\theta} &= C_{11}\varepsilon_{\theta\theta} + C_{12}(\varepsilon_{rr} + \varepsilon_{zz}) + R_{11}w_{\theta\theta} + R_{12}(w_{rr} + w_{zz})
\end{aligned}$$

$$\begin{aligned}
\sigma_{zz} &= C_{11}\varepsilon_{zz} + C_{12}(\varepsilon_{rr} + \varepsilon_{\theta\theta}) + R_{11}w_{zz} + R_{12}(w_{\theta\theta} + w_{rr}) \\
\sigma_{zr} &= \sigma_{rz} = 2C_{44}\varepsilon_{rz} + 2R_{44}w_{rz} \\
H_{rr} &= R_{11}\varepsilon_{rr} + R_{12}(\varepsilon_{\theta\theta} + \varepsilon_{zz}) + K_{11}w_{rr} + K_{12}(w_{\theta\theta} + w_{zz}) \\
H_{\theta\theta} &= R_{11}\varepsilon_{\theta\theta} + R_{12}(\varepsilon_{rr} + \varepsilon_{zz}) + K_{11}w_{\theta\theta} + K_{12}(w_{rr} + w_{zz}) \\
H_{zz} &= R_{11}\varepsilon_{zz} + R_{12}(\varepsilon_{rr} + \varepsilon_{\theta\theta}) + K_{11}w_{zz} + K_{12}(w_{rr} + w_{\theta\theta}) \\
H_{zr} &= H_{rz} = 2R_{44}\varepsilon_{rz} + 2K_{44}w_{rz}
\end{aligned} \tag{9.9-2}$$

and the equations of deformation geometry

$$\varepsilon_{ij} = \frac{1}{2}\left(\frac{\partial u_i}{\partial x_j} + \frac{\partial u_j}{\partial x_i}\right), \quad w_{ij} = \frac{1}{2}\left(\frac{\partial w_i}{\partial x_j} + \frac{\partial w_j}{\partial x_i}\right)$$

which are

$$\begin{aligned}
&\varepsilon_{rr} = \frac{\partial u_r}{\partial r}, \quad \varepsilon_{\theta\theta} = \frac{u_r}{r}, \quad \varepsilon_{zz} = \frac{\partial u_z}{\partial z} \\
&\varepsilon_{rz} = \varepsilon_{zr} = \frac{1}{2}\left(\frac{\partial u_r}{\partial z} + \frac{\partial u_z}{\partial r}\right) \\
&w_{rr} = \frac{\partial w_r}{\partial r}, \quad w_{\theta\theta} = \frac{w_r}{r}, \quad \varepsilon_{zz} = \frac{\partial w_z}{\partial z} \\
&w_{rz} = w_{zr} = \frac{1}{2}\left(\frac{\partial w_r}{\partial z} + \frac{\partial w_z}{\partial r}\right)
\end{aligned} \tag{9.9-3}$$

and the equations of equilibrium

$$\begin{aligned}
&\frac{\partial \sigma_{rr}}{\partial r} + \frac{\partial \sigma_{rz}}{\partial z} + \frac{\sigma_{rr} - \sigma_{\theta\theta}}{r} = 0 \\
&\frac{\partial \sigma_{zr}}{\partial r} + \frac{\partial \sigma_{zz}}{\partial z} + \frac{\sigma_{zr}}{r} = 0 \\
&\frac{\partial H_{rr}}{\partial r}\frac{\partial \sigma_{zr}}{\partial r} + \frac{\partial \sigma_{zz}}{\partial z} + \frac{\sigma_{zr}}{r} = 0 \\
&\frac{\partial H_{zr}}{\partial r} + \frac{\partial H_{zz}}{\partial z} + \frac{H_{zr}}{r} = 0
\end{aligned} \tag{9.9-4}$$

If all displacements and stresses can be expressed by the potential $F(r, z)$, which satisfies

$$\left[\frac{\partial^8}{\partial z^8} - b\left(\frac{\partial^2}{\partial r^2} + \frac{1}{r}\frac{\partial}{\partial r}\right)\frac{\partial^6}{\partial z^6} + c\left(\frac{\partial^2}{\partial r^2} + \frac{1}{r}\frac{\partial}{\partial r}\right)^2\frac{\partial^4}{\partial z^4}\right.$$
$$\left.-d\left(\frac{\partial^2}{\partial r^2} + \frac{1}{r}\frac{\partial}{\partial r}\right)^3\frac{\partial^2}{\partial z^2} + e\left(\frac{\partial^2}{\partial r^2} + \frac{1}{r}\frac{\partial}{\partial r}\right)^4\right]F = 0 \qquad (9.9\text{-}5)$$

then Eqs. (9.9-2)–(9.9-4) have been automatically satisfied.

Due to the length of these expressions which are not listed here.

As an application of the above theory and method, the solutions of the elastic field of cubic quasicrystal with a penny-shaped crack are discussed in the following.

Assume a penny-shaped crack with radius a in the centre of the cubic quasicrystal material, the size of the crack is much smaller than the solid, so the size of the material can be considered as infinite, and at infinity, the quasicrystal material is subjected to a tension p in the z-direction. The origin of the coordinate system is at the centre of the crack (as shown in Fig. 9.13).

From the symmetry of the problem, it is enough to study the upper half-space $z > 0$ or the lower half-space $z < 0$. In this case,

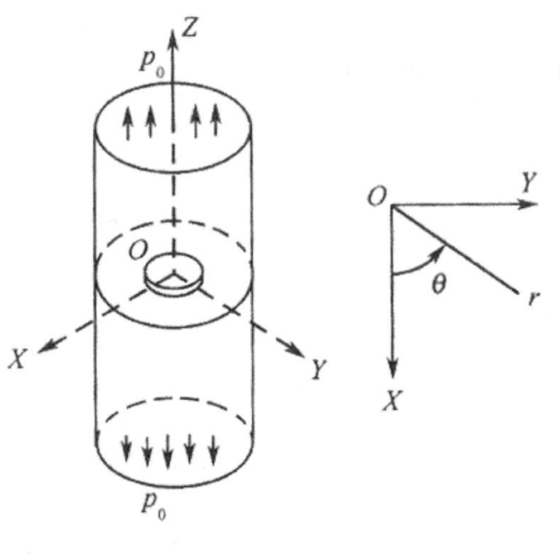

Fig. 9.13. Penny-shaped crack subject to a tension in a cubic quasicrystal.

for studying the upper half-space, the boundary conditions of the problem are described by

$$
\begin{aligned}
&\sqrt{r^2+z^2} \to \infty : \sigma_{zz} = p, \quad H_{zz} = 0, \sigma_{rz} = 0, H_{rz} = 0 \\
&z = 0, \quad 0 \le r \le a, \quad \sigma_{zz} = \sigma_{rz} = 0; \quad H_{zz} = H_{rz} = 0 \\
&z = 0, \quad r > a : \sigma_{rz} = 0, \quad u_z = 0; \quad H_{rz} = 0, \quad w_z = 0
\end{aligned} \tag{9.9-6}
$$

But the boundary conditions can be replaced by

$$
\begin{aligned}
&\sqrt{r^2+z^2} \to \infty : \sigma_{ij} = 0, \quad H_{ij} = 0 \\
&z = 0, \quad 0 \le r \le a, \quad \sigma_{zz} = -p_0, \quad \sigma_{rz} = 0; \quad H_{zz} = H_{rz} = 0 \\
&z = 0, \quad r > a : \sigma_{rz} = 0, \quad u_z = 0; \quad H_{rz} = 0, \ w_z = 0
\end{aligned} \tag{9.9-6$'$}
$$

which are equivalent to (9.9-6) in the sense of fracture mechanics, if $p = p_0$.

By taking the Hankel transform to Eq. (9.9-1) and boundary conditions (9.8-5), the solution at the transformed space such as

$$
\bar{F}(\xi, z) = A_1 e^{-\lambda_1 \xi z} + A_2 e^{-\lambda_2 \xi z} + A_3 e^{-\lambda_3 \xi z} + A_4 e^{-\lambda_4 \xi z} \tag{9.9-7}
$$

where A_i $(i = 1, 2, 3, 4)$ are the unknown functions of ξ to be determined and λ_i $(i = 1, 2, 3, 4)$ are the eigen roots obtained from the ordinary differential equation of $\bar{F}(\xi, r)$. According to the boundary conditions, $A_i(\xi)$ can be determined by solving the following dual integral equations:

$$
\begin{aligned}
&\int_0^\infty \xi A_i(\xi) J_0(\xi r) d\xi = M_i p_0, \quad 0 < r < a \\
&\int_0^\infty A_i(\xi) J_0(\xi r) d\xi = 0, \quad r > a
\end{aligned} \tag{9.9-8}
$$

and $i = 1, 2, 3, 4$, in which M_i are some constants that consists of elastic moduli and $J_0(\xi r)$ is the first kind of Bessel function of zero order.

According to the theory of dual integral equations (refer to Major Appendix), we obtain the solution of dual integral Eq. (9.9-8) as follows:

$$A_i(\xi) = 2a^2 M_i p(2\pi a\xi)^{-1/2}\xi^{-7} J_{3/2}(a\xi) \tag{9.9-9}$$

in which $J_{3/2}(a\xi)$ is the first kind of Bessel function with 3/2 order (refer to the Major Appendix for the detailed calculation).

After some calculation, stress intensity factor K_I, strain energy W_I and strain energy release rate G_I can be obtained as follows:

$$K_I = \frac{2}{\pi}\sqrt{\pi a}p, \quad W_I = Mp^2a^3, \quad G_I = \frac{1}{2\pi a}\frac{\partial W_I}{\partial a} = \frac{3Mp^2a}{2\pi} \tag{9.9-10}$$

where M is the constant composed of the elastic constants which is quite lengthy and so has not been included here.

The applications of the Fourier transform, Hankel transform and dual integral equations in Sections 7.6, 8.3, 9.2, 9.7 and 9.9 are developed from the work of Sneddon in the classical elasticity [27]; this shows that the Fourier analysis is a powerful tool in solving not only classical but also modern elasticity.

References

[1] Ding D H, Yang W G, Hu C Z and Wang R H, 1993, Generalized theory of elasticity of quasicrystals, *Phys. Rev. B*, **48**(10), 7003–7010.

[2] Hu C Z, Wang R H and Ding D H *et al.*, 1996, Point groups and elastic properties of two-dimensional quasicrystals, *Acta. Crystallog. A*, **52**(2), 251–256.

[3] Yang W G, Ding D H *et al.*, 1998, Atomtic model of dislocation in icosahedral quasicrystals, *Phil. Mag. A*, **77**(6), 1481–1497.

[4] Fan T Y and Guo L H, 2005, Final governing equation of plane elasticity of icosahedral quasicrystals, *Phys. Lett. A*, **341**(5), 235–239.

[5] Li L H and Fan T Y, 2006, Final governing equation of plane elasticity of icosahedral quasicrystals–stress potential method, *Chin. Phys. Lett.*, **24**(9), 2519–2521.

[6] Reynolds G A M, Golding B, Kortan A R *et al.*, 1990, Isotropic elasticity of the Al-Cu-Li quasicrystal, *Phys. Rev. B*, **41**(2), 1194–1195.

[7] Spoor P S, Maynard J D and Kortan A R, 1995, Elastic isotropy and anisotropy in quasicrystalline and cubic AlCuLi, *Phys. Rev. Lett.*, **75**(19), 3462–3465.

[8] Tanaka K, Mitarai and Koiwa M, 1996, Elastic constants of Al-based icosahedral quasicrystals, *Phil. Mag. A*, **73**(6), 1715–1723.
[9] Duquesne J-Y and Perrin B, 2002, Elastic wave interaction in icosahedral AlPdMn, *Phys. B*, **316**, 317–320.
[10] Foster K, Leisure R G, Shaklee A *et al.*, 1999, Elastic moduli of a Ti-Zr-Ni icosahedral quasicrystal and a 1/1 bcc crystal approximant, *Phys. Rev. B*, **59**(17), 11132–11135.
[11] Schreuer J, Steurer W, Lograsso T A *et al.*, 2004, Elastic properties of icosahedral i-$Cd_{84}Yb_{16}$ and hexagonal h-$Cd_{51}Yb_{14}$, *Phil. Mag. Lett.*, **84**(10), 643–653.
[12] Sterzel R, Hinkel C, Haas A *et al.*, 2000, Ultrasonic measurements on FCI Zn-Mg-Y single crystals, *Europhys. Lett.*, **49**(6), 742–747.
[13] Letoublon A, de Boissieu M, Boudard M *et al.*, 2001, Phason elastic constants of the icosahedral Al-Pd-Mn phase derived from diffuse scattering measurements, *Phil. Mag. Lett.*, **81**(4), 273–283.
[14] de Boissieu M, Francoual S, Kaneko Y *et al.*, 2005, Diffuse scattering and phason fluctuations in the Zn-Mg-Sc icosahedral quasicrystal and its Zn-Sc periodic approximant, *Phys. Rev. Lett.*, **95**(10), 105503/1–4.
[15] Edagawa K and So GI Y, 2007, Experimental evaluation of phonon-phason coupling in icosahedral quasicrystals, *Phil. Mag.*, **87**(1), 77–95.
[16] Zhu A Y and Fan T Y, 2007, Elastic field of a mode II Griffith crack in icosahedral quasicrystals, *Chin. Phys.*, **16**(4), 1111–1118.
[17] Zhu A Y, Fan T Y and Guo L H, 2007, A straight dislocation in an icosahedral quasicrystal, *J. Phys.: Condens. Matter*, **19**(23), 236216.
[18] Li X F and Fan T Y, 1998, New method for solving elasticity problems of some planar quasicrystals, *Chin. Phys. Lett.*, **15**(4), 278–280.
[19] Fan T Y, 1999, *Mathematical Theory of Elasticity of Quasicrystals and Its Applications*, Beijing Institute of Technology Press, Beijing (in Chinese).
[20] Guo Y C and Fan T Y, 2001, A mode II Griffith crack in decagonal quasicrystals, *Appl. Math. Mech.*, **22**(11), 1311–1317.
[21] Li L H and Fan T Y, 2008, Complex variable function method for solving Griffith crack in an icosahedral quasicrystal, *Sci. China*, G, **51**(6), 723–780.
[22] Li L H and Fan T Y, 2006, Complex function method for solving notch problem of point 10 two-dimensional quasicrystal based on the stress potential function, *J. Phys.: Condens. Matter*, **18**(47), 10631–10641.
[23] Zhou W M and Fan T Y, 2000, Axisymmetric elasticity problem of cubic quasicrystal, *Chin. Phys.*, **9**(4), 294–303.
[24] Fan T Y, Xie L Y, Fan L and Wang Q Z, 2011, Study on interface of quasicrystal-crystal, *Chin. Phys. B*, **20**(7), 076102.

[25] Zhu A Y and Fan T Y, 2009, Elastic analysis of a Griffith crack in icosahedral Al-Pd-Mn quasicrystal, *Int. J. Mod. Phys. B*, **23**(10), 1–16.
[26] Fan T Y, Tang Z Y, Li L H and Li W, 2010, The strict theory of complex variable function method of sextuple harmonic equation and applications, *J. Math. Phys.*, **51**(5), 053519.
[27] Sneddon I N, 1950, *Fourier Transforms*, New York, McGraw-Hill.

Chapter 10

Phonon–Phason Dynamics and Defect Dynamics of Solid Quasicrystals

Elastodynamics or phonon–phason dynamics of quasicrystals is a topic with different points of view. The focus of contradictions between different scholar circles lies in the role of phason variables in the dynamic process.

Lubensky *et al.* [1], Socolar and Lubensky [2] pointed out that phonon field **u** and phason field **w** play very different roles in the hydrodynamics of quasicrystals, and because **w** is insensitive to spatial translations, the phason modes represent the relative motion of the constituent density waves. They claimed that phasons are diffusive, not oscillatory, with very large diffusion times. On the other hand, according to Bak [3, 4], the phason describes particular structure disorders or structure fluctuations in quasicrystals, and it can be formulated based on a six-dimensional space description, as has been displayed in the previous chapters. Since there are six continuous symmetries, there exist six hydrodynamic vibration modes. Following this point of view **u** and **w** play similar roles in the dynamics, Bak called this dynamics as acoustics of quasicrystals. It is evident that the difference between the arguments of Lubensky *et al.* and Bak lies only in the dynamics. There is no difference between their arguments in statics (i.e., static equilibrium). Based on this reason in the discussions of the previous chapters, we need not distinguish the arguments of either Lubensky *et al.* or Bak.

Probably due to simpler mathematical formulation, many authors followed the Bak's argument, e.g., [5–12], in dynamic study. In this chapter, we present some results given in the references. These are Sections 10.1–10.4 which constitute the part one of this chapter. In the meantime, we introduce some other results [13, 14, 27] which are carried out by following the argument of Lubensky *et al.*; in this line of thinking, it appears that elastodynamics and hydrodynamics are combined in some extent, so it can be called as elasto-/hydro-dynamics of quasicrystals or named phonon–phason dynamics by Rochal and Lorman [16]; they also call this model the minimal model of phonon–phason dynamics. The discussions given in Sections 10.5–10.7 constitute the part two of this chapter. However, the hydrodynamics of solid quasicrystals created by Lubensky *et al.* will not be discussed in this chapter, because this is beyond the scope that we here studied. Their theory is partly concerned in Chapter 16 and introduced from an angle of additional derivation in Major Appendix of this book.

The results based on different hypothesis are presented to provide readers for their consideration and comparison. Though some researchers believe that the hydrodynamics based on the argument of Lubensky *et al.* is more fundamentally sound, the major shortcoming so far is lack of proper experimental data for confirmation. Recently, research interest in this respect is growing up [13–16], but the most important accomplishment shall still be the quantitative results.

However, Coddens [31], etc. put forward some different points of view to the work given by Lubensky *et al.*, which could not be discussed in this and other chapters of this book.

10.1 Acoustics of Quasicrystals Followed Bak's Argument

Ding *et al.* [5] first discussed the acoustics of quasicrystals. The basic equations in deformation geometry and generalized Hooke's law are the same as those of elastostatics, i.e.,

$$\varepsilon_{ij} = \frac{1}{2}\left(\frac{\partial u_i}{\partial x_j} + \frac{\partial u_j}{\partial x_i}\right), \quad w_{ij} = \frac{\partial w_i}{\partial x_j} \tag{10.1-1}$$

$$\begin{aligned} \sigma_{ij} &= C_{ijkl}\varepsilon_{kl} + R_{ijkl}w_{kl} \\ H_{ij} &= K_{ijkl}w_{kl} + R_{klij}\varepsilon_{kl} \end{aligned} \tag{10.1-2}$$

They claimed that the law of momentum conservation holds for both phonons and phasons, namely, for linear and small deformation cases, the equations of motion are

$$\frac{\partial \sigma_{ij}}{\partial x_j} = \rho \frac{\partial^2 u_i}{\partial^2 t}, \quad \frac{\partial H_{ij}}{\partial x_j} = \rho \frac{\partial^2 w_i}{\partial^2 t} \tag{10.1-3}$$

where ρ denotes the average mass density of the material.

This implies that they follow Bak's argument. After 7 years, Hu *et al.* [6] confirmed the point of view again. In fact, the final elastodynamic equations can be deduced by substituting (10.1-1) and (10.1-2) into (10.1-3). The mathematical structure of this theory is relatively simpler, and the formulations are similar to that of classical elastodynamics, so many authors take this formulation to develop the dynamics of quasicrystals and give applications in defect dynamics and thermodynamics. In the subsequent sections, we will present some examples of applications of the theory.

10.2 Acoustics of Anti-plane Elasticity for Some Quasicrystals

For three-dimensional icosahedral or cubic of one-dimensional hexagonal quasicrystals, in the anti-plane elasticity, the basic equations have similar form. First, we consider icosahedral quasicrystals

$$\begin{aligned}
\sigma_{zy} &= \sigma_{yz} = \mu \frac{\partial u_z}{\partial y} + R \frac{\partial w_z}{\partial y} \\
\sigma_{xz} &= \sigma_{zx} = \mu \frac{\partial u_z}{\partial x} + R \frac{\partial w_z}{\partial x} \\
H_{zy} &= (K_1 - K_2) \frac{\partial w_z}{\partial y} + R \frac{\partial u_z}{\partial y} \\
H_{zx} &= (K_1 - K_2) \frac{\partial w_z}{\partial x} + R \frac{\partial u_z}{\partial x}
\end{aligned} \tag{10.2-1}$$

Substituting (10.2-1) into equations of motion of (10.1-3) yields

$$\begin{aligned}
\mu \nabla^2 u_z + R \nabla^2 w_z &= \rho \frac{\partial^2 u_z}{\partial t^2} \\
R \nabla^2 u_z + (K_1 - K_2) \nabla^2 w_z &= \rho \frac{\partial^2 w_z}{\partial t^2}
\end{aligned} \tag{10.2-2}$$

If we define displacement functions ϕ and ψ such as

$$u_z = \alpha\phi - R\psi, \quad w_z = R\phi + \alpha\psi \tag{10.2-3}$$

then Eq. (10.2-2) reduce to the standard wave equations

$$\nabla^2\phi = \frac{1}{s_1^2}\frac{\partial^2\phi}{\partial t^2}, \quad \nabla^2\psi = \frac{1}{s_2^2}\frac{\partial^2\psi}{\partial t^2} \tag{10.2-4}$$

where

$$\alpha = \frac{1}{2}\left[\mu - (K_1 - K_2) + \sqrt{(\mu - (K_1 - K_2))^2 + 4R^2}\right] \tag{10.2-5}$$

and

$$s_j = \sqrt{\frac{\varepsilon_j}{\rho}}, \quad j = 1, 2 \tag{10.2-6}$$

$$\varepsilon_{1,2} = \frac{1}{2}\left[\mu + (K_1 - K_2) \pm \sqrt{(\mu - (K_1 - K_2))^2 + 4R^2}\right]$$

s_j can be understood as the speeds of wave propagation in anti-plane deformation of the material. It is obvious that the wave speeds are the results of phonon–phason coupling. If there is no coupling, i.e., $R \to 0$, then

$$s_1 \to \sqrt{\frac{\mu}{\rho}} \quad s_2 \to \sqrt{\frac{(K_1 - K_2)}{\rho}} \tag{10.2-7}$$

and $\sqrt{\frac{\mu}{\rho}}$ represents the speed of transverse wave of phonon field and $\sqrt{\frac{K_1-K_2}{\rho}}$ the speed of pure-phason elastic wave, requiring $K_1 - K_2 > 0$.

Substituting (10.2-3) into (10.2-1) yields

$$\sigma_{yz} = \sigma_{zy} = (\alpha\mu + R^2)\frac{\partial\phi}{\partial y} + R(\alpha - \mu)\frac{\partial\psi}{\partial y}$$

$$\sigma_{xz} = \sigma_{zx} = (\alpha\mu + R^2)\frac{\partial\phi}{\partial x} + R(\alpha - \mu)\frac{\partial\psi}{\partial x}$$

$$H_{zy} = R(\alpha + (K_1 - K_2))\frac{\partial\phi}{\partial y} + (\alpha(K_1 - K_2) - R^2)\frac{\partial\psi}{\partial y}$$

$$H_{zx} = R_3(\alpha + (K_1 - K_2))\frac{\partial\phi}{\partial x} + (\alpha(K_1 - K_2) - R^2)\frac{\partial\psi}{\partial x} \tag{10.2-8}$$

Formulas (10.2-3) and (10.2-8) give the expressions for displacements and stresses in terms of displacement functions ϕ and ψ, which satisfy the standard wave equations (10.2-4) for elastodynamics of the anti-plane elasticity of three-dimensional icosahedral quasicrystals.

The above discussion is valid for the anti-plane elasticity of three-dimensional cubic quasicrystals or one-dimensional quasicrystals too; the difference between these quasicrystals is only the material constants. If $\mu, K_1 - K_2$ and R are replaced by C_{44}, K_{44} and R_{44} (see Section 9.8) for cubic quasicrystals, or by C_{44}, K_2 and R_3 for one-dimensional hexagonal quasicrystals with the Laue classes $6/m_h$ and $6/m_h mm$ (see Section 7.1 or 8.1), one can find similar equations.

The solution of (10.2-4) can be found by using the method for solving pure wave equations in classical mathematical physics.

10.3 Moving Screw Dislocation in Anti-plane Elasticity

Assume a straight screw dislocation line parallel to the quasiperiodic axis which moves along one of the periodic axes, say, the x-axis in the periodic plane. For simplicity, consider the dislocation moves with a constant velocity V.

For the problem, a dislocation condition is assumed

$$\int_\Gamma du_z = \mathrm{b}_3^{\|}, \quad \int_\Gamma dw_z = \mathrm{b}_3^{\perp} \tag{10.3-1}$$

i.e., we assume that the dislocation has the Burgers vector $(0, 0, \mathrm{b}_3^{\|}, 0, \mathrm{b}_3^{\perp})$ and Γ denotes the Burgers circuit surroundings the core of the moving dislocation. Starting now, we denote the fixed coordinates as (x_1, x_2, t) and the moving ones as (x, y).

By introducing Galilean transformation

$$x = x_1 - Vt, \quad y = x_2 \tag{10.3-2}$$

The wave equations (10.2-4) reduce to the Laplace equations (i.e., $\left(\nabla^2 - \frac{1}{s_1^2}\frac{\partial^2}{\partial t^2}\right) \to \nabla_1^2$,

$$\left.\begin{array}{l}\left(\nabla^2 - \dfrac{1}{s_2^2}\dfrac{\partial^2}{\partial t^2}\right) \to \nabla_2^2, \quad \nabla^2 = \dfrac{\partial^2}{\partial x_1^2} + \dfrac{\partial^2}{\partial x_2^2}\Big) \\ \nabla_1^2\phi = 0, \quad \nabla_2^2\psi = 0\end{array}\right. \tag{10.3-3}$$

where

$$\nabla_1^2 = \frac{\partial^2}{\partial x^2} + \frac{\partial^2}{\partial y_1^2}, \quad \nabla_2^2 = \frac{\partial^2}{\partial x^2} + \frac{\partial^2}{\partial y_2^2} \tag{10.3-4a}$$

$$y_j = \beta_j y, \quad \beta_j = \sqrt{1 - V^2/c_j^2}, \quad j = 1, 2 \tag{10.3-4b}$$

Let complex variables z_j be

$$z_j = x + iy_j \quad (i = \sqrt{-1}) \tag{10.3-5}$$

the solution of Eq. (10.3-3) is

$$\phi = \mathrm{Im} F_1(z_1), \quad \psi = \mathrm{Im}\, F_2(z_2) \tag{10.3-6}$$

where $F_1(z_1)$ and $F_2(z_2)$ are the analytic functions of z_1 and z_2, respectively, and Im marks the imaginary part of a complex function.

The boundary condition (10.3-1) determines the analytic functions as

$$\phi(x, y_1) = \frac{A_1}{2\pi}\arctan\frac{y_1}{x}, \quad \psi(x, y_2) = \frac{A_2}{2\pi}\arctan\frac{y_2}{x} \tag{10.3-7a}$$

with constants

$$A_1 = \frac{\alpha b_3^{||} + R b_3^{\perp}}{\alpha^2 + R^2}, \quad A_2 = \frac{\alpha b_3^{||} - R b_3^{\perp}}{\alpha^2 + R^2} \tag{10.3-7b}$$

The displacement field is determined in the fixed coordinate system

$$u_z(x,y,t) = \frac{1}{2\pi(\alpha^2 + R^2)} \left[\left(\alpha^2 \arctan \frac{\beta_1 y}{x - Vt} + R^2 \arctan \frac{\beta_2 y}{x - Vt} \right) b_3^{||} + \left(\arctan \frac{\beta_1 y}{x - Vt} - \arctan \frac{\beta_2 y}{x - Vt} \right) \alpha R b_3^{\perp} \right] \tag{10.3-8a}$$

$$w_z(x,y,t) = \frac{1}{2\pi(\alpha^2 + R^2)} \left[\left(R^2 \arctan \frac{\beta_1 y}{x - Vt} + \alpha_3^2 \arctan \frac{\beta_2 y}{x - Vt} \right) b_3^{\perp} + \left(\arctan \frac{\beta_1 y}{x - Vt} - \arctan \frac{\beta_2 y}{x - Vt} \right) \alpha R b_3^{||} \right] \tag{10.3-8b}$$

The expressions for strains and stresses are omitted here due to limitation of space.

We give the evaluation on the energy of the moving dislocation. Denote energy W per unit length on the moving dislocation which consists of the kinetic energy W_k and potential energy W_p defined by the integrals

$$\begin{aligned} W_k &= \frac{1}{2}\rho \iint_\Omega \left[\left(\frac{\partial u_z}{\partial t} \right)^2 + \left(\frac{\partial w_z}{\partial t} \right)^2 \right] dx_1 dx_2, \\ W_p &= \frac{1}{2} \iint_\Omega \left[\sigma_{ij} \frac{\partial u_z}{\partial t} + H_{ij} \frac{\partial w_z}{\partial t} \right] dx_1 dx_2 \end{aligned} \tag{10.3-9}$$

respectively, where the integration should be taken over a ring $r_0 < r < R_0$, and r_0 denotes the size of the dislocation core and R_0 the size of the so-called dislocation net similar to those in conventional crystals. In general, $r_0 \sim 10^{-8}$ cm and $R_0 \sim 10^4 r_0$. Substituting the displacement formulas and corresponding stress formulas into (10.3-9), we obtain

$$W_k = \frac{k_k}{4\pi} \ln \frac{R_0}{r_0}, \quad W_p = \frac{k_p}{4\pi} \ln \frac{R_0}{r_0} \tag{10.3-10}$$

with

$$k_k = \frac{\rho V^2(\alpha^2 + R^2)}{2}\left(\frac{A_1^2}{\beta_1} + \frac{A_2^2}{\beta_2}\right)$$

$$\begin{aligned} k_p &= \frac{A_1^2}{2}(\mu\alpha^2 + (K_1 - K_2))R^2 + 2\alpha R^2)\left(\beta_1 + \frac{1}{\beta_1}\right) \\ &\quad + \frac{A_2^2}{2}(\mu R^2 + (K_1 - K_2)\alpha^2 - 2\alpha R^2)\left(\beta_2 + \frac{1}{\beta_2}\right) \end{aligned} \tag{10.3-11}$$

and A_1, A_2 given by (10.3-7). Therefore, the total energy is

$$W = \frac{k_k + k_p}{4\pi}\ln\frac{R_0}{r_0} \tag{10.3-12}$$

It is concluded that when $V \to s_2$, i.e., $\beta_2 \to 0$, this leads to the infinity of the energy and is invalid, and thus s_2 is the limit of the velocity of a moving dislocation. Additionally, if $V \ll s_2$, the total energy can be written in the following simple form:

$$W \approx W_0 + \frac{1}{2}\rho V^2[(\mathrm{b}_3^{\|})^2 + (\mathrm{b}_3^{\perp})^2]\frac{1}{4\pi}\ln\frac{R_0}{r_0} = W_0 + \frac{1}{2}m_0V^2 \tag{10.3-13}$$

where W_0 is the elastic energy per unit length of a rest screw dislocation, i.e.,

$$W_0 = \left[\mu(\mathrm{b}_3^{\|})^2 + R(\mathrm{b}_3^{\perp})^2 + 2\mathrm{b}_3^{\|}\mathrm{b}_3^{\perp}R\right]\frac{1}{4\pi}\ln\frac{R_0}{r_0} \tag{10.3-14}$$

and m_0 is the so-called "apparent" mass of the dislocation per unit length in the case considered

$$m_0 = [\mu(\mathrm{b}_3^{\|})^2 + R(\mathrm{b}_3^{\perp})^2 + 2\mathrm{b}_3^{\|}\mathrm{b}_3^{\perp}R]\frac{1}{4\pi}\ln\frac{R_0}{r_0} \tag{10.3-15}$$

It is evident that if $V = 0$, the solution reduces to that of static dislocation, which is given in 7.1.

Furthermore, if $\mathrm{b}_3^{\perp} = 0, R = 0$, then $\varepsilon_1 = \mu$, $\beta_1 = \sqrt{1 - V^2/c_2^2}$, $s_1 = c_2 = \sqrt{\mu/\rho}$ is the speed of transverse wave of a conventional crystal, and the above solution reduces to

$$
\begin{aligned}
u_z(x - Vt, y) &= \frac{\mathrm{b}}{2\pi} \arctan \frac{\beta_1 y}{x - Vt} \\
\sigma_{yz} = \sigma_{zy} &= \frac{\mathrm{b}}{2\pi} \frac{\mu\beta_1(x - Vt)}{(x - Vt)^2 + \beta_1^2 y^2} \\
\sigma_{xz} = \sigma_{zx} &= -\frac{\mathrm{b}}{2\pi} \frac{\mu\beta_1 y}{(x - Vt)^2 + \beta_1^2 y^2} \\
W &\approx \left(\mu\mathrm{b}^2 + \frac{1}{2}\rho V^2 \mathrm{b}^2\right) \frac{1}{4\pi} \ln \frac{R_0}{r_0} \\
m_0 &= \frac{\rho\mathrm{b}^2}{4\pi} \ln \frac{R_0}{r_0}
\end{aligned}
\tag{10.3-16}
$$

This is identical to the well-known Eshellby solution for crystals [17].

The above discussion is valid for the anti-plane elasticity of three-dimensional cubic or one-dimensional hexagonal quasicrystals, and only the material constants $\mu, K_1 - K_2$ and R need to be replaced by C_{44}, K_{44} and R_{44}, or by C_{44}, K_2 and R_3, respectively.

10.4 Mode III Moving Griffith Crack in Anti-plane Elasticity

As another application of the above acoustics theory, we study a moving Griffith crack of Mode III, which moves with constant speed V along x_1 (see Fig. 10.1).

Here, we also take the fixed coordinates (x_1, x_2, t) and moving coordinates (x, y). In the moving coordinates, the boundary conditions are

$$
\begin{aligned}
&\sqrt{x^2 + y^2} \to \infty : \sigma_{ij} = 0, H_{ij} = 0 \\
&y = 0, |x| < a : \sigma_{yz} = -\tau, H_{yz} = 0
\end{aligned}
\tag{10.4-1}
$$

Assume that the Laplace equations have the solution

$$
\phi(x_1, y_1) = \mathrm{Re}F_1(z_1), \quad \psi(x_1, y_2) = \mathrm{Re}\, F_2(z_2) \tag{10.4-2}
$$

where $F_1(z_1)$ and $F_2(z_2)$ are any analytic functions of z_1 and z_2.

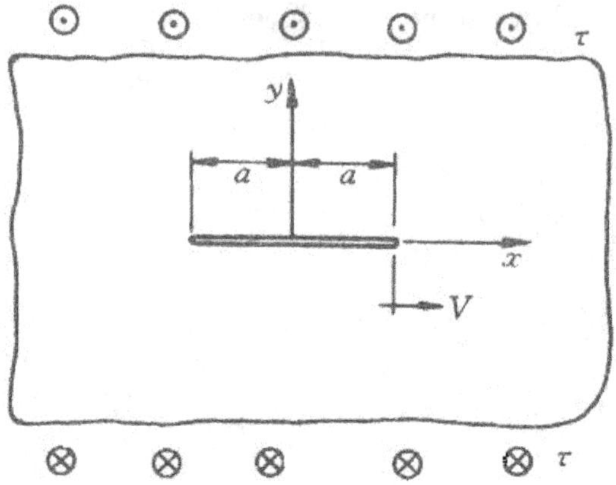

Fig. 10.1. Moving Griffith crack of Mode III.

Because the boundary conditions (10.4-1) are more complicated than those given in (10.3-1), we must use conformal mapping

$$z_1, z_2 = \omega(\zeta) = \frac{a}{2}(\zeta + \zeta^{-1}) \tag{10.4-3}$$

to solve the problem at the $\zeta(= \xi + i\eta)$-plane.

After some calculations, we find the solution

$$\begin{aligned} F_1(z_1) &= F_1[\omega(\zeta)] = G_1(\zeta) = \frac{i\Delta_1}{\Delta}\zeta, \\ F_2(z_2) &= F_2[\omega(\zeta)] = G_2(\zeta) = \frac{i\Delta_2}{\Delta}\zeta \end{aligned} \tag{10.4-4}$$

in which

$$\begin{aligned} \Delta &= \beta_1\beta_2[(\alpha\mu + R^2)(\alpha(K_1 - K_2) - R^2) \\ &\quad - R^2(\alpha + (K_1 - K_2))(\alpha - \mu)] \\ \Delta_1 &= \tau\alpha\beta_2(\alpha(K_1 - K_2) - R^2), \quad \Delta_2 = \tau\alpha\beta_1 R(\alpha + (K_1 - K_2)) \end{aligned} \tag{10.4-5}$$

There is the inverse mapping as

$$\zeta = \omega^{-1}(z_1) = \frac{z_1}{a} - \sqrt{\left(\frac{z_1}{a}\right)^2 - 1} = \omega^{-1}(z_2) = \frac{z_2}{a} - \sqrt{\left(\frac{z_2}{a}\right)^2 - 1} \tag{10.4-6}$$

The manipulation afterwards can also be done in z_1-plane/z_2-plane.

The corresponding stresses are

$$\sigma_{yz} = \sigma_{zy} = (\alpha\mu + R^2)\beta_1 \frac{\partial}{\partial y_1}\operatorname{Re} F_1(z_1) + R(\alpha - \mu)\beta_2 \frac{\partial}{\partial y_2}\operatorname{Re} F_2(z_2)$$

$$\sigma_{xz} = \sigma_{zx} = (\alpha\mu + R^2)\frac{\partial}{\partial x}\operatorname{Re} F_1(z_1) + R_3(\alpha - \mu)\beta_2 \frac{\partial}{\partial x}\operatorname{Re} F_2(z_2)$$

$$H_{zy} = R_3(\alpha + (K_1 - K_2))\beta_1 \frac{\partial}{\partial y_1}\operatorname{Re} F_1(z_1) + (\alpha(K_1 - K_2) - R^2)\frac{\partial}{\partial y_2}\operatorname{Re} F_2(z_2)$$

$$H_{zx} = R_3(\alpha + (K_1 - K_2))\beta_1 \frac{\partial}{\partial x}\operatorname{Re} F_1(z_1) + (\alpha(K_1 - K_2) - R^2)\frac{\partial}{\partial x}\operatorname{Re} F_2(z_2) \tag{10.4-7}$$

Substituting (10.4-6) into (10.4-4) and then into (10.4-7), the stresses can be evaluated in an explicit version, e.g.,

$$\begin{aligned}\sigma_{yz} = \sigma_{zy} = &-\frac{\tau}{\Delta}(\alpha\mu + R^2)\beta_1\beta_2(\alpha(K_1 - K_2) - R^2)\\ &\times \left[1 - \frac{d}{(d_1 d_2)^{\frac{1}{2}}}\cos\left(\theta - \frac{1}{2}\theta_1 - \frac{1}{2}\theta_2\right)\right]\\ &+ \frac{\tau}{\Delta}\beta_1\beta_2 R^2(\alpha + (K_1 - K_2))(\alpha - \mu)\\ &\times \left[1 - \frac{D}{(D_1 D_2)^{\frac{1}{2}}}\cos\left(\Theta - \frac{1}{2}\Theta_1 - \frac{1}{2}\Theta_2\right)\right]\end{aligned} \tag{10.4-8}$$

with

$$d = \sqrt{x^2 + y_1^2}, \quad d_1 = \sqrt{(x-a)^2 + y_1^2}, \quad d_2 = \sqrt{(x+a)^2 + y_1^2}$$

$$D = \sqrt{x^2 + y_2^2}, \quad D_1 = \sqrt{(x-a)^2 + y_2^2}, \quad D_2 = \sqrt{(x+a)^2 + y_2^2}$$

$$\theta = \arctan\left(\frac{y_1}{x}\right), \quad \theta_1 = \arctan\left(\frac{y_1}{x-a}\right), \quad \theta_2 = \arctan\left(\frac{y_1}{x+a}\right)$$

$$\Theta = \arctan\left(\frac{y_2}{x}\right), \quad \Theta_1 = \arctan\left(\frac{y_2}{x-a}\right), \quad \Theta_2 = \arctan\left(\frac{y_2}{x+a}\right) \tag{10.4-9}$$

It is easy to prove that (10.4-8) satisfies the relevant boundary conditions and is the exact solution.

Similarly, $\sigma_{xz} = \sigma_{zx}$, H_{zx} and H_{zy} can also be expressed explicitly. From (10.4-8), as $y = 0$, it yields

$$\sigma_{yz}(x,0) = \begin{cases} \dfrac{x\tau}{\sqrt{x^2-a^2}} - \tau, & |x| > a \\ -\tau, & |x < a| \end{cases} \tag{10.4-10}$$

The stress presents the singularity of order $(x-a)^{-1/2}$ as $x \to a$.

The stress intensity factor for Mode III for a phonon field is

$$K_I^{\parallel} = \lim_{x \to a^+} \sqrt{\pi(x-a)}\sigma_{yz}(x,0) = \sqrt{\pi a}\tau \tag{10.4-11}$$

This is identical to the classical Yoffe solution [18]; there the stress intensity factor is also independent of crack moving speed V.

Now, we calculate the energy of the moving crack, which is defined by

$$\begin{aligned} W &= 2\int_0^a [\sigma_{zy}(x,0) \oplus H_{zy}(x,0)][u_z(x,0) \oplus w_z(x,0)]dx \\ &= \frac{1}{\Delta}(\Delta_1\alpha - \Delta_2 R)\tau\pi a = \frac{1}{\Delta}[\alpha\beta_2(\alpha(K_1-K_2) - R^2) \\ &\quad - \beta_1 R^2(\alpha + (K_1 - K_2))]\pi a^2\tau \end{aligned} \tag{10.4-12}$$

The crack energy release rate is

$$\begin{aligned} G = \frac{1}{2}\frac{\partial W}{\partial a} &= \frac{1}{2\Delta}[\alpha\beta_2(\alpha(K_1-K_2) - R^2) \\ &\quad - \beta_1 R^2(\alpha + (K_1 - K_2)](K_I^{\parallel})^2 \end{aligned} \tag{10.4-13}$$

The above discussion, results and conclusions are hold too for the anti-plane elasticity of three-dimensional cubic or one-dimensional hexagonal quasicrystals, and only the material constants μ, $K_1 - K_2$ and R need to be replaced by C_{44}, K_{44} and R_{44} or by C_{44}, K_2 and R_3, respectively.

10.5 Two-dimensional Phonon–Phason Dynamics, Fundamental Solution

In contrast to Eq. (10.1-3) based on Bak's argument, we here use formulas of phonon–phason dynamics which are originated from the hydrodynamics of Lubensky *et al.* (for details refer to Fan *et al.* [13])

$$\rho\frac{\partial^2 u_i}{\partial t^2} = \frac{\partial \sigma_{ij}}{\partial x_j}, \quad \kappa\frac{\partial w_i}{\partial t} = \frac{\partial H_{ij}}{\partial x_j} \tag{10.5-1}$$

where $\kappa = 1/\Gamma_w$, and Γ_w is dissipation coefficient of phason field. The first equation is the phonon elastodynamic equation (wave equation) and the second equation is the equation of motion of phasons; a diffusion equation comes from the hydrodynamics of solid quasicrystals but after simplification. For coupling phonon–phason systems of quasicrystals, these two equations are coupled to each other. The detailed hydrodynamics of solid quasicrystals can be found in Major Appendix of this book; we here do not concern the aspect, which is beyond the scope discussed in this chapter.

Li [19] studies the phonon–phason dynamics for two-dimensional decagonal quasicrystals starting from Eq. (10.5-1) and obtains the following wave equations in the plane deformation case:

$$\begin{aligned} &M\nabla^2 u_x + (L+M)\frac{\partial}{\partial x}\left(\frac{\partial u_x}{\partial x} + \frac{\partial u_y}{\partial y}\right) \\ &\quad + R\left[\left(\frac{\partial^2}{\partial x^2} - \frac{\partial^2}{\partial y^2}\right) w_x + 2\frac{\partial^2 w_y}{\partial x \partial y}\right] = \rho\frac{\partial^2 u_x}{\partial t^2} \end{aligned} \tag{10.5-2}$$

$$M\nabla^2 u_y + (L+M)\frac{\partial}{\partial y}\left(\frac{\partial u_x}{\partial x} + \frac{\partial u_y}{\partial y}\right)$$

$$+R\left[\left(\frac{\partial^2}{\partial x^2} - \frac{\partial^2}{\partial y^2}\right) w_y - 2\frac{\partial^2 w_x}{\partial x \partial y}\right] = \rho \frac{\partial^2 u_y}{\partial t^2} \tag{10.5-3}$$

$$K_1 \nabla^2 w_x + R\left[\left(\frac{\partial^2}{\partial x^2} - \frac{\partial^2}{\partial y^2}\right) u_x - 2\frac{\partial^2 u_y}{\partial x^2}\right] = \kappa \frac{\partial w_x}{\partial t} \tag{10.5-4}$$

$$K_1 \nabla^2 w_y + R\left[\left(\frac{\partial^2}{\partial x^2} - \frac{\partial^2}{\partial y^2}\right) u_y + 2\frac{\partial^2 u_x}{\partial x \partial y}\right] = \kappa \frac{\partial w_y}{\partial t} \tag{10.5-5}$$

where $\nabla^2 = \partial_1^2 + \partial_2^2$. If we introduce a function Y

$$u_1 = L_1 Y, \quad u_2 = L_2 Y \tag{10.5-6}$$

in which $L_j (j = 1, 2)$ are two unknown linear operators, will be determined as follows.

In order to find the explicit expressions of $L_j (j = 1, 2)$, substituting (10.5-6) into (10.5-4) and (10.5-5) yields

$$K_1 \nabla^2 w_1 + R\left[\left(\frac{\partial^2}{\partial x^2} - \frac{\partial^2}{\partial y^2}\right) L_1 Y - 2\frac{\partial^2}{\partial x \partial y} L_2 Y\right] = \kappa \dot{w}_1 \tag{10.5-7}$$

$$K_1 \nabla^2 w_2 + R\left[\left(\frac{\partial^2}{\partial x^2} - \frac{\partial^2}{\partial y^2}\right) L_2 Y + 2\frac{\partial^2}{\partial x \partial y} L_1 Y\right] = \kappa \dot{w}_2 \tag{10.5-8}$$

Thus, there are

$$w_1 = -R\left[\left(\frac{\partial^2}{\partial x^2} - \frac{\partial^2}{\partial y^2}\right) L_1 - 2\frac{\partial^2 L_2}{\partial x \partial y}\right] Z, \tag{10.5-9}$$

$$w_2 = -R\left[\left(\frac{\partial^2}{\partial x^2} - \frac{\partial^2}{\partial y^2}\right) L_2 + 2\frac{\partial^2 L_1}{\partial x \partial y}\right] Z, \tag{10.5-10}$$

$$Y = K_1 \nabla^2 Z - \kappa \dot{Z}, \tag{10.5-11}$$

Then, (10.5-7) and (10.5-8) will be satisfied automatically, where Z represents a new unknown function. Substituting (10.5-9)–(10.5-11) and (10.5-6) into (10.5-2) and (10.5-3), respectively, leads to

$$\left\{\left(K_1\nabla^2-\kappa\frac{\partial}{\partial t}\right)\left[(L+M)\frac{\partial^2}{\partial x^2}+M\nabla^2-\rho\frac{\partial^2}{\partial t^2}\right]-R^2\nabla^2\nabla^2\right\}L_1Z$$
$$+(L+M)\left(K_1\nabla^2-\kappa\frac{\partial}{\partial t}\right)\frac{\partial^2}{\partial x\partial y}L_2Z=0 \tag{10.5-12}$$
$$\left\{\left(K_1\nabla^2-\kappa\frac{\partial}{\partial t}\right)\left[(L+M)\frac{\partial^2}{\partial y^2}+M\nabla^2-\rho\frac{\partial^2}{\partial t^2}\right]-R^2\nabla^2\nabla^2\right\}L_2Z$$
$$+(L+M)\left(K_1\nabla^2-\kappa\frac{\partial}{\partial t}\right)\frac{\partial^2}{\partial x\partial y}L_1Z=0 \tag{10.5-13}$$

Putting

$$L_1Z=-\left(K_1\nabla^2-\kappa\frac{\partial}{\partial t}\right)\frac{\partial^2}{\partial x\partial y}F \tag{10.5-14}$$
$$L_2Z=\frac{1}{L+M}\left\{\left(K_1\nabla^2-\kappa\frac{\partial}{\partial t}\right)\left[(L+M)\frac{\partial^2}{\partial x^2}+M\nabla^2-\rho\frac{\partial^2}{\partial t^2}\right]\right.$$
$$\left.-R^2\nabla^2\nabla^2\right\}F \tag{10.5-15}$$

one can find that (10.5-12) has been automatically satisfied, while (10.5-13) reduces to

$$\left\{\left(K_1\nabla^2-\kappa\frac{\partial}{\partial t}\right)\left[(2M+L)\nabla^2-\rho\frac{\partial^2}{\partial t^2}\right]-R^2\nabla^2\nabla^2\right\}$$
$$\left[\left(K_1\nabla^2-\kappa\frac{\partial}{\partial t}\right)\left(M\nabla^2-\rho\frac{\partial^2}{\partial t^2}\right)-R^2\nabla^2\nabla^2\right]F=0 \tag{10.5-16}$$

In (10.5-13) and (10.5-14) L_1, L_2 denote operators. Similarly if taking L_1Z and L_2Z in the following version:

$$L_1Z = \frac{1}{L+M}\left\{\left(K_1\nabla^2 - \kappa\frac{\partial}{\partial t}\right)\left[(L+M)\frac{\partial^2}{\partial y^2} + M\nabla^2 - \rho\frac{\partial^2}{\partial t^2}\right] - R^2\nabla^2\nabla^2\right\}F \tag{10.5-17}$$

$$L_2Z = -\left(K_1\nabla^2 - \kappa\frac{\partial}{\partial t}\right)\frac{\partial^2}{\partial x\partial y}F \tag{10.5-18}$$

then it reduces to the final governing equation (10.5-16). As a check, in the static case, (10.5-15) reduces to

$$\nabla^2\nabla^2\nabla^2\nabla^2 F = 0 \tag{10.5-19}$$

This is Eq. (6.2-7) in Chapter 6, and the correctness of the above derivation is proved.

The equations obtained above can be simplified further. Function F is decomposed into $F = F_1 + F_2$ so that $F_j (j = 1, 2)$ satisfies, respectively,

$$\left(K_1\nabla^2 - \kappa\frac{\partial}{\partial t}\right)\left[(2M+L)\nabla^2 - \rho\frac{\partial^2}{\partial t^2}\right]F_1 - R^2\nabla^2\nabla^2 F_1 = 0 \tag{10.5-20}$$

$$\left(K_1\nabla^2 - \kappa\frac{\partial}{\partial t}\right)\left(M\nabla^2 - \rho\frac{\partial^2}{\partial t^2}\right)F_2 - R^2\nabla^2\nabla^2 F_2 = 0 \tag{10.5-21}$$

or

$$\left[K_1(2M+L) - R^2\right]\nabla^2\nabla^2 F_1 - \kappa(2M+L)\nabla^2\frac{\partial}{\partial t}F_1 - \rho K_1\nabla^2\frac{\partial^2}{\partial t^2}F_1 + \kappa\rho\frac{\partial^3}{\partial t^3}F_1 = 0 \tag{10.5-22}$$

$$\left(K_1M - R^2\right)\nabla^2\nabla^2 F_2 - \kappa M\nabla^2\frac{\partial}{\partial t}F_2 - \rho K_1\nabla^2\frac{\partial^2}{\partial t^2}F_2 + \kappa\rho\frac{\partial^3}{\partial t^3}F_2 = 0 \tag{10.5-23}$$

These two equations are the final governing equations, which are "conjugated" through phonon–phason coupling constant R, and describe the interaction between wave propagation and diffusion. Otherwise, if $R = 0$, then

$$F_1 = \xi + \zeta \tag{10.5-24}$$

$$F_2 = \eta + \zeta \tag{10.5-25}$$

and ξ and η satisfy

$$(2M + L)\nabla^2\xi = \rho\frac{\partial^2}{\partial t^2}\xi \tag{10.5-26}$$

$$M\nabla^2\eta = \rho\frac{\partial^2}{\partial t^2}\eta \tag{10.5-27}$$

This is entirely identical to the wave equations of the Lamé potentials. If the material is isotropic then $L = \lambda$, 和 $M = \mu$ and λ and μ are the Lamé constants. In this case, ζ satisfies

$$K_1\nabla^2\zeta = \kappa\partial_t\zeta \tag{10.5-28}$$

which is a classical diffusion equation.

Once potentials $F_j (j = 1, 2)$ are determined, the displacement field can be evaluated. For example, denoting

$$\begin{aligned} \phi &= -\left(K_1\nabla^2 - \kappa\partial_t\right)(\partial_1 + \partial_2) F_1, \\ \psi &= -\left(K_1\nabla^2 - \kappa\partial_t\right)(\partial_1 - \partial_2) F_2 \end{aligned} \tag{10.5-29}$$

then

$$u_x = \left(K_1\nabla^2 - \kappa\frac{\partial}{\partial t}\right)\left(\frac{\partial}{\partial x}\phi + \frac{\partial}{\partial y}\psi\right) \tag{10.5-30}$$

$$u_y = \left(K_1\nabla^2 - \kappa\frac{\partial}{\partial t}\right)\left(\frac{\partial}{\partial y}\phi - \frac{\partial}{\partial x}\psi\right) \tag{10.5-31}$$

$$w_x = R\left(\Pi_2\frac{\partial}{\partial x}\varphi - \Pi_1\frac{\partial}{\partial y}\psi\right) \tag{10.5-32}$$

$$w_y = -R\left(\Pi_1\frac{\partial}{\partial y}\varphi + \Pi_2\frac{\partial}{\partial x}\psi\right) \tag{10.5-33}$$

where

$$\Pi_1 = 3\frac{\partial^2}{\partial x^2} - \frac{\partial^2}{\partial y^2}, \quad \Pi_2 = 3\frac{\partial^2}{\partial y^2} - \frac{\partial^2}{\partial x^2}$$

Furthermore, the stresses are

$$\sigma_{xx} = \left[\left(K_1\nabla^2 - \kappa\frac{\partial}{\partial t}\right)\left(L\nabla^2 + 2M\frac{\partial^2}{\partial x^2}\right) - R^2\nabla^2\left(\frac{\partial^2}{\partial x^2} - \frac{\partial^2}{\partial y^2}\right)\right]\phi$$

$$+2\frac{\partial^2}{\partial x\partial y}\left[M\left(K_1\nabla^2 - \kappa\frac{\partial}{\partial t}\right) - R^2\nabla^2\right]\psi \tag{10.5-34}$$

$$-2\frac{\partial^2}{\partial x\partial y}\left[M\left(K_1\nabla^2 - \kappa\frac{\partial}{\partial t}\right) - R^2\nabla^2\right]\psi \tag{10.5-35}$$

$$\sigma_{xy} = 2\frac{\partial^2}{\partial x\partial y}\left[M\left(K_1\nabla^2 - \kappa\frac{\partial}{\partial t}\right) - R^2\nabla^2\right]\phi$$

$$-\left(\frac{\partial^2}{\partial x^2} - \frac{\partial^2}{\partial y^2}\right)\left[M\left(K_1\nabla^2 - \kappa\frac{\partial}{\partial t}\right) - R^2\nabla^2\right]\psi \tag{10.5-36}$$

$$H_{xx} = R\left[K_1\frac{\partial^2}{\partial x^2}\Pi_2 - K_2\frac{\partial^2}{\partial y^2}\Pi_1 + \left(K_1\nabla^2 - \kappa\frac{\partial}{\partial t}\right)\left(\frac{\partial^2}{\partial x^2} - \frac{\partial^2}{\partial y^2}\right)\right]\phi$$

$$-R\frac{\partial^2}{\partial x\partial y}\left[K_1\Pi_1 + K_2\Pi_2 - 2\left(K_1\nabla^2 - \kappa\frac{\partial}{\partial t}\right)\right]\psi \tag{10.5-37}$$

$$H_{yy} = -R\left[K_1\frac{\partial^2}{\partial y^2}\Pi_1 - K_2\frac{\partial^2}{\partial x^2}\Pi_2 - \left(K_1\nabla^2 - \kappa\frac{\partial}{\partial t}\right)\left(\frac{\partial^2}{\partial x^2} - \frac{\partial^2}{\partial y^2}\right)\right]\phi$$

$$-R\frac{\partial^2}{\partial x\partial y}\left[K_1\Pi_2 + K_2\Pi_1 - 2\left(K_1\nabla^2 - \kappa\frac{\partial}{\partial t}\right)\right]\psi \tag{10.5-38}$$

$$H_{12} = R\frac{\partial^2}{\partial x\partial y}\left[K_1\Pi_2 + K_2\Pi_1 - 2\left(K_1\nabla^2 - \kappa\frac{\partial}{\partial t}\right)\right]\phi$$

$$-R\left[K_1\frac{\partial^2}{\partial y^2}\Pi_1 - K_2\frac{\partial^2}{\partial x^2}\Pi_2 - \left(K_1\nabla^2 - \kappa\frac{\partial}{\partial t}\right)\left(\frac{\partial^2}{\partial x^2} - \frac{\partial^2}{\partial y^2}\right)\right]\psi \tag{10.5-39}$$

$$H_{21} = -R\frac{\partial^2}{\partial x \partial y}\left[K_1\Pi_1 + K_2\Pi_2 - 2\left(K_1\nabla^2 - \kappa\frac{\partial}{\partial t}\right)\right]\phi$$
$$-R\left[K_1\frac{\partial^2}{\partial x^2}\Pi_2 - K_2\frac{\partial^2}{\partial y^2}\Pi_1 + \left(K_1\nabla^2 - \kappa\frac{\partial}{\partial t}\right)\left(\frac{\partial^2}{\partial x^2} - \frac{\partial^2}{\partial y^2}\right)\right]\psi \quad (10.5\text{-}40)$$

It can be verified that the all equations in phonon–phason dynamics are satisfied for the plane field of two-dimensional quasicrystals with point group 10 mm; the further work is solving these equations under prescribed boundary conditions.

Li [20] develops the work introduced above, extends it into the dynamics of decagonal qyasicrystals of point groups $10, \overline{10}$ and $10/m$, determines relevant wave speeds and analyzes the wave propagation. Figures 10.2 and 10.3 show his calculating results partly.

10.6 Phonon–Phason Dynamics and Solutions of Two-dimensional Decagonal Quasicrystals

In this section, we would like to give a detailed description on the solution of two-dimensional quasicrystals based on the phonon–phason dynamics formulation.

The equations of deformation geometry and the generalized Hooke's law are the same as before which are not listed again here.

By using the dynamic equations (10.5-1) in Section 5, i.e., the so-called phonon–phason equations for quasicrystals, we can create the formalism for two-dimensional quasicrystals for the new dynamics. As an application of the formulation, some dynamic crack solutions are given in this section.

10.6.1 *The mathematical formalism of dynamic crack problems of decagonal quasicrystals*

In over 200 quasicrystals observed to date, there are over 70 two-dimensional decagonal quasicrystals, so these kinds of solid phases play an important role in the material. For simplicity, only point group 10 mm two-dimensional decagonal quasicrystals will be

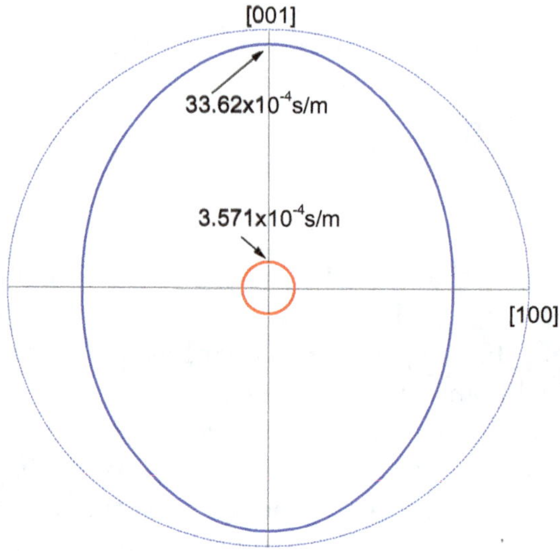

Fig. 10.2. A section of slow surface of acoustic wave propagation in xz-plane, material constants are $L = 30$ GPa, $M = 40$ GPa, $K_1 = 300$ MPa, $R = -0.05\mu$, $K_2 = -0.52K_1$ and $K_3 = 0.5K_1$.

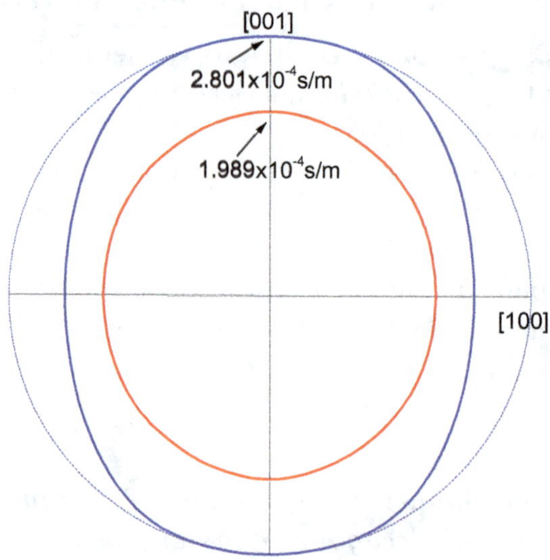

Fig. 10.3. A section of slow surface of acoustic wave propagation in xz-plane, material constants are $L = 75$ GPa, $M = 65$ GPa, $K_1 = 81$ GPa, $R = 0.4K_1$, $K_2 = -0.52K_1$ and $K_3 = 0.5K_1$.

considered herein. We denote the periodic direction as the z axis and the quasiperiodic plane as the xy-plane. Assume a Griffith crack in the solid along the periodic direction, i.e., the z-axis. It is obvious that the elastic field induced by a uniform tensile stress at upper and lower surfaces of the specimen is independent of z, so $\partial/\partial z = 0$. In this case, the stress–strain relations are reduced to

$$\begin{aligned}
\sigma_{xx} &= L(\varepsilon_{xx} + \varepsilon_{yy}) + 2M\varepsilon_{xx} + R(w_{xx} + w_{yy}) \\
\sigma_{yy} &= L(\varepsilon_{xx} + \varepsilon_{yy}) + 2M\varepsilon_{yy} - R(w_{xx} + w_{yy}) \\
\sigma_{xy} &= \sigma_{yx} = 2M\varepsilon_{xy} + R(w_{yx} - w_{xy}) \\
H_{xx} &= K_1 w_{xx} + K_2 w_{yy} + R(\varepsilon_{xx} - \varepsilon_{yy}) \\
H_{yy} &= K_1 w_{yy} + K_2 w_{xx} + R(\varepsilon_{xx} - \varepsilon_{yy}) \\
H_{xy} &= K_1 w_{xy} - K_2 w_{yx} - 2R\varepsilon_{xy} \\
H_{yx} &= K_1 w_{yx} - K_2 w_{xy} + 2R\varepsilon_{xy}
\end{aligned} \tag{10.6-1}$$

where $L = C_{12}$ and $M = (C_{11} - C_{12})/2$ are the phonon elastic constants, K_1 and K_2 are the phason elastic constants and R is phonon–phason coupling elastic constant.

Substituting (10.6-1) into (10.5-1), we obtain the equations of motion of decagonal quasicrystals as follows:

$$\begin{aligned}
\frac{\partial^2 u_x}{\partial t^2} &= c_1^2 \frac{\partial^2 u_x}{\partial x^2} + (c_1^2 - c_2^2)\frac{\partial^2 u_y}{\partial x \partial y} + c_2^2 \frac{\partial^2 u_x}{\partial y^2} \\
&\quad + c_3^2 \left(\frac{\partial^2 w_x}{\partial x^2} + 2\frac{\partial^2 w_y}{\partial x \partial y} - \frac{\partial^2 w_x}{\partial y^2} \right) \\
\frac{\partial^2 u_y}{\partial t^2} &= c_2^2 \frac{\partial^2 u_y}{\partial x^2} + (c_1^2 - c_2^2)\frac{\partial^2 u_x}{\partial x \partial y} + c_1^2 \frac{\partial^2 u_y}{\partial y^2} \\
&\quad + c_3^2 \left(\frac{\partial^2 w_y}{\partial x^2} - 2\frac{\partial^2 w_x}{\partial x \partial y} - \frac{\partial^2 w_y}{\partial y^2} \right) \\
\frac{\partial w_x}{\partial t} &= d_1^2 \left(\frac{\partial^2 w_x}{\partial x^2} + \frac{\partial^2 w_x}{\partial y^2} \right) + d_2^2 \left(\frac{\partial^2 u_x}{\partial x^2} - 2\frac{\partial^2 u_y}{\partial x \partial y} - \frac{\partial^2 u_x}{\partial y^2} \right) \\
\frac{\partial w_y}{\partial t} &= d_1^2 \left(\frac{\partial^2 w_y}{\partial x^2} + \frac{\partial^2 w_y}{\partial y^2} \right) + d_2^2 \left(\frac{\partial^2 u_y}{\partial x^2} + 2\frac{\partial^2 u_x}{\partial x \partial y} - \frac{\partial^2 u_y}{\partial y^2} \right)
\end{aligned} \tag{10.6-2}$$

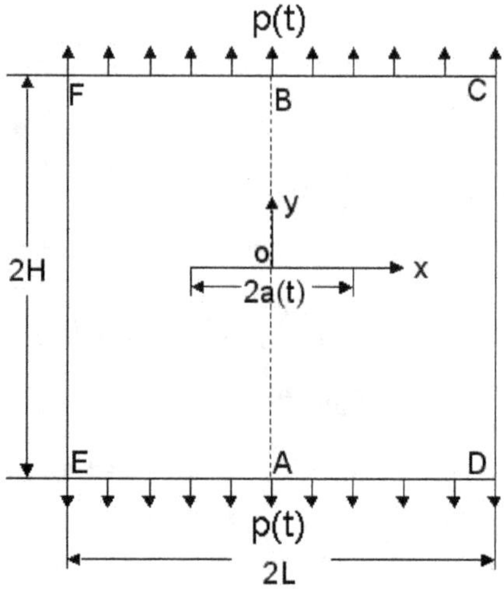

Fig. 10.4. The specimen with a central crack.

where

$$c_1 = \sqrt{\frac{L+2M}{\rho}}, \quad c_2 = \sqrt{\frac{M}{\rho}}, \quad c_3 = \sqrt{\frac{R}{\rho}}, \quad d_1 = \sqrt{\frac{K_1}{\kappa}} \quad \text{and}$$

$$d_2 = \sqrt{\frac{R}{\kappa}}, \quad d_3 = \sqrt{\frac{K_2}{\kappa}} \qquad (10.6\text{-}3)$$

Note that constants c_1, c_2 and c_3 have the meaning of elastic wave speeds, while d_1^2, d_2^2 and d_3^2 do not represent wave speed; they are diffusive coefficients.

The decagonal quasicrystal with a crack is shown in Fig. 10.4. It is a rectangular specimen with a central crack of length $2a(t)$ subjected to a dynamic or static tensile stress at its ends ED and FC, in which $a(t)$ represents the crack length being a function of time, and for dynamic initiation of crack growth, the crack is stable, so $a(t) = a_0 =$ constant, and for fast crack propagation, $a(t)$ varies with time. At first, we consider dynamic initiation of crack growth, and then study crack fast propagation. Due to the symmetry of the specimen, only the upper right quarter is considered.

Referring to the upper right part and considering a fix grips case, the following boundary conditions should be satisfied:

$$\begin{array}{lll}
u_x = 0,\ \sigma_{yx} = 0,\ w_x = 0,\ H_{yx} = 0 & \text{on } x = 0 & \text{for } 0 \le y \le H \\
\sigma_{xx} = 0,\ \sigma_{yx} = 0,\ H_{xx} = 0,\ H_{yx} = 0 & \text{on } x = L & \text{for } 0 \le y \le H \\
\sigma_{yy} = p(t),\ \sigma_{xy} = 0,\ H_{yy} = 0,\ H_{xy} = 0 & \text{on } y = H & \text{for } 0 \le x \le H \\
\sigma_{yy} = 0,\ \sigma_{xy} = 0,\ H_{yy} = 0,\ H_{xy} = 0 & \text{on } y = 0 & \text{for } 0 \le x \le a(t) \\
u_y = 0,\ \sigma_{xy} = 0,\ w_y = 0,\ H_{xy} = 0 & \text{on } y = 0 & \text{for } a(t) \le x \le L
\end{array} \tag{10.6-4}$$

in which $p(t) = p_0 f(t)$ is a dynamic load if $f(t)$ varies with time, otherwise it is a static load (i.e., if $f(t) = \text{const}$), and $p_0 = \text{const}$ with the stress dimension.

The initial conditions are

$$\begin{array}{ll}
u_x(x,y,t)\,|_{t=0} = 0 & u_y(x,y,t)\,|_{t=0} = 0 \\
w_x(x,y,t)\,|_{t=0} = 0 & w_y(x,y,t)\,|_{t=0} = 0 \\
\dfrac{\partial u_x(x,y,t)}{\partial t}\,|_{t=0} = 0 & \dfrac{\partial u_y(x,y,t)}{\partial t}\,|_{t=0} = 0
\end{array} \tag{10.6-5}$$

For implementation of finite difference, all field variables in governing equations (10.6-2) and boundary-initial conditions (10.6-4) and (10.6-5) must be expressed by displacements and their derivatives. This can be done through the constitutive equations (10.6-1). The details of the finite difference scheme are given in Appendix A of this chapter.

For the related parameters in this section, the experimentally determined mass density for decagonal Al–Ni–Co quasicrystal $\rho = 4.186 \times 10^{-3}\ \text{g}\cdot\text{mm}^{-3}$ is used and phonon elastic moduli are $C_{11} = 2.3433$, $C_{12} = 0.5741$ ($10^{12}\text{dyn/cm}^2 = 10^2$ GPa) which are obtained by resonant ultrasound spectroscopy [21], and we have also chosen phason elastic constants $K_1 = 1.22$ and $K_2 = 0.24$ (10^{12} dyn/cm^2 = 10^2 GPa) estimated by the Monte Carlo simulation [22] and $\Gamma_w = 1/\kappa = 4.8 \times 10^{-19}\ \text{m}^3\cdot\text{s/kg} = 4.8 \times 10^{-10}\ \text{cm}^3\cdot\mu\text{s/g}$ [23]. The coupling constant R has been measured for some special cases recently; see Chapters 6 and 9, respectively. In computation, we take $R/M = 0.01$ for the coupling case corresponding to quasicrystals, and $R/M = 0$ for the decoupled case in which the latter corresponds to crystals.

10.6.2 *Examination on the physical model*

In order to verify the correctness of the suggested model and the numerical simulation, we first explore the specimen without a crack. We know that there are the fundamental solutions characterizing time variation natures based on the wave propagation of phonon field and on the motion of diffusion according to mathematical physics

$$\begin{cases} u \sim e^{i\omega(t-x/c)} \\ w \sim \dfrac{1}{\sqrt{t-t_0}} e^{-(x-x_0)^2/\Gamma_w(t-t_0)} \end{cases} \tag{10.6-6}$$

where ω is a frequency and c a speed of the wave, t, the time and t_0 a special value of t, x a distance, x_0 a special value of x and Γ_w kinetic coefficient of phason defined previously.

Comparison results are shown in Figs. 10.5(a)–10.5(c), in which the solid line represents the numerical solution of quasicrystals and the dotted line represents the fundamental solution of Eq. (10.6-6). From Figs. 10.5(a) and 10.5(b), we can see that both displacement components of phonon field are an excellent agreement to the fundament solutions. However, there are some differences because the phonon field is influenced by the phason field and the phonon–phason coupling effect. From Fig. 10.5(c), in the phason field, we find that the phason mode presents diffusive nature in the overall tendency, but because of the influence of the phonon and phonon–phason coupling, it can also have some character of fluctuation. So, the model describes the dynamic behaviour of phonon field and phason field in deed. This also shows that the mathematical modelling of this work is valid.

10.6.3 *Testing the Scheme and the Computer Program*

10.6.3.1 *Stability of the scheme*

The stability of the scheme is the core problem of finite difference method which depends upon the choice of parameter $\alpha = c_1\tau/h$, which is the ratio between time step and space step substantively. The choice is related to the ratio c_1/c_2, i.e., the ratio between speeds of elastic longitudinal and transverse waves of the phonon

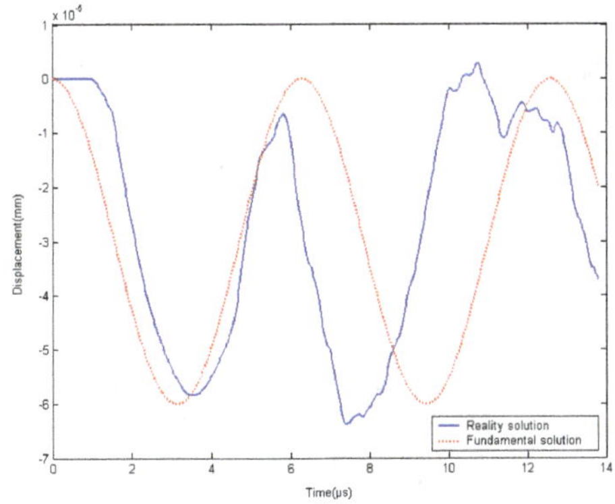

(a)

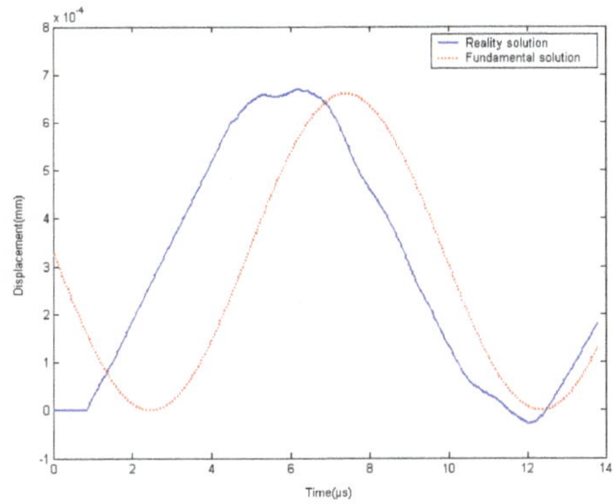

(b)

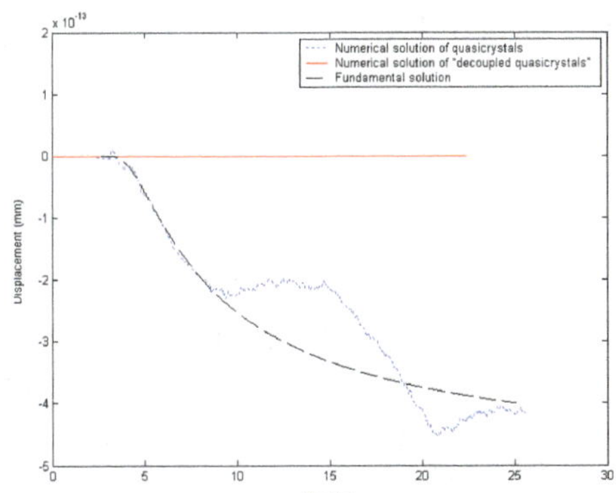

(c)

Fig. 10.5. (a) Displacement component of phonon field u_x versus time. (b) Displacement component of phonon field u_y versus time. (c) Displacement component of phason field w_x versus time.

field. To determine the upper bound for the ration to guarantee the stability, according to our computational practice and considering the experiences of computations for conventional materials, we choose $\alpha = 0.8$ in all cases and the results are stable.

10.6.3.2 *Accuracy test*

The stability is only a necessary condition for successful computation. We must check the accuracy of the numerical solution. This can be realized through some comparison with some well-known classical solutions (analytic as well as numerical solutions). For this purpose, the material constants in the computation are chosen as $c_1 = 7.34$, $c_2 = 3.92$ (mm/μs) and $\rho = 5 \times 10^3\,\text{kg/m}^3$, $p_0 = 1$ MPa which are the same with those in Refs. [24–26] (but are different from those listed in Section 10.6.1). At first, the comparison to the classical exact analytic solution is carried out, and in this case, we put $w_x = w_y = 0$ (i.e., $K_1 = K_2 = R = 0$) for the numerical solution. The comparison has been done with the key physical quantity — dynamic stress intensity factor, which is defined by

$$K_{\mathrm{I}}(t) = \lim_{x \to a_0^+} \sqrt{\pi(x - a_0)}\sigma_{yy}(x, 0, t) \qquad (10.6\text{-}7)$$

The normalized dynamic stress intensity factor can be denoted as $K_I(t)/K_I^{\text{static}}$, in which K_I^{static} is the corresponding static stress intensity factor, whose value here is taken as $\sqrt{\pi a_0}p_0$ (for the reason of this, refer to Section 10.6.3.3). For the dynamic initiation of crack growth in classical fracture dynamics, there is the only exact analytic solution — the Maue's solution [24], but the configuration of whose specimen is quite different from that of our specimen. Maue studied a semi-infinite crack in an infinite body, subjected to a Heaviside impact loading at the crack surface. While our specimen is a finite size rectangular plate with a central crack, the applied stress is around the external boundary of the specimen. Generally, the Maue model cannot describe the interaction between wave and external boundary. However, consider a very short time interval, i.e., during the period between the stress wave from the external boundary arriving at the crack tip (this time is denoted by t_1) and before the reflecting

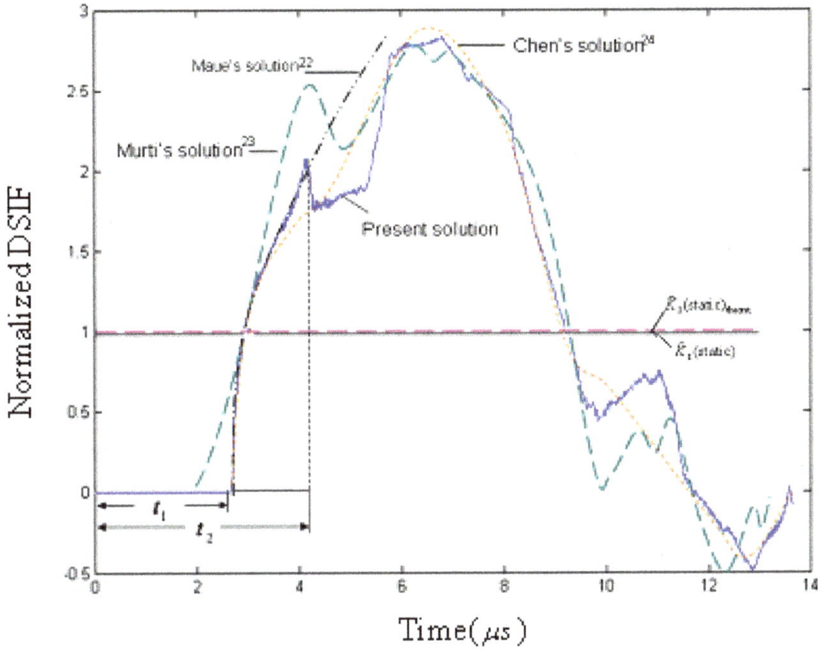

Fig. 10.6. Comparison of the present solution with analytic solution and other numerical solution for conventional structural materials given by other authors.

by external boundary stress wave emanating from the crack tip in the finite size specimen (the time is marked as t_2). During this special very short time interval, our specimen can be seen as an "infinite specimen". The comparison given in Fig. 10.6 shows that the numerical results are in excellent agreement with those of Maue's solution within the short interval in which the solution is valid.

Our solution corresponding to the case of $w_x = w_y = 0$ is also compared with the numerical solutions of conventional crystals, e.g., Murti's solution [25] and Chen's solutions [26], which are also shown in Fig. 10.6; it is evident that our solution presents very high precison.

10.6.3.3 *Influence of mesh size (space step)*

The mesh size or the space step of the algorithm can influence the computational accuracy too. To check the accuracy of the algorithm,

Table 10.1. The normalized static S.I.F. of quasicrystals for different space steps.

H	$a_0/10$	$a_0/15$	$a_0/20$	$a_0/30$	$a_0/40$
$\bar{K}$	0.9259	0.94829	0.96229	0.97723	0.99516
Errors	7.410%	5.171%	3.771%	2.277%	0.484%

we take different space steps shown in Table 10.1, which indicates if $h = a_0/40$ the accuracy is good enough. The check is carried out through static solution, because the static crack problem in infinite body of decagonal quasicrystals has exact solution, given by Li and Fan in Section 3 of Chapter 8, the normalized static intensity factor is equal to unit. In the static case, there is no wave propagation effect, $L/a_0 \geq 3, H/a_0 \geq 3$, and the effect of boundary to solution is very weak, and for our present specimen $L/a_0 \geq 4$, $H/a_0 \geq 8$, which may be seen as an infinite specimen, so the normalized static stress intensity factor is approximate but with a high precise equal to unit. The table shows that the algorithm is quite highly accuracy when $h = a_0/40$.

10.6.4 *Results of dynamic initiation of crack growth*

The dynamic crack problem presents two "phases" in the process: the dynamic initiation of crack growth and fast crack propagation. In the phase of dynamic initiation of crack growth, the length of the crack is constant, assuming $a(t) = a_0$. The specimen with stationary crack that is subjected to a rapidly varying applied load $p(t) = p_0 f(t)$, where p_0 is a constant with stress dimension and $f(t)$ is taken as the Heaviside function. It is well known that the coupling effect between phonon and phason is very important, which reveals the distinctive physical properties including mechanical properties and makes quasicrystals distinguish the periodic crystals. So, studying the coupling effect is significant.

The dynamic stress intensity factor $K_{\mathrm{I}}(t)$ for quasicrystals has the same definition given by (10.6-7), whose numerical results are plotted in Fig. 10.6, where the normalized dynamics stress

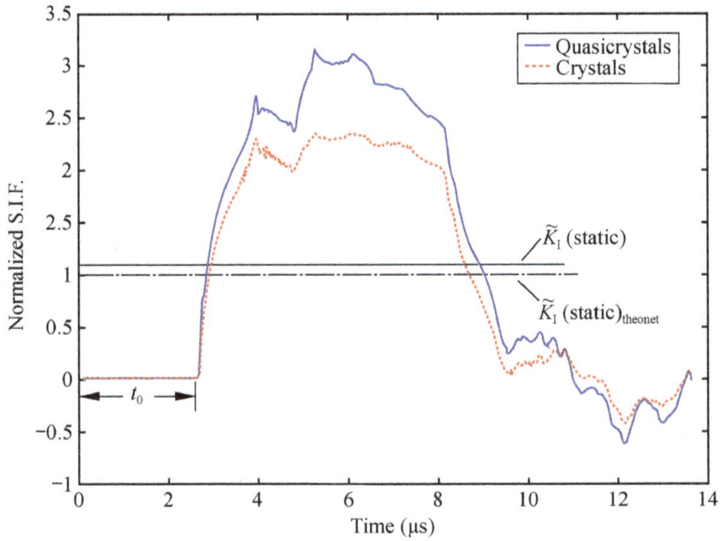

Fig. 10.7. Normalized dynamics stress intensity factor (DSIF) versus time.

intensity factor $K_{\rm I}(t)/\sqrt{\pi a_0}p_0$ is used. There are two curves in the figure: one represents quasicrystal, i.e., $R/M = 0.01$, and the other describes periodic crystals corresponding to $R/M = 0$; the two curves of the figure are apparently different, though they are similar to some extent. Because of the phonon–phason coupling effect, the mechanical properties of the quasicrystals are obviously different from the classical crystals. Thus, the coupling effect plays an important role.

In Fig. 10.7, t_0 represents the time that the wave from the external boundary propagates to the crack surface, in which $t_0 = 2.6735\,\mu$s. So, the velocity of the wave propagation is $\nu_0 = H/t_0 = 7.4807$ km/s, which is just equal to the longitudinal wave speed $c_1 = \sqrt{(L+2M)/\rho}$. This indicates that for the complex system of wave propagation motion of diffusion coupling, the phonon wave propagation presents dominating role.

There are many oscillations in the figure, especially the stress intensity factor. These oscillations characterize the reflection and diffraction between waves coming from the crack surface and the specimen boundary surfaces. The oscillations are influenced by the material constants and specimen geometry including the shape and size very much.

10.6.5 *Results of the fast crack propagation*

In this section, we focus on the discussion for the "phase" of fast crack propagation. To explore the inertia effect caused by the fast crack propagation, the specimen is designed under the action of constant load $P(t) = p$ rather than time-varying load, but the crack grows with high speed in this case. The problem for fast crack propagation is a nonlinear problem because for one part of the boundary, the crack is with unknown length beforehand. For this moving boundary problem, we must give additional condition for the solution to be definite. That is, we must give a criterion checking crack propagation or crack arrest at the growing crack tip. This criterion can be imposed in different ways, e.g., the critical stress criterion or critical energy criterion. The stress criterion is used in this paper: $\sigma_{yy} < \sigma_c$ represents crack arrest, $\sigma_{yy} = \sigma_c$ represents critical state and $\sigma_{yy} > \sigma_c$ represents crack propagation. Here, we take $\sigma_c = 450$ MPa for decagonal Al–Ni–Co quasicrystals, which was obtained by referring the measured value by Meng *et al.* for decagonal Al–Cu–Co quasicrystals, refer to Ref. [2] in Chapter 8; the modification by referring the hardness of alloys Al–Ni–Co and Al–Cu–Co and the hardness on decagonal Al–Ni–Co can be found in the paper given by Takeuchi *et al.* [28]. The simulation of a fracturing process runs as follows:

Given the specimen geometry and its material constants, we first solve the initial dynamic problem in the way previously described. When the stress σ_{yy} reaches a prescribed critical value σ_c, the crack is extended by one grid interval. The crack now continues to grow, by one grid interval at a time, as long as the σ_{yy} stress level ahead of the propagating crack tip reaches the value of σ_c. During the propagation stage, the time that elapses between the two sequential extensions is recorded and the corresponding velocity is evaluated.

The crack velocity for quasicrystals and periodic crystals is constructed in Fig. 10.8, from which, we observe that the velocity in

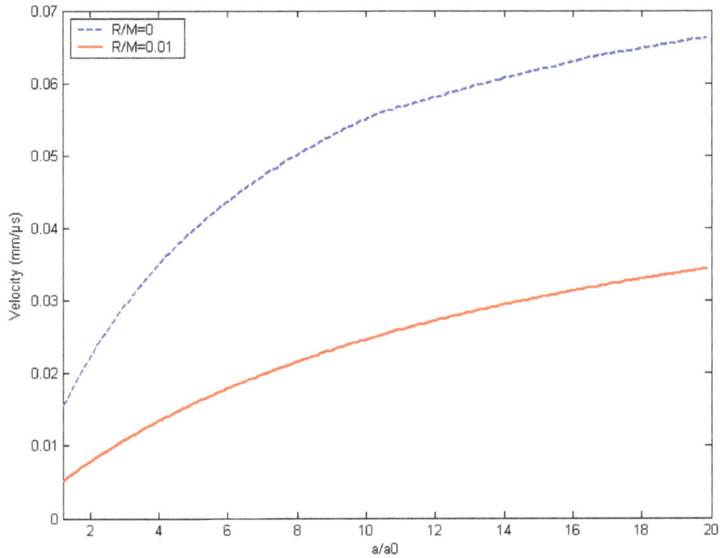

Fig. 10.8. The crack velocity versus normalized crack size with different phonon–phason coupling elastic constants.

quasicrystals is lower than that of the periodic crystals; the phonon–phason coupling makes the quasicrystals different from periodic crystals. The reason for this is not so clear.

We find that the fast crack propagating velocity is obviously different in quasicrystals compared to the crystalline and conventional engineering materials. Next, we will explore the velocity under different loads in quasicrystals. The above described procedure was conducted, keeping the same geometry and material constants. With various loads, the relation between velocity and crack growth is constructed in Fig. 10.8. The crack velocity increases smoothly with increasing applied load. It is understandable as the load increases, the time to reach the critical stress is less, so the velocity increases.

As shown in Figs. 10.9 and 10.10, the calculated crack propagation results show some roughness as the load increases. Currently, there is no experimental observation for fast crack propagation, though Ebert *et al.* [27] made some observation by scanning tunneling microscopy

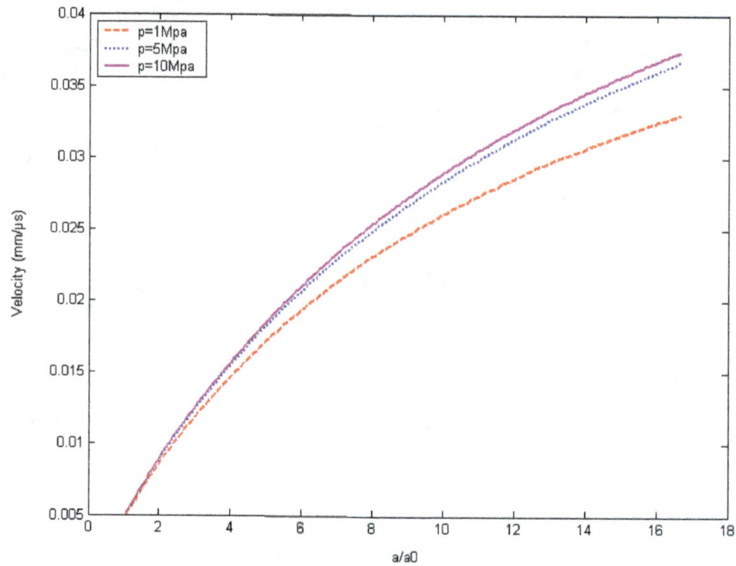

Fig. 10.9. Variation of crack velocity versus normalized crack size for different load levels.

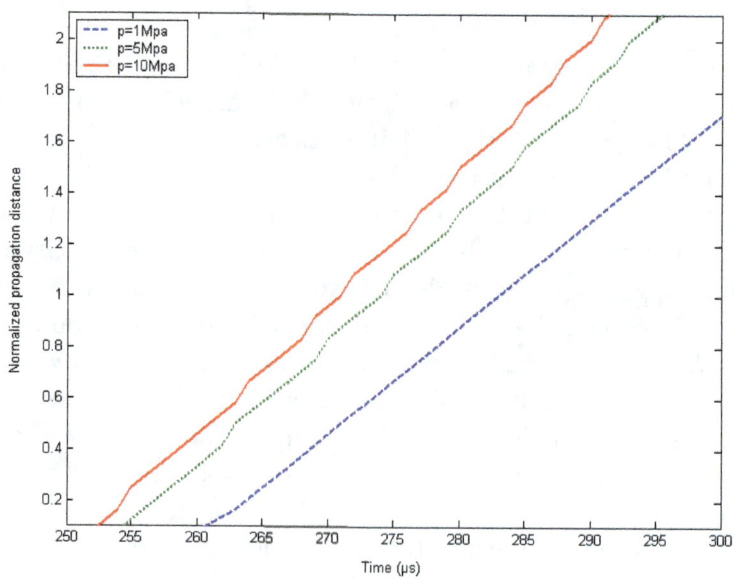

Fig. 10.10. Normalized crack growth size $(a - a_0)/a_0$ of crack tip versus time for different load levels.

for quasistatic crack growth. Because the fast crack propagation and quasistatic crack growth belong to two different regimes, the comparison cannot be easily made.

10.7 Phonon–Phason Dynamics and Applications to Fracture Dynamics of Icosahedral Quasicrystals

10.7.1 *Basic Equations, Boundary and Initial Conditions*

The elasto-/hydro-dynamics of icosahedral Al–Pd–Mn quasicrystals is a more interesting topic than that of decagonal Al–Ni–Co quasicrystals, especially a rich set of experimental data for elastic constants can be used for the computation described here. From the previous section, we know there are lack of measured data for phason elastic constants, which are obtained by Monte-Carlo simulation; this makes some undetermined factors for computational results for decagonal quasicrystals. This shows the discussion on icosahedral quasicrystals is more necessary, and the formalism and numerical results are presented in this section.

If considering only the plane problem, especially for the crack problems, there are much similarities with those discussed in the previous section. We present herein only the parts that are different.

For the plane problem, i.e.,

$$\frac{\partial}{\partial z} = 0 \tag{10.7-1}$$

the linearized elasto-/hydro-dynamics of icosahedral quasicrystals has non-zero displacements u_z, w_z apart from u_x, u_y, w_x, w_y, so in the strain tensors

$$\varepsilon_{ij} = \frac{1}{2}\left(\frac{\partial u_i}{\partial x_j} + \frac{\partial u_j}{\partial x_i}\right) \quad w_{ij} = \frac{\partial w_i}{\partial x_j}$$

it increases some non-zero components compared with those in two-dimensional quasicrystals. In connecting with this, in the stress tensors, the non-zero components increase too relatively to two-dimensional ones. With these reasons, the stress–strain relation presents different nature with that of the decagonal quasicrystals though the generalized Hooke's law has the same form with that in one- and two-dimensional quasicrystals, i.e.,

$$\sigma_{ij} = C_{ijkl}\varepsilon_{kl} + R_{ijkl}w_{kl} \quad H_{ij} = R_{klij}\varepsilon_{kl} + K_{ijkl}w_{kl}$$

In particular, the elastic constants are quite different from those discussed in the previous sections, in which the phonon elastic constants can be expressed such as

$$C_{ijkl} = \lambda\delta_{ij}\delta_{kl} + \mu(\delta_{ik}\delta_{jl} + \delta_{il}\delta_{jk}) \tag{10.7-2}$$

and the phason elastic constant matrix [K] and phonon–phason coupling elastic one [R] are defined by formula (9.1-6) in Chapter 9, which are not listed here again.

Substituting these non-zero stress components into the equations of motion

$$\rho\frac{\partial^2 u_i}{\partial t^2} = \frac{\partial\sigma_{ij}}{\partial x_j}, \quad \kappa\frac{\partial w_i}{\partial t} = \frac{\partial H_{ij}}{\partial x_j} \tag{10.7-3}$$

and through the generalized Hooke's law and strain–displacement relation, we obtain the final dynamic equations as follows:

$$\begin{aligned}
\frac{\partial^2 u_x}{\partial t^2} + \theta\frac{\partial u_x}{\partial t} &= c_1^2\frac{\partial^2 u_x}{\partial x^2} + (c_1^2 - c_2^2)\frac{\partial^2 u_y}{\partial x\partial y} + c_2^2\frac{\partial^2 u_x}{\partial y^2} \\
&\quad + c_3^2\left(\frac{\partial^2 w_x}{\partial x^2} + 2\frac{\partial^2 w_y}{\partial x\partial y} - \frac{\partial^2 w_x}{\partial y^2}\right) \\
\frac{\partial^2 u_y}{\partial t^2} + \theta\frac{\partial u_y}{\partial t} &= c_2^2\frac{\partial^2 u_y}{\partial x^2} + (c_1^2 - c_2^2)\frac{\partial^2 u_x}{\partial x\partial y} + c_1^2\frac{\partial^2 u_y}{\partial y^2} \\
&\quad + c_3^2\left(\frac{\partial^2 w_y}{\partial x^2} - 2\frac{\partial^2 w_x}{\partial x\partial y} - \frac{\partial^2 w_y}{\partial y^2}\right) \\
\frac{\partial^2 u_z}{\partial t^2} + \theta\frac{\partial u_z}{\partial t} &= c_2^2\left(\frac{\partial^2}{\partial x^2} + \frac{\partial^2}{\partial y^2}\right)u_z + c_3^2\left(\frac{\partial^2 w_x}{\partial x^2} - \frac{\partial^2 w_x}{\partial y^2}\right. \\
&\quad \left. - 2\frac{\partial^2 w_y}{\partial x\partial y} + \frac{\partial^2 w_z}{\partial x^2} + \frac{\partial^2 w_z}{\partial y^2}\right) \\
\frac{\partial w_x}{\partial t} + \theta w_x &= d_1\left(\frac{\partial^2}{\partial x^2} + \frac{\partial^2}{\partial y^2}\right)w_x + d_2\left(\frac{\partial^2}{\partial x^2} - \frac{\partial^2}{\partial y^2}\right)w_z
\end{aligned}$$

$$+d_3\left(\frac{\partial^2 u_x}{\partial x^2}-2\frac{\partial^2 u_y}{\partial x\partial y}-\frac{\partial^2 u_x}{\partial y^2}+\frac{\partial^2 u_z}{\partial x^2}-\frac{\partial^2 u_z}{\partial y^2}\right)$$

$$\frac{\partial w_y}{\partial t}+\theta w_y=d_1\left(\frac{\partial^2}{\partial x^2}+\frac{\partial^2}{\partial y^2}\right)w_y-d_2\frac{\partial^2 w_z}{\partial x\partial y}$$

$$+d_3\left(\frac{\partial^2 u_y}{\partial x^2}+2\frac{\partial^2 u_x}{\partial x\partial y}-\frac{\partial^2 u_y}{\partial y^2}-2\frac{\partial^2 u_z}{\partial x\partial y}\right)$$

$$\frac{\partial w_z}{\partial t}+\theta w_z=(d_1-d_2)\left(\frac{\partial^2}{\partial x^2}+\frac{\partial^2}{\partial y^2}\right)w_z$$

$$+d_2\left(\frac{\partial^2 w_x}{\partial x^2}-\frac{\partial^2 w_x}{\partial y^2}-2\frac{\partial^2 w_y}{\partial x\partial y}\right)$$

$$+d_3\left(\frac{\partial^2}{\partial x^2}+\frac{\partial^2}{\partial y^2}\right)u_z \tag{10.7-4}$$

in which

$$c_1=\sqrt{\frac{\lambda+2\mu}{\rho}},\quad c_2=\sqrt{\frac{\mu}{\rho}},\quad c_3=\sqrt{\frac{R}{\rho}},\quad d_1=\frac{K_1}{\kappa},\quad d_2=\frac{K_2}{\kappa},$$

$$d_3=\frac{R}{\kappa}, \tag{10.7-5}$$

Note that constants c_1, c_2 and c_3 have the meaning of elastic wave speeds, while d_1, d_2 and d_3 do not represent wave speed, but are diffusive coefficients and parameter θ may be understood as a manmade damping coefficient as in the previous section.

Consider an icosahedral quasicrystal specimen with a Griffith crack shown in Fig. 10.11; all parameters of geometry and loading are the same with those given in the previous, but in the boundary conditions, there are some different points, which are given as follows:

$$u_x=0,\ \sigma_{yx}=0,\ \sigma_{zx}=0,\ w_x=0,\ H_{yx}=0,\ H_{zx}=0 \text{ on } x=0 \text{ for } 0\le y\le H$$

$$\sigma_{xx}=0,\ \sigma_{yx}=0,\ \sigma_{zx}=0,\ H_{xx}=0,\ H_{yx}=0,\ H_{zx}=0 \text{ on } x=L \text{ for } 0\le y\le H$$

$$\sigma_{yy}=p(t),\ \sigma_{xy}=0,\ \sigma_{zy}=0,\ H_{yy}=0,\ H_{xy}=0,\ H_{zy}=0 \text{ on } y=H \text{ for } 0\le x\le L$$

$$\sigma_{yy} = 0,\ \sigma_{xy} = 0,\ \sigma_{zy} = 0,\ H_{yy} = 0,\ H_{xy} = 0,\ H_{zy} = 0 \text{ on } y = 0$$
$$\text{for } 0 \le x \le a(t)$$
$$u_y = 0,\ \sigma_{xy} = 0,\ \sigma_{zy} = 0,\ w_y = 0,\ H_{xy} = 0,\ H_{zy} = 0 \text{ on } y = 0$$
$$\text{for } a(t) < x \le L. \quad (10.7\text{-}6)$$

The initial conditions are

$$u_x(x,y,t)\,|_{t=0} = 0 \quad u_y(x,y,t)\,|_{t=0} = 0 \quad u_z(x,y,t)\,|_{t=0} = 0$$
$$w_x(x,y,t)\,|_{t=0} = 0 \quad w_y(x,y,t)\,|_{t=0} = 0 \quad w_z(x,y,t)\,|_{t=0} = 0$$
$$\frac{\partial u_x(x,y,t)}{\partial t}\,|_{t=0} = 0 \quad \frac{\partial u_y(x,y,t)}{\partial t}\,|_{t=0} = 0 \quad \frac{\partial u_z(x,y,t)}{\partial t}\,|_{t=0} = 0 \quad (10.7\text{-}7)$$

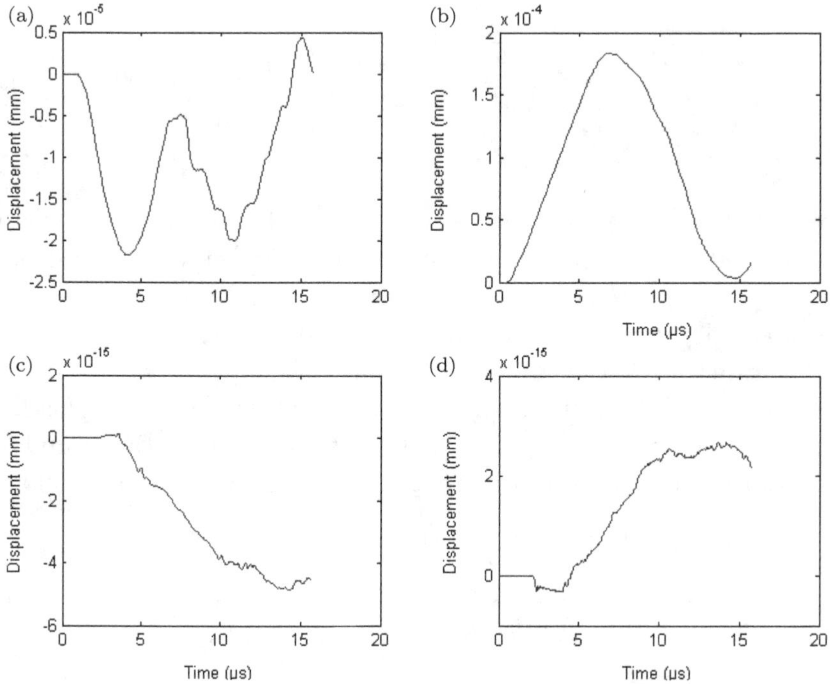

Fig. 10.11. Displacement components of quasicrystals versus time. (a) Displacement component u_x; (b) displacement component u_y; (c) displacement component w_x; (d) displacement component w_y.

10.7.2 *Some results*

We now concentrate on investigating the phonon and phason fields in the icosahedral Al–Pd–Mn quasicrystal, in which we take $\rho = 5.1\text{g/cm}^3$ and $\lambda = 74.2, \mu = 70.4(\text{GPa})$ of the phonon elastic moduli, for phason ones $K_1 = 72, K_2 = -37(\text{MPa})$ (refer to Chapter 9) and the constant relevant to diffusion coefficient of phason is $\Gamma_w = 1/\kappa = 4.8 \times 10^{-19}\text{m}^3 \cdot \text{s/kg} = 4.8 \times 10^{-10}\ \text{cm}^3 \cdot \mu\text{s/g}$ [23]. On the phonon–phason coupling constant, there is no measured result for icosahedral quasicrystals so far, thus we take $R/\mu = 0.01$ for quasicrystals, and $R/\mu = 0$ for "decoupled quasicrystals" or crystals.

The problem is solved by the finite difference method, and the principle, scheme and algorithm are illustrated as those in the previous section and shall not be repeated here. The testing for the physical model, scheme, algorithm and computer programme are similar to those given in Section 10.6.

The numerical results for the dynamic initiation of the crack growth problem and the phonon and phason displacements are shown in Fig. 10.12.

The dynamic stress intensity factor $K_{\text{I}}(t)$ is defined by

$$K_{\text{I}}(t) = \lim_{x \to a_0^+} \sqrt{\pi(x - a_0)}\sigma_{yy}(x, 0, t)$$

and the normalized dynamics stress intensity factor (S.I.F.) $\tilde{K}_{\text{I}}(t) = K_{\text{I}}(t)/\sqrt{\pi a_0}p_0$ is used, and the results are illustrated in Fig. 10.13, in which the comparison with those of crystals are shown, and one can see that the effects of phason and phonon–phason coupling are evident very much.

For the fast crack propagation problem, the primary results are listed only the dynamic stress intensity factor versus time as given in Fig. 10.13:

Details of this work are reported in Ref. [29].

10.7.3 *Conclusion and discussion*

In Sections 10.7.1–10.7.2, a new model on dynamic response of quasicrystals based on the argument of Bak as well as the argument of Lubensky *et al.* is formulated. This model is regarded as an

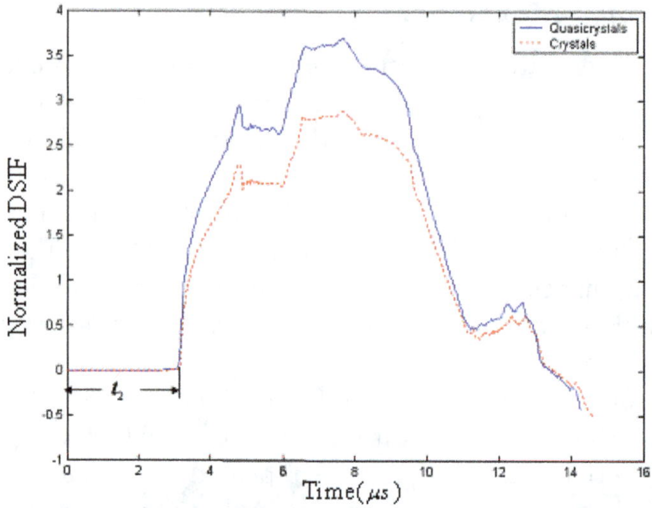

Fig. 10.12. Normalized dynamic stress intensity factor of central crack specimen under impact loading versus time.

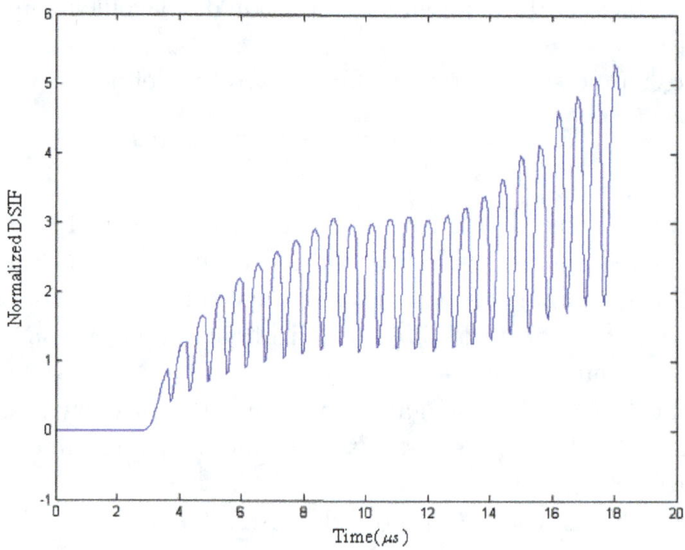

Fig. 10.13. Normalized stress intensity factor of propagating crack with constant crack speed versus time.

elasto-/hydro-dynamics model for the material or as a collaborating model of wave propagation and diffusion. This model is more complex than the pure wave propagation model for conventional crystals, and the analytic solution is very difficult to obtain, except maybe for a few simple examples given in Section 10.5. The numerical procedure based on a finite difference algorithm is developed. Computed results confirm the validity of the wave propagation behaviour of a phonon field and the behaviour of the diffusion of a phason field. The interaction between phonons and phasons are also recorded.

The finite difference formalism is applied to analyze the dynamic initiation of crack growth and crack fast propagation for two-dimensional decagonal Al–Ni–Co and three-dimensional icosahedral Al–Pd–Mn quasicrystals; the displacement and stress fields around the tip of the stationary and propagating cracks are revealed, and the stress presents singularity with order $r^{-1/2}$, in which r denotes the distance measured from the crack tip. For the fast crack propagation, which is a nonlinear problem — moving boundary problem — one must provide additional condition for determining the solution. For this purpose, we give a criterion for checking the crack propagation/crack arrest based on the critical stress criterion. The application of this additional condition for determining the solution has helped us achieve the numerical simulation of the moving boundary value problem and reveal crack length-time evolution. However, more important and difficult problems are left open for further study.

10.8 Appendix A — The Detail of Finite Difference Scheme

Equations (10.6-2) subjected to conditions (10.6-4) and (10.6-5) are very complicated, and the analytic solution for the boundary-initial value problem is not available at present, which has to be solved by numerical method. Here, we extend the method of finite difference of Shmuely and Alterman [30] scheme for analysing the crack problem for conventional engineering materials to quasicrystalline materials.

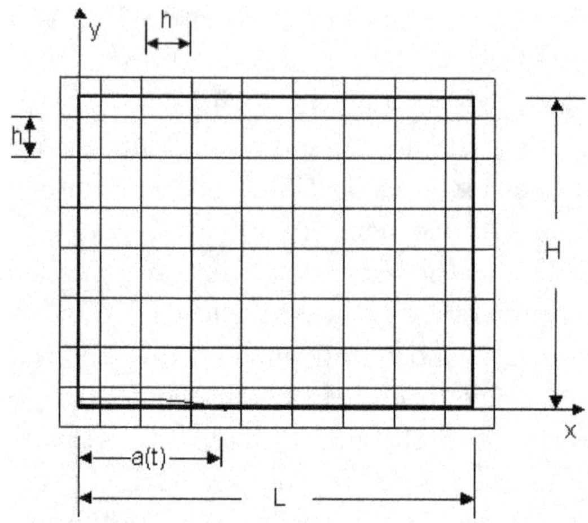

Fig. 10.14. Scheme of grid used.

A grid is imposed on the upper right of the specimen as shown in Fig. 10.14. For convenience, the mesh size h is taken to be the same in both x and y directions. The grid is extended beyond the half-step by adding four special grid lines $x = -h/2$, $x = L + h/2$, $y = -h/2$ and $y = H + h/2$, which form the grid boundaries.

Denoting the time step by τ and using central difference approximations, the finite difference formulations of Eq. (10.6-2), valid at the inner part of the grids are

$$\begin{aligned}
u_x(x, y, t+\tau) &= 2u_x(x, y, t) - u_x(x, y, t-\tau) + \left(\frac{\tau}{h}c_1\right)^2 \\
&\quad \times [u_x(x+h, y, t) - 2u_x(x, y, t) + u_x(x-h, y, t)] \\
&\quad + \left(\frac{\tau}{h}\right)^2 (c_1^2 - c_2^2) [u_y(x+h, y+h, t) - u_y(x+h, y-h, t) \\
&\quad - u_y(x-h, y+h, t) + u_y(x-h, y-h, t)] + \left(\frac{\tau}{h}c_2\right)^2 \\
&\quad \times [u_x(x, y+h, t) - 2u_x(x, y, t) + u_x(x, y-h, t)] \\
&\quad + \left(\frac{\tau}{h}c_3\right)^2 [w_x(x+h, y, t) - 2w_x(x, y, t) + w_x(x-h, y, t)] \\
&\quad + 2\left(\frac{\tau}{h}\right)^2 c_3^2 [w_y(x+h, y+h, t) - w_y(x+h, y-h, t)
\end{aligned}$$

$$
\begin{aligned}
&- w_y(x-h, y+h, t) + w_y(x-h, y-h, t)] - \left(\frac{\tau}{h}c_3\right)^2 \\
&\times [w_x(x, y+h, t) - 2w_x(x, y, t) + w_x(x, y-h, t)] \\
u_y(x, y, t+\tau) &= 2u_y(x, y, t) - u_y(x, y, t-\tau) + \left(\frac{\tau}{h}c_2\right)^2 \\
&\times [u_y(x+h, y, t) - 2u_y(x, y, t) + u_y(x-h, y, t)] \\
&+ \left(\frac{\tau}{2h}\right)^2 (c_1^2 - c_2^2)[u_x(x+h, y+h, t) - u_x(x+h, y-h, t) \\
&- u_x(x-h, y+h, t) + u_x(x-h, y-h, t) \\
&+ \left(\frac{\tau}{h}c_1\right)^2 [u_y(x, y+h, t) - 2u_y(x, y, t) + u_y(x, y-h, t)] \\
&+ \left(\frac{\tau}{h}c_3\right)^2 [w_y(x+h, y, t) - 2w_y(x, y, t) + w_y(x-h, y, t)] \\
&- 2\left(\frac{\tau}{2h}\right)^2 c_3^2[w_x\,(x+h_s y+h, t) - w_x(x+h, y-h, t) \\
&- w_x(x-h, y+h, t) + w_x(x-h, y-h, t)] - \left(\frac{\tau}{h}c_2\right)^2 \\
&\times [w_y(x, y+h, t) - 2w_y(x, y, t) + w_y(x, y-h, t)] \\
w_x(x, y, t+\tau) &= w_x(x, y, t) + d_2^2\frac{\tau}{h^2} \\
&\times [u_x(x+h, v, t) - 2u_x(x, y, t) + u_x(x-h, y, t)] \\
&+ d_1^2\frac{\tau}{h^2}[w_x(x+h, y, t) + w_x(x-h, y, t) - 4w_x(x, y, t) \\
&+ w_x(x, y+h, t) + w_x(x, y-h, t)] - 2d_2^2\frac{\tau}{(2h)^2} \\
&\times [u_y(x+h, y+h, t) - u_y(x+h, y-h, t) - u_y(x-h, y+h, t) \\
&+ u_y(x-h, y-h, t)] - d_2^2\frac{\tau}{h^2}[u_x(x, y+h, t) - 2u_x(x, y, t) \\
&+ u_x(x, y-h, t)] \\
w_y(x, y, t+\tau) &= w_y(x, y, t) + d_2^2\frac{\tau}{h^2} \\
&\times [u_y(x+h, y, t) - 2u_y(x, y, t) + u_y(x-h, y, t)] \\
&+ d_1\frac{\tau}{h^2}[w_y(x+h, y, t) + w_y(x-h, y, t) - 4w_y(x, y, t)
\end{aligned}
$$

$$
\begin{aligned}
&+ w_y(x, y+h, t) + w_y(x, y-h, t)] + 2d_2^2 \frac{\tau}{(2h)^2} \\
&\times [u_x(x+h, y+h, t) - u_x(x+h, y-h, t) - u_x(x-h, y+h, t) \\
&+ u_x(x-h, y-h, t)] - d_2 \frac{\tau}{h^2}[u_y(x, y+h, t) - 2u_y(x, y, t) \\
&+ u_y(x, y-h, t)] \qquad (10.8\text{-}1)
\end{aligned}
$$

The displacements at mesh points located at the special lines are determined by satisfying the boundary conditions, and we obtain respectively for points on the grid lines $x = -h/2$ and $x = L + h/2$.

$$
\begin{aligned}
u_x\left(\substack{-\frac{h}{2}\\ L+\frac{h}{2}}, y, t\right) &= u_x\left(\substack{-\frac{h}{2}\\ L-\frac{h}{2}}, y, t\right) \pm \frac{1}{2}\frac{d_1^2(c_1^2 - 2c_2^2) + c_3^2 d_2^2}{c_1^2 d_1^2 - c_3^2 d_2^2} \\
&\quad \times \begin{bmatrix} u_y\left(\substack{\frac{h}{2}\\ L-\frac{h}{2}}, y+h, t\right) \\ -u_y\left(\substack{\frac{h}{2}\\ L-\frac{h}{2}}, y-h, t\right) \end{bmatrix} \pm \frac{1}{2}\frac{c_3^2(d_1^2 - d_3^2)}{c_1^2 d_1^2 - c_3^2 d_2^2} \\
&\quad \times \left[w_y\left(\substack{\frac{h}{2}\\ L-\frac{h}{2}}, y+h, t\right) - w_y\left(\substack{\frac{h}{2}\\ L-\frac{h}{2}}, y-h, t\right)\right]
\end{aligned}
\qquad (10.8\text{-}2a)
$$

$$
\begin{aligned}
w_x\left(\substack{-\frac{h}{2}\\ L+\frac{h}{2}}, y, t\right) &= w_x\left(\substack{-\frac{h}{2}\\ L-\frac{h}{2}}, y, t\right) \pm 2\frac{d_2^2(c_1^2 - 2c_2^2)}{c_3^2 d_2^2 - c_1^2 d_1^2}\left[u_y\left(\substack{\frac{h}{2}\\ L-\frac{h}{2}}, y+h, t\right)\right. \\
&\quad \left. - u_y\left(\substack{\frac{h}{2}\\ L-\frac{h}{2}}, y-h, t\right)\right] \pm \frac{1}{2}\frac{c_3^2 d_2^2 - c_1^2 d_3^2}{c_3^2 d_2^2 - c_1^2 d_1^2} \\
&\quad \times \left[w_y\left(\substack{\frac{h}{2}\\ L-\frac{h}{2}}, y+h, t\right) - w_y\left(\substack{\frac{h}{2}\\ L-\frac{h}{2}}, y-h, t\right)\right]
\end{aligned}
\qquad (10.8\text{-}2b)
$$

$$
\begin{aligned}
u_y\left(\substack{-\frac{h}{2}\\ L+\frac{h}{2}}, y, t\right) &= u_y\left(\substack{-\frac{h}{2}\\ L-\frac{h}{2}}, y, t\right) \pm \frac{1}{2}\left[u_x\left(\substack{\frac{h}{2}\\ L-\frac{h}{2}}, y+h, t\right)\right. \\
&\quad \left. - u_x\left(\substack{\frac{h}{2}\\ L-\frac{h}{2}}, y-h, t\right)\right] \pm \frac{1}{2}\frac{c_3^2(d_1^2 - d_3^2)}{c_2^2 d_1^2 - c_3^2 d_2^2} \\
&\quad \times \left[w_x\left(\substack{\frac{h}{2}\\ L-\frac{h}{2}}, y+h, t\right) - w_x\left(\substack{\frac{h}{2}\\ L-\frac{h}{2}}, y-h, t\right)\right]
\end{aligned}
\qquad (10.8\text{-}2c)
$$

$$w_y\left(\substack{-\frac{h}{2}\\L+\frac{h}{2}}, y, t\right) = w_y\left(\substack{\frac{h}{2}\\L-\frac{h}{2}}, y, t\right) \pm \frac{1}{2}\frac{c_3^2 d_2^2 - c_2^2 d_3^2}{c_2^2 d_1^2 - c_3^2 d_2^2} \times \left[w_x\left(\substack{\frac{h}{2}\\L-\frac{h}{2}}, y+h, t\right) - w_x\left(\substack{\frac{h}{2}\\L-\frac{h}{2}}, y-h, t\right)\right] \tag{10.8-2d}$$

where equations (10.8-2a) and (10.8-2b) as related to $x = -h/2$ which is not valid. From the first condition of (10.6-5), at $x = 0$, $u_x = 0$ and $w_x = 0$. To satisfy the condition, the displacements u_x and w_x at $x = -h/2$ are approximated by

$$\begin{aligned} u_x(x, -h/2, t) &= -u_x(x, h/2, t) \\ w_x(x, -h/2, t) &= -w_x(x, h/2, t) \end{aligned} \tag{10.8-3}$$

On the grid line $y = -h/2$ and $y = H + h/2$, we obtain

$$u_x\left(x, \substack{-\frac{h}{2}\\L+\frac{h}{2}}, t\right) = u_x\left(x, \substack{-\frac{h}{2}\\L-\frac{h}{2}}, t\right) \pm \frac{1}{2}\left[u_y\left(x+h, \substack{\frac{h}{2}\\L-\frac{h}{2}}, t\right) - u_y\left(x-h, \substack{\frac{h}{2}\\L-\frac{h}{2}}, t\right)\right] \pm \frac{1}{2}\frac{c_3^2(d_1^2 - d_3^2)}{c_2^2 d_1^2 - c_3^2 d_2^2} \times \left[w_y\left(x+h, \substack{\frac{h}{2}\\L-\frac{h}{2}}, t\right) - w_y\left(x-h, \substack{\frac{h}{2}\\L-\frac{h}{2}}, t\right)\right] \tag{10.8-4a}$$

$$w_x\left(x, \substack{-\frac{h}{2}\\L+\frac{h}{2}}, t\right) = w_x\left(x, \substack{-\frac{h}{2}\\L-\frac{h}{2}}, t\right) \pm \frac{1}{2}\frac{c_3^2 d_2^2 - c_2^2 d_3^2}{c_2^2 d_1^2 - c_3^2 d_2^2} \times \left[w_y\left(x+h, \substack{\frac{h}{2}\\L-\frac{h}{2}}, t\right) - w_y\left(x-h, \substack{\frac{h}{2}\\L-\frac{h}{2}}, t\right)\right] \tag{10.8-4b}$$

$$u_y\left(x, \substack{-\frac{h}{2}\\L+\frac{h}{2}}, t\right) = u_y\left(x, \substack{\frac{h}{2}\\L-\frac{h}{2}}, t\right) \pm \frac{1}{2}\frac{c_3^2 d_2^2 + d_1^2(c_1^2 - 2c_2^2)}{c_1^2 d_1^2 - c_3^2 d_2^2} \times \left[u_x\left(x+h, \substack{\frac{h}{2}\\L-\frac{h}{2}}, t\right) - u_x\left(x-h, \substack{\frac{h}{2}\\L-\frac{h}{2}}, t\right)\right] \pm \frac{1}{2}\frac{c_3^2(d_3^2 - d_1^2)}{c_1^2 d_1^2 - c_3^2 d_2^2}\left[w_x\left(x+h, \substack{\frac{h}{2}\\L-\frac{h}{2}}, t\right) - w_x\left(x-h, \substack{\frac{h}{2}\\L-\frac{h}{2}}, t\right)\right] \tag{10.8-4c}$$

$$
\begin{aligned}
w_y\left(x,{}^{-\frac{h}{2}}_{L+\frac{h}{2}},t\right) &= w_y\left(x,{}^{\frac{h}{2}}_{L-\frac{h}{2}},t\right) \pm \frac{d_2^2(c_1^2-c_2^2)}{c_1^2d_1^2-c_3^2d_2^2}\left[u_x\left(x+h,{}^{\frac{h}{2}}_{L-\frac{h}{2}},t\right)\right. \\
&\quad \left. - u_x\left(x-h,{}^{\frac{h}{2}}_{L-\frac{h}{2}},t\right)\right] \pm \frac{1}{2}\frac{c_1^2d_3^2-c_3^2d_2^2}{c_1^2d_1^2-c_3^2d_2^2} \\
&\quad \times \left[w_x\left(x+h,{}^{\frac{h}{2}}_{L-\frac{h}{2}},t\right) - w_x\left(x-h,{}^{\frac{h}{2}}_{L-\frac{h}{2}},t\right)\right]
\end{aligned}
\tag{10.8-4d}
$$

in which, Eq. (10.8-4c) and (10.8-4) as related to $y = -h/2$ are valid only along the crack surface, namely, only for $x \leq a - h/2$, at $y = 0$, in which the crack terminates. From the last condition of (10.6-5), at $y = 0$ and the ahead of the crack, $u_y = 0, w_y = 0$. To satisfy this condition, the displacements u_y and w_y at $y = -h/2$ are approximated by

$$
\begin{aligned}
u_y(x,-h/2,t) &= -u_y(x,h/2,t) \\
w_y(x,-h/2,t) &= -w_y(x,h/2,t)
\end{aligned}
\tag{10.8-5}
$$

In constructing the approximation (10.8-2–10.8-5), we follow a method proposed by Shmuely and Peretz [31] which was also successfully employed in Ref. [30] for conventional engineering materials. According to this method, derivatives perpendicular to the boundary are proposed by uncentered differences and derivatives parallel to the boundary by centered difference. The real boundary can be considered as located at a distance of half the mesh size from the grid boundaries.

The four grid corners require a special treatment. Difference methods of handling the discontinuities at such points have been proposed in the past. Here, we found that satisfactory results are obtained when the displacements sought are extrapolated from those given along both sides of the corner in question. Accordingly, the components u_x, u_y, w_x, w_y at $(-h/2, -h/2)$ are given by

$$
\begin{aligned}
{}^{u_x}_{u_y}(-h/2,-h/2,t) &= {}^{u_x}_{u_y}(h/2,-h/2,t) + {}^{u_x}_{u_y}(-h/2,h/2,t) \\
&\quad -0.5\left[{}^{u_x}_{u_y}(3h/2,-h/2,t) + {}^{u_x}_{u_y}(-h/2,3h/2,t)\right]
\end{aligned}
$$

$$\begin{aligned}\begin{matrix}w_x\\w_y\end{matrix}(-h/2,-h/2,t) &= \begin{matrix}w_x\\w_y\end{matrix}(h/2,-h/2,t)+\begin{matrix}w_x\\w_y\end{matrix}(-h/2,h/2,t)\\ &\quad -0.5\left[\begin{matrix}w_x\\w_y\end{matrix}(3h/2,-h/2,t)+\begin{matrix}w_x\\w_y\end{matrix}(-h/2,3h/2,t)\right]\end{aligned} \tag{10.8-6}$$

Similar expressions are used for deriving the displacement components at $(-h/2, H+h/2)$, $(L+h/2, L+h/2)$ and $(L+h/2, -h/2)$.

By following relevant stability criterion of the scheme, the computation is always stable and achieves high precision. Discussions on this aspect are omitted here due to space limitation.

References

[1] Lubensky T C, Ramaswamy S and Toner J, 1985, Hydrodynamics of icosahedral quasicrystals, *Phys. Rev. B*, **32**(11), 7444–7452.

[2] Socolar J E S, Lubensky T C and Steinhardt P J, 1986, Phonons, phasons and dislocations in quasicrystals, *Phys. Rev. B*, **34**(5), 3345–3360.

[3] Bak P, 1985, Phenomenological theory of icosahedral in commensurate (quasiperiodic)order in Mn-Al alloys, *Phys. Rev. Lett.*, **54**(14), 1517–1519.

[4] Bak P, 1985, Symmetry, stability and elastic properties of icosahedral in commensurate crystals, *Phys. Rev. B*, **32**(9), 5764–5772.

[5] Ding D H, Yang W G, Hu C Z *et al.*, 1993, Generalized elasticity theory of quasicrystals, *Phys. Rev. B*, **48**(10), 7003–7010.

[6] Hu C Z, Wang R H and Ding D H, 2000, Symmetry groups, physical property tensors, elasticity and dislocations in quasicrystals, *Rep. Prog. Phys.*, **63**(1), 1–39.

[7] Fan T Y, Li X F and Sun Y F, 1999, A moving screw dislocation in one-dimensional hexagonal quasicrystal, *Acta Phys. Sin. (Overseas Ed.)*, **8**(3), 288–295.

[8] Fan T Y, 1999, A study on special heat of one-dimensional hexagonal quasicrystals, *J. Phys.: Condense. Matter*, **11**(45), L513–L517.

[9] Fan T Y and Mai Y W, 2003, Partition function and state equation of point group 12mm two-dimensional quasicrystals, *Euro. Phys. J. B*, **31**(1), 25–27.

[10] Fan T Y and Mai Y W, 2004, Elasticity theory, fracture mechanics and some relevant thermal properties of quasicrystalline materials, *Appl. Mech. Rev.*, **57**(5), 325–344.

[11] Li C L and Liu Y Y, 2001, Phason-strain influences on low-temperature specific heat of the decagonal Al-Ni-Co quasicrystal, *Chin. Phys. Lett.* **18**(4), 570–572.
[12] Li C L and Liu Y Y, 2001, Low-temperature lattice excitation of icosahedral Al-Mn-Pd quasicrystals, *Phys. Rev. B*, **63**(6), 064203.
[13] Fan T Y, Wang X F, Li W, Zhu A Y, 2009, Elasto-hydrodynamics of quasicrystals, *Phil. Mag.*, **89**(6), 501–512.
[14] Zhu A Y and Fan T Y, 2008, Dynamic crack propagation in a decagonal Al-Ni-Co quasicrystal, *J. Phys. Condens. Matter*, **20**(29), 295217.
[15] Rochal S B and Lorman V L, 2000, Anisotropy of acoustic-phonon properties of an icosahedral quasicrystal at high temperature due to phonon-phason coupling, *Phys. Rev. B*, **62**(2), 874.
[16] Rochal S B and Lorman V L, 2002, Minimal model of the phonon-phason dynamics on icosahedral quasicrystals and its application for the problem of internal friction in the i-AIPdMn alloys, *Phys. Rev. B*, **66**(14), 144204.
[17] On the Eshelby's solution one can refer to Hirth J P and Lorthe J, 1982, *Theory of Dislocations*, 2nd edition, John Wiely & Sons, New York.
[18] Yoffe E H, 1951, Moving Griffith crack, *Phil. Mag.*, **43**(10), 739–750.
[19] Li X F, 2011, A general solution of elasto-hydrodynamics of two-dimensional quasicrystals, *Philos. Mag. Lett.*, **91**(4), 313–320.
[20] Li X F, 2013, Elastohydrodynamic problems in quasicrystal elasticity theory and wave propagation, *Philos. Mag. Lett.*, **93**(13), 1500–1519.
[21] Chernikov M A, Ott H R, Bianchi A *et al.*, 1998, Elastic moduli of a single quasicrystal of decagonal Al-Ni-Co: evidence for transverse elastic isotropy, *Phys. Rev. Lett.*, **80**(2), 321–324.
[22] H. C. Jeong and P. J. Steinhardt, 1993, Finite-temperature elasticity phase transition in decagonal quasicrystals, *Phys. Rev. B*, **48**(13), 9394–9403.
[23] Walz C, 2003, Zur Hydrodynamik in Quasikristallen, Diplomarbeit, Universitaet Stuttgart.
[24] Maue A W, 1954, Die entspannungswelle bei ploetzlischem Einschnitt eines gespannten elastischen Koepores, *Zeitschrift fuer angewandte Mathematik und Mechanik*, **14**(1), 1–12.
[25] Murti V and Vlliappan S, 1982, The use of quarter point element in dynamic crack analysis, *Eng. Fract. Mech.*, **23**(3), 585–614.
[26] Chen Y M, 1975, Numerical computation of dynamic stress intensity factor s by a Lagrangian finite-difference method (the HEMP code), *Eng. Fract. Mech.*, **7**(8), 653–660.

[27] Ebert Ph, Feuerbacher M, Tamura N *et al.*, 1996, Evidence for a cluster-based on structure of AlPdMn single quasicrystals, *Phys. Rev. Lett.*, **77**(18), 3827–3830.

[28] Takeuchi S, Iwanaga H and Shibuya T, 1991, Hardness of quasicystals, *Jpn. J. Appl. Phys.*, **30**(3), 561–562.

[29] Wang X F, Fan T Y and Zhu A Y, 2009, Dynamic behaviour of the icosahedral Al-Pd-Mn quasicrystal with a Griffith crack, *Chin Phys B*, **18**(2), 709–714. (in Section 10.7 a results are introduced from Zhu A Y and Fan T Y's unpublished work "Fast crack propagation in three-dimensional icosahedral Al-Pd-Mn quasicrystals", 2007 or referring to Zhu A Y: Study on analytic solutions in elasticity of three-dimensional quasicrystals and elastodynamics of two- and three-dimensional quasicrystals, Dissertation, Beijing Institute of Technology, 2009, in Chinese).

[30] Shmuely M and Alterman Z S, 1973, Crack propagation analysis by finite differences, *J. Appl. Mech.*, **40**(4), 902–908.

[31] Shmuely M and Peretz D, 1976, Static and dynamic analysis of the DCB problem in fracture mechanics, *Int. J. Solids Struct.*, **12**(1), 67–67.

Chapter 11

Complex Analysis Method for Elasticity of Quasicrystals

In Chapters 7–9, we frequently used the complex analysis method to solve problems of the elasticity of quasicrystals, and many exact analytic solutions were obtained by the method. In those chapters, we only provided the results, and the underlying principle and details of the method could not be discussed. Considering the relatively new feature and particular effect of the method, it is helpful to attempt a further discussion in-depth. Of course, this may lead to a slight repletion with relevant contents of Chapters 7–9.

It is well-known that the so-called complex potential method in elasticity is effective, in general, only for solving harmonic and biharmonic partial differential equations in the classical theory of elasticity. For these equations, the solutions can be expressed by analytic functions of a single complex variable $z = x + iy$, $i = \sqrt{-1}$. In addition, in the classical elasticity, quasi-biharmonic partial differential equations can be solved by analytic functions of some different complex variables such as $z_1 = x + \alpha_1 y, z_2 = x + \alpha_2 y, \ldots$ in which $\alpha_1, \alpha_2, \ldots$ are complex constants. The study of the elasticity of quasicrystals has led to the discovery of some multi-harmonic and multi-quasiharmonic equations, which cover quite a wide range of partial differential equations appearing in the field to date and have been introduced in Chapters 5–9. The discussion on the complex analysis of these equations is significant. We know that the

Muskhelishvili complex analysis method for classical plane elasticity [1], which solves mainly the biharmonic equation, and the complex potential method developed by Lekhlitzkii [2] for classical anisotropic plane elasticity, which solves mainly the quasi-biharmonic equation, made great contributions for quite a wide range of fields in science and engineering. The present formulation and solutions of the complex analysis, e.g., quadruple and sextuple harmonic equations and quadruple quasiharmonic equation, are a new development of the complex analysis method used for classical elasticity. Though the new method is used to solve elasticity problems of quasicrystals at present, it may be extended into other disciplines of science and technology in future.

At first, we simply review the complex analysis method for harmonic and biharmonic equations and then focus on those for quadruple and sextuple harmonic equations and quadruple quasi-harmonic equations, with discussions in detail, presenting their new features from the angle of elasticity as well as complex potential method.

11.1 Harmonic and Biharmonic in Anti-plane Elasticity of One-dimensional Quasicrystals

The final governing equations of the elasticity of one-dimensional quasicrystals present the following two kinds discussed in Chapter 5:

$$
\begin{aligned}
&c_{44}\nabla^2 u_z + R_3\nabla^2 w_z = 0\\
&R_3\nabla^2 u_z + K^2\nabla^2 w_z = 0
\end{aligned} \tag{11.1-1}
$$

$$
\left(c_1\frac{\partial^4}{\partial x^4} + c_2\frac{\partial^4}{\partial x^3\partial y} + c_3\frac{\partial^4}{\partial x^2\partial y^2} + c_4\frac{\partial^4}{\partial x\partial y^3} + c_5\frac{\partial^4}{\partial y^4}\right) G = 0 \tag{11.1-2}
$$

in which Eq. (11.1-1) are actually two decoupled harmonic equations of u_z and w_z, whose complex variable function method was introduced in Sections 8.1 and 8.2. Here, we do not repeat any more.

Equation (11.1-2) is a quasi-biharmonic equation which describes the phonon–phason coupling elasticity field for some kinds of one-dimensional quasicrystal systems; refer to Chapter 5. Some solutions of them in terms of the complex variable function method, whose origin comes from the classical work of Lekhlitskii [2], readers can find some beneficial hints in the monograph.

11.2 Biharmonic Equations in Plane Elasticity of Point Group 12mm Two-dimensional Quasicrystals

From Chapter 6, we know that in the elasticity of dodecagonal quasicrystals, the phonon and phason fields are decoupled from each other. For whose plane elasticity we have the final governing equations as follows:

$$\nabla^2\nabla^2 F = 0, \quad \nabla^2\nabla^2 G = 0 \tag{11.2-1}$$

The complex representation of the solution of (11.2-1) is

$$\left.\begin{aligned} F(x,y) &= \mathrm{Re}[\bar{z}\varphi_1(z) + \int \psi_1(z)dz] \\ G(x,y) &= \mathrm{Re}[\bar{z}\pi_1(z) + \int \chi_1(z)dz] \end{aligned}\right\} \tag{11.2-2}$$

where $\phi_1(z)$, $\psi_1(z)$, $\pi_1(z)$ and $\chi_1(z)$ are any analytic functions of a complex variable $z = x + iy$ $(i = \sqrt{-1})$. For these kinds of biharmonic equations, Muskhelishvili [1] developed the systematical complex variable function method; readers can find some details in the well-known monograph, and we need not discuss those any more. Muskhelishvili's method has some developments in China, e.g., Lu [3] and Fan [4].

11.3 The Complex Analysis of Quadruple Harmonic Equations and Applications in Two-dimensional Quasicrystals

As discussed in Chapters 6–8, for point groups 5m and 10mm or point groups 5, $\overline{5}$ and 10, $\overline{10}$ quasicrystals, either by the displacement potential formulation or by the stress potential formulation, we obtain that the final governing equation is a quadruple harmonic equation, whose complex variable function method is newly created by Liu and Fan [5,6] based on the displacement potential formulation and by Li and Fan [7, 8] based on the stress potential formulation. This complex potential method greatly develops the methodology used in the classical elasticity. It is necessary to give some further discussions in-depth. For simplicity, the following discussion is based on the stress potential formulation only, and solutions are given only for point groups 5, $\overline{5}$ and 10, $\overline{10}$ quasicrystals because the point groups $5m$ and $10mm$ quasicrystals can be seen as a special case of the former.

11.3.1 *Complex representation of solution of the governing equation*

Because it is relatively simpler for the case of point groups $5m$ and $10mm$, which belong to the special case of point groups 5, $\overline{5}$ and point groups 10, $\overline{10}$, we here discuss only the final governing equation of plane elasticity of a pentagonal of point groups 5, $\overline{5}$ and decagonal quasicrystals of point groups 10, $\overline{10}$

$$\nabla^2\nabla^2\nabla^2\nabla^2 G = 0 \tag{11.3-1}$$

where $G(x, y)$ is the stress potential function. The solution of Eq. (11.3-1) is

$$G = 2\,\mathrm{Re}\left[g_1(z) + \bar{z}g_2(z) + \frac{1}{2}\bar{z}^2 g_3(z) + \frac{1}{6}\bar{z}^3 g_4(z)\right] \tag{11.3-2}$$

where $g_j(z)$ $(j = 1, \ldots, 4)$ are four analytic functions of a single complex variable $z \equiv x + iy = re^{i\theta}$. The bar denotes the complex conjugate hereinafter, i.e., $\bar{z} = x - iy = re^{-i\theta}$. We call these functions the complex stress potentials or the complex potentials in brief.

11.3.2 *Complex representation of the stresses and displacements*

Section 8.4 shows that, from fundamental solution (11.3-2), one can find the complex representation of the stresses as follows:

$$\begin{aligned}
\sigma_{xx} &= -32c_1 \operatorname{Re}(\Omega(z) - 2g_4'''(z)) \\
\sigma_{yy} &= 32c_1 \operatorname{Re}(\Omega(z) + 2g_4'''(z)) \\
\sigma_{xy} &= \sigma_{yx} = 32c_1 \operatorname{Im}\Omega(z) \\
H_{xx} &= 32R_1 \operatorname{Re}(\Theta'(z) - \Omega(z)) - 32R_2 \operatorname{Im}(\Theta'(z) - \Omega(z)) \\
H_{xy} &= -32R_1 \operatorname{Im}(\Theta'(z) + \Omega(z)) - 32R_2 \operatorname{Re}(\Theta'(z) + \Omega(z)) \\
H_{yx} &= -32R_1 \operatorname{Im}(\Theta'(z) - \Omega(z)) - 32R_2 \operatorname{Re}(\Theta'(z) - \Omega(z)) \\
H_{yy} &= -32R_1 \operatorname{Re}(\Theta'(z) + \Omega(z)) + 32R_2 \operatorname{Im}(\Theta'(z) + \Omega(z)
\end{aligned} \tag{11.3-3}$$

where

$$\begin{aligned}
\Theta(z) &= g_2^{(\mathrm{IV})}(z) + \bar{z}g_3^{(\mathrm{IV})}(z) + \frac{1}{2}\bar{z}^2 g_4^{(\mathrm{IV})}(z) \\
\Omega(z) &= g_3^{(\mathrm{IV})}(z) + \bar{z}g_4^{(\mathrm{IV})}(z)
\end{aligned} \tag{11.3-4}$$

in which the prime, two prime, three prime and superscript (IV) denote the first- to fourth-order differentiation of $g_j(z)$ to the variable z, and in addition, $\Theta'(z) = d\Theta(z)/dz$. It is evident that $\Theta(z)$ and $\Omega(z)$ are not analytic functions.

By derivation from (11.3-3), we have the complex representation of the displacements such as

$$\begin{aligned}
u_x + iu_y = {} & 32(4c_1c_2 - c_3 - c_1c_4)g_4''(z) \\
& - 32(c_1c_4 - c_3)(\overline{g_3'''(z)} + z\overline{g_4'''(z)})
\end{aligned} \tag{11.3-5}$$

$$w_x + iw_y = \frac{32(R_1 - iR_2)}{K_1 - K_2}\overline{\Theta(z)} \tag{11.3-6}$$

with constants

$$c = M(K_1 + K_2) - 2(R_1^2 + R_2^2), \quad c_1 = \frac{c}{K_1 - K_2} + M,$$

$$c_2 = \frac{c + (L + M)(K_1 + K_2)}{4(L + M)c} \quad c_3 = \frac{R_1^2 + R_2^2}{c}, \quad c_4 = \frac{K_1 + K_2}{c} \tag{11.3-7}$$

11.3.3 *The complex representation of boundary conditions*

In the following, we consider only the stress boundary value problem, i.e., at the boundary curve where L_t the tractions (T_x, T_y) and generalized tractions (h_x, h_y) are given, and there are the stress boundary conditions such as

$$\begin{aligned} &\sigma_{xx} \cos(\mathbf{n}, x) + \sigma_{xy} \cos(\mathbf{n}, y) = T_x, \\ &\sigma_{xy} \cos(\mathbf{n}, x) + \sigma_{yy} \cos(\mathbf{n}, y) = T_y, \quad (x, y) \in L_t \end{aligned} \tag{11.3-8}$$

$$\begin{aligned} &H_{xx} \cos(\mathbf{n}, x) + H_{xy} \cos(\mathbf{n}, y) = h_x, \\ &H_{xy} \cos(\mathbf{n}, x) + H_{yy} \cos(\mathbf{n}, y) = h_y, \quad (x, y) \in L_t \end{aligned} \tag{11.3-9}$$

where T_x, T_y and h_x, h_y are tractions and generalized tractions at the boundary L_t where the stresses are prescribed.

From (11.3-8) and after some derivation, the phonon stress boundary condition can be reduced to the equivalent form

$$g_4''(z) + \overline{g_3'''(z)} + z\overline{g_4'''(z)} = \frac{i}{32c_1} \int (T_x + iT_y) ds, \quad z \in L_t \tag{11.3-10}$$

From Eqs. (11.3-9), (11.3-3) and (11.3-4), we have

$$(R_2 - iR_1)\Theta(z) = i \int (h_x + ih_y) ds, \quad z \in L_t \tag{11.3-11}$$

11.3.4 *Structure of complex potentials*

11.3.4.1 *Arbitrariness in the definition of the complex potentials*

For simplicity, we introduce the following new symbols:

$$g_2^{(\mathrm{IV})}(z) = h_2(z), \quad g_3'''(z) = h_3(z), \quad g_4''(z) = h_4(z) \tag{11.3-12}$$

Then Eq. (11.3-3) can be rewritten as follows:

$$\sigma_{xx} + \sigma_{yy} = 128c_1 \operatorname{Re} h_4'(z) \tag{11.3-13}$$

$$\sigma_{yy} - \sigma_{xx} + 2i\sigma_{xy} = 64c_1\Omega(z) = 64c_1[h_3'(z) + \bar{z}h_4''(z)] \tag{11.3-14}$$

$$H_{xy} - H_{yx} - i(H_{xx} + H_{yy}) = 64(iR_1 - R_2)\Omega(z) \tag{11.3-15}$$

$$(H_{xx} - H_{yy}) - i(H_{xy} + H_{yx}) = 64(R_1 + R_2)\Theta'(z) \tag{11.3-16}$$

Similar to the classical elasticity, from Eqs. (11.3-13)–(11.3-16), it is obvious that a state of phonon and phason stresses is not altered, if one replaces

$$h_4(z) \quad \text{by } h_4(z) + Diz + \gamma \tag{11.3-17}$$

$$h_3(z) \quad \text{by } h_3(z) + \gamma' \tag{11.3-18}$$

$$h_2(z) \quad \text{by } h_2(z) + \gamma'' \tag{11.3-19}$$

where D is a real constant and γ, γ', γ'' are arbitrary complex constants.

Now, consider how these substitutions affect the displacement components which were determined by formulas (11.3-5) and (11.3-6). Direct substitution shows that

$$\begin{aligned} u_x + iu_y &= 32(4c_1c_2 - c_3 - c_1c_4)h_4(z) - 32(c_1c_4 - c_3)(\overline{h_3(z)} + z\overline{h_4'(z)}) \\ &\quad + 32(4c_1c_2 - 2c_3)Diz \\ &\quad + [32(4c_1c_2 - c_3 - c_1c_4)\gamma - 32(c_1c_4 - c_3)\overline{\gamma'}] \end{aligned} \tag{11.3-20}$$

$$\begin{aligned} w_x + iw_y &= \frac{32(R_1 - iR_2)}{K_1 - K_2}\left[\overline{h_2(z)} + z\overline{h_3'(z)} + \frac{1}{2}z^2\overline{h_4''(z)}\right] \\ &\quad + \frac{32(R_1 - iR_2)}{K_1 - K_2}\overline{\gamma''} \end{aligned} \tag{11.3-21}$$

Formulas (11.3-20) and (11.3-21) show that a substitution of the forms (11.3-17) and (11.3-19) will affect the displacement unless

$$D = 0, \quad \gamma = \frac{c_1c_4 - c_3}{4c_1c_2 - c_3 - c_1c_4}\overline{\gamma'}, \quad \overline{\gamma''} = 0$$

11.3.4.2 *General formulas for finite multiply connected regions*

Consider now the case when the region S, occupied by the quasicrystal, is multiple connected. In general, the region is bounded by several simple closed contours $s_1, s_2, \ldots, s_m, s_{m+1}$. The last of these contours is to contain all the others, depicted in Fig. 11.1, i.e., a plate with holes. We assume that the contours do not intersect themselves and have no points in common. Sometimes, we call $s_1, s_2, \ldots, s_m$ as inner boundaries and s_{m+1} as the outer boundary of the region. It is evident that the points $z_1, z_2, \ldots, z_m$ are fixed points in the holes but are located out of the material.

Similar to the discussion of the classical elasticity theory (refer to [1]), we can obtain

$$h_4'(z) = \sum_{k=1}^{m} A_k \ln(z - z_k) + h_{4*}'(z) \tag{11.3-22}$$

$$h_4(z) = \sum_{k=1}^{m} A_k z \ln(z - z_k) + \sum_{k=1}^{m} \gamma_k \ln(z - z_k) + h_{4*}(z) \tag{11.3-23}$$

$$h_3(z) = \sum_{k=1}^{m} \gamma_k' \ln(z - z_k) + h_{3*}(z) \tag{11.3-24}$$

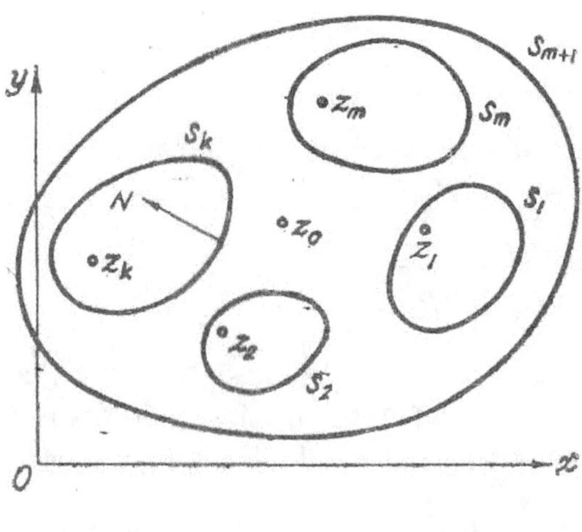

Fig. 11.1. Finite multi-connected region.

Recalling z_k, denote fixed points outside the region S; $h_{3*}(z)$, $h_{4*}(z)$ are holomorphic (analytic and single-valued, refer to Major Appendix) in region S, A_k real constants and γ_k, γ_k' complex constants.

By substituting (11.3-22)–(11.3-24) into (11.3-16), one can find that

$$h_2(z) = \sum_{k=1}^{m} \gamma_k'' \ln(z - z_k) + h_{2*}(z) \tag{11.3-25}$$

$h_{2*}(z)$ is holomorphic in S, and γ_k'' are complex constants.

Consideration will be given to the condition of single-valuedness of phonon displacements. From Eq. (11.3-5), one has

$$\begin{aligned} u_x + iu_y &= 32(4c_1c_2 - c_3 - c_1c_4)h_4(z) \\ &\quad - 32(c_1c_4 - c_3)(\overline{h_3(z)} + z\overline{h_4'(z)}) \end{aligned} \tag{11.3-26}$$

Substituting (11.3-23)–(11.3-25) into (11.3-26), it is immediately seen that

$$\begin{aligned} [u_x + iu_y]_{s_k} &= 2\pi i \Big\{ [32(4c_1c_2 - c_3 - c_1c_4) + 32(c_1c_4 - c_3)]A_k z \\ &\quad +32(4c_1c_2 - c_3 - c_1c_4)\gamma_k + \overline{\gamma_k'(z)} \Big\} \end{aligned} \tag{11.3-27}$$

in which $[]_k$ denotes the increase undergone by the expression in brackets for one anti-clockwise circuit of the contour s_k. Hence, it is necessary and sufficient for the single-valuedness of phonon displacements in formulas (11.3-22)–(11.3-25)

$$A_k = 0, \quad 32(4c_1c_2 - c_3 - c_1c_4)\gamma_k + \overline{\gamma_k'} = 0 \tag{11.3-28}$$

Similar to the above-mentioned discussion, from Eq. (11.3-6), one has

$$[w_x + iw_y]_{s_k} = \frac{32(R_1 - iR_2)}{K_1 - K_2}(-2\pi i)\overline{\gamma_k''} \tag{11.3-29}$$

Hence, it is necessary and sufficient for the single-valuedness of phason displacements

$$\gamma_k'' = 0 \tag{11.3-30}$$

It will now be shown that the quantities γ_k, γ_k' may be very simply expressed in terms of X_k, Y_k, where (X_k, Y_k) denotes the resultant vector of the external stresses, exerted on the contour s_k. By (11.3-10), applying it to the contour s_k, one has

$$-32c_1 i[h_4(z) + \overline{h_3(z)} + z\overline{h_4'(z)}]_{s_k} = X_k + iY_k \tag{11.3-31}$$

with

$$X_k = \int_{S_k} T_x ds, \quad Y_k = \int_{S_k} T_y ds$$

In the present case, the normal vector **n** must be directed outwards with respect to the region s_k. Consequently, the contour s_k must be traversed in the clockwise direction. Taking this fact into consideration, one obtains

$$-2\pi i(\gamma_k - \overline{\gamma_k'}) = \frac{i}{32c_1}(X_k + iY_k) \tag{11.3-32}$$

By Eqs. (11.3-28), (11.3-31) and (11.3-32), one has

$$\begin{aligned} &A_k = 0 \\ &\gamma_k = d_1(X_k + iY_k), \quad \gamma_k' = d_2(X_k - iY_k) \end{aligned} \tag{11.3-33}$$

where

$$\begin{aligned} d_1 &= \frac{1}{64c_1\pi[32(4c_1c_2 - c_3 - c_1c_4) + 1]}, \\ d_2 &= -\frac{4c_1c_2 - c_3 - c_1c_4}{2c_1\pi[32(4c_1c_2 - c_3 - c_1c_4) + 1]} \end{aligned} \tag{11.3-34}$$

which are independent from the suffix k. So,

$$\begin{aligned} h_4(z) &= d_1 \sum_{k=1}^{m} (X_k + iY_k)\ln(z - z_k) + h_{4*}(z) \\ h_3(z) &= d_2 \sum_{k=1}^{m} (X_k - iY_k)\ln(z - z_k) + h_{3*}(z) \\ h_2(z) &= h_{2*}(z) \end{aligned} \tag{11.3-35}$$

We can conclude that the complex functions $h_2(z)$, $h_3(z)$, $h_4(z)$ must be expressed by formulas (11.3-35) to assure the single-valuedness of stresses and displacements, where $h_{2*}(z)$, $h_{3*}(z)$, $h_{4*}(z)$ are holomorphic in region S.

11.3.4.3 *Case of infinite regions*

From the point of view of the application, the consideration of infinite regions is likewise of major interest. We assume that the contour s_{m+1} has entirely moved to infinity.

Due to Eqs. (11.3-13) and (11.3-14) being similar to the classical elasticity theory, we have

$$\begin{aligned} h_4(z) &= d_1(X+iY)\ln z + (B+iC)z + h_4^0(z) \\ h_3(z) &= d_2(X-iY)\ln z + (B'+iC')z + h_3^0(z) \end{aligned} \tag{11.3-36}$$

where B, C, B', C' are unknown real constants to be determined, and

$$X = \sum_{k=1}^{m} X_k, \quad Y = \sum_{k=1}^{m} Y_k$$

$h_3^0(z)$, $h_4^0(z)$ are functions, holomorphic in region S, including the point at infinity, i.e., for sufficiently large $|z|$, they may be expanded into series of the form

$$h_4^0(z) = a_0 + \frac{a_1}{z} + \frac{a_2}{z^2} + \cdots, \quad h_3^0(z) = a_0' + \frac{a_1'}{z} + \frac{a_2'}{z^2} + \cdots \tag{11.3-37}$$

On the basis of (11.3-2), the state of phonon and phason, stresses will not be altered by assuming

$$a_0 = a_0' = 0$$

By the theorem of Laurent, the function $h_{2*}(z)$ may be represented in region S including the point at infinity by the series

$$h_{2*}(z) = \sum_{-\infty}^{+\infty} c_n z^n \tag{11.3-38}$$

Substituting Eqs. (11.3-36) and (11.3-38) into Eq. (11.3-16), one has

$$\begin{aligned}&(H_{xx} - H_{yy}) - i(H_{xy} + H_{yx}) \\ &= 2 \times 32(R_1 + R_2)\left[\sum_{-\infty}^{+\infty} c_n n z^{n-1} + \bar{z}\left(-\frac{d_2}{z^2} + h_3^{0\prime\prime}(z)\right)\right. \\ &\left.+\frac{1}{2}\bar{z}^2\left(\frac{2d_1}{z^3} + h_4^{0\prime\prime\prime}(z)\right)\right] \end{aligned} \quad (11.3\text{-}39)$$

Hence it follows that for the stresses to remain finite as $|z| \to \infty$, one must have

$$c_n = 0 \quad (n \geq 2)$$

It is obvious that the phonon and phason stresses will be bounded, if these conditions are satisfied. Hence, one has finally

$$\begin{aligned} h_4(z) &= d_1(X + iY)\ln z + (B + iC)z + h_4^0(z) \\ h_3(z) &= d_2(X - iY)\ln z + (B' + iC')z + h_3^0(z) \\ h_2(z) &= (B'' + iC'')z + h_2^0(z) \end{aligned} \quad (11.3\text{-}40)$$

where B'', C'' are unknown real constants to be determined and $h_2^0(z)$ is a function, holomorphic in region S, including the point at infinity, thus it has the form similar to that of (11.3-37):

$$h_2^0(z) = a_0'' + \frac{a_1''}{z} + \frac{a_2''}{z^2} + \cdots \quad (11.3\text{-}41)$$

We have assumed that $a_0 = a_0' = 0$ already, now further assume $a_0'' = 0$, i.e.,

$$h_4^0(\infty) = h_3^0(\infty) = h_2^0(\infty) = 0$$

Then from (11.3-40) and (11.3-13)–(11.3-16), one can determine

$$\begin{aligned} &B = \frac{\sigma_{xx}^{(\infty)} + \sigma_{yy}^{(\infty)}}{128c_1}, \quad B' = \frac{\sigma_{xx}^{(\infty)} - \sigma_{yy}^{(\infty)}}{64c_1}, \quad C' = \frac{\sigma_{xy}^{(\infty)}}{32c_1} \\ &B'' = \frac{R_2(H_{xy}^{(\infty)} - H_{yx}^{(\infty)}) - R_1(H_{xx}^{(\infty)} + H_{yy}^{(\infty)})}{64(R_1^2 - R_2^2)}, \\ &C'' = \frac{R_1(H_{xy}^{(\infty)} - H_{yx}^{(\infty)}) - R_2(H_{xx}^{(\infty)} + H_{yy}^{(\infty)})}{64(R_1^2 - R_2^2)} \end{aligned} \quad (11.3\text{-}42)$$

and C has no usage, and so we put it to be zero, in which $\sigma_{ij}^{(\infty)}$ and $H_{ij}^{(\infty)}$ represent the applied stresses at the point of infinity.

11.3.5 *Conformal mapping*

If we constrain our discussion only for the case of stress boundary value problems, then the problems will be solved under boundary conditions (11.3-10) and (11.3-11). For some complicated regions, solutions of the problems cannot be directly obtained in the physical plane (i.e., the z-plane). We must use a conformal mapping

$$z = \omega(\zeta) \tag{11.3-43}$$

to transform the region studied in the plane onto the interior of unit circle γ in the mapping plane (e.g., ζ-plane).

Substituting (11.3-43) into (11.3-40), we have

$$\begin{aligned}
h_4(z) &= \Phi_4(\zeta) = d_1(X+iY)\ln\omega(\zeta) + B\omega(\zeta) + \Phi_4^0(\zeta)\\
h_3(z) &= \Phi_3(\zeta) = d_2(X-iY)\ln\omega(\zeta) + (B'+iC')\omega(\zeta) + \Phi_3^0(\zeta)\\
h_2(z) &= \Phi_2(\zeta) = (B''+iC'')\omega(\zeta) + \Phi_2^0(\zeta)
\end{aligned} \tag{11.3-44}$$

where

$$\Phi_j(\zeta) = h_j[\omega(\zeta)], \quad \Phi_j^0(\zeta) = h_j^0[\omega(\zeta)], \quad j = 1, \ldots, 4$$

In addition,

$$h_i'(z) = \frac{\Phi_i'(\zeta)}{\omega'(\zeta)}$$

At the mapping plane, the boundary conditions (11.3-10) and (11.3-11) stand for

$$\Phi_4(\sigma) + \overline{\Phi_3(\sigma)} + \omega(\sigma)\frac{\overline{\Phi_4'(\sigma)}}{\overline{\omega'(\sigma)}} = \frac{i}{32c_1}\int (T_x + iT_y)ds \tag{11.3-10$'$}$$

$$(R_2 - iR_1)\Theta(\sigma) = i\int (h_x + ih_y)ds \tag{11.3-11$'$}$$

where $\sigma = e^{i\varphi}$ represents the value of ζ at the unit circle (i.e., $\rho = 1$). From these boundary value equations, we can determine the unknown functions $\Phi_j(\zeta)$ $(j = 2, 3, 4)$.

11.3.6 *Reduction of the boundary value problem to function equations*

Due to $\Phi_1(\zeta) = 0$, we now have three unknown functions $\Phi_i(\zeta)$ $(i = 2, 3, 4)$. Taking conjugate of (11.3-10′) yields

$$\overline{\Phi_4(\sigma)} + \Phi_3(\sigma) + \overline{\omega(\sigma)}\frac{\Phi_4'(\sigma)}{\omega'(\sigma)} = -\frac{i}{32c_1}\int (T_x - iT_y)ds \quad (11.3\text{-}10'')$$

Substituting the first one of Eq. (11.3-4) into (11.3-11′) and then multiplying $\frac{1}{2\pi i}\frac{d\sigma}{\sigma-\zeta}$ to both sides of (11.3-10′), (11.3-10″) and (11.3-11′) leads to

$$\frac{1}{2\pi i}\int_\gamma \frac{\Phi_4(\sigma)d\sigma}{\sigma-\zeta} + \frac{1}{2\pi i}\int_\gamma \frac{\overline{\Phi_3(\sigma)}d\sigma}{\sigma-\zeta} + \frac{1}{2\pi i}\int_\gamma \frac{\omega(\sigma)}{\overline{\omega'(\sigma)}}\frac{\overline{\Phi_4'(\sigma)}d\sigma}{\sigma-\zeta}$$
$$= \frac{1}{32c_1}\frac{1}{2\pi i}\int_\gamma \frac{\mathbf{t}d\sigma}{\sigma-\zeta}$$

$$\frac{1}{2\pi i}\int_\gamma \frac{\overline{\Phi_4(\sigma)}d\sigma}{\sigma-\zeta} + \frac{1}{2\pi i}\int_\gamma \frac{\Phi_3(\sigma)d\sigma}{\sigma-\zeta} + \frac{1}{2\pi i}\int_\gamma \frac{\overline{\omega(\sigma)}}{\omega'(\sigma)}\frac{\Phi_4'(\sigma)d\sigma}{\sigma-\zeta}$$
$$= \frac{1}{32c_1}\frac{1}{2\pi i}\int_\gamma \frac{\bar{\mathbf{t}}d\sigma}{\sigma-\zeta}$$

$$\frac{1}{2\pi i}\int_\gamma \frac{\Phi_2(\sigma)d\sigma}{\sigma-\zeta} + \frac{1}{2\pi i}\int_\gamma \frac{\overline{\omega(\sigma)}}{\omega'(\sigma)}\frac{\Phi_3'(\sigma)d\sigma}{\sigma-\zeta} + \frac{1}{2\pi i}\left[\int_\gamma \frac{\overline{\omega(\sigma)}^2}{[\omega'(\sigma)]^2}\frac{\Phi_4''(\sigma)d\sigma}{\sigma-\zeta}\right.$$
$$\left. - \int_\gamma \frac{\overline{\omega(\sigma)}^2\omega''(\sigma)}{[\omega'(\sigma)]^3}\frac{\Phi_4'(\sigma)d\sigma}{\sigma-\zeta}\right] = \frac{1}{R_1 - iR_2}\frac{1}{2\pi i}\int_\gamma \frac{\mathbf{h}d\sigma}{\sigma-\zeta}$$
$$(11.3\text{-}45)$$

where $\mathbf{t} = i\int(T_x + iT_y)ds$, $\bar{\mathbf{t}} = -i\int(T_x - iT_y)ds$, $\mathbf{h} = i\int(h_1 + ih_2)ds$ in Eq. (11.3-45), which are the function equations to determine complex potentials $\Phi_i(\zeta)$ which are analytic in the interior of the unit circle γ and satisfy boundary value conditions (11.3-45) at the unit circle.

11.3.7 *Solution of the function equations*

According to Cauchy's integral formula (refer to Major Appendix I),

$$\frac{1}{2\pi i}\int_\gamma \frac{\Phi_i(\sigma)}{\sigma-\zeta}d\sigma = \Phi_i(\zeta), \frac{1}{2\pi i}\int_\gamma \frac{\overline{\Phi_i(\sigma)}}{\sigma-\zeta}d\sigma = \overline{\Phi_i(0)}, \quad |\zeta|<1$$

So, (11.3-45) are reduced to

$$\begin{aligned}
&\Phi_4(\zeta)+\overline{\Phi_3(0)}+\frac{1}{2\pi i}\int_\gamma \frac{\omega(\sigma)}{\overline{\omega(\sigma)}}\frac{\overline{\Phi_4'(\sigma)}d\sigma}{\sigma-\zeta} = \frac{i}{32c_1}\frac{1}{2\pi i}\int_\gamma \frac{\mathbf{t}d\sigma}{\sigma-\zeta}\\
&\overline{\Phi_4(0)}+\Phi_3(\zeta)+\frac{1}{2\pi i}\int_\gamma \frac{\overline{\omega(\sigma)}}{\omega(\sigma)}\frac{\Phi_4'(\sigma)d\sigma}{\sigma-\zeta} = -\frac{i}{32c_1}\frac{1}{2\pi i}\int_\gamma \frac{\bar{\mathbf{t}}d\sigma}{\sigma-\zeta}\\
&\Phi_2(\zeta)+\frac{1}{2\pi i}\int_\gamma \frac{\overline{\omega(\sigma)}}{\omega'(\sigma)}\frac{\Phi_3'(\sigma)d\sigma}{\sigma-\zeta}+\frac{1}{2\pi i}\left[\int_\gamma \frac{\overline{\omega(\sigma)}^2}{[\omega'(\sigma)]^2}\frac{\Phi_4''(\sigma)d\sigma}{\sigma-\zeta}\right.\\
&\qquad\left.-\int_\gamma \frac{\overline{\omega(\sigma)}^2\omega''(\sigma)}{[\omega'(\sigma)]^3}\frac{\Phi_4'(\sigma)d\sigma}{\sigma-\zeta}\right] = \frac{i}{R_1-iR_2}\frac{1}{2\pi i}\int_\gamma \frac{\mathbf{h}d\sigma}{\sigma-\zeta}
\end{aligned} \tag{11.3-46}$$

The calculation of integrals in (11.3-46) depends upon the configuration of the sample, so the mapping function is $\omega(\zeta)$ and the applied stresses are $\mathbf{t}$ and $\mathbf{h}$. In the following, we will give a concrete solution for a given configuration and applied traction.

11.3.8 *Example 1: Elliptic Notch/Crack Problem and Solution*

We calculate the stress and displacement field induced by an elliptic notch L: $\left(\frac{x^2}{a^2}+\frac{y^2}{b^2}=1\right)$ in an infinite plane of decagonal quasicrystal, see Fig. 11.2, the edge of which is subjected to a uniform pressure p. Though the problem was solved in Section 8.4, figuring out its outline from the general formulation is meaningful.

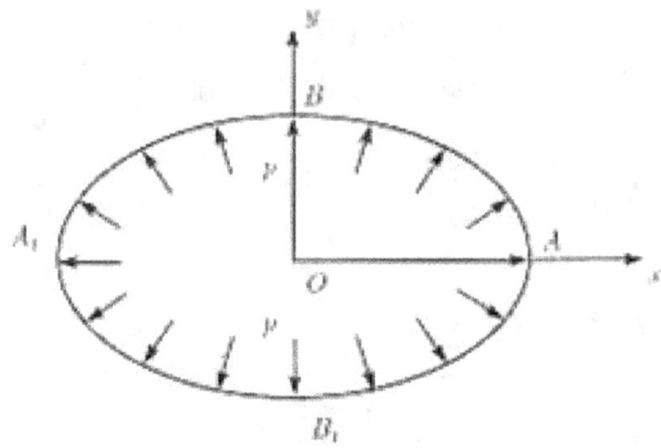

Fig. 11.2. An elliptic notch in a decagonal quasicrystal.

The boundary conditions can be expressed by Eqs. (11.3-10) and (11.3-11); for simplicity, we assume $h_x = h_y = 0$. Thus,

$$i\int (T_x + iT_y)ds = i\int (-p\cos(\mathbf{n},x) - ip\cos(\mathbf{n},y))ds = -pz = -p\omega(\sigma)$$
$$i\int (h_x + ih_y)ds = 0 \tag{11.3-47}$$

In addition, in this case in formulas (11.3-44)

$$\begin{aligned} &X = Y = 0 \\ &B = 0, \quad B' = C' = 0, \quad B'' = C'' = 0 \end{aligned} \tag{11.3-48}$$

So $\Phi_j(\zeta) = \Phi_j^0(\zeta)$, but in the following, we omit the superscript of the functions $\Phi_i^0(\zeta)$ for simplicity.

The conformal mapping is

$$z = \omega(\zeta) = R_0\left(\frac{1}{\zeta} + m\zeta\right) \tag{11.3-49}$$

to transform the region containing ellipse at the z-plane onto the interior of the unit circle at the ζ-plane, refer to Fig. 11.3, where $\zeta = \xi + i\eta = \rho e^{i\varphi}$ and

$$R_0 = \frac{a+b}{2}, \quad m = \frac{a-b}{a+b}$$

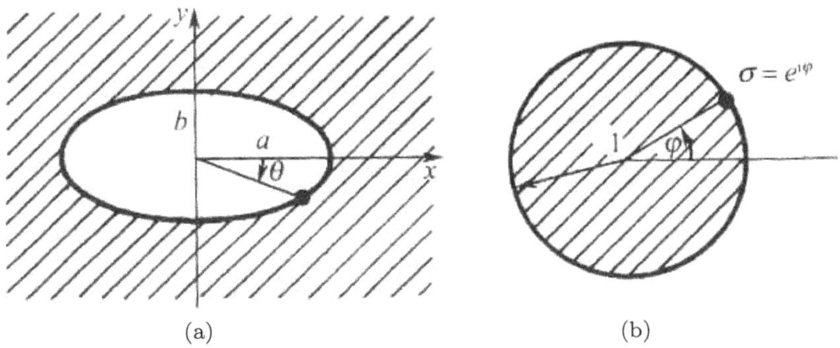

Fig. 11.3. Conformal mapping from the region at z-plane (left) with an elliptic hole onto the interior of the unit circle at ζ-plane (right).

Substituting (11.3-48) and (11.3-49) into function Eq. (11.3-46), one obtains

$$\begin{aligned}\Phi_3(\zeta) &= \frac{pR_0}{32c_1}\frac{(1+m^2)\zeta}{m\zeta^2-1}\\ \Phi_4(\zeta) &= -\frac{pR_0}{32c_1}m\zeta \qquad (11.3\text{-}50)\\ \Phi_2(\zeta) &= \frac{pR_0}{32c_1}\frac{\zeta(\zeta^2+m)[(1+m^2)(1+m\zeta^2)-(\zeta^2+m)]}{(m\zeta^2-1)^3}\end{aligned}$$

If take $m = 1$, from (11.3-50), we can obtain the solution of the Griffith crack, in particular, the explicit solution at z-plane can be explored by taking inversion $\zeta = \omega^{-1}(z) = z/a - \sqrt{z^2/a^2-1}$ (as $m = 1$) into the relevant formulas.

The concrete results are given in Section 8.4, which are omitted here.

11.3.9 *Example 2: Infinite plane with an elliptic hole subjected to a tension at infinity*

In this case,

$$\begin{aligned}&X = Y = 0, \quad T_x = T_y = 0, \quad B = \frac{p}{64c_1}, \quad B' = C' = 0\\ &B'' = C'' = 0, \quad \mathbf{t} = \bar{\mathbf{t}} = \mathbf{h} = 0 \qquad (11.3\text{-}51)\end{aligned}$$

so from (11.3-44),

$$\begin{aligned} h_4(z) &= \Phi_4(\zeta) = B\omega(\zeta) + \Phi_4^0(\zeta) \\ h_3(z) &= \Phi_3(\zeta) = \Phi_3^0(\zeta) \\ h_2(z) &= \Phi_2(\zeta) = \Phi_2^0(\zeta) \end{aligned} \tag{11.3-52}$$

Substituting (11.3-52) into (11.3-45), we obtain similar equations on functions $\Phi_j^0(\zeta)$ $(j = 2, 3, 4)$, so the solution is similar to (11.3-50).

11.3.10 *Example 3: Infinite plane with an elliptic hole subjected to a distributed pressure at a part of surface of the hole*

The problem is shown in Fig. 11.4. We here use the conformal mapping

$$z = \omega(\zeta) = R_0\left(\zeta + \frac{m}{\zeta}\right) \tag{11.3-53}$$

to transform the region at z-plane onto the exterior of the unit circle γ at ζ-plane, see Fig. 11.5.

In terms of the similar procedure, we find that solution [9] is as follows:

$$\begin{aligned} \Phi_4(\zeta) = &\frac{1}{32c_1} \cdot \frac{p}{2\pi i} \cdot \left[-\frac{mR_0}{\zeta} \ln \frac{\sigma_2}{\sigma_1} \right. \\ &\left. + z \ln \frac{\sigma_2 - \zeta}{\sigma_1 - \zeta} + z_1 \ln(\sigma_1 - \zeta) - z_2 \ln(\sigma_2 - \zeta) \right] \\ &+ ip(d_1 - d_2)(z_1 - z_2) \ln \zeta \end{aligned}$$

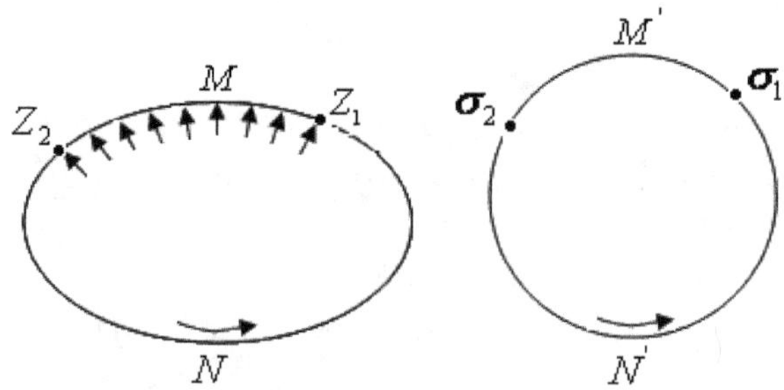

Fig. 11.4. Infinite plane with an elliptic hole subjected to a distributed pressure at a part of surface of the hole and its conformal mapping at ζ-plane.

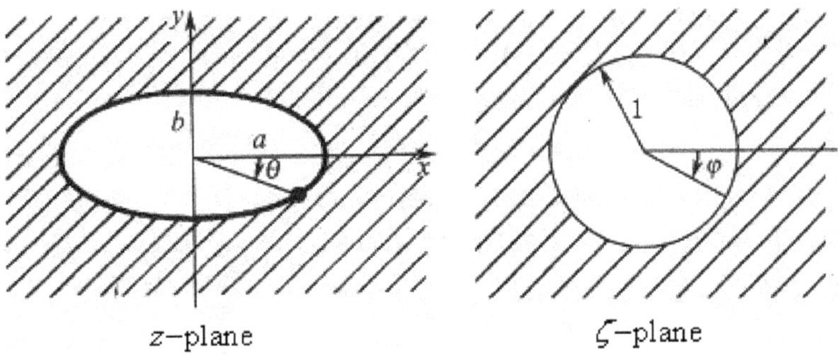

Fig. 11.5. Conformal mapping from the region at z-plane with an elliptic hole onto the exterior of the unit circle at ζ-plane.

$$\begin{aligned}\Phi_3(\zeta) &= \frac{1}{32c_1}\cdot\frac{p}{2\pi i}\cdot\left[-\frac{(1+m^2)R_0\zeta}{(\zeta^2-m)}\ln\frac{\sigma_2}{\sigma_1}+\frac{R_0(\sigma_1-\sigma_2)(1+m\zeta^2)}{(\zeta^2-m)}\right.\\ &\quad\left.-\overline{z_2}\ln(\sigma_2-\zeta)+\overline{z_1}\ln(\sigma_1-\zeta)\right]\\ &\quad-ip(d_1+d_2)\cdot\left[(\overline{z_1}-\overline{z_2})\ln\zeta+(z_1-z_2)\frac{(1+m^2)}{\zeta^2-m}\right]\\ \Phi_2(\zeta) &= \frac{1}{32c_1}\cdot\frac{pR_0}{2\pi i}\cdot\frac{(m\zeta^2+1)(\zeta^2+m)}{(\zeta^2-m)^3}\left(\ln\frac{\sigma_2}{\sigma_1}+\frac{\sigma_2-\sigma_1}{(\sigma_2-\zeta)(\sigma_1-\zeta)}\right)\\ &\quad+\frac{1}{32c_1}\cdot\frac{p}{2\pi i}\cdot\frac{(m\zeta^2+1)}{(\zeta^2-m)^2}\\ &\quad\times\left\{2\operatorname{Re}z_2\cdot\frac{\sigma_2-\sigma_1}{(\sigma_2-\zeta)(\sigma_1-\zeta)}+\left[z_2-R_0\left(\zeta-\frac{m}{\zeta}\right)\right]\right.\\ &\quad\left.\cdot\left[\frac{(\sigma_2-\zeta)(\sigma_1-\zeta)+(\sigma_2+\sigma_1-2\zeta)(\sigma_2-\sigma_1)}{(\sigma_2-\zeta)(\sigma_1-\zeta)}\right]\right\}\\ &\quad\frac{(m\zeta^2+1)(\zeta^2+m)}{(\zeta^2-m)^3}ip\left\{d_1(\overline{z_1}-\overline{z_2}-z_1+z_2)\frac{1}{\zeta-\sigma_1}\right.\\ &\quad\left.+(d_2-d_1)(z_1-z_2)\left[\frac{1}{\zeta^2}+\frac{1}{\zeta}+\frac{1}{(\zeta-\sigma_1)^2}\right]\right\}\end{aligned}\tag{11.3-54}$$

where

$$z_1=R_0\left(\sigma_1+\frac{m}{\sigma_1}\right),\quad z_2=R_0\left(\sigma_2+\frac{m}{\sigma_2}\right)$$

11.4 Complex Analysis for Sextuple Harmonic Equation and Applications to Three-dimensional Icosahedral Quasicrystals

Plane elasticity of icosahedral quasicrystals has been reduced to a sextuple harmonic equation to solve in Chapter 9, where we have shown the solution procedure of the equation for a notch/crack problem by complex variable function method, and we here provide further in-depth discussion from the point of complex function theory. The aim is to develop the complex potential method for higher-order multi-harmonic equations. Though there are some similar natures in the following description with that introduced in the preceding section, the discussion here is necessary because the governing equation and boundary conditions for icosahedral quasicrystals are quite different from those for decagonal quasicrystals.

11.4.1 *The complex representation of stresses and displacements*

In Section 9.5, by the stress potential, we obtain the final governing equation under the approximation $R^2/\mu K_1 \ll 1$

$$\nabla^2\nabla^2\nabla^2\nabla^2\nabla^2\nabla^2 G = 0 \tag{11.4-1}$$

The fundamental solution of Eq. (11.4-1) can be expressed in six analytic functions of complex variable z, i.e.,

$$G(x, y) = \operatorname{Re}[g_1(z) + \bar{z}g_2(z) + \bar{z}^2 g_3(z) + \bar{z}^3 g_4(z) + \bar{z}^4 g_5(z) + \bar{z}^5 g_6(z)] \tag{11.4-2}$$

where $g_i(z)$ are arbitrary analytic functions of $z = x + iy$, and the bar denotes the complex conjugate.

By Eqs. (11.4-1), (11.4-2) and (9.5-2), (9.5-3), the stresses can be expressed as follows:

$$\sigma_{xx} + \sigma_{yy} = 48c_2c_3R\operatorname{Im}\Gamma'(z)$$

$$\sigma_{yy} - \sigma_{xx} + 2i\sigma_{xy} = 8ic_2c_3R(12\overline{\Psi'(z)} - \Omega'(z))$$

$$\sigma_{zy} - i\sigma_{zx} = -960c_3c_4f_6'(z) \quad \sigma_{zz} = \frac{24\lambda R}{(\mu+\lambda)}c_2c_3\operatorname{Im}\Gamma'(z)$$

$$
\begin{aligned}
&H_{xy} - H_{yx} - i(H_{xx} + H_{yy}) = -96c_2c_5\overline{\Psi'(z)} - 8c_1c_2R\Omega'(z) \\
&H_{yx} + H_{xy} + i(H_{xx} - H_{yy}) = -480c_2c_5\overline{f_6'(z)} - 4c_1c_2R\Theta'(z) \\
&H_{yz} + iH_{xz} = 48c_2c_6\Gamma'(z) - 4c_2R^2(2K_2 - K_1)\overline{\Omega'(z)} \\
&H_{zz} = \frac{24R^2}{(\mu+\lambda)}c_2c_3 \operatorname{Im}\Gamma'(z)
\end{aligned}
\tag{11.4-3}
$$

where

$$
\begin{aligned}
&\Psi(z) = f_5(z) + 5\bar{z}f_6'(z) \\
&\Gamma(z) = f_4(z) + 4\bar{z}f_5'(z) + 10\bar{z}^2 f_6''(z) \\
&\Omega(z) = f_3(z) + 3\bar{z}f_4'(z) + 6\bar{z}^2 f_5''(z) + 10\bar{z}^3 f_6'''(z) \\
&\Theta(z) = f_2(z) + 2\bar{z}f_3'(z) + 3\bar{z}^2 f_4''(z) + 4\bar{z}^3 f_5'''(z) + 5\bar{z}^4 f_6^{(\mathrm{IV})}(z) \\
&c_1 = \frac{R(2K_2 - K_1)(\mu K_1 + \mu K_2 - 3R^2)}{2(\mu K_1 - 2R^2)}, \\
&c_3 = \frac{1}{R}K_2(\mu K_2 - R^2) - R(2K_2 - K_1) \\
&c_2 = \mu(K_1 - K_2) - R^2 - \frac{(\mu K_2 - R^2)^2}{\mu K_1 - 2R^2}, \\
&c_4 = c_1R + \frac{1}{2}c_3\left(K_1 + \frac{\mu K_1 - 2R^2}{\lambda + \mu}\right) \\
&c_5 = 2c_4 - c_1R, \quad c_6 = (2K_2 - K_1)R^2 - 4c_4\frac{\mu K_2 - R^2}{\mu K_1 - 2R^2}
\end{aligned}
\tag{11.4-4}
$$

In the above expressions, the function $g_1(z)$ is not used and is to be assumed as $g_1(z) = 0$, so $f_1(z) = 0$. For simplicity, we have introduced the following new symbols:

$$
\begin{aligned}
&g_2^{(9)}(z) = f_2(z), \quad g_3^{(8)}(z) = f_3(z), \quad g_4^{(7)}(z) = f_4(z) \\
&g_5^{(6)}(z) = f_5(z), \quad g_6^{(5)}(z) = f_6(z)
\end{aligned}
\tag{11.4-5}
$$

where $g_i^{(n)}$ denotes the nth derivative to the argument z. Similar to manipulation in the previous section, the complex representations of displacement components can be written as follows (here we have omitted the rigid body displacements):

$$\begin{aligned}
u_y + iu_x &= -6c_3R\left(\frac{2c_2}{\mu+\lambda} + c_7\right)\overline{\Gamma(z)} - 2c_3c_7R\Omega(z)\\
u_z &= \frac{4}{\mu(K_1+K_2)-3R^2}(240c_{10}\operatorname{Im} f_6(z) + c_1c_2R^2\operatorname{Im}(\Theta(z)\\
&\quad - 2\Omega(z) + 6\Gamma(z) - 24\Psi(z)))\\
w_y + iw_x &= -\frac{R}{c_1(\mu K_1 - 2R^2)}(24c_9\overline{\Psi(z)} - c_8\Theta(z))\\
w_z &= \frac{4(\mu K_2 - R^2)}{(K_1-2K_2)R(\mu(K_1+K_2)-3R^2)}\\
&\quad \times (240c_{10}\operatorname{Im} f_6(z)) + c_1c_2R^2\operatorname{Im}(\Theta(z) - 2\Omega(z)\\
&\quad + 6\Gamma(z) - 24\Psi(z))) \qquad (11.4\text{-}6)
\end{aligned}$$

in which

$$c_7 = \frac{c_2K_1 + 2c_1R}{\mu K_1 - 2R^2},\quad c_8 = c_1c_3R(\mu(K_1-K_2) - R^2)$$

$$c_9 = c_8 + 2c_3c_4\left(c_2 - \frac{(\mu K_2 - R^2)^2}{\mu K_1 - 2R^2}\right),\quad c_{10} = c_1c_3R^2 - c_4(c_3R - c_2K_1) \qquad (11.4\text{-}7)$$

11.4.2 *The complex representation of boundary conditions*

The boundary conditions of plane elasticity of icosahedral quasicrystals can be expressed as follows:

$$\sigma_{xx}l + \sigma_{xy}m = T_x,\quad \sigma_{yx}l + \sigma_{yy}m = T_y,\quad \sigma_{zx}l + \sigma_{zy}m = T_z \qquad (11.4\text{-}8)$$

$$H_{xx}l + H_{xy}m = h_x,\quad H_{yx}l + H_{yy}m = h_y,\quad H_{zx}l + H_{zy}m = h_z \qquad (11.4\text{-}9)$$

for $(x, y) \in L$ which represents the boundary of a multi-connected quasicrystalline material, and

$$l = \cos(\mathbf{n}, x) = \frac{dy}{ds},\quad m = \cos(\mathbf{n}, y) = -\frac{dx}{ds}$$

$\mathbf{T} = (T_x, T_y, T_z)$ and $\mathbf{h} = (h_x, h_y, h_z)$ denote the surface traction vector and generalized surface traction vector and $\mathbf{n}$ represents the outward unit normal vector of any point of the boundary, respectively.

Utilizing Eq. (11.4-3) and the first two formulas of Eq. (11.4-8), one has

$$\begin{aligned}&-4c_2c_3R[3(f_4(z)+4\bar{z}f_5'(z)+10\bar{z}^2f_6''(z))-(\overline{f_3(z)}+3z\overline{f_4'(z)}\\&\quad+6z^2\overline{f_5''(z)}+10z^3\overline{f_6'''(z)})]\\&=i\int(T_x+iT_y)ds,\quad z\in L\end{aligned}\tag{11.4-10}$$

Taking conjugate on both sides of Eq. (11.4-10) yields

$$\begin{aligned}&-4c_2c_3R[3(\overline{f_4(z)}+4z\overline{f_5'(z)}+10z^2\overline{f_6''(z)})-(f_3(z)+3\bar{z}f_4'(z)\\&\quad+6z^2f_5'(z)+10\bar{z}^3f_6'''(z))]\\&=-i\int(T_x-iT_y)ds,\quad z\in L\end{aligned}\tag{11.4-11}$$

Similarly, from Eq. (11.4-3) and the first two formulas of (11.4-9), one obtains

$$48c_2(2c_4-c_1R)\overline{\Psi(z)}+2c_1c_2R\Theta(z)=i\int(h_x+ih_y)ds,\quad z\in L\tag{11.4-12}$$

Furthermore, we assume

$$T_z=h_z=0\tag{11.4-13}$$

For simplicity and by the third equation in (11.4-8) and (11.4-9) and the formulas of (11.4-3) and (11.4-13), one has

$$\begin{cases}f_6(z)+\overline{f_6(z)}=0\\4c_{11}\operatorname{Re}[f_5(z)+5\bar{z}f_6'(z)]+(2K_2-K_1)R\operatorname{Re}[f_4(z)+4\bar{z}f_5'(z)\\\quad+10\bar{z}^2f_6''(z)+20f_6(z)]=0\end{cases}$$

$$z\in L\tag{11.4-14}$$

in which

$$c_{11} = (2K_2 - K_1)R - \frac{4c_4(\mu K_2 - R^2)}{(\mu K_1 - 2R^2)R} \tag{11.4-15}$$

As we have shown in the previous section, complex analytic functions (i.e., the complex potentials) must be determined by boundary value equations; the discussion is as follows.

11.4.3 *Structure of complex potentials*

11.4.3.1 *The arbitrariness of the complex potentials*

For explicit description, Eq. (11.4-3) can be written as follows:

$$\begin{aligned}
&\sigma_{zy} - i\sigma_{zx} = -960c_3c_4f_6'(z)\\
&c_1(\sigma_{yy} - \sigma_{xx} - 2i\sigma_{xy}) + ic_2[H_{xy} - H_{yx} + i(H_{xx} + H_{yy})]\\
&\quad = -192ic_2c_3c_4\Psi'(z)\\
&2c_1(H_{zy} + iH_{zx}) - R(2K_2 - K_1)[H_{xy} - H_{yx} + i(H_{xx} + H_{yy})]\\
&\quad = 96c_3cR(2K_2 - K_1)\Psi'(z) + 96c_1c_3c_6\Gamma'(z)\\
&c_5(\sigma_{yy} - \sigma_{xx} + 2i\sigma_{xy}) + ic_2R[H_{xy} - H_{yx} - i(H_{xx} + H_{yy})]\\
&\quad = -16ic_2c_3c_4\Omega'(z)\\
&H_{yx} + H_{xy} + i(H_{xx} - H_{yy}) = -480c_2c_5\overline{f_6'(z)} - 4c_1c_2R\Theta'(z)
\end{aligned} \tag{11.4-16}$$

Similar to the discussion of two-dimensional quasicrystals, from the equations, it is obvious that a state of phonon and phason stresses is not altered, if one replaces

$$f_i(z) \quad \text{by } f_i(z) + \gamma_i \quad (i = 2, \ldots, 6) \tag{11.4-17}$$

where γ_i are arbitrary complex constants.

Now, consider how these substitutions affect the components of the displacement vectors which were determined by the formula (11.4-6). Substituting (11.4-13) into (11.4-8))–(11.4-12) shows that if the complex constants $\gamma_i (i = 2, \ldots, 6)$ satisfy

$$\begin{aligned} & 3\left(\frac{2c_2}{\mu+\lambda}+c_7\right)\overline{\gamma_4}+c_7\gamma_3=0 \\ & 24c_9\overline{\gamma_5}-c_8\gamma_2=0 \\ & 40c_{10}\gamma_6-c_1c_3R^2\left[4\left(1-\frac{c_9}{c_8}\right)\overline{\gamma_5}-\frac{2c_2}{(\mu+\lambda)c_7}\gamma_4\right]=0 \end{aligned} \tag{11.4-18}$$

then the substitution (11.4-17) will not affect the displacements.

11.4.3.2 *General formulas for finite multiply connected region*

Consider now the case when the region S, occupied by the body, is multiply connected, see Fig. 11.1.

Since the stress must be single-valued and Eq. (11.4-16)

$$\sigma_{zy} - i\sigma_{zx} = -960c_3c_4f_6'(z) \tag{11.4-19}$$

We know that $f_6'(z)$ is holomorphic and hence single-valued in the region inside contour s_{m+1}, so the complex function can be expressed as

$$f_6(z) = \int_{z_0}^{z} f_6'(z)dz + \text{constant} \tag{11.4-20}$$

where z_0 denotes the fixed point. By Eq. (11.4-20), we have

$$f_6(z) = b_k \ln(z - z_k) + f_{6*}(z) \tag{11.4-21}$$

$f_{6*}(z)$ is holomorphic in the region with contour s_{m+1}.

Substituting (11.4-21) into the second formula of Eq. (11.4-16), i.e.,

$$\begin{aligned} & c_1(\sigma_{yy}-\sigma_{xx}-2i\sigma_{xy})+ic_2[H_{xy}-H_{yx}+i(H_{xx}+H_{yy})] \\ & \quad = -192ic_2c_3c_4\Psi'(z) \end{aligned}$$

shows that $f_5'(z)$ is holomorphic in the region enclosed by contour s_{m+1}, so one has

$$f_5(z) = c_k \ln(z - z_k) + f_{5*}(z) \tag{11.4-22}$$

$f_{5*}(z)$ is holomorphic in the region of interior of contour s_{m+1}.

Similar to the above-mentioned discussion, by Eqs. (11.4-16)–(11.4-18), the complex functions $f_i(i = 2, 3, 4)$ can be written as

$$\begin{aligned} f_4(z) &= d_k \ln(z - z_k) + f_{4*}(z) \\ f_3(z) &= e_k \ln(z - z_k) + f_{3*}(z) \\ f_2(z) &= t_k \ln(z - z_k) + f_{2*}(z) \end{aligned} \tag{11.4-23}$$

where d_k, e_k and t_k are complex constants and $f_{i*}(z)$ $(i = 2, 3, 4)$ is holomorphic in the region inside contour s_{m+1}.

By substituting (11.4-21)–(11.4-23) into the complex expressions of displacements, the condition of single-valuedness of displacements will be given as follows:

$$\begin{aligned} &-3\left(\frac{2c_2}{\mu+\lambda} + c_7\right)\overline{d_k} + c_7 e_k = 0 \\ &24c_9\overline{c_k} + c_8 t_k = 0 \\ &240c_{10}b_k + c_1c_3R^2(t_k - 2e_k + 6d_k - 24c_k) = 0 \end{aligned} \tag{11.4-24}$$

Applying the boundary conditions given above to the contour s_k and by Eq. (11.4-24), we know that the above complex constants may be very simply expressed in terms of surface traction and generalized surface traction as

$$\begin{aligned} b_k &= \frac{c_1c_3R^2}{240c_{10}}\left[\frac{12c_2}{(\mu+\lambda)c_7}\overline{d_k} + 24\left(1 + \frac{c_9}{c_8}\right)c_k\right] \\ c_k &= \frac{c_8}{-96\pi[c_3c_8(2c_4 - c_1R) - c_1c_3R]}(h_x - ih_y) \\ t_k &= \frac{c_8}{4\pi[c_3c_8(2c_4 - c_1R) - c_1c_3R]}(h_x + ih_y) \\ d_k &= \frac{(\mu+\lambda)c_7}{24\pi c_2c_3R(2c_2 + (\mu+\lambda)c_7)}(T_x + iT_y) \\ e_k &= -\frac{2c_2 + (\mu+\lambda)c_7}{16\pi c_2^2c_3R}(T_x - iT_y) \end{aligned} \tag{11.4-25}$$

We can easily extend the above results to the case where there are m inner boundaries.

11.4.4 *Case of infinite regions*

From the point of view of the application, the consideration of infinite regions is likewise of major interest. We assume that the contour s_{m+1} has entirely moved to infinity.

Similar to the discussion of two-dimensional quasicrystal, we have

$$f_6(z) = \sum_{k=1}^{m} b_k \ln z + f_{6**}(z), \quad f_5(z) = \sum_{k=1}^{m} c_k \ln z + f_{5**}(z)$$

$$f_4(z) = \sum_{k=1}^{m} d_k \ln z + f_{4**}(z), \quad f_3(z) = \sum_{k=1}^{m} e_k \ln z + f_{3**}(z)$$

$$f_2(z) = \sum_{k=1}^{m} t_k \ln z + f_{2**}(z) \tag{11.4-26}$$

$f_{j**}(z)$ $(j = 2, \ldots, 6)$ are functions, holomorphic outside s_{m+1}, not including the point at infinity. By the theorem of Laurent, the function $h_{2*}(z)$ may be represented outside s_{m+1} by the series

$$f_{ji**}(z) = \sum_{-\infty}^{+\infty} a_{jn} z^n \quad (j = 2, \ldots, 6) \tag{11.4-27}$$

Substituting the first equation of (11.4-26) and (11.4-27) into the first one of Eq. (11.4-16), one has

$$\sigma_{zy} - i\sigma_{zx} = -960 c_3 c_4 \left(\sum_{k=1}^{m} b_k \frac{1}{z} + \sum_{-\infty}^{\infty} n a_{6n} z^{n-1} \right) \tag{11.4-28}$$

Hence it follows that for the stress to remain finite as $|z| \to \infty$, one must have

$$a_{6n} = 0 \quad (n \geq 2) \tag{11.4-29}$$

Similarly, from Eqs. (11.4-15)–(11.4-18), to make the stresses be bounded, the following conditions are also satisfied:

$$a_{jn} = 0 \quad (n \geq 2, \ j = 2, \ldots, 5) \tag{11.4-30}$$

So, we can obtain the expressions of the complex function $f_i(z)(i = 2, \ldots, 6)$ for the stresses to remain finite as $|z| \to \infty$, for example,

$$f_6(z) = \sum_{k=1}^{m} b_k \ln z + (B + iC)z + f_6^0(z) \tag{11.4-31}$$

where B, C are unknown real constants to be determined and $f_6^0(z)$ is a function, holomorphic outside s_{m+1}, including the point at infinity. The determination of unknown constants B, C, etc. is similar to that given in Section 11.3.4, but the details are omitted here due to limitation of space.

11.4.5 *Conformal mapping and function equations at ζ-plane*

We now have five equations of boundary value (11.4-10)–(11.4-12) and (11.4-14), from which the unknown functions $f_j(z)$ $(j = 2, \ldots, 6)$ will be determined; in addition, we have assumed that $f_1(z) = 0$ because it has no usage. For some complicated regions, the function equations cannot be directly solved at the physical plane (i.e., the z-plane), and the conformal mapping is particularly meaningful in the case.

Assume that a conformal mapping

$$z = \omega(\zeta) \tag{11.4-32}$$

is used to transform the region at z-plane onto the interior of the unit circle γ at ζ-plane. Under the mapping, the unknown functions $f_j(z)$ become

$$f_j(z) = f_j[\omega(\zeta)] = \Phi_j(\zeta) \quad (j = 2, \ldots, 6) \tag{11.4-33}$$

Substituting (11.4-32) and (11.4-33) into the first relation of boundary conditions, (11.4-14) yields

$$\frac{1}{2\pi i}\int_\gamma \frac{\Phi_6(\sigma)}{\sigma - \varsigma} d\sigma + \frac{1}{2\pi i}\int_\gamma \frac{\overline{\Phi_6(\sigma)}}{\sigma - \varsigma} d\sigma = 0$$

This shows that

$$\Phi_6(\varsigma) = 0 \tag{11.4-34}$$

according to the Cauchy integral formula.

Substitution of (11.4-32), (11.4-33) and (11.4-34) into boundary conditions (11.4-10)–(11.4-12) and the second one of conditions (11.4-14) leads to boundary value equations for determining the unknown functions $\Phi_j(\zeta)$ $(j = 2, \ldots, 5)$ at ζ-plane, i.e.,

$$\begin{aligned}
&\frac{3}{2\pi i}\int_\gamma \frac{\Phi_4(\sigma)}{\sigma-\varsigma}d\sigma + \frac{4}{2\pi i}\int_\gamma \frac{\overline{\omega(\sigma)}}{\omega'(\sigma)}\frac{\Phi_5'(\sigma)}{\sigma-\varsigma}d\sigma - \frac{1}{2\pi i}\int_\gamma \frac{\overline{\Phi_3(\sigma)}}{\sigma-\varsigma}d\sigma\\
&\quad - 3\frac{1}{2\pi i}\int_\gamma \frac{\omega(\sigma)}{\overline{\omega'(\sigma)}}\frac{\overline{\Phi_4'(\sigma)}}{\sigma-\varsigma}d\sigma - 6\frac{1}{2\pi i}\int_\gamma\left[\frac{[\omega(\sigma)]^2\overline{\Phi_5''(\sigma)}}{\overline{\omega'(\sigma)}^2}\right.\\
&\quad\left. - \frac{[\omega(\sigma)]^2\overline{\omega''(\sigma)}}{\overline{\omega'(\sigma)}^3}\overline{\Phi_5'(\sigma)}\right]\frac{d\sigma}{\sigma-\varsigma} = \frac{1}{4c_2c_3}\frac{1}{2\pi i}\int_\gamma \frac{\mathbf{t}}{\sigma-\varsigma}d\sigma
\end{aligned}\tag{11.4-35}$$

$$\begin{aligned}
&\frac{3}{2\pi i}\int_\gamma \frac{\overline{\Phi_4(\sigma)}}{\sigma-\varsigma}d\sigma + \frac{4}{2\pi i}\int_\gamma \frac{\overline{\omega(\sigma)}}{\omega'(\sigma)}\frac{\Phi_4'(\sigma)}{\sigma-\varsigma}d\sigma - \frac{1}{2\pi i}\int_\gamma \frac{\Phi_3(\sigma)}{\sigma-\zeta}d\sigma\\
&\quad - 3\frac{1}{2\pi i}\int_\gamma \frac{\overline{\omega(\sigma)}}{\omega'(\sigma)}\frac{\Phi_3(\sigma)}{\sigma-\zeta}d\sigma - 6\frac{1}{2\pi i}\int_\gamma\left[\frac{\overline{\omega(\sigma)}^2\Phi_5''(\sigma)}{[\omega'(\sigma)]^2}\right.\\
&\quad\left. - \frac{\overline{\omega(\sigma)}^2\omega''(\sigma)\Phi_5'(\sigma)}{[\omega'(\sigma)]^3}\right]\frac{d\sigma}{\sigma-\varsigma} = \frac{1}{4c_2c_3R}\frac{1}{2\pi i}\int_\gamma \frac{\bar{\mathbf{t}}}{\sigma-\varsigma}d\sigma
\end{aligned}\tag{11.4-36}$$

$$\begin{aligned}
&\frac{1}{2\pi i}\int_\gamma \frac{\Phi_2(\sigma)}{\sigma-\varsigma}d\sigma + 2\frac{1}{2\pi i}\int_\gamma \frac{\overline{\omega(\sigma)}}{\omega'(\sigma)}\frac{\Phi_3'(\sigma)}{\sigma-\varsigma}d\sigma + 3\frac{1}{2\pi i}\int_\gamma\left[\frac{\overline{\omega(\sigma)}^2\Phi_4''(\sigma)}{[\omega'(\sigma)]^2}\right.\\
&\quad\left. - \frac{\overline{\omega(\sigma)}^2\omega''(\sigma)\Phi_4'(\sigma)}{[\omega'(\sigma)]^3}\right]\frac{d\sigma}{\sigma-\varsigma} + 4\frac{1}{2\pi i}\int_\gamma\left[\frac{\overline{\omega(\sigma)}^2\Phi_5'''(\sigma)}{[\omega'(\sigma)]^3}\right.\\
&\quad - 3\frac{\overline{\omega(\sigma)}^3\omega''(\sigma)\Phi_5''(\sigma)}{[\omega'(\sigma)]^4} + 3\frac{\overline{\omega(\sigma)}^3\omega''(\sigma)\Phi_5'(\sigma)}{[\omega'(\sigma)]^5}\\
&\quad\left. - \frac{\overline{\omega(\sigma)}^3\omega'''(\sigma)\Phi_5'(\sigma)}{[\omega'(\sigma)]^4}\right]\frac{d\sigma}{\sigma-\varsigma} = \frac{1}{2\pi i}\int \frac{\mathbf{h}}{\sigma-\varsigma}d\sigma
\end{aligned}\tag{11.4-37}$$

$$\begin{aligned}
&\frac{4c_{11}}{2\pi i}\int_\gamma \frac{\Phi_5(\sigma)}{\sigma-\varsigma}d\sigma + \frac{(2K_2-K_1)R}{2\pi i}\\
&\quad\times\int_\gamma\left[\frac{\Phi_4(\sigma)}{\sigma-\varsigma} + 4\frac{\overline{\omega(\sigma)}}{\omega'(\sigma)}\frac{\Phi_5'(\sigma)}{\sigma-\varsigma}\right]d\sigma = 0
\end{aligned}\tag{11.4-38}$$

in which $\mathbf{t} = i\int (T_x + iT_y)ds$, $\bar{\mathbf{t}} = -i\int (T_x - iT_y)ds$, $\mathbf{h} = i\int(h_1 + ih_2)ds$. For a given configuration and applied stresses, we can obtain the solution by solving these function equations.

11.4.6 *Example: Elliptic notch problem and solution*

We consider an icosahedral quasicrystal solid with an elliptic notch, which penetrates through the medium along the z-axis direction, the edge of the elliptic notch subjected to the uniform pressure p, similar to Fig. 11.2.

Since the measurement of generalized traction has not been reported so far, for simplicity, we assume that $h_x = 0$, $h_y = 0$.

However, the calculation cannot be completed at the z-plane owing to the complicity, so we have to employ the conformal mapping

$$z = \omega(\zeta) = R_0\left(\frac{1}{\zeta} + m\zeta\right) \tag{11.4-38}$$

to transform the exterior of the ellipse at the z-plane onto the interior of the unit circle γ at the ζ-plane, in which

$$R_0 = (a+b)/2, \quad m = (a-b)/(a+b)$$

Let

$$f_j(z) = f_j[\omega(\zeta)] = \Phi_j(\zeta) \quad (j = 2, \ldots, 6) \tag{11.4-39}$$

Substituting (11.4-38) into the first formula of (11.4-25) and then multiplying both sides of equations by $d\sigma/[2\pi i(\sigma - \zeta)]$ (σ represents the value of ζ at the unit circle γ(i.e., $\rho = 1$)) yields

$$\frac{1}{2\pi i}\int_\gamma \frac{\Phi_6(\sigma)}{\sigma - \varsigma}d\sigma + \frac{1}{2\pi i}\int_\gamma \frac{\overline{\Phi_6(\sigma)}}{\sigma - \varsigma}d\sigma = 0 \tag{11.4-40}$$

By means of Cauchy integral formula, we have

$$\Phi_6(\varsigma) = 0 \tag{11.4-41}$$

Substituting (11.4-38) and (11.4-41) into (11.4-22)–(11.4-24), then multiplying both sides of the equations by $d\sigma/[2\pi i(\sigma-\zeta)]$ (σ represents the value of ζ at the unit circle γ (i.e., $\rho=1$)) and integrating around the unit circle γ yield

$$\begin{aligned}&\frac{3}{2\pi i}\int_\gamma\frac{\Phi_4(\sigma)}{\sigma-\varsigma}d\sigma+\frac{4}{2\pi i}\int_\gamma\frac{\overline{\omega(\sigma)}}{\omega'(\sigma)}\frac{\Phi'_5(\sigma)}{\sigma-\varsigma}d\sigma-\frac{1}{2\pi i}\int_\gamma\frac{\overline{\Phi_3(\sigma)}}{\sigma-\varsigma}d\sigma\\&\quad-3\frac{1}{2\pi i}\int_\gamma\frac{\omega(\sigma)}{\overline{\omega''(\sigma)}}\frac{\overline{\Phi'_4(\sigma)}}{\sigma-\varsigma}d\sigma-6\frac{1}{2\pi i}\int_\gamma\left[\frac{[\omega(\sigma)]^2\overline{\Phi''_5(\sigma)}}{\overline{[\omega'(\sigma)]}^2}\right.\\&\quad\left.-\frac{[\omega(\sigma)]^2\overline{\omega''(\sigma)}}{\overline{[\omega'(\sigma)]}^3}\overline{\Phi'_5(\sigma)}\right]\frac{d\sigma}{\sigma-\varsigma}=\frac{p}{4c_2c_3}\int_\gamma\frac{\omega(\sigma)}{\sigma-\varsigma}d\sigma\end{aligned}\tag{11.4-42}$$

$$\begin{aligned}&\frac{3}{2\pi i}\int_\gamma\frac{\overline{\Phi_4(\sigma)}}{\sigma-\varsigma}d\sigma+\frac{4}{2\pi i}\int_\gamma\frac{\omega(\sigma)}{\overline{\omega'(\sigma)}}\frac{\Phi'_5(\sigma)}{\sigma-\varsigma}d\sigma-\frac{1}{2\pi i}\int_\gamma\frac{\Phi_3(\sigma)}{\sigma-\zeta}d\sigma\\&\quad-3\frac{1}{2\pi i}\int_\gamma\frac{\overline{\omega(\sigma)}}{\omega'(\sigma)}\frac{\Phi'_4(\sigma)}{\sigma-\zeta}d\sigma-6\frac{1}{2\pi i}\int_\gamma\left[\frac{\overline{\omega(\sigma)}^2\Phi''_5(\sigma)}{[\omega'(\sigma)]^2}\right.\\&\quad\left.-\frac{\overline{\omega(\sigma)}^2\omega''(\sigma)\Phi'_5(\sigma)}{[\omega'(\sigma)]^3}\right]\frac{d\sigma}{\sigma-\varsigma}=\frac{p}{4c_2c_3R}\frac{1}{2\pi i}\int_\gamma\frac{\overline{\omega(\sigma)}}{\sigma-\varsigma}d\sigma\end{aligned}\tag{11.4-43}$$

$$\begin{aligned}&\frac{1}{2\pi i}\int_\gamma\frac{\Phi_2(\sigma)}{\sigma-\varsigma}d\sigma+2\frac{1}{2\pi i}\int_\gamma\frac{\overline{\omega(\sigma)}}{\omega'(\sigma)}\frac{\Phi'_3(\sigma)}{\sigma-\varsigma}d\sigma+3\frac{1}{2\pi i}\int_\gamma\left[\frac{\overline{\omega(\sigma)}^2\Phi''_4(\sigma)}{[\omega'(\sigma)]^2}\right.\\&\quad\left.-\frac{\overline{\omega(\sigma)}^2\omega''(\sigma)\Phi'_4(\sigma)}{[\omega'(\sigma)]^3}\right]\frac{d\sigma}{\sigma-\varsigma}+4\frac{1}{2\pi i}\int_\gamma\left[\frac{\overline{\omega(\sigma)}^3\Phi'''_5(\sigma)}{[\omega'(\sigma)]^3}\right.\\&\quad-3\frac{\overline{\omega(\sigma)}^3\omega''(\sigma)\Phi''_5(\sigma)}{[\omega'(\sigma)]^4}+3\frac{\overline{\omega(\sigma)}^3\omega''(\sigma)\Phi'_5(\sigma)}{[\omega'(\sigma)]^5}\\&\quad\left.-\frac{\overline{\omega(\sigma)}^3\omega'''(\sigma)\Phi'_5(\sigma)}{[\omega'(\sigma)]^4}\right]\frac{d\sigma}{\sigma-\varsigma}=0\end{aligned}\tag{11.4-44}$$

$$\begin{aligned}&\frac{4c_{11}}{2\pi i}\int_\gamma\frac{\Phi_5(\sigma)}{\sigma-\varsigma}d\sigma+\frac{(2K_2-K_1)R}{2\pi i}\\&\quad\times\int_\gamma\left[\frac{\Phi_4(\sigma)}{\sigma-\varsigma}+4\frac{\overline{\omega(\sigma)}}{\omega'(\sigma)}\frac{\Phi'_5(\sigma)}{\sigma-\varsigma}\right]d\sigma=0\end{aligned}\tag{11.4-45}$$

Because

$$\frac{\overline{\omega(\sigma)}}{\omega'(\sigma)} = \sigma\frac{\sigma^2+m}{m\sigma^2-1}$$

and

$$\zeta\frac{\zeta^2+m}{m\zeta^2-1}\Phi_5'(\zeta) = \zeta\frac{\zeta^2+m}{m\zeta^2-1}(\alpha_1+2\alpha_2\zeta+3\alpha_3\zeta^2+\cdots)$$

are analytic in $|\zeta|<1$ and continuous in the unit circle γ, by means of the Cauchy integral formula, from Eq. (11.4-42), we have

$$\frac{1}{2\pi i}\int_\gamma \frac{\Phi_4(\sigma)}{\sigma-\varsigma}d\sigma = \Phi_4(\zeta)$$

$$\frac{1}{2\pi i}\int_\gamma \sigma\frac{\sigma^2+m}{m\sigma^2-1}\frac{\Phi_5'(\sigma)}{\sigma-\varsigma}d\sigma = \zeta\frac{\zeta^2+m}{m\zeta^2-1}\Phi_5'(\zeta)$$

Substituting

$$\frac{\omega(\sigma)}{\overline{\omega'(\sigma)}} = -\frac{1}{\sigma}\frac{m\sigma^2+1}{\sigma^2-m}, \quad \frac{\omega(\sigma)^2\overline{\omega''(\sigma)}}{\overline{\omega'(\sigma)}^3} = \frac{2\sigma(m\sigma^2+1)^2}{(\sigma^2-m)^3}$$

into Eq. (11.4-42) noting that

$$-\frac{1}{\zeta}\frac{m\zeta^2+1}{\zeta^2-m}\overline{\Phi_4'(\zeta)} = -\frac{1}{\zeta}\frac{m\zeta^2+1}{\zeta^2-m}\left(\overline{\beta_1}+2\frac{\overline{\beta_2}}{\zeta}+3\frac{\overline{\beta_3}}{\zeta^2}+\cdots\right)$$

$$\frac{2\zeta(m\zeta^2+1)^2}{(\zeta^2-m)^3}\overline{\Phi_5'(\zeta)} = \frac{2\zeta(m\zeta^2+1)^2}{(\zeta^2-m)^3}\left(\overline{\alpha_1}+2\frac{\overline{\alpha_2}}{\zeta}+3\frac{\overline{\alpha_3}}{\zeta^2}+\cdots\right)$$

are analytic in $|\zeta|>1$ and continuous in the unit circle γ, by means of the Cauchy integral formula and analytic extension of the complex variable function theory, from Eq. (11.4-42), we obtain

$$\frac{1}{2\pi i}\int_\gamma \frac{\overline{\Phi_3(\sigma)}}{\sigma-\varsigma}d\sigma = 0, \quad \frac{1}{2\pi i}\int_\gamma \frac{\omega(\sigma)}{\overline{\omega'(\sigma)}}\frac{\overline{\Phi_4'(\sigma)}}{\sigma-\varsigma}d\sigma = 0$$

$$\frac{1}{2\pi i}\int_\gamma \left[\frac{\omega(\sigma)^2\overline{\Phi_5''(\sigma)}}{\overline{\omega'(\sigma)}^2} - \frac{\omega(\sigma)^2\overline{\omega''(\sigma)}}{\overline{\omega'(\sigma)}^3}\overline{\Phi_5'(\sigma)}\right]\frac{d\sigma}{\sigma-\varsigma} = 0$$

Substituting the above results into Eq. (11.4-42), with the help of Eq. (11.4-45), one has

$$\begin{aligned}\Phi_4(\zeta) &= \frac{R_0}{12c_2c_3R}pm\zeta - \frac{(2K_2-K_1)R_0}{2c_2c_3C_{11}}\frac{pm\zeta(\zeta^2+m)}{(m\zeta^2-1)} \\ \Phi_5(\zeta) &= -\frac{(2K_2-K_1)R_0}{48c_2c_3C_{11}}pm\zeta\end{aligned} \tag{11.4-46}$$

Similar to the above discussion, from Eqs. (11.4-43) and (11.4-44), one has

$$\begin{aligned}\Phi_2(\zeta) &= -\frac{R_0}{2c_2c_3R}\frac{p\zeta(\zeta^2+m)(m^3\zeta^2+1)}{(m\zeta^2-1)^3} + \frac{(2K_2-K_1)R_0}{2c_2c_3C_{11}} \\ &\quad \times \frac{pm\zeta^3(\zeta^2+m)[m^2\zeta^6-(m^3+4m)\zeta^4+(2m^4+4m^2+5)\zeta^2+m]}{(m\zeta^2-1)^5} \\ \Phi_3(\zeta) &= -\frac{R_0}{4c_2c_3R}\frac{p\zeta(m^2+1)}{(m\zeta^2-1)} - \frac{(2K_2-K_1)R_0}{12c_2c_3C_{11}} \\ &\quad \times \frac{pm\zeta^3(\zeta^2+m)(m\zeta^2-m^2-2)}{(m\zeta^2-1)^3}\end{aligned} \tag{11.4-47}$$

The elliptic notch problem is solved. The solution of the Griffith crack subjected to a uniform pressure can be obtained corresponding to the case $m=1$, $R_0=a/2$ of the above solution. The solution of the crack can be expressed explicitly in the z-plane, and refer to Section 9.7 in Chapter 9 for the concrete results.

11.5 Complex Analysis of Generalized Quadruple Harmonic Equation

In Chapters 6–8, we have known that the plane elasticity of octagonal quasicrystals is governed by the final equation

$$(\nabla^2\nabla^2\nabla^2\nabla^2 - 4\varepsilon\nabla^2\nabla^2\Lambda^2\Lambda^2 + 4\varepsilon\Lambda^2\Lambda^2\Lambda^2\Lambda^2)F = 0 \tag{11.5-1}$$

either by the displacement potential or by the stress potential, in which

$$\left.\begin{aligned}&\nabla^2=\frac{\partial^2}{\partial x^2}+\frac{\partial^2}{\partial y^2},\quad \Lambda^2=\frac{\partial^2}{\partial x^2}-\frac{\partial^2}{\partial y^2}\\&\varepsilon=\frac{R^2(L+M)(K_2+K_3)}{[M(K_1+K_2+K_3)-R^2][(L+2M)K_1-R^2]}\end{aligned}\right\}\tag{11.5-2}$$

Due to the appearance of operator Λ^2, it seems there is no connection with complex variable functions in solving Eq. (11.5-1). But if we rewrite it as

$$\begin{aligned}&\left[\frac{\partial^8}{\partial x^8}+4(1-4\varepsilon)\frac{\partial^8}{\partial x^6\partial y^2}+2(3+16\varepsilon)\frac{\partial^8}{\partial x^4\partial y^4}\right.\\&\quad\left.+4(1-4\varepsilon)\frac{\partial^8}{\partial x^2\partial y^6}+\frac{\partial^8}{\partial y^8}\right]F=0\end{aligned}\tag{11.5-3}$$

then we find that this is one of the typical multi-quasiharmonic partial differential equations with a quadruple, and there is complex representation of solution such as

$$F(x,y)=2\,\mathrm{Re}\sum_{k=1}^{4}F_k(z_k),\qquad z_k=x+\mu_k y\tag{11.5-4}$$

in which functions $F_k(z_k)$ are analytic functions of the complex variable z_k $(k=1,\ldots,4)$ and $\mu_k=\alpha_k+i\beta_k$ $(k=1,\ldots,4)$ are complex parameters and are determined by the roots of the following eigenvalue equation:

$$\mu^8+4(1-4\varepsilon)\mu^6+2(3+16\varepsilon)\mu^4+4(1-4\varepsilon)\mu^2+1=0\tag{11.5-5}$$

We have shown in Chapters 7 and 8 that some solutions of dislocations (based on displacement potential formulation) and notchs/cracks (based on stress potential formulation) can be found in terms of this complex analysis. In the procedure, it must carry out some calculations on the determinants of fourth order, so the solution expressions are quite lengthy, but are analytic substantively.

11.6 Conclusion and Discussion

The discovery of quadruple and sextuple harmonic equations is significant for modern elasticity as well as for partial differential equations, and provides an applying a field for complex analysis. This chapter gives a comprehensive discussion on the complex analysis for solving the equations, we think the study is primary.

The complex potential approach mentioned above is a new development of the Muskhelishvili's approach to the classical elasticity; it extends greatly the scope of the method. We believe that the quadruple and sextuple harmonic equations are not only useful in quasicrystals but probably also for other disciplines of science and engineering. So, the complex analysis method can be used for other studies.

Apart from the development to extend the scope of the complex potential theory and method, we also developed the Muskhelishvili method for the conformal mapping. According to the monograph [1], the conformal mapping is limited within the rational function class. But we extended it into the transcendental function class, and some exact analytic solutions for more complicated cracked configurations are achieved, see, e.g., Chapter 8.

The method is effective not only for solving elasticity problems but also for solving plasticity problems, see, e.g., [11–13]. The new summarization of the method can be found in article [14] and other references [15, 16].

11.7 Appendix: Basic Formulas of Complex Analysis

It is enlightening that Muskhelishvili [1] gave extensive descriptions in detail on the complex analysis in due presentation of elasticity in his classical monograph; this is very beneficial to readers. However, there is no possibility for this book. We provide here some points only of the function theory which were frequently cited in the text. These can be referred for readers who are advised to read the books of I. I. Privalov [17] and M. A. Lavrentjev and B. A. Schabat [18] for further details. Other knowledge has been provided in due succession of the text in Chapters 7–9 and 11. The present contents can also be seen as a supplement in reading the material given in Chapters 7–9 and 11 if it is needed. The importance of complex analysis is not only in deriving the solutions by the complex potential formulation but also in dealing with the solutions by integral transforms and dual integral equations which are to be discussed in Appendix B of Major Appendix of this book.

11.7.1 *Complex functions, analytic functions*

Usually $z = x+iy$ is denoted as a complex variable in which $\sqrt{-1} = i$, or $z = re^{i\theta}$, and $r = \sqrt{x^2+y^2}$, called the modulus of the complex number, $\theta = \arctan\left(\frac{y}{x}\right)$, the argument angle of z. Assume $f(z)$ be a function of one complex variable, or a complex function in abbreviation, and denote

$$f(z) = P(x, y) + iQ(x, y) \tag{11.7-1}$$

in which both $P(x, y)$ and $Q(x, y)$ are functions with real variables and are called the real and imaginary parts, respectively, and marked by

$$P(x, y) = \operatorname{Re} f(z), \quad Q(x, y) = \operatorname{Im} f(z)$$

There is a sort of complex functions called analytic functions (or regular functions, and single-valued analytic functions are called holomorphic functions) which have important applications in many branches of mathematics, physics and engineering. The concepts related to this are discussed as follows.

The complex function $f(z)$ is analytic in a given region; this means that it can be expanded in the neighbourhood of any point z_0 of the region into a non-negative integer power series (i.e., the Taylor series) of the form

$$f(z) = \sum_{n=0}^{\infty} a_n(z - z_0)^n \tag{11.7-2}$$

in which a_n is a constant (in general, a complex number). The concept will be frequently used in later calculations.

Another definition of an analytic function is that if the complex function $f(z)$ given in the region, whose real part $P(x, y)$ and imaginary part $Q(x, y)$ are single valued, has continuous partial derivatives of the first order and satisfies the Cauchy–Riemann condition such as

$$\frac{\partial P}{\partial x} = \frac{\partial Q}{\partial y}, \quad \frac{\partial P}{\partial y} = -\frac{\partial Q}{\partial x} \tag{11.7-3}$$

in the region.

This kind of function, P and Q are named mutually conjugate harmonic ones. From (11.7-3), it follows that

$$\nabla^2 P = \left(\frac{\partial^2}{\partial x^2} + \frac{\partial^2}{\partial y^2}\right) P = 0, \quad \nabla^2 Q = \left(\frac{\partial^2}{\partial x^2} + \frac{\partial^2}{\partial y^2}\right) Q = 0$$

This concept will also be often used in the following.

An analytic function can also be defined in integral form. Assuming $f(z)$ is a complex function in a certain complex number region D and Γ is any simple smooth closed curve (sometimes called a simple curve for simplicity) in D, we can obtain that $f(z)$ is analytic in the region if

$$\int_\Gamma f(z)dz = 0 \tag{11.7-4}$$

The result is known as Cauchy's integral theorem (or simply called Cauchy's theorem) which will be frequently used later.

The theory of complex functions proves that the above definitions are mutually equivalent.

11.7.2 *Cauchy's formula*

An important result of Cauchy's theorem is the so-called Cauchy's formula, i.e., if $f(z)$ is analytic in a single-connected region D bounded by a closed curve Γ and continuous in $D^+ + \Gamma$ (Fig. 11.6), then

$$\frac{1}{2\pi i}\int_\Gamma \frac{f(t)}{t - z}dt = f(z) \tag{11.7-5}$$

in which z is an arbitrary point in D^+.

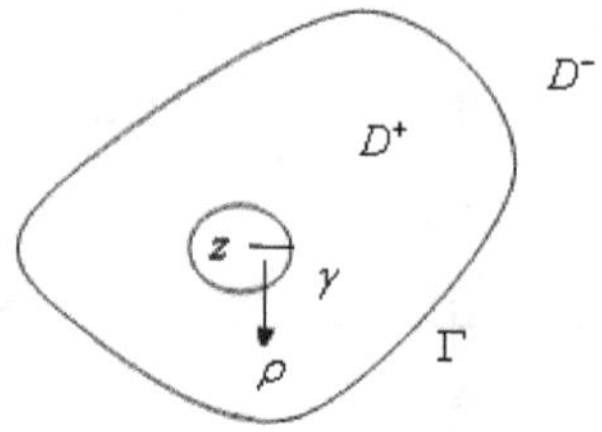

Fig. 11.6. A finite region D^+.

Proof. Taking z as the centre and ρ as the radius make a small circle γ in D^+. According to Cauchy's theorem (11.7-4),

$$\int_\Gamma \frac{f(t)}{t-z}dt = \int_\gamma \frac{f(t)}{t-z}dt \tag{11.7-6}$$

As $f(z)$ is analytic in D^+ and continuous in $D^+ + \Gamma$, there is a small number $\varepsilon > 0$, for any point t and γ, if ρ is sufficiently small, such as

$$|f(t) - f(z)| < \varepsilon$$

and note that $|t - z| = \rho$, hence

$$\lim_{\varepsilon \to 0} \int_\gamma \frac{f(t)}{t-z}dt = \int_\gamma \frac{f(z)}{t-z}dt \tag{11.7-7}$$

Just as mentioned previously, $f(z)$ is analytic in D^+, and the value of the integral

$$\int_\gamma \frac{f(z)}{t-z}dt$$

will not be changed when ρ is reducing. Therefore, the limit mark in the left-hand side of (11.7-7) can be removed. In addition,

$$\int_\gamma \frac{f(z)}{t-z}dt = f(z)\int_\gamma \frac{dt}{t-z} = f(z)\int_0^{2\pi} \frac{\rho e^{i\theta}}{\rho e^{i\theta}}d\theta = 2\pi i f(z)$$

Based on (11.7-6) and this result, formula (11.7-5) is proved.

In formula (11.7-5), if z is taken as its values in a region D^- consisting of the points lying outside Γ (see Fig. 11.6), then

$$\frac{1}{2\pi i}\int_\Gamma \frac{f(t)}{t-z}dt = 0 \tag{11.7-8}$$

In fact, this is a direct consequence of Cauchy's theorem because in this case the integrand $f(z)/(\zeta - z)$ as function of ζ is analytic in region D^+, where ζ denotes the point in the region D^+.

Suppose all conditions are the same as those for (11.7-5), then

$$\frac{1}{2\pi i}\int_\Gamma \frac{\overline{f(t)}}{t-z}dt = \overline{f(0)} \tag{11.7-9}$$

Proof. For simplicity, here the proof is given for the case Γ being a circle. Being analytic in the region D^+, $f(z)$ may be expanded as a non-negative integer power series, in which, taking $z_0 = 0$, such that

$$f(z) = a_0 + a_1 z + a_2 z^2 + \cdots = f(0) + f'(0)z + \frac{1}{2!}f''(0)z^2 + \cdots$$

The function $\overline{f(z)}$ in formula (11.7-9) is the value of $\bar{f}\left(\frac{1}{z}\right)$ at the circle Γ, here

$$\bar{f}\left(\frac{1}{z}\right) = \overline{f(0)} + \overline{f'(0)}\frac{1}{z} + \frac{1}{2!}\overline{f''(0)}\frac{1}{z^2} + \cdots$$

is an analytic function in D^-. From Cauchy's formula,

$$\frac{1}{2\pi i}\int_\Gamma \frac{dt}{t^k(t-z)} = \begin{cases} 1 & k = 0 \\ 0 & k > 0 \end{cases}$$

such that (11.7-9) is proved.

In contrast to the above, the current function $f(z)$ is analytic in D^- (including $z = \infty$), then

$$\frac{1}{2\pi i}\int_\Gamma \frac{f(z)}{t-z}dt = \begin{cases} -f(z) + f(\infty) & z \in D^- \\ f(\infty) & z \in D^+ \end{cases} \tag{11.7-10}$$

The proof of this formula can be offered in a similar manner adopted for (11.7-5), but the following points must be noted:

(i) The analytic function $f(z)$ in D^- (including $z = \infty$) may be expanded as the following series:

$$f(z) = c_0 + c_1\frac{1}{z} + c_2\frac{1}{z^2} + \cdots$$

(ii) $\displaystyle\frac{1}{2\pi i}\int_\Gamma \frac{c_0}{t-z}dt = \begin{cases} 0 & z \in D^- \\ c_0 & z \in D^+ \end{cases}$
where $c_0 = f(\infty) \neq 0$.

All conditions are the same as that for formula (11.7-10), and there exists

$$\frac{1}{2\pi i}\int_\Gamma \frac{\overline{f(t)}}{t-z}dt = 0 \tag{11.7-11}$$

11.7.3 *Poles*

Suppose a finite point in z-plane (i.e., z is not a point at infinity), and in the neighbourhood of the point, the function presents the form as follows:

$$f(z) = G(z) + f_0(z) \tag{11.7-12}$$

in which $f_0(z)$ is an analytic function in the neighbourhood of point a and

$$G(z) = \frac{A_0}{z-a} + \frac{A_1}{(z-a)^2} + \cdots \frac{A_m}{(z-a)^m} \tag{11.7-13}$$

where $A_1, A_2, \ldots, A_m$ are constants such that $f(z)$ is called having a pole with order m and $z = a$ is the pole.

If a is a point at infinity, $f_0(z)$ in (11.7-12) is regular at point at infinity (i.e., $f(t) = c_0 + c_1 z^{-1} + c_2 z^{-2} + \cdots$), while at $z = \infty$,

$$G(z) = A_0 + A_1 z + \cdots + A_m z^m \tag{11.7-14}$$

then we say that $f(z)$ has a pole of order m at $z = \infty$.

11.7.4 *Residual theorem*

If the function $f(z)$ has pole a with order m, its integral may be evaluated simply by computing residual.

What is the meaning of the residual? Suppose $f(z)$ is analytic in the neighbourhood of point $z = a$, except $z = a$, and infinite at $z = a$. In this case, the point $z = a$ is named isolated singular point. The residual of the function $f(z)$ at point $z = a$ is the value of the integral

$$\frac{1}{2\pi i}\int_{\Gamma} f(z)dz$$

in which Γ represents any closed contour enclosing point $z = a$. For a residual, we will use the resignation as $\mathrm{Re}\, s f(a)$.

If $z = a$ is an m-order pole of $f(z)$, its residual may be evaluated from the following formula:

$$\mathrm{Re}\, s f(a) = \frac{1}{(m-1)!}\lim_{z\to a}\frac{d^{m-1}}{dz^{m-1}}\{(z-a)^m f(z)\} \tag{11.7-15}$$

Obviously, the integral is

$$\int_{\Gamma} f(z)dz = 2\pi i\, \mathrm{Re}\, s f(a)$$

So, the evaluation of integrals may be reduced to the calculation of derivatives, and it is greatly simplified. If $z = a$ is a first-order pole, then

$$\mathrm{Re}\, s f(a) = \lim_{z\to a}(z-a)f(z) \tag{11.7-16}$$

the calculation is much simpler.

What follows the residual theorem is introduced as follows: let the function $f(z)$ be analytic in region D and continuous in $D+\Gamma$ except at finite isolated poles $a_1, a_2, \ldots, a_n$, then

$$\int_{\Gamma} f(z)dz = 2\pi i\sum_{k=1}^{n}\mathrm{Re}\, s f(a_k) \tag{11.7-17}$$

where Γ represents the boundary of region D.

Almost all integrals in the text can be evaluated by the residual theorem.

Example. Calculate the integral

$$\frac{1}{2\pi}\int_{-\infty}^{\infty}\frac{1}{-m\omega^2+k}e^{-i\omega t}d\omega = I \tag{11.7-18}$$

in terms of the residual theorem, where m and k are positive constants.

Though the integral is a real integral, it is difficult to evaluate because the integration limit is infinite and there are two singular points at the integration path but are easily completed by using the residual theorem. At first, we extend the real variable ω to a complex one, i.e., put $\omega = \omega_1 + i\omega_2$, and ω_1, ω_2 are real variables. At the complex plane ω, a half-circle with origin $(0, 0)$ and radius $R \to \infty$ is taken as an additional integral path, referring to Fig. 11.7. Along the real axis, the integrand of the integral has two poles $(-\sqrt{k/m}, 0)$ and $(\sqrt{m/k}, 0)$, and the value of the integral is equal to

$$\frac{1}{2\pi}\int_{-\infty}^{\infty}\frac{1}{-m\omega^2+k}e^{-i\omega t}d\omega$$

$$= I_1 = \lim_{R\to\infty, r\to 0}\left(\int_{C_R}+\int_1+\int_2+\int_3+\int_{C_1}+\int_{C_2}\right) \tag{11.7-19}$$

where the first integral on the right-hand side of (11.7-19) is carried out on the path of the grand half-circle, the second to

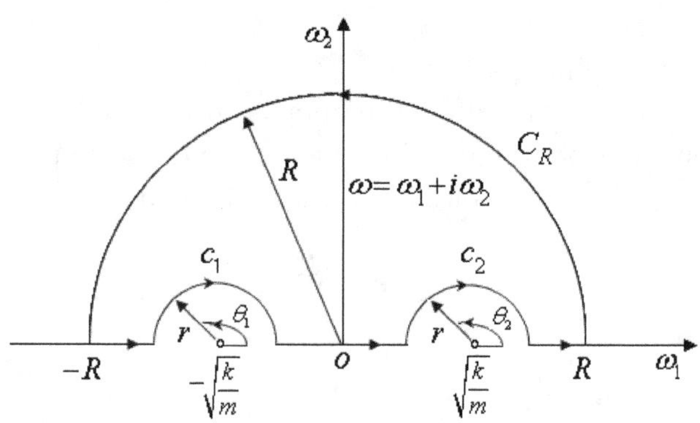

Fig. 11.7. Integration path at ω-plane.

fourth ones are on the path along the real axes except intervals $(-r-\sqrt{k/m}, -\sqrt{k/m}+r)$ and $(-r+\sqrt{k/m}, \sqrt{k/m}+r)$, and the fifth and sixth ones are on two small half-circle arcs C_1 and C_2 with origins $(-\sqrt{k/m}, 0)$ and $(\sqrt{k/m}, 0)$ and radius r, respectively. Because the integrand in the interior enclosing by the integration path in (11.7-19) is analytic, according to the Cauchy theorem (referring to formula (11.7-3)),

$$I_1 = 0 \tag{11.7-20}$$

Based on the bebaviour of the integrand and the Jordan lemma, the first one on the right-hand side of (11.7-19) must be zero. So

$$\lim_{R\to\infty, r\to 0}\left(\int_1 + \int_2 + \int_3 + \int_{C_1} + \int_{C_2}\right) = 0$$

and

$$\lim_{R\to\infty, r\to 0}\left(\int_1 + \int_2 + \int_3\right) = I = -\lim_{r\to 0}\left(\int_{C_1} + \int_{C_2}\right)$$

At arc $C_1{:}\omega + \sqrt{k/m} = re^{i\theta_1}$, $d\omega = ire^{i\theta_1}d\theta_1$ and at arc $C_2{:}\omega - \sqrt{k/m} = re^{i\theta_2}$, $d\omega = ire^{i\theta_2}d\theta_2$. Substituting these into the above integrals and after some simple calculations, we obtain

$$I = \frac{\pi}{m\sqrt{k/m}} \sin\sqrt{k/m}t \tag{11.7-21}$$

The inversion of some integral transforms even the solution of certain integral equations, many key calculations are completed by the similar procedure exhibited above, which are shown in Section 11.7.10 and in the Major Appendix B of this book.

11.7.5 *Analytic extension*

A function $f_1(z)$ is analytic at region D_1, if one can construct another function $f_2(z)$ analytic at region D_2, and D_1 and D_2 are not mutually intersected regions but with common bounding Γ, further more

$$f_1(z) = f_2(z) \quad z \in \Gamma$$

We can say that $f_1(z)$ and $f_2(z)$ are analytic extensions to each other, and we can also say that the function

$$F(z) = \begin{cases} f_1(z) & \text{as } z \in D_1 \\ f_2(z) & \text{as } z \in D_2 \end{cases}$$

analytic at $D = D_1 + D_2$ is an analytic extension of $f_1(z)$ as well as $f_2(z)$.

11.7.6 *Conformal mapping*

In the text of Chapters 7–9 and 11, by using one or several analytic functions which are also named complex potentials, we have expressed the solutions of harmonic, biharmonic, quadruple harmonic, sextuple harmonic equations and quasi-biharmonic and quasi-quadruple harmonic equations, and this is the complex representation of solutions. We can see that the complex representation is only the first step for solving the boundary value problems. For some problems with complicated boundaries, one must utilize the conformal mapping to transform the problem onto the mapping plane, and the corresponding boundaries can be simplified to a unit circle or straight line, and the calculation can be put forward; in some cases, exact analytic solutions are available.

The so-called conformal mapping is that the complex variable $z = x + iy$ and another one $\zeta = \xi + i\eta$ can be connected by

$$z = \omega(\zeta) \tag{11.7-22}$$

in which $\omega(\zeta)$ is a single-valued analytic function of $\zeta = \xi + i\eta$ in some region. Except certain points, the inversion of mapping (11.7-22) exists. If for a certain region the mapping is single-valued, we say it is a single-valued conformal mapping. In general, the mapping is single-valued, but the inversion $\zeta = \omega^{-1}(z)$ is impossible single-valued. It has the following properties:

(1) An angle at point $z = z_0$ after the mapping becomes an angle at point $\zeta = \zeta_0$, but the both angles have the same value of the argument, and the rotation is either in the same direction, which is the first kind of conformal mapping (e.g., shown in Fig. 11.5),

or in counter direction, which is the second kind of conformal mapping (e.g., depicted in Fig. 11.5).

(2) If $\omega(\zeta)$ is analytic and single-valued in region Ω and transforms the region into region D, then the inversion $\zeta = \omega^{-1}(z)$ is analytic and single-valued in region D and maps D onto Ω.

(3) If D is a region and c is a simple closed curve in it, and its interior belongs to D, and if $\omega^{-1}(z)$ is analytic and maps c onto a closed curve γ at Ω region bilaterally single-valued, then $\omega(\zeta)$ is analytic and single-valued in the region and maps D onto the interior of Ω.

In the text, we mainly used the following two kinds of conformal mapping:

(1) Rational function conformal mapping, e.g.,

$$\omega(\zeta) = \frac{c}{\zeta} + a_0 + a_1\zeta + \cdots + a_n\zeta^n \tag{11.7-23}$$

or

$$\omega(\zeta) = R\zeta + b_0 + b_1\frac{1}{\zeta} + \cdots + b_n\frac{1}{\zeta^n} \tag{11.7-24}$$

in which c, $a_0, a_1, \ldots, a_n$, R, $b_0, b_1, \ldots, b_n$ are constants. These mappings can be used in studying the infinite region with a crack at the physical plane onto the interior of the unit circle at the mapping plane. In the monograph of Muskhelishvili [1], he postulated that his method is only suitable for this kind of mapping function. Fan [4] extended it to transcendental mapping functions and achieved exact analytic solutions for crack problems for complicated configuration.

(2) Transcendental functions, e.g.,

$$\omega(\zeta) = \frac{H}{\pi}\ln\left[1 + \frac{(1+\zeta)^2}{(1-\zeta)^2}\right] \tag{11.7-25}$$

and

$$\omega(\zeta) = \frac{2W}{\pi}\arctan\left\{\sqrt{1-\zeta^2}\tan\left(\frac{\pi a}{2W}\right)\right\} - a \tag{11.7-26}$$

which can be used to transform a finite specimen with a crack onto the interior of unit circle at mapping plane, where H, W and a represent sample sizes and the crack size. These transcendental conformal mapping functions have been used by Fan and his group to construct exact analytic solutions, e.g., [14–16], which breaks through the limitation of the Muskhelishvili method.

References

[1] Muskhelishvili N I, 1956, *Some Basic Problems of the Mathematical Theory of Elasticity*, P. Noordhoff, Groningen.

[2] Lekhnitskii S G, 1963, *Theory of Elasticity of an Anisotropic Body*, Holden-Day, San Francisco.

[3] Lu J K, 2000, *Complex Variable Function Method of Plane Elasticity*, Science Press, Beijing (in Chinese).

[4] Fan T Y, 1990, Semi-infinite crack in a strip, *Chin. Phys. Lett.*, **8**(9), 401–404; Foundation of Fracture Theory, Science Press, Beijing, 2003 (in Chinese).

[5] Liu G T, 2004, The complex variable function method of the elastic theory of quasicrystals and defects and auxiliary equation method for solving some nonlinear evolution equations, Dissertation, Beijing Institute of Technology (in Chinese).

[6] Liu G T and Fan T Y, 2003,The complex method of the plane elasticity in 2D quasicrystals point group 10mm ten-fold rotation symmetry notch problems, *Science in China, Series E*, **46**(3), 326–336.

[7] Li L H and Fan T Y, 2006, Final governing equation of plane elasticity of icosahedral quasicrystals — stress potential method, *Chin. Phys. Lett.*, **24**(9), 2519–2521.

[8] Li L H and Fan T Y, 2006, Complex function method for solving notch problem of point group 10 and $\overline{10}$ two-dimensional quasicrystal based on the stress potential function, *J. Phys.: Condens. Matter*, **18**(47), 10631–10641.

[9] Li L H and Fan T Y, 2008, Complex function method for notch problem of plane elasticity of icosahedral quasicrystals, *Science in China G*, **51**(6), 773–780.

[10] Li W and Fan T Y, 2009, Study on elastic analysis of crack problem of two-dimensional decagonal quasicrystals of point group 10, $\overline{10}$, *Int. Mod. Phys. Lett. B*, in press.

[11] Fan T Y and Fan L, 2008, Plastic fracture of quasicrystals, *Phil. Mag.*, **88**(4), 323–335.

[12] Li W and Fan T Y, 2011, Plastic analysis of crack problem of two-dimensional decagonal Al-Ni-Co quasicrystalline materials of point group $10, \overline{10}$, *Chin. Phys. B*, **20**(3), 036101.
[13] Li W and Fan T Y, 2009, Plastic solution of crack in three-dimensional icosahedral Al-Pd-Mn quasicrystals, *Phili. Mag.*, **89**(31), 2823–2832.
[14] Fan T Y, Tang Z Y, Li L H and Li W, 2010, The strict theory of complex variable function method of sextuple harmonic equation and applications, *J. Math. Phys.*, **51**(5), 053519.
[15] Shen D W and Fan T Y, 2003, Tow collinear semi-infinite cracks in a strip, *Eng. Fract. Mech.*, **70**(8), 813–822.
[16] Li W, 2011, Analytic solutions of a finite width strip with a single edge crack of two-dimensional quasicrystals, *Chin. Phys. B*, **20**(11), 116201.
[17] Privalov I I, 1983, *Introduction to Theory of Complex Variable Functions*, Science, Moscow (in Russian).
[18] Lavrentjev M A, Schabat B A, 1986, *Method of Complex Variable Function Theory*, National Technical-Theoretical Literature Press, Moscow (in Russian).

Chapter 12

Variational Principle of Elasticity of Quasicrystals, Numerical Analysis and Applications

From Chapters 5–11, we developed analytic theories and methods. The elasticity problems of quasicrystals were reduced to boundary value or initial-boundary value problems of some partial differential equations to solve, in which complex analysis and conformal mapping method, integral transform and integral equation method, etc. were used. For some boundary value problems, these methods are extremely powerful, even capable of obtaining exact analytic solutions. In Chapter 14 and Major Appendix, we will further develop the analytic method in studying some problems, such as nonlinear deformation. The analytic solutions are very beautiful, simple and explicit, which indicates the power of the methods.

However, there are limitations to these analytic methods. In general, they can only treat some problems with simple configurations and simple boundary conditions, while for more complicated problems, the methods cannot display their power.

Those solved by these analytic methods directly are partial differential equations, the solutions of which hold for the neighbourhood of any point in the region considered if the solutions are constructed. In this sense, the solutions are exact, which belong to the classical solutions in mathematical physics, while the numerical methods are methods connected with discretization. Among them, the finite difference method displayed in Chapter 10 is one discretizing scheme. And the procedure of the finite element method is another discretizing scheme. The strict formulation of the finite element method can utilize variational principle, but it is not always necessary. It has been shown that solutions obtained by these two discrete methods can approach exact solutions as the size of the discrete mesh (or element) tends to be infinitesimal. By collaborating with computers, the methods can solve problems with very complicated configurations, boundary conditions and material structures. This shows the power of numerical methods, which are modernized and systematized. In contrast with the so-called analytic (classical) solutions, modern theory on partial differential equations proposed the so-called generalized (weak) solutions, and the above numerical methods are a tool to implement weak solutions. The finite difference method has been discussed in Chapter 10, and in this chapter, only the finite element method and its basis — variational principle — are discussed. Further mathematical principle on the weak solutions is developed in Chapter 13.

12.1 Review of Basic Relations of Elasticity of Icosahedral Quasicrystals

Because of the importance of icosahedral quasicrystals, we consider the finite element analysis only on this kind of matter. For this purpose, we here recall the basic relations of elasticity of icosahedral quasicrystals. There are

$$\varepsilon_{ij} = \frac{1}{2}\left(\frac{\partial u_i}{\partial x_j} + \frac{\partial u_j}{\partial x_i}\right), \quad w_{ij} = \frac{\partial w_i}{\partial x_j} \tag{12.1-1}$$

In this case, the generalized Hooke's law stands for

$$\begin{aligned} \sigma_{ij} &= C_{ijkl}\varepsilon_{kl} + R_{ijkl}w_{kl} \\ H_{ij} &= K_{ijkl}w_{kl} + R_{klij}\varepsilon_{kl} \end{aligned} \tag{12.1-2}$$

where

$$C_{ijkl} = \lambda\delta_{ij}\delta_{kl} + \mu(\delta_{ik}\delta_{jl} + \delta_{il}\delta_{jk}) \tag{12.1-3}$$

$$K_{ijkl} = K_1\delta_{ik}\delta_{jl} + K_2(\delta_{ij}\delta_{kl} - \delta_{il}\delta_{jk}) \tag{12.1-4}$$

$$R_{ijkl} = R(\delta_{i1} - \delta_{i2})(\delta_{ij}\delta_{kl} - \delta_{ik}\delta_{jl} + \delta_{il}\delta_{jk}) \tag{12.1-5}$$

λ and μ are Lamé coefficients, K_1 and K_2 are the phason elastic constants and R is the phonon–phason coupling elastic constant, respectively.

The stress components satisfy the equilibrium equations

$$\frac{\partial \sigma_{ij}}{\partial x_j} + f_i = 0, \quad \frac{\partial H_{ij}}{\partial x_j} + g_i = 0 \tag{12.1-6}$$

The above formulas hold in any interior point of region Ω, and at boundary S_t, the stresses satisfy the boundary conditions

$$\begin{aligned} \sigma_{ij} n_j &= T_i \quad (x_1, x_2, x_3) \in S_t \\ H_{ij} n_j &= h_i \end{aligned} \tag{12.1-7}$$

and at boundary S_u, the displacements satisfy the boundary conditions

$$\left.\begin{aligned} u_i &= \bar{u}_i \\ w_i &= \bar{w}_i \end{aligned}\right. \quad (x_1, x_2, x_3) \in S_u \tag{12.1-8}$$

where T_i is the traction vector, h_i is the generalized traction vector at boundary S_t, $\bar{u}_i$ and $\bar{w}_i$ are the given displacements at boundary S_u, n_i is the unit outward normal vector at any point of the boundary and $S = S_u + S_t$.

In the following, only the static problems are studied, and the initial value conditions will not be concerned. For the dynamic problems, the initial value conditions must be used, which have been discussed in Chapter 10.

12.2 General Variational Principle for Static Elasticity of Quasicrystals

The variational principle in mathematical physics is one of the basic principles, which reveals that the extreme value (or stationary value) of the energy functional of a system is equivalent to the governing equations and the corresponding boundary value (or initial-boundary value) conditions of the system. Accordingly, solutions of the initial-boundary value problem of the partial differential equations can be converted to determine the extreme value of the corresponding energy functional. And the latter will be implemented by a discretization procedure; one among them is the finite element method.

We here extend the minimum potential energy principle of classical elasticity [1] to describe the elasticity of quasicrystals.

Theorem 12.2.1 (Variational principle of elasticity of quasicrystals). *For sufficient smooth boundary, if all u_i and w_i satisfy the equations of deformation geometry* (12.1-1) *and displacement boundary conditions* (12.1-8), *let the energy functional of quasicrystals*

$$\Pi = \int_\Omega F d\Omega + \int_\Omega (f_i u_i + g_i w_i) d\Omega + \int_{\Gamma_t} (T_i u_i + h_i w_i) dS \quad (12.2\text{-}1)$$

take a minimum value, then they will be the solution satisfying the equilibrium equations (12.1-6) *and the stress boundary conditions* (12.1-7), *in which F is defined by*

$$\begin{aligned} F &= \int_0^{\varepsilon_{ij}} \sigma_{ij} d\varepsilon_{ij} + \int_0^{w_{ij}} H_{ij} dw_{ij} = F_u + F_w + F_{uw} \\ F_u &= \frac{1}{2} C_{ijkl} \varepsilon_{ij} \varepsilon_{kl} \\ F_w &= \frac{1}{2} K_{ijkl} w_{ij} w_{kl} \\ F_{uw} &= R_{ijkl} \varepsilon_{ij} w_{kl} \end{aligned} \quad (12.2\text{-}2)$$

or by (4.4-1), *Ω is the region occupied by the quasicrystal and S is the boundary of Ω.*

In addition, the conditions in the theorem are sufficient and necessary.

Proof.

(i) Necessity

Assume that the functional Π takes its extreme value, i.e., $\delta\Pi = 0$. From (12.2-1),

$$\begin{aligned} \delta\Pi = & \int_\Omega \frac{\partial F}{\partial \varepsilon_{ij}} \delta\varepsilon_{ij} + \frac{\partial F}{\partial w_{ij}} \delta w_{ij} - (f_i \delta u_i + g_i \delta w_i) d\Omega \\ & - \int_{\Gamma_t} (T_i \delta u_i + h_i \delta w_i) dS = 0 \end{aligned} \quad (12.2\text{-}3)$$

where

$$\sigma_{ij} = \frac{\partial F}{\partial \varepsilon_{ij}}, \quad H_{ij} = \frac{\partial F}{\partial w_{ij}} \tag{12.2-4}$$

which have been introduced in Chapter 4.

Noting that the suffixes of quantities $\frac{\partial F}{\partial \varepsilon_{ij}}$ are symmetric, we obtain

$$\int_\Omega \frac{\partial F}{\partial \varepsilon_{ij}} \delta \varepsilon_{ij} d\Omega = \int_\Omega \frac{\partial F}{\partial \varepsilon_{ij}} \delta \left(\frac{\partial u_i}{\partial x_j} \right) d\Omega$$

Making use of the Green formula to above formula yields

$$\begin{aligned}\int_\Omega \frac{\partial F}{\partial \varepsilon_{ij}} \delta \varepsilon_{ij} d\Omega &= \int_\Omega \frac{\partial}{\partial x_j} \left(\frac{\partial F}{\partial \varepsilon_{ij}} \delta u_i \right) d\Omega - \int_\Omega \frac{\partial}{\partial x_j} \left(\frac{\partial F}{\partial \varepsilon_{ij}} \right) \delta u_i) d\Omega \\ &= \int_{S_u + S_t} \frac{\partial F}{\partial \varepsilon_{ij}} n_j \delta u_i dS - \int_\Gamma \frac{\partial}{\partial x_j} \left(\frac{\partial F}{\partial \varepsilon_{ij}} \right) \delta u_i d\Omega\end{aligned}$$

Because the displacements are given at the boundary S_u, at which $\delta u_i = \delta \bar{u}_i = 0$, the above formula has been reduced to

$$\int_\Omega \frac{\partial F}{\partial \varepsilon_{ij}} \delta \varepsilon_{ij} d\Omega = \int_{S_t} \frac{\partial F}{\partial \varepsilon_{ij}} n_j \delta u_i dS - \int_\Omega \frac{\partial}{\partial x_j} \left(\frac{\partial F}{\partial \varepsilon_{ij}} \right) \delta u_i d\Omega \tag{12.2-5}$$

Due to $w_{ij} = \partial w_i / \partial x_j$, a similar analysis to what was adopted just above gives rise to

$$\int_\Omega \frac{\partial F}{\partial w_{ij}} \delta w_{ij} d\Omega = \int_{\Gamma_t} \frac{\partial F}{\partial w_{ij}} n_j \delta w_i d\Gamma - \int_\Omega \frac{\partial}{\partial x_j} \left(\frac{\partial F}{\partial w_{ij}} \right) \delta u_i d\Omega \tag{12.2-6}$$

Substituting (12.2-5) and (12.2-6) into (12.2-3) leads to

$$\begin{aligned}\delta \Pi = &- \int_\Omega \left\{ \left[\frac{\partial}{\partial x_j} \left(\frac{\partial F}{\partial \varepsilon_{ij}} \right) + f_i \right] \delta u_i d\Omega + \left[\frac{\partial}{\partial x_j} \left(\frac{\partial F}{\partial w_{ij}} \right) + g_i \right] \delta w_i \right\} d\Omega \\ &+ \int_{S_t} \left\{ \left[\left(\frac{\partial F}{\partial \varepsilon_{ij}} \right) n_j - T_i \right] \delta u_i + \left[\left(\frac{\partial F}{\partial w_{ij}} \right) n_j - h_i \right] \delta w_i \right\} dS = 0\end{aligned} \tag{12.2-7}$$

Since δu_i and δw_i are of arbitrary and independent variation at region Ω and boundary S, the validity of (12.2-7) must be

$$\frac{\partial}{\partial x_j}\left(\frac{\partial F}{\partial \varepsilon_{ij}}\right) + f_i = 0, \quad \frac{\partial}{\partial x_j}\left(\frac{\partial F}{\partial w_{ij}}\right) + g_i = 0, \quad (x_1, x_2, x_3) \in \Omega$$

$$\left(\frac{\partial F}{\partial w_{ij}}\right) n_j - h_i = 0, \quad \left(\frac{\partial F}{\partial \varepsilon_{ij}}\right) n_j - T_i = 0, \quad (x_1, x_2, x_3) \in S_t$$

Substituting (12.2-4) into the above formula yields just the equilibrium equations and stress boundary conditions. This shows that u_i and w_i satisfy equations of deformation geometry, stress–strain relations and displacement boundary conditions and making energy functional to have minimum value should be the solution satisfying the equilibrium equations and stress boundary conditions.

(ii) Sufficiency

The sufficiency of the conditions given by the theorem means that, if u_i and w_i satisfy relations of deformation geometry and displacement boundary and make the equilibrium equations and stress boundary conditions to be satisfied, then they should do the energy functional to be minimum.

Suppose that quantities u_i, ε_{ij}, w_i and w_{ij} obey the stress–strain relations (12.1-2) and satisfy the displacement boundary conditions (12.1-8) and set

$$\begin{aligned} \varepsilon_{ij}^* &= \varepsilon_{ij} + \delta\varepsilon_{ij}, u_i^* = u_i + \delta u_i \\ w_{ij}^* &= w_{ij} + \delta w_{ij}, w_i^* = w_i + \delta w_i \end{aligned} \tag{12.2-8}$$

through the displacement–strain relations, what follows is that

$$\begin{aligned} \delta\varepsilon_{ij} &= \frac{1}{2}(\delta u_{i,j} - \delta u_{j,i}) \\ \delta w_{ij} &= \delta w_{i,j} \end{aligned} \tag{12.2-9}$$

where $u_{i,j} = \partial u_i/\partial x_j$, etc.

The free energy $F(\varepsilon_{ij}^*, w_{ij}^*)$ can be expanded the into the Taylor series, as follows:

$$\begin{aligned}F(\varepsilon_{ij}^*, w_{ij}^*) &= F(\varepsilon_{ij}+\delta\varepsilon_{ij}, w_{ij}+\delta w_{ij})\\ &= F(\varepsilon_{ij}, w_{ij}) + \frac{\partial F}{\partial \varepsilon_{ij}}\delta\varepsilon_{ij} + \frac{\partial F}{\partial w_{ij}}\delta w_{ij}\\ &\quad + \frac{1}{2}\frac{\partial^2 F}{\partial\varepsilon_{ij}\partial\varepsilon_{kl}}\delta\varepsilon_{ij}\delta\varepsilon_{kl} + \frac{1}{2}\frac{\partial^2 F}{\partial w_{ij}\partial w_{kl}}\delta w_{ij}\delta w_{kl}\\ &\quad + \frac{\partial^2 F}{\partial\varepsilon_{ij}\partial w_{kl}}\delta\varepsilon_{ij}\delta w_{kl} + \cdots\end{aligned} \tag{12.2-10}$$

If the free energy is a homogeneous quantity of strain components of second order, then the expansion of energy functional corresponding to (12.2-10) does not contain terms higher than third order, i.e.,

$$\Pi^* = \Pi + \delta\Pi + \delta^2\Pi + O(\delta^3) \tag{12.2-11}$$

in which

$$\begin{aligned}\delta\Pi &= \int_\Omega \frac{\partial F}{\partial\varepsilon_{ij}}\delta\varepsilon_{ij} + \frac{\partial F}{\partial w_{ij}}\delta w_{ij} - f_i\delta u_i - g_i\delta w_i)d\Omega\\ &\quad - \int_{S_t}(T_i\delta u_i + h_i\delta w_i)dS\end{aligned} \tag{12.2-12}$$

$$\begin{aligned}\delta^2\Pi &= \int_\Omega\left\{\frac{1}{2}\frac{\partial^2 F}{\partial\varepsilon_{ij}\partial\varepsilon_{kl}}\delta\varepsilon_{ij}\delta\varepsilon_{kl} + \frac{1}{2}\frac{\partial^2 F}{\partial w_{ij}\partial w_{kl}}\delta w_{ij}\delta w_{kl}\right.\\ &\quad \left. + \frac{1}{2}\frac{\partial^2 F}{\partial\varepsilon_{ij}\partial w_{kl}}\delta\varepsilon_{ij}\delta w_{kl}\right\} d\Omega\end{aligned} \tag{12.2-13}$$

Applying the Green formula to (12.2-12) leads to

$$\delta\Pi = 0$$

u_i and w_i (through the corresponding σ_{ij} and H_{ij}) satisfy the equilibrium equations and stress boundary conditions. This means that the energy functional takes extreme values.

According to the discussion in Ref. [2], we do some extension, i.e., $\lambda+\mu > 0$, $\mu > 0$, $K_1 > 0$, $K_2 > 0$, $\mu K_1 > R^2$, then the elasticity of quasicrystals presents stability, in the case the stress–strain elastic matrix should be positive determination, so

$$\delta^2\Pi > 0$$

This guarantees that $\delta\Pi = 0$ takes not only an extreme value but also the minimum value of the energy functional.

But recent experimental results and some simulation show that there may be $K_2 < 0$; this does not influence the energy-taking extreme value, but the extreme value may not be a minimum value.

Collaborating variational principle and theory of functional analysis, we can prove the existence, uniqueness and stability of the solution of the boundary value problem (12.1-3) and (12.1-8)–(12.1-11). This is concerned not only with the numerical implementation but also with other topics; a detailed discussion of this is given in Chapter 13 or see Guo and Fan [2].

The variational principle can be extended to the dynamic case, in which it is needed only to extend the energy functional (12.1-1) to be as follows:

$$\Pi = \int_{\Omega} F d\Omega + \int_{\Omega} [(f_i - \rho\ddot{u}_i)u_i + (g_i - \kappa\dot{w}_i)w_i)d\Omega + \int_{S_t} (T_i u_i + h_i w_i)dS \tag{12.2-14}$$

where the meaning of ρ and κ can be found in Chapter 10. From (12.2-14), we can obtain the corresponding variational equation similar to (12.2-3), which is equivalent to the equations of phonon–phason dynamics and related boundary and initial conditions of quasicrystals. Further discussion about this is omitted here.

12.3 Finite Element Method for Elasticity of Icosahedral Quasicrystals

Finite element method is a method discreting variational equations and region Ω. Dividing the quasicrystal body into M subregions or M elements $\Omega^{(m)}$; the superscript m denotes the number of an element, and $m = 1, \ldots, M$. For any element $\Omega^{(m)}$; the phonon and phason displacements are expressed by $u_i^{(m)}$ and $w_i^{(m)}$

$$\begin{cases} u_i^{(m)} = \sum_{\alpha=1}^{n} I_\alpha u_{i\alpha}^{(m)} \\ w_i^{(m)} = \sum_{\alpha=1}^{n} I_\alpha w_{i\alpha}^{(m)} \end{cases} \quad (x, y, z) \in \Omega^{(m)} \tag{12.3-1}$$

in which n is the amount of mth element, and sub index α is the number of nodes of element m, I_α is the interpolating function of node α and $u_{i\alpha}^{(m)}$ and $w_{i\alpha}^{(m)}$ are the phonon and phason ith displacement components of node α. In the interior of every element, the displacements $u_i^{(m)}$ are $w_i^{(m)}$ are continuous and single-valued, and at the interface between the elements, the displacements are continuous, i.e.,

$$\begin{cases} u_i^{(m)} = u_i^{(m')} \\ w_i^{(m)} = w_i^{(m')} \end{cases} \quad (x, y, z) \in S^{(mm')} \tag{12.3-2}$$

where $S^{(mm')}$ represents the interface between elements m and m'. At the boundary S_u on which the displacements are given, the displacements satisfy the displacement boundary conditions (12.1-8).

Under satisfying these conditions, the discretization form of energy functional Π takes

$$\begin{aligned} \Pi^* = \sum_{m=1}^{M} \Bigg[\int_{\Omega^{(m)}} \Bigg(& \frac{1}{2} C_{ijkl} \varepsilon_{ij}^{(m)} \varepsilon_{kl}^{(m)} + \frac{1}{2} K_{ijkl} w_{ij}^{(m)} w_{kl}^{(m)} \\ & + R_{ijkl} \varepsilon_{ij}^{(m)} w_{kl}^{(m)} - f_i^{(m)} u_i^{(m)} - g_i^{(m)} w_i^{(m)} \Bigg) d\Omega \\ & - \int_{S_t^{(m)}} (T_i^{(m)} u_i^{(m)} + h_i^{(m)} w_i^{(m)}) dS \Bigg] \end{aligned} \tag{12.3-3}$$

According to the following order, the strain components can be arranged as a vector

$$\begin{aligned} \{\varepsilon_{ij}, w_{ij}\}^{(m)T} = \{ & \varepsilon_{11}, \varepsilon_{22}, \varepsilon_{33}, \gamma_{23}, \gamma_{31}, \gamma_{12}, w_{11}, w_{22}, w_{33}, \\ & w_{23}, w_{31}, w_{12}, w_{32}, w_{13}, w_{21}\}^{(m)} \end{aligned} \tag{12.3-4}$$

where $\gamma_{ij} = 2\varepsilon_{ij}$ $(i \neq j)$ and the superscript T denotes the transpose of the matrix. The stresses can also be expressed as a vector

$$\{\sigma_{ij}, H_{ij}\}^{(m)T} = \{\sigma_{11}, \sigma_{22}, \sigma_{33}, \sigma_{23}, \sigma_{31}, \sigma_{12}, H_{11}, H_{22}, H_{33}, H_{23}, H_{31}, H_{12}, H_{32}, H_{13}, H_{21}\}^{(m)} \quad (12.3\text{-}5)$$

Utilizing (12.3-4) and (12.3-5), the relation (12.1-2) between stresses and strains stands for

$$\{\sigma_{ij}, H_{ij}\}^{m} = [D]\{\varepsilon_{ij}, w_{ij}\}^{m} \quad (12.3\text{-}6)$$

where $[D]$ is the elastic constant matrix, namely,

$$[D] = \begin{bmatrix} [C] & [R] \\ [R]^T & [K] \end{bmatrix} \quad (12.3\text{-}7)$$

with the submatrices as

$$[C] = \begin{bmatrix} \lambda+2\mu & \lambda & \lambda & 0 & 0 & 0 \\ \lambda & \lambda+2\mu & \lambda & 0 & 0 & 0 \\ \lambda & \lambda & \lambda+2\mu & 0 & 0 & 0 \\ 0 & 0 & 0 & \mu & 0 & 0 \\ 0 & 0 & 0 & 0 & \mu & 0 \\ 0 & 0 & 0 & 0 & 0 & \mu \end{bmatrix}$$

$$[R] = \begin{bmatrix} R & R & R & 0 & 0 & 0 & 0 & R & 0 \\ -R & -R & R & 0 & 0 & 0 & 0 & -R & 0 \\ 0 & 0 & -2R & 0 & 0 & 0 & 0 & 0 & 0 \\ 0 & 0 & 0 & 0 & 0 & -R & R & 0 & -R \\ R & -R & 0 & 0 & R & 0 & 0 & 0 & 0 \\ 0 & 0 & 0 & -R & 0 & -R & 0 & 0 & R \end{bmatrix}$$

$$[K] = \begin{bmatrix} K_1 & 0 & 0 & 0 & K_2 & 0 & 0 & K_2 & 0 \\ 0 & K_1 & 0 & 0 & -K_2 & 0 & 0 & K_2 & 0 \\ 0 & 0 & K_2 + K_1 & 0 & 0 & 0 & 0 & 0 & 0 \\ 0 & 0 & 0 & K_1 - K_2 & 0 & K_2 & 0 & 0 & -K_2 \\ K_2 & -K_2 & 0 & 0 & K_1 - K_2 & 0 & 0 & 0 & 0 \\ 0 & 0 & 0 & K_2 & 0 & K_1 & -K_2 & 0 & 0 \\ 0 & 0 & 0 & 0 & 0 & -K_2 & K_1 - K_2 & 0 & -K_2 \\ K_2 & K_2 & 0 & 0 & 0 & 0 & 0 & K_1 - K_2 & 0 \\ 0 & 0 & 0 & -K_2 & 0 & 0 & -K_2 & 0 & K_1 \end{bmatrix}$$

The vector representation of phonon and phason displacements can be written as

$$\{\bar{u}^{(m)}\}^T = \{u_1, u_2, u_3, w_1, w_2, w_3\}^{(m)} \tag{12.3-8}$$

In accordance with the strain–displacement relation (12.1-1), Eq. (12.3-4) can be expressed by

$$\{\varepsilon_{ij}, w_{ij}\}^{(m)} = [L]\{\bar{u}^{(m)}\} \tag{12.3-9}$$

where $[L]$ represents the differential operator matrix of element strain, i.e.,

$$[L]^T = \begin{bmatrix} \partial_1 & 0 & 0 & 0 & \partial_3 & \partial_2 & 0 & 0 & 0 & 0 & 0 & 0 & 0 & 0 & 0 \\ 0 & \partial_2 & 0 & \partial_3 & 0 & \partial_1 & 0 & 0 & 0 & 0 & 0 & 0 & 0 & 0 & 0 \\ 0 & 0 & \partial_3 & \partial_2 & \partial_1 & 0 & 0 & 0 & 0 & 0 & 0 & 0 & 0 & 0 & 0 \\ 0 & 0 & 0 & 0 & 0 & 0 & \partial_1 & 0 & 0 & 0 & 0 & \partial_2 & 0 & \partial_3 & 0 \\ 0 & 0 & 0 & 0 & 0 & 0 & 0 & \partial_2 & 0 & \partial_3 & 0 & 0 & 0 & 0 & \partial_1 \\ 0 & 0 & 0 & 0 & 0 & 0 & 0 & 0 & \partial_3 & 0 & \partial_1 & 0 & \partial_2 & 0 & 0 \end{bmatrix}. \quad (12.3\text{-}10)$$

in which $\partial_j = \frac{\partial}{\partial x_j}$ $(j = 1, 2, 3)$. Substituting (12.3-9) into (12.3-6), it follows that

$$\{\sigma_{ij}, H_{ij}\}^{(m)} = [D][L]\{\bar{u}^{(m)}\} \quad (12.3\text{-}11)$$

Inserting (12.3-8), (12.3-9) and (12.3-11) into the discreting form of the energy functional (12.3-3) yields

$$\Pi^* = \sum_{m=1}^{M} \left\{ \int_{\Omega^{(m)}} \frac{1}{2} \{\bar{u}^{(m)}\}^T [L]^T [D][L] \{\bar{u}^{(m)}\} d\Omega - \{\bar{u}^{(m)}\}^T \right. \\ \left. \times \left[\int_{\Omega_m} \begin{Bmatrix} f^{(m)} \\ g^{(m)} \end{Bmatrix} d\Omega + \int_{S_t^{(m)}} \begin{Bmatrix} T^{(m)} \\ h^{(m)} \end{Bmatrix} dS \right] \right\} \quad (12.3\text{-}12)$$

In the interior of element m, the displacement vector $\{\bar{u}^{(m)}\}$ can be expressed by the displacement vector of every node

$$\{\bar{u}^{(m)}\} = [I]\{\tilde{u}^{(m)}\} \quad (12.3\text{-}13)$$

where $[I] = [[I_1], [I_2], \ldots, [I_n]]$ is the matrix of interpolating functions:

$$[I_\alpha] = \begin{bmatrix} I_\alpha & 0 & 0 & 0 & 0 & 0 \\ 0 & I_\alpha & 0 & 0 & 0 & 0 \\ 0 & 0 & I_\alpha & 0 & 0 & 0 \\ 0 & 0 & 0 & I_\alpha & 0 & 0 \\ 0 & 0 & 0 & 0 & I_\alpha & 0 \\ 0 & 0 & 0 & 0 & 0 & I_\alpha \end{bmatrix}, \quad \alpha = 1, 2, \ldots, n \quad (12.3\text{-}14)$$

$\{\tilde{u}^{(m)}\} = \{\{\tilde{u}_1^{(m)}\}, \{\tilde{u}_2^{(m)}\}, \ldots \{\tilde{u}_\alpha^{(m)}\}, \ldots \{\tilde{u}_n^{(m)}\}\}$ consists of displacement vectors of every node within the element and $\{\tilde{u}_\alpha^{(m)}\}$ the displacement vector of node α, i.e.,

$$\{\tilde{u}_\alpha^{(m)}\} = \{u_{1\alpha}, u_{2\alpha}, u_{3\alpha}, w_{1\alpha}, w_{2\alpha}, w_{3\alpha}\}^{(m)} \tag{12.3-15}$$

Substituting (12.3-13) into (12.3-12), then (12.3-12) can be calculated as follows:

$$\delta\Pi^* = 0 \tag{12.3-16}$$

then we have obtained the finite element scheme

$$[K]\{\tilde{u}\} = \{\mathbf{R}\} \tag{12.3-17}$$

in which

$$[K] = \sum_{m=1}^{M} \int_{\Omega^{(m)}} [B]^T [D][B] d\Omega, \quad [B_i] = [L][I_i]$$

$$\{\mathbf{R}\} = \sum_{m=1}^{M} \left(\int_{\Omega^{(m)}} [I]^T \begin{Bmatrix} f^{(m)} \\ g^{(m)} \end{Bmatrix} d\Omega + \int_{S_t^{(m)}} [I]^T \begin{Bmatrix} T^{(m)} \\ h^{(m)} \end{Bmatrix} dS \right) \tag{12.3-18}$$

$[K]$ denotes the total stiffness matrix, $\{\mathbf{R}\}$ the vector of equivalent node force, $\{\tilde{u}\} = \{\{\tilde{u}^{(1)}\}, \{\tilde{u}^{(2)}\}, \ldots, \{\tilde{u}^{(N)}\}\}$ the displacement vector of all nodes within region Ω and N the amount of nodes after discretization. From (12.3-17), one obtains the solution $\{\tilde{u}\}$, which further gets strains and stresses. The stiffness matrix $[K]$ in integral form can be evaluated by the Gauss-integrating method.

The stresses at the Gauss integral point in element m of quasicrystals can be obtained through the following version:

$$\{\sigma_{ij}, H_{ij}\}^{(m)} = [D][B]\{\tilde{u}^{(m)}\} \tag{12.3-19}$$

12.4 Numerical Results

12.4.1 *Test example — An icosahedral Al–Pd–Mn quasicrystal bar subjected to uniaxial tension*

Consider an icosahedral Al–Pd–Mn quasicrystal bar with the height H subjected to uniaxial tensile force F as shown in Fig. 12.1.

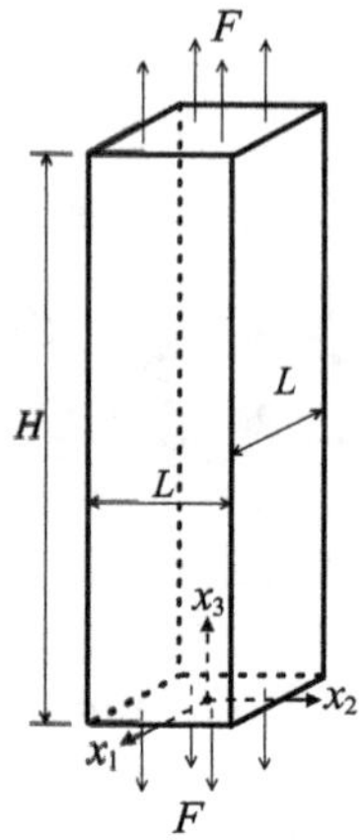

Fig. 12.1. An icosahedral quasicrystal bar subjected to uniaxial tension.

A Cartesian coordinate system (x_1, x_2, x_3) is introduced with the bottom and top surfaces lying in the planes $x_3 = 0$ and $x_3 = H$, respectively. The coordinate origin is located at the centre of the bottom of the cuboid. The area of the cross-section of the cuboid is L^2.

According to the loading of bar, it is obvious that the stress components yield

$$\sigma_{33} = \frac{F}{L^2}, \quad \sigma_{11} = \sigma_{22} = \sigma_{23} = \sigma_{31} = \sigma_{12} = 0, \quad H_{ij} = 0$$

Inserting Eq. (31) into Eq. (7) leads to the strain components in the x_3 direction:

$$\varepsilon_{11} = \varepsilon_{22} = \frac{-F\left(\lambda + \frac{2R^2}{K_1+K_2}\right)}{2L^2\left[2\mu^2 + 3\lambda\mu - \frac{R^2(9\lambda+6\mu)}{K_1+K_2}\right]},$$

$$\varepsilon_{33} = \frac{F\left(\lambda + \mu - \frac{R^2}{K_1+K_2}\right)}{L^2\left[2\mu^2 + 3\lambda\mu - \frac{R^2(9\lambda+6\mu)}{K_1+K_2}\right]}$$

$$w_{33} = \frac{FR}{L^2[\mu(K_1 + K_2) - 3R^2]}, \quad \text{other } \varepsilon_{ij} = w_{ij} = 0$$

The displacement boundary conditions for this problem can be written as

$$u_3 = w_3 = 0, \quad for\ x_3 = 0$$

$$u_1 = u_2 = w_1 = w_2 = 0, \quad for\ x_1 = x_2 = x_3 = 0$$

Further, tractions in both phonon and phason are zero at all other surfaces. The displacement components are

$$u_1 = \frac{-F\left(\lambda + \frac{2R^2}{K_1+K_2}\right)x_1}{2L^2\left[2\mu^2 + 3\lambda\mu - \frac{R^2(9\lambda+6\mu)}{K_1+K_2}\right]},$$

$$u_2 = \frac{-F\left(\lambda + \frac{2R^2}{K_1+K_2}\right)x_2}{2L^2\left[2\mu^2 + 3\lambda\mu - \frac{R^2(9\lambda+6\mu)}{K_1+K_2}\right]}$$

$$u_3 = \frac{F\left(\lambda + \mu - \frac{R^2}{K_1+K_2}\right)x_3}{L^2\left[2\mu^2 + 3\lambda\mu - \frac{R^2(9\lambda+6\mu)}{K_1+K_2}\right]}, \quad w_3 = \frac{FRx_3}{L^2(\mu K_1 + \mu K_2 - 3R^2)}$$

If there isn't any coupling between phonon and phason fields, i.e., $R = 0$, one can have

$$u_1 = \frac{-F\lambda x_1}{2L^2(2\mu^2 + 3\lambda\mu)}, \quad u_2 = \frac{-F\lambda x_2}{2L^2(2\mu^2 + 3\lambda\mu)}$$

$$u_3 = \frac{F(\lambda + \mu)x_3}{L^2(2\mu^2 + 3\lambda\mu)}, \quad w_3 = 0$$

coinciding with the displacements in the classical elastic theory of crystals.

Next, the displacements calculated numerically will be compared to the analytical results. The material parameters of a three-dimensional icosahedral Al–Pd–Mn quasicrystal are given as follows: see [3–6] or Chapter 9:

$$\lambda = 74.9\,\text{GPa}, \quad \mu = 72.4\,\text{GPa}, \quad K_1 = 72\,\text{MPa}, \quad K_2 = -37\,\text{MPa}$$

The coupling constant R of Al–Pd–Mn has not been measured so far. In the computations, we respectively assume $R/\mu = 0$, 0.001, 0.002, 0.004, 0.006, 0.008 and 0.01. The height H of the cuboid is

4 cm, and the area of the cross-section is $L^2 = 1\,\text{cm}^2$. The tensile force F is 1 kN. Eight-node brick elements are used to mesh this model. The calculated domain is meshed by 32 0.5 cm × 0.5 cm × 0.5 cm 8-node hexahedron elements.

The displacements of the point (0.5, 0.5, 4) at the top surface of the cuboid are shown in Table 12.1. The results show that the numerical and analytical solutions are the same. With increasing R/μ, the absolute values of displacements u_1, u_3 and w_3 become larger. For $R/\mu = 0.01$, the displacement w_3 is about six times u_3 demonstrating the coupling effect of this quasicrystal. Of course, the influence of R/μ on the displacements in the phonon field depends on K_1 and K_2.

The displacements along the loading axis, with x_1 or x_2 as the tensile axis shown in Fig. 12.2, are briefly given as follows:

Table 12.1. Comparison of the displacements of the point $(0.5, 0.5, 4)/10^{-2}$cm.

Displacement	R/μ	0	0.001	0.002	0.004	0.006	0.008	0.01
u_1	Analytical	0.00070	0.00071	0.00073	0.00083	0.00103	0.00146	0.00258
	Numerical	0.00070	0.00071	0.00073	0.00083	0.00103	0.00146	0.00258
u_3	Analytical	0.02202	0.02214	0.02249	0.02405	0.02732	0.03416	0.05215
	Numerical	0.02202	0.02214	0.02249	0.02405	0.02732	0.03416	0.05215
w_3	Analytical	0	0.11500	0.23439	0.50754	0.88298	1.51665	3.01205
	Numerical	0.00000	0.11500	0.23439	0.50754	0.88298	1.51665	3.01205

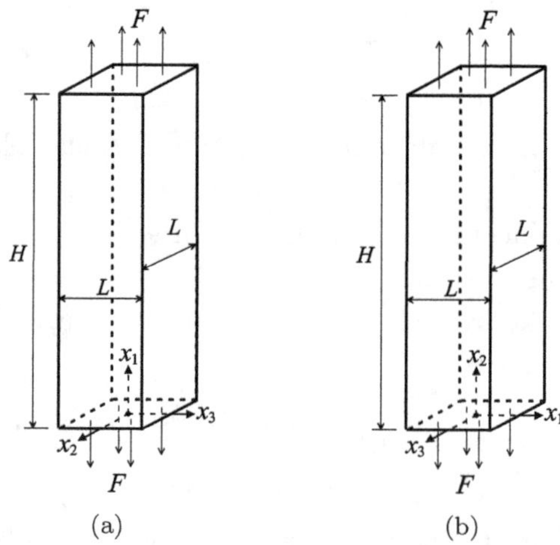

Fig. 12.2. An icosahedral quasicrystal bar subjected to uniaxial tension.

(a) when x_1 is the tensile axis, there are

$$u_1 = \frac{F\left(\lambda + \mu - \frac{R^2}{K_1+K_2}\right)x_1}{L^2\left[2\mu^2 + 3\lambda\mu - \frac{R^2(9\lambda+6\mu)}{K_1+K_2}\right]},$$

$$w_1 = -\frac{FRx_1}{2L^2(\mu K_1 + \mu K_2 - 3R^2)}$$

(b) when x_2 is the tensile axis, there are

$$u_2 = \frac{F\left(\lambda + \mu - \frac{R^2}{K_1+K_2}\right)x_2}{L^2\left[2\mu^2 + 3\lambda\mu - \frac{R^2(9\lambda+6\mu)}{K_1+K_2}\right]},$$

$$w_2 = \frac{FRx_2}{2L^2(\mu K_1 + \mu K_2 - 3R^2)}$$

It can be seen that the phonon displacements are the same for all three kinds of tensile axes, whereas the phason displacements are different from each other. This shows the three-dimensional anisotropic characteristic of icosahedral quasicrystals.

This checks the computer programme to be correct and effective.

12.4.2 *Specimen of icosahedral Al–Pd–Mn quasicrystal with a crack under tension*

An icosahedral Al–Pd–Mn quasicrystal plate containing a penetrating crack subjected to a uniform load P is shown in Fig. 12.3, where $a = 5\,\text{mm}$, $H = 50\,\text{mm}$, $L = 60\,\text{mm}$ and $P/\mu = 0.001$. The thickness of the plate is 1 mm, which is 10% of the crack length $2a$ and 1% of the plate height $2H$, so the analytic solutions [6–8] can also be used in the case. The point O is the coordinate origin, and the front and back faces of the plate are lying in plane $x_3 = 0$ and $x_3 = -1\,\text{mm}$. The elastic constants of the quasicrystal are the same as those in Section 4. We assume $R/\mu = 0.005$ for the coupled case and $R/\mu = 0$ for the decoupled case.

Due to the symmetry of the specimen, only the upper right 1/4 part $0 \leq x_1 \leq L$, $0 \leq x_2 \leq H$ has to be modelled. The following

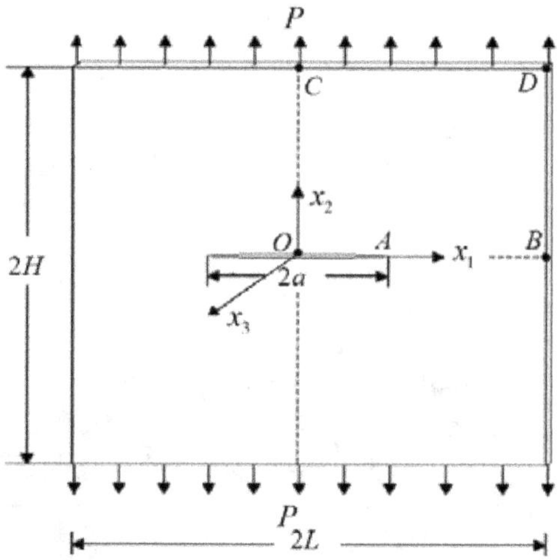

Fig. 12.3. Icosahedral quasicrystal plate with a crack under tension.

boundary conditions must be satisfied:

$$
\begin{aligned}
&a \leq x_1 \leq L, \quad x_2 = 0\text{:} \quad u_2 = 0, \quad w_2 = 0 \\
&x_1 = 0, \quad 0 \leq x_2 \leq H\text{:} \quad u_1 = 0, \quad w_1 = 0 \\
&x_3 = -0.5\,\text{mm:} \quad u_3 = 0, \quad w_3 = 0 \\
&0 \leq x_1 \leq L, \quad x_2 = H\text{:} \quad \sigma_{22} = P, \text{ other } \sigma_{ij}n_j = 0 \text{ and } H_{ij}n_j = 0 \\
&0 \leq x_1 \leq a, \quad x_2 = 0\text{:} \quad \sigma_{ij}n_j = 0 \text{ and } H_{ij}n_j = 0 \\
&x_1 = L, \quad 0 \leq x_2 \leq H\text{:} \quad \sigma_{ij}n_j = 0 \text{ and } H_{ij}n_j = 0 \\
&x_3 = 0 \text{ and } -1\text{mm:} \quad \sigma_{ij}n_j = 0 \text{ and } H_{ij}n_j = 0
\end{aligned}
$$

with the unit normal n_j at the boundaries. The domain $0 \leq x_1 \leq L$, $0 \leq x_2 \leq H$ is meshed by 20-node brick elements with a thickness of 1 mm along the x_3-axis. The discretization in the $x_3 = 0$ plane is shown in Fig. 12.4; there are 766 elements and 5775 nodes. A layer of elements with 0.1 mm height is placed at the bottom of the model in order to conveniently analyse the stress intensity factor of the crack. The zone around the crack is meshed by elements with 0.1 mm height and 0.1 mm width, and the width is 2% of the crack

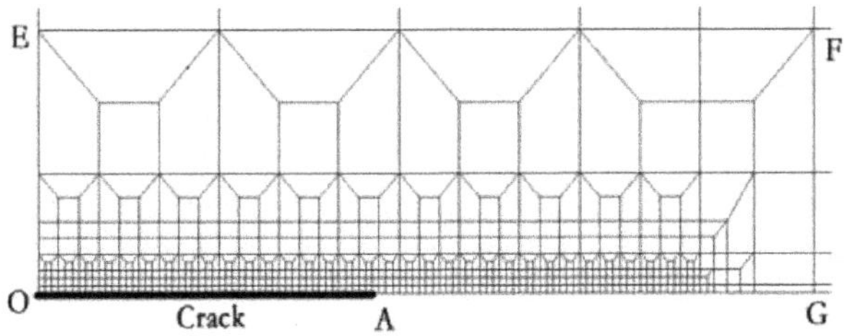

(a)

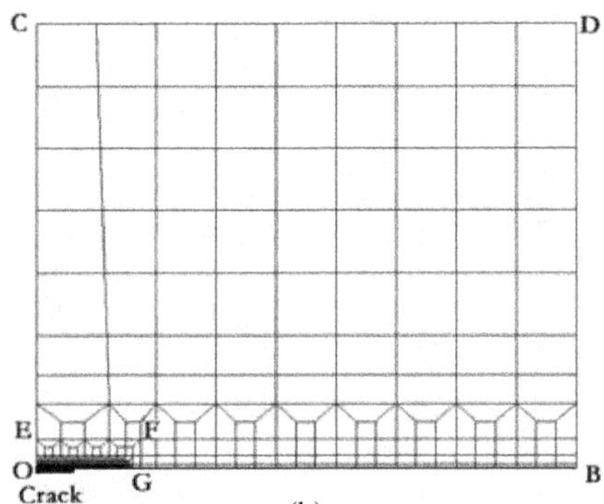

(b)

Fig. 12.4. Finite element discretization in the plane $x_3 = 0$.

length a. To model the displacements and stresses near the crack tip accurately, the elements related to the crack tip are degenerated in terms of singular quarter–point brick elements [9, 10].

For the plate, the thickness is very small compared to the height and width of the plate, the load P is parallel to the x_1–x_2 plane, and the differences of stresses in Gauss integral points of the finite element model along the x_3-direction are very small. Therefore, the stresses will be averaged along the x_3-axis.

The normal phonon stress ratio σ_{22}/P at the integration points of the element layer in the crack plane is given in Fig. 12.5, in which $r = \sqrt{(x_1 - a)^2 + x_2^2}$. The normal stress ratio at the crack tip when $R/\mu = 0.005$ is 35.667, while it is 34.187 when $R/\mu = 0$. Along with the increase in the ratio r/a, the stress ratios of the two cases are rapidly consistent with each other, finally approaching 1.0 satisfying the stress boundary condition shown according to Eq. (38).

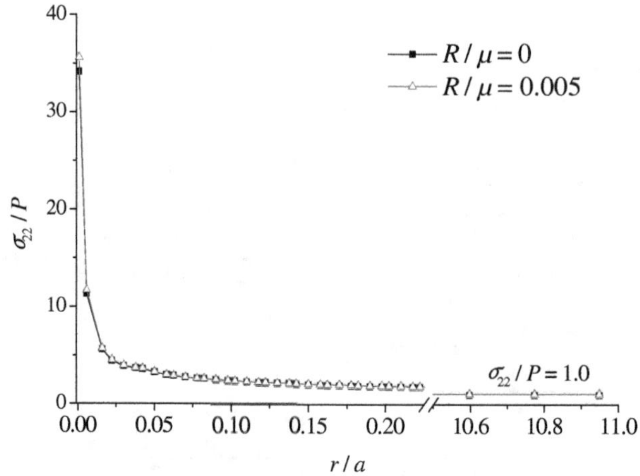

Fig. 12.5. Normal phonon stress ration σ_{22}/P versus r/a.

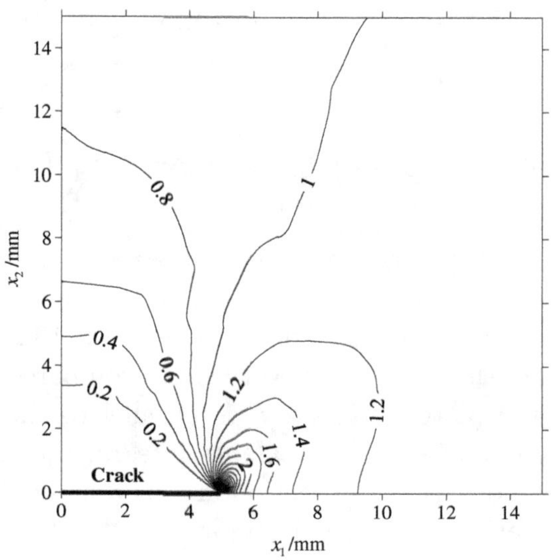

Fig. 12.6. Contours of normalized phonon stress σ_{22}/P near the crack tip for $R/\mu = 0.005$.

The contours of the normal stress ratio σ_{22}/P for $R/\mu = 0.005$ are shown in Fig. 12.6. It can be seen that there is a stress concentration at the crack tip line in classical elastic fracture mechanics. Far away from the crack, the stress ratio approaches 1.0, again demonstrating the consistency with the stress boundary condition according to the known analytic solutions.

To characterize crack tip loading, stress intensity factors are well-established quantities for linear fracture problems. From the well-known results, the phonon stress intensity factor for a Mode I crack, there is [8]

$$K^{\|} = \lim_{x_1 \to a^+} \left[\sqrt{2\pi(x_1 - a)} \sigma_{22}(x_1, 0) \right] = \sqrt{\pi a} P f(a/L, a/H)$$

with the assumption $R^2/(\mu K_1) \leq 1$, where $f(a/L, a/H)$ characterizes the geometry factor of a specimen with finite size. In other words, for different geometries of specimens, there will be different values of $f(a/L, a/H)$ which can be determined by the present numerical computation. Because the size of the specimen is quite greater than that of the crack, $f(a/L, a/H) \approx 1$ here.

An extrapolation method based on the stresses σ_{22} at element Gauss integration points is implemented to calculate the stress intensity factors from numerical solutions. The stresses at the Gaussian points of elements in the front of the crack tip and adjacent to the crack plane are almost identical to the stresses at $x_2 = 0$. With the position x_{1s} of the corresponding normal stress $(\sigma_{22})_s$ and $r_s = x_{1s} - a$, $K_s^{\|}$ reads

$$K_s^{\|} = \sqrt{2\pi r_s} (\sigma_{22})_s$$

where the subscript s denotes the number of Gauss point. We choose Q Gauss points, i.e., $s = 1, \ldots, Q$. Figure 12.7 gives the relations of $K_s^{\|}$ and r_s for the two cases $R/\mu = 0$ and $R/\mu = 0.005$. It shows that the stress intensity factors $K_s^{\|}$ of $R/\mu = 0.005$ are almost identical to those of $R/\mu = 0$. Near the crack tip, results are inaccurate due to the singularity approaching a straight line as r_s increases. Therefore, we can apply the least-squares method to the values of the straight lines. The stress intensity factor $K^{\|}$ finally is obtained as the intersection of the straight line with the ordinate which can be calculated as follows:

$$K^{\|} \approx \frac{\sum r_s \sum r_s K_s^{\|} - \sum r_s^2 \sum K_s^{\|}}{(\sum r_s)^2 - Q \sum r_s^2}$$

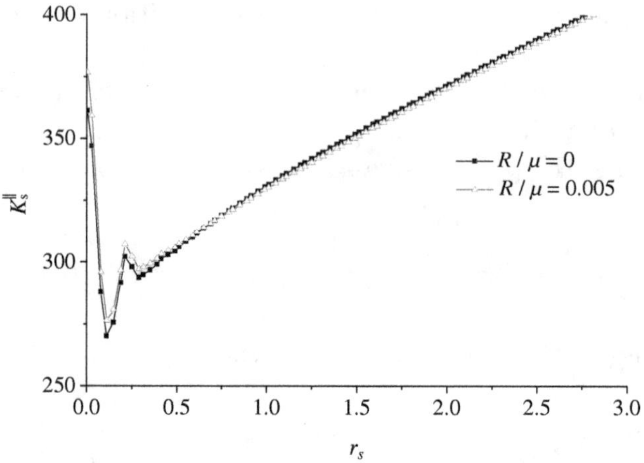

Fig. 12.7. Stress intensity factors $K_s^{\parallel}$ from different Gauss points versus $r_s = x_{1s} - a$.

The Gauss points in the domain $0.5 < r_s < 3$ are chosen to determine the stress intensity factors $K^{\parallel}$ and further $Q = 76$. The intensity factors for the two cases are obtained as 290.61 and 290.21 $\text{MPa}\sqrt{mm}$, respectively, for $R/\mu = 0$ and $R/\mu = 0.005$. Substituting $a = 5\,\text{mm}$ and P such that $P/\mu = 0.001$ into the last term of Eq. (39), the stress intensity factor is $286.95f$ $\text{MPa}\sqrt{mm}$. The numerical solutions for geometry factors f for $R/\mu = 0$ and $R/\mu = 0.005$ respectively are 1.013 and 1.011.

Figure 12.8 gives the opening displacement u_2 of the crack OA in the plane $x_3 = 0$. It shows that the crack opening displacement for $R/\mu = 0.005$ is larger than that for $R/\mu = 0$. At $x_1 = 0$, the opening displacement for $R/\mu = 0.005$ is $0.00441\,\text{mm}$, while for $R/\mu = 0$, $u_2 = 0.00402\,\text{mm}$ is obtained. It can be seen that the crack opening is obviously influenced by the coupling between the phonon and phason fields.

Figures 12.5, 12.7 and 12.8 show the coupling effect between the phonon and phason fields; it has a strong influence on the displacement fields, however, very weak influence on the stress fields. This phenomenon is due to the way of loading our problem.

The finite element analysis and computation are given by Yang *et al.* [11].

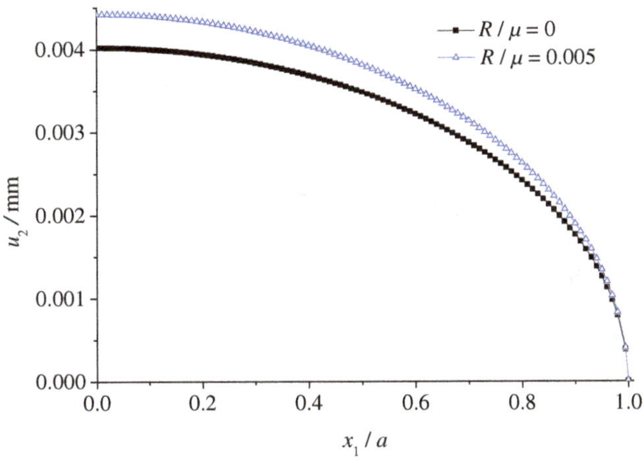

Fig. 12.8. The opening displacement u_2 of the crack at OA.

12.5 Conclusion

The variational principle is set up for three-dimensional quasicrystals, in fact, it is suitable for all quasicrystals if the constitutive law is given a slight modification. The example of cracked specimen given in Section 12.4 discussed a finite size body; the solution is very close to analytic solutions given by Fourier transform and complex analysis because the crack size is much smaller compared to the sample size; in this sense, the power of numerical method has not been fully performed; we need to do more computations with practical meanings. The advantages of the finite element method lie in exploring the effects of complex geometry, complex boundary conditions and complex material structures.

References

[1] Hu H C, 1981, *Variational Principles of Elasticity*, Beijing, Science Press (in Chinese).

[2] Guo L H and Fan T Y, 2007, Solvability of boundary value problems of elasticity theory of three-dimensional quasicrystals, *Appl. Math. Mech.*, **28**(8), 1061–1070.

[3] Newman M E, Henley C L, 1995, Phason elasticity of a three-dimensional quasicrystal: A transfer-matrix method. *Phys. Rev. B*, **52**: 6386–6399.

[4] Capitan M J, Calvayrac Y, Quivy A *et al.*, 1999, X-ray diffuse scattering from icosahedral Al–Pd–Mn quasicrystals. *Phys. Rev. B*, **60**: 6398–6404.

[5] Zhu W J and Henley C L, 1999, Phonon-phason coupling in icosahedral quasicrystals. *Euro. phys. Lett.*, **46**, 748–754.

[6] Zhu A Y and Fan T Y, 2009, Elastic analysis of a Griffith crack in icosahedral Al–Pd–Mn quasicrystal, *Int. J. Mod. Phys. B*, **23**(10), 1–16.

[7] Fan T Y, 2003, *Foundation of Fracture Theory*, Science Press, Beijing (in Chinese).

[8] Li L H and Fan T Y, 2008, Complex variable method for plane elasticity of icosahedral quasicrystals and elliptic notch problem, *Sci. China Ser. G*, **51**(10), 773–780.

[9] Henshell R D and Shaw K G, 1975, Crack tip finite elements are unnecessary, *Int. J. Num. Meth. Eng.*, **9**, 495–507.

[10] Barsoum R S, 1976, On the Use of isoparametric finite elements in linear fracture mechanics, *Int. J. Num. Meth. Eng.*, **10**, 25–37.

[11] Yang L Z, Ricoeur A, He F M and Gao Y, 2014, A finite element algorithm for static problems of icisahedral quasicrystals, *Chin. Phys. B*, **23**(5), 056102.

Chapter 13

Some Mathematical Principles on Solutions of Elasticity of Solid Quasicrystals

Starting from Chapter 4, we studied several mathematical models of the elasticity and generalized dynamics of quasicrystals and gave different kinds of solutions [1]. A further discussion on some common features of solutions of mathematical physics [2] is offered in this chapter.

13.1 Uniqueness of Solution of Elasticity of Quasicrystals

Assume a quasicrystal occupied by the region Ω is in equilibrium under the boundary conditions

$$\left.\begin{array}{ll} \sigma_{ij}n_j = T_i, H_{ij}n_j = h_i, & x_i \in S_t \\ u_i = \bar{u}_i, w_i = \bar{w}_i, & x_i \in S_u \end{array}\right\} \tag{13.1-1}$$

i.e., if the equations

$$\frac{\partial \sigma_{ij}}{\partial x_j} + f_i = 0, \quad \frac{\partial H_{ij}}{\partial x_j} + g_i = 0 \quad x_i \in \Omega \tag{13.1-2}$$

coupled by

$$\begin{aligned} \sigma_{ij} &= C_{ijkl}\varepsilon_{kl} + R_{ijkl}w_{kl} \\ H_{ij} &= R_{klij}\varepsilon_{kl} + K_{ijkl}w_{kl} \end{aligned} \tag{13.1-3}$$

and

$$\varepsilon_{ij} = \frac{1}{2}\left(\frac{\partial u_i}{\partial x_j} + \frac{\partial u_j}{\partial x_i}\right), \quad w_{ij} = \frac{\partial w_i}{\partial x_j} \tag{13.1-4}$$

satisfy the boundary conditions (13.1-1), then the boundary value problem (13.1-1)–(13.1-4) has a unique solution.

Proof. If the conclusion is not true, suppose there are two solutions of equations (13.1-2)–(13.1-4) under boundary conditions (13.1-1):

$$u_i^{(1)} \oplus w_i^{(1)}, u_i^{(2)} \oplus w_i^{(2)}$$
$$\varepsilon_{ij}^{(1)} \oplus w_{ij}^{(1)}, \varepsilon_{ij}^{(2)} \oplus w_{ij}^{(2)}$$
$$\sigma_{ij}^{(1)} \oplus H_{ij}^{(1)}, \sigma_{ij}^{(2)} \oplus H_{ij}^{(2)}$$

both solutions satisfy the equations of deformation geometry, stress–strain relationship, equilibrium equations and boundary conditions.

According to the linear superposition principle, the difference between these two solutions

$$(u_i^{(1)} - u_i^{(2)}) \oplus (w_i^{(1)} - w_i^{(2)}) = \Delta u_i \oplus \Delta w_i$$
$$(\sigma_{ij}^{(1)} - \sigma_{ij}^{(2)}) \oplus (H_{ij}^{(1)} - H_{ij}^{(2)}) = \Delta\sigma_{ij} \oplus \Delta H_{ij}$$

should be the solution of the problem too. The "differences" should satisfy zero boundary conditions. The work of external forces calculated by them is

$$\begin{aligned} 0 = &\int_\Omega (\Delta\mathbf{f} \oplus \Delta\mathbf{g}) \cdot (\Delta\mathbf{u} \oplus \Delta\mathbf{w}) d\Omega \\ &+ \int_\Gamma (\Delta\mathbf{T} \oplus \Delta\mathbf{h}) \cdot (\Delta\mathbf{u} \oplus \Delta\mathbf{w}) d\Gamma = 2\int_\Omega \Delta U d\Omega \end{aligned} \tag{13.1-5}$$

where ΔU represents the strain energy density corresponding to the differences between the two sets of solutions and the derivation of the last step of (13.1-5) used the Gauss theorem.

Because ΔU is the positive definite quadratic form of $\Delta\varepsilon_{ij}$ and Δw_{ij}, there is

$$\Delta U \geq 0 \tag{13.1-6}$$

Considering the left-hand side of (13.1-5) be zero, it flows that

$$\Delta U = 0 \tag{13.1-7}$$

Based on the positive definite property of ΔU, $\Delta\varepsilon_{ij}$ and Δw_{ij} must be zero, apart from only rigid displacements, so that $\varepsilon_{ij}^{(1)} = \varepsilon_{ij}^{(2)}$, $w_{ij}^{(1)} = w_{ij}^{(2)}$, etc.

At the same time, at boundary Γ_u, one has

$$\Delta\overline{u} = 0, \Delta\overline{w} = 0$$

This means the rigid displacements cannot exist; it must be

$$u_i^{(1)} = u_i^{(2)}, w_i^{(1)} = w_i^{(2)}$$

The uniqueness theorem is a powerful tool in the study of the elasticity of quasicrystals, e.g., if the solution satisfies all equations of elasticity and all boundary conditions, we confirm that the solution must be unique. All solutions offered in Chapters 7–9 and analytic solutions given in Chapter 10 exhibited this character. But the solution of the Saint-Venant problem in Section 6.9 is not unique.

13.2 Generalized Lax–Milgram Theorem

Consider a quasicrystal described in Eqs. (13.1-2)–(13.1-4) and boundary conditions (13.1-1). In this case, the displacements satisfy the following partial differential equations:

$$\left.\begin{aligned} -\left[C_{ijkl}\frac{\partial^2 u_k}{\partial x_j \partial x_l} + R_{ijkl}\frac{\partial^2 w_k}{\partial x_j \partial x_l}\right] &= f_i(x,y,z) \\ -\left[R_{klij}\frac{\partial^2 u_k}{\partial x_j \partial x_l} + K_{ijkl}\frac{\partial^2 w_k}{\partial x_j \partial x_l}\right] &= g_i(x,y,z) \end{aligned}\right\} \quad (x,y,z)\in\Omega \tag{13.2-1}$$

At boundary Γ_u, they satisfy homogenous conditions

$$u_i\,|_{\Gamma_u} = 0, w_i\,|_{\Gamma_u} = 0 \tag{13.2-2}$$

And at boundary Γ_t, the corresponding stresses satisfy non-homogenous conditions

$$\left.\begin{aligned} \left(C_{ijkl}\frac{\partial u_k}{\partial x_l} + R_{ijkl}\frac{\partial w_k}{\partial x_l}\right) n_j\,|_{\Gamma_t} &= T_i(x,y,z) \\ \left(R_{klij}\frac{\partial u_k}{\partial x_l} + C_{ijkl}\frac{\partial w_k}{\partial x_l} +\right) n_j\,|_{\Gamma_t} &= h_i(x,y,z) \end{aligned}\right\} \tag{13.2-3}$$

in which $x_1 = x$, $x_2 = y$, $x_3 = z$, $\Gamma = \Gamma_u + \Gamma_t$.

If $u_i, w_i \in C^2(\Omega)\cap C^1(\partial\Omega)$ and satisfy Eq. (13.2-1) and boundary conditions (13.2-2) and (13.2-3), then they will be called the classical solution of boundary value problem (13.2-1)–(13.2-4), where $\partial\Omega = \Gamma = \Gamma_u + \Gamma_t$.

The boundary value problem (13.2-1)–(13.2-3) can be reduced to the corresponding variational problem. For this purpose, we introduce the norm

$$\left.\begin{aligned} \|u_i\|_{1,\Omega}^2 &= \int_\Omega \left[\left(\frac{\partial u_i}{\partial x}\right)^2 + \left(\frac{\partial u_i}{\partial y}\right)^2 + \left(\frac{\partial u_i}{\partial z}\right)^2\right] d\Omega \\ \|w_i\|_{1,\Omega}^2 &= \int_\Omega \left[\left(\frac{\partial w_i}{\partial x}\right)^2 + \left(\frac{\partial w_i}{\partial y}\right)^2 + \left(\frac{\partial w_i}{\partial z}\right)^2\right] d\Omega \end{aligned}\right\} \tag{13.2-4}$$

This norm is suitable only for the case of homogeneous boundary conditions (13.2-2), otherwise the norm will not be the present form.

After defining the norm (13.2-4), the space defining $u_i(x, y, z)$ and $w_i(x, y, z)$ is denoted as $H'(\Omega)$. If we introduce an inner product

$$\left.\begin{aligned} (u_i^{(1)}, u_i^{(2)}) &= \int_\Omega \left(\frac{\partial u_i^{(1)}}{\partial x}\frac{\partial u_i^{(2)}}{\partial x} + \frac{\partial u_i^{(1)}}{\partial y}\frac{\partial u_i^{(2)}}{\partial y} + \frac{\partial u_i^{(1)}}{\partial z}\frac{\partial u_i^{(2)}}{\partial z}\right) d\Omega \\ (w_i^{(1)}, w_i^{(2)}) &= \int_\Omega \left(\frac{\partial w_i^{(1)}}{\partial x}\frac{\partial w_i^{(2)}}{\partial x} + \frac{\partial w_i^{(1)}}{\partial y}\frac{\partial w_i^{(2)}}{\partial y} + \frac{\partial w_i^{(1)}}{\partial z}\frac{\partial w_i^{(2)}}{\partial z}\right) d\Omega \end{aligned}\right\} \tag{13.2-5}$$

then $H'(\Omega)$ is a Hilbert space, and it is also named as Sobolev space.

We can define space V such as

$$V = \{(u_i, w_i) \in H'(\Omega), \quad (u_i)_{\Gamma_u} = 0, \quad (w_i)_{\Gamma_u} = 0\} \tag{13.2-6}$$

If defining the inner product at space V by (13.2-5), then V is also a Hilbert space, and

$$V \subset H'(\Omega) \tag{13.2-7}$$

Define space

$$\begin{aligned} H = \{X = (u_1, u_2, u_3, w_1, w_2, w_3) \in (H'(\Omega))_1^6, \\ (u_i)_{\Gamma_u} = 0, \quad (w_i)_{\Gamma_u} = 0\} \end{aligned} \tag{13.2-8}$$

and set up a norm

$$\|X\|_{1,\Omega} = \left[\sum_{i=1}^{3}\left[\|u_i\|_{1,\Omega}^2 + \|w_i\|_{1,\Omega}^2\right)\right]^{1/2} \tag{13.2-9}$$

For any $X = (u_1, u_2, u_3, w_1, w_2, w_3) \in (H'(\Omega))_1^6$, according to the strain–displacement relations

$$\varepsilon_{ij}(X) = \varepsilon_{ji}(X) = \frac{1}{2}\left(\frac{\partial u_i}{\partial x_j} + \frac{\partial u_j}{\partial x_i}\right)$$

$$w_{ij}(X) = \frac{\partial w_i}{\partial x_j}$$

and the stress–strain relations of quasicrystals

$$\sigma_{ij}(X) = \sigma_{ji}(X) = C_{ijkl}\varepsilon_{kl}(X) + R_{ijkl}w_{kl}(X)$$

$$H_{ij} = R_{klij}\varepsilon_{kl}(X) + K_{ijkl}w_{kl}(X)$$

define bilinear functional

$$B(X^{(1)}, X^{(2)}) = \int_\Omega [\sigma_{ij}(X^{(1)})\varepsilon_{ij}(X^{(2)}) + H_{ij}(X^{(1)})w_{ij}(X^{(2)})]d\Omega \tag{13.2-10}$$

in which $B(X^{(1)}, X^{(2)})$ presents symmetry, continuity and positive definite properties.

Take a linear functional

$$l(X) = \int_\Omega [f_i u_i + g_i w_i]d\Omega + \int_{\Gamma_t} [T_i u_i + h_i w_i]dS \tag{13.2-11}$$

where $(f_1, f_2, f_3, g_1, g_2, g_3) \in (L^2(\Omega))^6$, $(T_1, T_2, T_3, h_1, h_2, h_3) \in (L^2(\Omega))^6$ are the given functions at Ω and S, respectively. One can prove that $l(X)$ is a continuous functional of X at the region Ω, and L^2 represents the Lebesgue square integrable meaning.

There are the following theorems.

Theorem 1. *Variational problem corresponding to the boundary value problem (13.2-1)–(13.2-3) is concluded for obtaining $X \in H$ so that X*

$$B(X, Y) = l(Y) \quad \forall Y \in H \tag{13.2-12}$$

in which $l(Y)$ is a linear functional defined by (13.2-11).

(Proof is omitted)

Theorem 2. *If $X = (u_1, u_2, u_3, w_1, w_2, w_3) \in (C^2(\Omega))^6$ is a classical solution of the boundary value problem (13.2-1)–(13.2-3), then X must be the solution of the variational problem (13.2-12), i.e., X is also a generalized solution (or weak solution). Alternatively, if X is a solution of the variational problem (13.2-12), and $X \in (C^2(\Omega))^6$, then it is also a classical solution of the boundary value problem (13.2-1)–(13.2-3).*

(Proof is omitted)

Theorem 3 (Generalized Lax–Milgram theorem). *Assume that H is the Hilbert space defined above; for elasticity of quasi-crystals, $B(X, Y)$ is a bilinear functional at $H \times H$ and satisfies*

$$B(X, Y) = f(X) \quad \forall Y \in H \tag{13.2-13}$$

admits a unique solution $X \in H$ and

$$\|X\| \leq \frac{1}{\alpha}\|f\|_{(H)'} \tag{13.2-14}$$

where $(H)'$ is the dual space of H consisting of all bounded linear functionals and equipped with the norm

$$\|f\|_{(H)'} = \sup_{Y \in H, Y \neq 0} \frac{f(Y)}{\|Y\|}$$

This theorem gives an extension of the Lax–Milgram theorem [3, 4].

(Proof is omitted)

Theorem 4. *Assume*

$$J(X) = \frac{1}{2}B(X, X) - l(X) \tag{13.2-15}$$

and H is the same as mentioned previously, $B(X, Y)$ is a bilinear functional at $H \times H$ defined in (13.2-10), then the variational problem for having $X \in H$ so that $J(X)$ being minimum

$$J(X_0) = \min_{X \in H} J(X) \tag{13.2-16}$$

in which, we have the following:

(1) *There exists solution, and the number of solutions does not exceed one;*

(2) *If there is a solution to problem (13.2-16), which must be the solution of problem (13.2-12) and vice versa, ((13.2-12) is called as the Galerkin variational problem, and (13.2-16) is named as the Ritz variational problem);*

(3) If X^* is their solution, then

$$J(X) - J(X^*) = \frac{1}{2}B(X - X^*, X - X^*) \quad \forall X \in H \quad (13.2\text{-}17)$$

(Proof is omitted)

The discussion of this section may provide preparedness for the following text.

13.3 Matrix Expression of Elasticity of Three-dimensional Quasicrystals

In the previous section, the variational problem on plane elasticity of two-dimensional quasicrystals is discussed in which the concept of weak solution (generalized solution) is mentioned. In the following sections, we will deal with the weak solution of elasticity of three-dimensional icosahedral quasicrystals; for simplicity, a matrix expression for the elasticity will be used. From the discussion of previous chapters, we know that the basic equations for elasticity of quasicrystals except cubic quasicrystals

$$\begin{cases} \dfrac{\partial \sigma_{ij}}{\partial x_j} + f_i = \rho \dfrac{\partial^2 u_i}{\partial^2 t}, \quad \dfrac{\partial H_{ij}}{\partial x_j} + g_i = \rho \dfrac{\partial^2 w_i}{\partial^2 t} \left(\text{or } \kappa \dfrac{\partial w_i}{\partial t} \right), & (13.3\text{-}1) \\ \varepsilon_{ij} = \dfrac{1}{2} \left(\dfrac{\partial u_i}{\partial x_j} + \dfrac{\partial u_j}{\partial x_i} \right), \quad w_{ij} = \dfrac{\partial w_i}{\partial x_j}, & (13.3\text{-}2) \\ \qquad x_i \in \Omega, \ i, \quad j = 1,2,3 \ t > 0 & \\ \sigma_{ij} = C_{ijkl}\varepsilon_{kl} + R_{ijkl}w_{kl}, \quad H_{ij} = K_{ijkl}w_{kl} + R_{klij}\varepsilon_{kl}, & (13.3\text{-}3) \end{cases}$$

where $u = (u_x, u_y, u_z)$ represents the phonon displacement vector, $w = (w_x, w_y, w_z)$ the phason one, ε_{ij}, w_{ij} the phonon and phason strains, σ_{ij}, H_{ij} the corresponding phonon and phason stresses,

$C_{ijkl}, K_{ijkl}, R_{ijkl}$ the elastic constants, f_i, g_i the body and generalized body force densities, ρ the mass density and $\kappa = 1/\Gamma_w$, respectively.

Denoting $\partial\Omega = (\partial\Omega)_u + (\partial\Omega)_\sigma$ to express the boundary of the region occupied by the quasicrystal, there are the boundary conditions

$$x \in (\partial\Omega)_u : u_i = u_i^0, \quad w_i = w_i^0 \tag{13.3-4}$$

$$x \in (\partial\Omega)_\sigma : \sigma_{ij}n_j = T_i, \quad H_{ij}n_j = h_i \tag{13.3-5}$$

where u_i^0 and w_i^0 represent the known functions at boundary $(\partial\Omega)_u, T_i$ and h_i the prescribed traction and generalized traction density at boundary $(\partial\Omega)_\sigma$ and n_j the unit outward normal vector at any point of the boundary.

There are initial conditions for dynamic problem:

$$\begin{aligned} & u_i \,|_{t=0} = a_i\,(x), \quad w_i\,|_{t=0} = b_i(x), \quad \dot{u}_i\,|_{t=0} = c_i\,(x), \\ & \quad \dot{w}_i\,|_{t=0} = d_i\,(x) \\ & (\text{or } u_i\,|_{t=0} = a_i\,(x), \quad w_i\,|_{t=0} = b_i(x), \quad \dot{u}_i\,|_{t=0} = c_i\,(x)) \end{aligned} \tag{13.3-6}$$

where a_i, b_i, c_i, d_i are known functions, $x = (x_1, x_2, x_3) \in \Omega$.

Putting

$$\tilde{U}^T = (u_1\, u_2\, u_3\, w_1\, w_2\, w_3)_{1\times 6},$$

$$\tilde{F}^T = (f_i\; g_i)_{1\times 6} = (f_1\, f_2\, f_3\, g_1\; g_2\, g_3)_{1\times 6}$$

$$\begin{aligned} \tilde{\sigma}^T = (&\sigma_{11}\,\sigma_{22}\,\sigma_{33}\,\sigma_{12}\,\sigma_{23}\,\sigma_{31} \\ &H_{11}\,H_{22}\,H_{33}\,H_{12}\,H_{23}\,H_{31}\,H_{13}\,H_{21}\,H_{32})_{1\times 15} \end{aligned}$$

$$\begin{aligned} \tilde{\varepsilon}^T = (&\varepsilon_{11}\,\varepsilon_{22}\,\varepsilon_{33}\;\; 2\varepsilon_{12}\;\, 2\varepsilon_{23}\;\, 2\varepsilon_{31} \\ &w_{11}\,w_{22}\,w_{33}\,w_{12}\,w_{23}\,w_{31}\,w_{13}\,w_{21}\,w_{32})_{1\times 15} \end{aligned}$$

$$\tilde{\partial} = \begin{bmatrix} \tilde{\partial}^{(1)} & 0 \\ 0 & \tilde{\partial}^{(2)} \end{bmatrix}, \quad \tilde{\partial}^{(1)} = \begin{bmatrix} \partial_1 & 0 & 0 \\ 0 & \partial_2 & 0 \\ 0 & 0 & \partial_3 \\ \partial_2 & \partial_1 & 0 \\ 0 & \partial_3 & \partial_2 \\ \partial_3 & 0 & \partial_1 \end{bmatrix},$$

$$\tilde{\partial}^{(2)} = \begin{bmatrix} \partial_1 & 0 & 0 \\ 0 & \partial_2 & 0 \\ 0 & 0 & \partial_3 \\ \partial_2 & 0 & 0 \\ 0 & \partial_3 & 0 \\ 0 & 0 & \partial_1 \\ \partial_3 & 0 & 0 \\ 0 & \partial_1 & 0 \\ 0 & 0 & \partial_2 \end{bmatrix}, \quad \left(\partial_i = \frac{\partial}{\partial x_i}\right)$$

$\tilde{U}^T, \tilde{F}^T, \partial^T, \tilde{\sigma}^T, \tilde{\varepsilon}^T$ denote the transpose of $\tilde{U}, \tilde{F}, \partial, \tilde{\sigma}, \tilde{\varepsilon}$,$\sigma^i = (\sigma_{i1}, \sigma_{i2}, \sigma_{i3})$ the i-th row of matrix $(\sigma_{ij})_{3\times3}$ and $H^i = (H_{i1}, H_{i2}, H_{i3})$ the i-th row of matrix $(H_{ij})_{3\times3}$.

So that (13.3-2) can be rewritten as the matrix form:

$$\tilde{\varepsilon} = \partial\tilde{U} \tag{13.3-2$'$}$$

Note that

$$\left(\frac{\partial\sigma_{1j}}{\partial x_j} \quad \frac{\partial\sigma_{2j}}{\partial x_j} \quad \frac{\partial\sigma_{3j}}{\partial x_j} \quad \frac{\partial H_{1j}}{\partial x_j} \quad \frac{\partial H_{2j}}{\partial x_j} \quad \frac{\partial H_{3j}}{\partial x_j}\right)^T = \partial^T\tilde{\sigma},$$

then (13.3-1) can be rewritten by

$$\partial^T\tilde{\sigma} + F = \rho\ddot{\tilde{U}} \tag{13.3-1$'$}$$

Putting

$$C = \begin{bmatrix} & C_{11ij} & \\ & C_{22ij} & \\ & C_{33ij} & \\ \cdots & C_{12ij} & \cdots \\ & C_{23ij} & \\ & C_{31ij} & \end{bmatrix}_{6\times6}$$

$$= \begin{bmatrix} C_{1111} & C_{1122} & C_{1133} & C_{1112} & C_{1123} & C_{1131} \\ C_{2211} & C_{2222} & C_{2233} & C_{2212} & C_{2223} & C_{2231} \\ C_{3311} & C_{3322} & C_{3333} & C_{3312} & C_{3323} & C_{3331} \\ C_{1211} & C_{1222} & C_{1233} & C_{1212} & C_{1223} & C_{1231} \\ C_{2311} & C_{2322} & C_{2333} & C_{2312} & C_{2323} & C_{2331} \\ C_{3111} & C_{3122} & C_{3133} & C_{3112} & C_{3123} & C_{3131} \end{bmatrix}_{6\times6}$$

$$K = \begin{bmatrix} & K_{11ij} & \\ & K_{22ij} & \\ & K_{33ij} & \\ \cdots & K_{12ij} & \cdots \\ & K_{23ij} & \\ & K_{31ij} & \\ & K_{13ij} & \\ & K_{21ij} & \\ & K_{32ij} & \end{bmatrix}$$

$$= \begin{bmatrix} K_{1111} & K_{1122} & K_{1133} & K_{1112} & K_{1123} & K_{1131} & K_{1113} & K_{1121} & K_{1132} \\ K_{2211} & K_{2222} & K_{2233} & K_{2212} & K_{2223} & K_{2231} & K_{2213} & K_{2221} & K_{2232} \\ K_{3311} & K_{3322} & K_{3333} & K_{3312} & K_{3323} & K_{3331} & K_{3313} & K_{3321} & K_{3332} \\ K_{1211} & K_{1222} & K_{1233} & K_{1212} & K_{1223} & K_{1231} & K_{1213} & K_{1221} & K_{1232} \\ K_{2311} & K_{2322} & K_{2333} & K_{2312} & K_{2323} & K_{2331} & K_{2313} & K_{2321} & K_{2332} \\ K_{3111} & K_{3122} & K_{3133} & K_{3112} & K_{3123} & K_{3131} & K_{3113} & K_{3121} & K_{3132} \\ K_{1311} & K_{1322} & K_{1333} & K_{1312} & K_{1323} & K_{1331} & K_{1313} & K_{1321} & K_{1332} \\ K_{2111} & K_{2122} & K_{2133} & K_{2112} & K_{2123} & K_{2131} & K_{2113} & K_{2121} & K_{2132} \\ K_{3211} & K_{3222} & K_{3233} & K_{3212} & K_{3223} & K_{3231} & K_{3213} & K_{3221} & K_{3232} \end{bmatrix}_{9\times 9}$$

$$R = \begin{bmatrix} & R_{11ij} & \\ & R_{22ij} & \\ & R_{33ij} & \\ \cdots & R_{12ij} & \cdots \\ & R_{23ij} & \\ & R_{31ij} & \end{bmatrix}_{6\times 9}$$

$$= \begin{bmatrix} R_{1111} & R_{1122} & R_{1133} & R_{1112} & R_{1123} & R_{1131} & R_{1113} & R_{1121} & R_{1132} \\ R_{2211} & R_{2222} & R_{2233} & R_{2212} & R_{2223} & R_{2231} & R_{2213} & R_{2221} & R_{2232} \\ R_{3311} & R_{3322} & R_{3333} & R_{3312} & R_{3323} & R_{3331} & R_{3313} & R_{3321} & R_{2232} \\ R_{1211} & R_{1222} & R_{1233} & R_{1212} & R_{1223} & R_{1231} & R_{1213} & R_{1221} & R_{1232} \\ R_{2311} & R_{2322} & R_{2333} & R_{2312} & R_{2323} & R_{2331} & R_{2313} & R_{2321} & R_{2332} \\ R_{3111} & R_{3122} & R_{3133} & R_{3112} & R_{3123} & R_{3131} & R_{3113} & R_{3121} & R_{3132} \end{bmatrix}_{6\times 9}$$

then

$$D = (d_{ij})_{15\times 15} = \begin{bmatrix} C & R \\ R^T & K \end{bmatrix}$$

Here, the order of index i, j of C is the same with those of the phonon strain tensor, the order of index i, j of K, R is the same with those of phason strain tensor and R^T is the transpose of R. From the above expressions, one can find that due to the symmetry of C and K (see, e.g., (4.4-3) and (4.4-5)), the matrix D is symmetry.

The generalized Hooke's law (13.3-3) can be rewritten as

$$\tilde{\sigma} = D\tilde{\varepsilon} \tag{13.3-3$'$}$$

and (13.3-1′) and (13.3-2′) can be collected as follows:

$$\partial^T D \partial \tilde{U} + \tilde{F} = \rho \ddot{\tilde{U}} \tag{13.3-7}$$

Put

$$A(x) = \begin{bmatrix} a_1(x) \\ a_2(x) \\ a_3(x) \\ b_1(x) \\ b_2(x) \\ b_3(x) \end{bmatrix}_{6\times 1}, \quad B(x) = \begin{bmatrix} c_1(x) \\ c_2(x) \\ c_3(x) \\ d_1(x) \\ d_2(x) \\ d_3(x) \end{bmatrix}_{6\times 1}$$

$$\tilde{U}^0 = \begin{bmatrix} u_1^0 \\ u_2^0 \\ u_3^0 \\ w_1^0 \\ w_2^0 \\ w_3^0 \end{bmatrix}_{6\times 1}, \quad \tilde{\sigma}^0 = \begin{bmatrix} T_1 \\ T_2 \\ T_3 \\ h_1 \\ h_2 \\ h_3 \end{bmatrix}_{6\times 1}, \quad \tilde{\partial}_n = \begin{bmatrix} \tilde{\partial}_n^{(1)} & 0 \\ 0 & \tilde{\partial}_n^{(2)} \end{bmatrix}$$

$$\tilde{\partial}_n^{(1)} = \begin{bmatrix} \cos(n,x_1) & 0 & 0 \\ 0 & \cos(n,x_2) & 0 \\ 0 & 0 & \cos(n,x_3) \\ \cos(n,x_2) & \cos(n,x_1) & 0 \\ 0 & \cos(n,x_3) & \cos(n,x_2) \\ \cos(n,x_3) & 0 & \cos(n,x_1) \end{bmatrix}$$

$$\tilde{\partial}_n^{(2)} = \begin{bmatrix} \cos(n,x_1) & 0 & 0 \\ 0 & \cos(n,x_2) & 0 \\ 0 & 0 & \cos(n,x_3) \\ \cos(n,x_2) & 0 & 0 \\ 0 & \cos(n,x_3) & 0 \\ 0 & 0 & \cos(n,x_1) \\ \cos(n,x_3) & 0 & 0 \\ 0 & \cos(n,x_1) & 0 \\ 0 & 0 & \cos(n,x_2) \end{bmatrix}$$

where $\tilde{\partial}_n, \tilde{\partial}_n^{(1)}, \tilde{\partial}_n^{(2)}$ are obtained from the differential operator matrixes $\tilde{\partial}, \tilde{\partial}^{(1)}, \tilde{\partial}^{(2)}$ through a replacement ∂_i by $\cos(n, x_i)$.

Equation (13.3-4) can be rewritten as

$$\tilde{U}(x,t) = \tilde{U}^0, \quad x \in (\partial\Omega)_u \tag{13.3-4$'$}$$

Considering the similarity of the left-hand side of (13.3-5) with the first term of (13.3-1), then (13.3-5) can be rewritten as $\tilde{\partial}_n^T \tilde{\sigma} = \tilde{\sigma}^0, x \in (\partial\Omega)_\sigma$. In addition, by using (13.3-2′) and (13.3-3′) there is

$$\tilde{\partial}_n^T D \tilde{\partial} \tilde{U} = \tilde{\sigma}^0, \quad x \in (\partial\Omega)_\sigma \tag{13.3-5$'$}$$

If the quasicrystal is in static equilibrium: $\rho \ddot{\tilde{U}} = 0$ (i.e., the inertia forces vanish) in this case, it need not be the initial conditions, and the boundary value problem of quasicrystals is interpreted as follows:

$$\begin{cases} -\tilde{\partial}^T D \tilde{\partial} \tilde{U} = \tilde{F}, \quad x \in \Omega, t > 0 & (13.3\text{-}8) \\ \tilde{U}(x,t)\Big|_{(\partial\Omega)_u} = \tilde{U}^0, \quad \tilde{\partial}_n^T D \tilde{\partial} \tilde{U}(x,t)\Big|_{(\partial\Omega)_\sigma} = \tilde{\sigma}^0 & (13.3\text{-}9) \end{cases}$$

where

$$\partial\Omega = (\partial\Omega)_u + (\partial\Omega)_\sigma.$$

13.4 The Weak Solution of Boundary Value Problem of Elasticity of Quasicrystals

For simplicity, only the displacement boundary problem (or say the Dirichlet problem) is dealt with in the following. Suppose that $\tilde{F} \in$

$(L^2(\Omega))^6$, if $\tilde{U}(x) \in (C^2(\tilde{\Omega}))^6$ is the solution of problems (13.3-8) and (13.3-9), then for any vector function

$$\eta = \begin{bmatrix} \eta^1_{3\times 1} \\ \eta^2_{3\times 1} \end{bmatrix} = \begin{bmatrix} \eta^1_1 & \eta^1_2 & \eta^1_3 & \eta^2_1 & \eta^2_2 & \eta^2_3 \end{bmatrix}^T_{1\times 6} \in (C_0^\infty(\Omega))^6,$$

both the sides of (13.3-8) are multiplied by η (by making a scalar product), and then integrating along Ω, we have

$$\iiint_\Omega (-\tilde{\partial}^T D\tilde{\partial}\tilde{U}) \cdot \eta dx = \iiint_\Omega \tilde{F} \cdot \eta dx \tag{13.4-1}$$

From (13.3-3) and Chapter 4 known $\sigma_{ij} = \sigma_{ji}$, in terms of Gauss formula, (13.3-2′ and (13.3-3′), there exists

$$\begin{aligned}
&\iiint_\Omega (-\tilde{\partial}^T D\tilde{\partial}\tilde{U}) \cdot \eta dx \\
&= -\iiint_\Omega \left(\frac{\partial \sigma_{ij}}{\partial x_j}\eta^1_i + \frac{\partial H_{ij}}{\partial x_j}\eta^2_i\right) dx \\
&= -\iiint_\Omega \left[\frac{\partial}{\partial x_j}(\sigma_{ij}\eta^1_i + H_{ij}\eta^2_i) - \left(\sigma_{ij}\frac{\partial \eta^1_i}{\partial x_j} + h_{ij}\frac{\partial \eta^2_i}{\partial x_j}\right)\right] dx \\
&= -\iint_{\partial\Omega} (\sigma_{ij}\eta^1_i + H_{ij}\eta^2_i) n_j dS + \iiint_\Omega \left(\sigma_{ij}\frac{\partial \eta^1_i}{\partial x_j} + H_{ij}\frac{\partial \eta^2_i}{\partial x_j}\right) dx \\
&= \iiint_\Omega \left(\frac{1}{2}\sigma_{ij}\frac{\partial \eta^1_i}{\partial x_j} + \frac{1}{2}\sigma_{ij}\frac{\partial \eta^1_j}{\partial x_i} + H_{ij}\frac{\partial \eta^2_i}{\partial x_j}\right) dx \\
&= \iiint_\Omega [(\sigma_{ij}(\tilde{U})\varepsilon_{ij}(\eta^1) + H_{ij}(\tilde{U})w_{ij}(\eta^2)] dx \\
&= \iiint_\Omega \tilde{\sigma}(\tilde{U}) \cdot \tilde{\varepsilon}(\eta) dx = \iiint\limits_\Omega (\tilde{\partial}\eta)^T D\tilde{\partial}\tilde{U} dx,
\end{aligned} \tag{13.4-2}$$

and (13.4-1) and (13.4-2) yield

$$\iiint_\Omega (\tilde{\partial}\eta)^T D\tilde{\partial}\tilde{U} dx = \iiint_\Omega \tilde{\sigma}(\tilde{U}) \cdot \tilde{\varepsilon}(\eta) dx = \iiint_\Omega \tilde{F} \cdot \eta dx \tag{13.4-3}$$

Because C_0^∞ is dense in $H_0^1(\Omega)$, (13.4-3) holds too $\forall \eta(x) \in (H_0^1(\Omega))^6$.

In contrast, if $\tilde{U}(x) \in (C^2(\bar{\Omega}))^6$, and 13.4-3) is valid $\forall \eta(x) \in (H_0^1(\Omega))^6$, we can do derivation in counter-order of the above procedure and find (13.4-1) through the fundamental lemma of the variational method [5]. So we have the following:

Definition. Assume $\tilde{F} \in (L^2(\Omega))^6$, if $\tilde{U}(x) \in (H_0^1(\Omega))^6$, and (13.4-3) holds for $\forall \eta(x) \in (H_0^1(\Omega))^6$, then say $\tilde{U}(x)$ is the weak solution (or generalized solution) of the boundary value problem

$$\begin{cases} -\tilde{\partial}^T D\tilde{\partial}\tilde{U}(x) = \tilde{F}(x), & (13.3\text{-}8') \\ \tilde{U}(x)\,|_{\partial\Omega} = 0 & (13.3\text{-}9') \end{cases}$$

13.5 The Uniqueness of Weak Solution

Making use of $(\cdot,\cdot)$ to express the inner product in $L^2(\Omega)$, the corresponding norm is $\|\cdot\| : \|\mathrm{v}\| = (\int_\Omega \mathrm{v}^2 dx)^{1/2}$, for the scalar function $\mathrm{v} \in L^2(\Omega)$. In terms of $(\cdot,\cdot)_1$ to denote the inner product in $H_0^1(\Omega)$, the corresponding norm is $\|\cdot\|_1 : \|\mathrm{v}\|_1 = [\int_\Omega \mathrm{v}^2 dx + \sum_{k=1}^3 \int_\Omega (\frac{\partial \mathrm{v}}{\partial x_k})^2 dx]^{1/2}$, and the semi-norm is $|\cdot|_1 : |\mathrm{v}|_1 = [\sum_{k=1}^3 \int_\Omega (\frac{\partial \mathrm{v}}{\partial x_k})^2 dx]^{1/2}$, for the scalar function $\mathrm{v} \in L^2(\Omega)$.

Note 1: Norm $\|\cdot\|_1$is equivalent to semi-norm$|\cdot|_1$.

Note 2: The norm and semi-norm of the vector function $\mathrm{v} = (\mathrm{v}_1, \mathrm{v}_2, \ldots, \mathrm{v}_n) \in (H_0^1(\Omega))^n$ (it is marked by $H_0^1(\Omega)$ some times) are denoted as $\|\cdot\|_1$ and $|\cdot|_1$ such as

$$\|\mathrm{v}\|_1^2 = \sum_{i=1}^n \|\mathrm{v}_i\|_1^2 = \sum_{i=1}^n \int_\Omega \mathrm{v}_i^2 dx + \sum_{i=1}^n \sum_{k=1}^3 \int_\Omega \left(\frac{\partial \mathrm{v}_i}{\partial x_k}\right)^2 dx$$

$$= \int_\Omega |\mathrm{v}|^2\, dx + \int_\Omega |\mathrm{v}_x|^2\, dx$$

$$|\mathrm{v}|_1^2 = \sum_{i=1}^n \sum_{k=1}^3 \int_\Omega \left(\frac{\partial \mathrm{v}_i}{\partial x_k}\right)^2 dx$$

where $|\mathrm{v}|^2 = \sum_{i=1}^n \mathrm{v}_i^2$, $|\mathrm{v}_x|^2 = \sum_{i=1}^n \sum_{k=1}^3 (\frac{\partial \mathrm{v}_i}{\partial x_k})^2$, $\frac{\partial \mathrm{v}}{\partial x_k} = (\frac{\partial \mathrm{v}_1}{\partial x_k}, \frac{\partial \mathrm{v}_2}{\partial x_k}, \ldots, \frac{\partial \mathrm{v}_n}{\partial x_k})$. Obviously, Note 1 holds for the vector function **v** too.

Lemma (Korn inequality [6–8]). *Assume Ω is a bounded region with the boundary $\partial\Omega$ of sufficient smooth in $\mathbf{R^n}$, and $\forall v = (v_1, v_2, \ldots, v_n) \in H_0^1(\Omega)$, there is*

$$\sum_{i,k=1}^{n} \int_\Omega \left(\frac{\partial v_i}{\partial x_k} + \frac{\partial v_k}{\partial x_i}\right)^2 dx \geq c_1 \|v\|_1^2$$

in which the positive constant c_1 is dependent only on Ω.

Theorem. *Suppose Ω is a bounded region in $\mathbf{R^3}$ and with sufficient smooth boundary $\partial\Omega$; if real symmetric matrix $D = (d_{ij})$ satisfies the inequality*

$$\lambda_1 \sum_{i=1}^{15} \xi_i^2 \leq \sum_{i,j=1}^{15} \xi_i d_{ij} \xi_j \leq \lambda_2 \sum_{i=1}^{15} \xi_i^2$$

where λ_1, λ_2 are positive constants, then for any $\tilde{F} \in (L^2(\Omega))^6$, displacement boundary value problems (13.3-8′) and (13.3-9′) exists unique weak solution (or generalized solution).

Proof. Put $\lfloor \tilde{U}, \eta \rfloor = \iiint_\Omega (\partial\eta)^T D \partial\tilde{U} dx$, then(13.4-3) can be rewritten as

$$\left\lfloor \tilde{U}, \eta \right\rfloor = (\tilde{F}, \eta), \quad \forall \eta \in (H_0^1(\Omega))^6 \tag{13.5-1}$$

At first, we prove $\lfloor \cdot, \cdot \rfloor$ is a new inner product at $(H_0^1(\Omega))^6$. For this purpose, it needs to prove: $\lfloor \tilde{U}, \tilde{U} \rfloor \geq 0$, and $\lfloor \tilde{U}, \tilde{U} \rfloor = 0 \Leftrightarrow \tilde{U} = 0$, $\forall \tilde{U} \in (H_0^1(\Omega))^6$.

In the following, we give only an outline of the proof; the details are omitted. In addition, $\lfloor \tilde{U}, \eta \rfloor$ in (13.5-1) is a positive definite bilinear functional at $H_0^1(\Omega)$; the proof can be done from the Lax–Milgram theorem (see Section 13.2).

Due to the assumption, matrix $D = (d_{ij})_{15\times15}$ being positive definite, matrix $D = (d_{ij})_{15\times15}$ and unit matrix I are in contract, i.e., there exists a reversible matrix $\mathbf{C}$ such that $D = \mathbf{C^T C}$ (note that here $\mathbf{C}$ is not the phonon elastic constant matrix).

Then,

$$\lfloor \tilde{U}, \tilde{U} \rfloor = \iiint_\Omega (\tilde{\partial}\tilde{U})^T D\tilde{\partial}\tilde{U}dx = \iiint_\Omega (\tilde{\partial}\tilde{U})^T (\mathbf{C}^T\mathbf{C})\tilde{\partial}\tilde{U}dx$$
$$= \iiint_\Omega (\mathbf{C}\tilde{\partial}\tilde{U})^T (\mathbf{C}\tilde{\partial}\tilde{U})dx \geq 0;$$
$$\lfloor \tilde{U}, \tilde{U} \rfloor = 0 \Leftrightarrow \iiint_\Omega (\mathbf{C}\tilde{\partial}\tilde{U})^T (\mathbf{C}\tilde{\partial}\tilde{U})dx = 0 \Leftrightarrow \mathbf{C}\tilde{\partial}\tilde{U} = 0$$

Because $\mathbf{C}$ is reversible, $\tilde{\partial}\tilde{U} = 0$, i.e.,

$$\frac{\partial u_i}{\partial x_i} = 0, \quad \frac{\partial u_i}{\partial x_j} + \frac{\partial u_j}{\partial x_i} = 0 \ (i \neq j), \quad \frac{\partial w_i}{\partial x_j} = 0, \quad i, j = 1, 2, 3$$

It follows $\frac{\partial w_i}{\partial x_j} = 0$ that w_i should be constants; besides, $\tilde{U}\mid_{\partial\Omega} = 0$, and $w_i = 0$ at the boundary. In a similar analysis, we find that $u_i = 0$ at the boundary. Thus, $\tilde{U} = 0$ at the boundary.

In this way, we have proved that $\lfloor \cdot, \cdot \rfloor$ is a new inner product at $(H_0^1(\Omega))^6$, and the corresponding norm is $\|\tilde{U}\|_{(1)} = \lfloor \tilde{U}, \tilde{U} \rfloor^{1/2}$.

Second, at $(H_0^1(\Omega))^6$, the new norm $\|\cdot\|_{(1)}$ is equivalent to the initial norm $\|\cdot\|_1$. We are going to give the proof about this.

From the Cauchy inequality, the assumption of the theorem and Note 1, there is

$$\|\tilde{U}\|_{(1)}^2 = \iiint_\Omega \sum_{i,j=1}^{15} (\tilde{\partial}\tilde{U})_i d_{ij} (\tilde{\partial}\tilde{U})_j dx \leq \lambda_2 \iiint_\Omega (\tilde{\partial}\tilde{U})^T (\tilde{\partial}\tilde{U})dx$$
$$= \lambda_2 \iiint_\Omega \left[\sum_{i=1}^{3} \left(\frac{\partial u_i}{\partial x_i}\right)^2 + \sum_{\substack{i,j=1 \\ i<j}}^{3} \left(\frac{\partial u_i}{\partial x_j} + \frac{\partial u_j}{\partial x_i}\right)^2 + \sum_{i,j=1}^{3} \left(\frac{\partial w_i}{\partial x_j}\right)^2 \right] dx$$
$$\leq \lambda_2 \iiint_\Omega \left\{ \sum_{i=1}^{3} \left(\frac{\partial u_i}{\partial x_i}\right)^2 + 2 \sum_{\substack{i,j=1 \\ i<j}}^{3} \left[\left(\frac{\partial u_i}{\partial x_j}\right)^2 + \left(\frac{\partial u_j}{\partial x_i}\right)^2 \right] \right.$$
$$\left. + \sum_{i,j=1}^{3} \left(\frac{\partial w_i}{\partial x_j}\right)^2 \right\} dx$$

$$\leq 2\lambda_2 \iiint_\Omega \sum_{i,j=1}^{3} \left[\left(\frac{\partial u_i}{\partial x_j}\right)^2 + \left(\frac{\partial w_i}{\partial x_j}\right)^2 \right] dx$$

$$= 2\lambda_2 \left|\tilde{U}\right|_1^2 \leq 2c \left\|\tilde{U}\right\|_1^2$$

In other hand, from the assumption of the theorem, the Korn inequality and Note 1, there is

$$\|\tilde{U}\|_{(1)}^2 = \iiint_\Omega \sum_{i,j=1}^{15} (\tilde{\partial}\tilde{U})_i d_{ij} (\tilde{\partial}\tilde{U})_j dx \geq \lambda_1 \iiint_\Omega (\tilde{\partial}\tilde{U})^T (\tilde{\partial}\tilde{U}) dx$$

$$= \lambda_1 \iiint_\Omega \left[\sum_{i=1}^{3} \left(\frac{\partial u_i}{\partial x_i}\right)^2 + \sum_{\substack{i,j=1 \\ i<j}}^{3} \left(\frac{\partial u_i}{\partial x_j} + \frac{\partial u_j}{\partial x_i}\right)^2 + \sum_{i,j=1}^{3} \left(\frac{\partial w_i}{\partial x_j}\right)^2 \right] dx$$

$$= \lambda_1 \iiint_\Omega \left[\frac{1}{4} \sum_{i,j=1}^{3} \left(\frac{\partial u_i}{\partial x_j} + \frac{\partial u_j}{\partial x_i}\right)^2 + \sum_{i,j=1}^{3} \left(\frac{\partial w_i}{\partial x_j}\right)^2 \right] dx$$

$$\geq \frac{1}{4}\lambda_1 c_2 \iiint_\Omega \left(\frac{\partial u_i}{\partial x_j}\right)^2 dx + \lambda_1 \sum_{i,j=1}^{3} \iiint_\Omega \left(\frac{\partial w_i}{\partial x_j}\right)^2 dx$$

$$\geq \min\left\{\frac{1}{4}\lambda_1 c_2, \lambda_1\right\} |\tilde{U}|_1^2 \geq c\left\|\tilde{U}\right\|_1^2$$

Consequently, we have proved the equivalency between the new norm $\|\cdot\|_{(1)}$and the initial norm $\|\cdot\|_1$.

Finally, for $\tilde{F} \in (L^2(\Omega))^6$, by using Schwarz inequality and the fact that the embedding $H_0^1(\Omega) \subset_\rightarrow L^2(\Omega)$ is a compact embedding, we have

$$\left| \iiint_\Omega \tilde{F} \cdot \eta dx \right| \leq \|\tilde{F}\| \cdot \|\eta\| \leq M \|\tilde{F}\| \cdot \|\eta\|_1$$

$$\leq M_1 \|\tilde{F}\| \cdot \|\eta\|_1 \quad \forall \eta \in (H_0^1(\Omega))^6$$

i.e., $\eta \mapsto \iiint_\Omega \tilde{F} \cdot \eta dx \quad (\forall \eta \in (H_0^1(\Omega))^6)$ is a unique continuous linear functional at $(H_0^1(\Omega))^6$).

Therefore, from Riesz theorem, there must be a unique $y_{\tilde{F}} \in (H_0^1(\Omega))^6$ so that

$$\iiint_\Omega \tilde{F} \cdot \eta dx = \lfloor y_{\tilde{F}}, \eta \rfloor \quad \forall \eta \in \left(H_0^1(\Omega)\right)^6$$

Thus, (13.5-1) is rewritten as

$$\left\lfloor \tilde{U}, \eta \right\rfloor = \lfloor y_{\tilde{F}}, \eta \rfloor \quad \forall \eta \in (H_0^1(\Omega))^6$$

This shows the $\tilde{U} = y_{\tilde{F}}$ is the unique weak solution (generalized solution) of the displacement value problems (13.3-8′) and (13.3-9′). □

In the above proof, using the Korn inequality is crucial (the second Korn inequality will be used for the stress boundary value problem, but this discussion is not included here).

13.6 Conclusion and Discussion

To some extent, the discussion Sections 13.3–13.5 are a development of Section 13.2. These discussions provide a basis for the numerical methods developed in Chapters 10 and 12 from the point of a view of a weak solution. Of course, for the stress boundary value problems and dynamic problems, the discussion needs to do some extensions of the previous discussions. Due to the limitation of space the additional extensions are not be given any more.

It is evident that the discussion in Sections 13.3–13.5 is valid for any systems of one-, two- and three-dimensional quasicrystals, except cubic quasicrystals, for which $w_{ij} = \partial w_i/\partial x_j$ should be replaced by $w_{ij} = (\partial w_i/\partial x_j + \partial w_j/\partial x_i)/2$ in the equations (13.3-2) only. The difference in the formulation for different quasicrystal systems lies in the elastic constant matrix D, one can directly use the above formulation for any quasicrystal systems only by substituting the concrete matrix D into the relevant formulas.

References

[1] Fan T Y and Mai Y W, 2004, Elasticity theory, fracture mechanics and some relevant thermal properties of quasicrystalline materials, *Appl. Mech. Rev.*, **57**(5), 325–344.

[2] Courant R and Hilbert D, 1955, *Methods of Mathematical Physics*, Interscience Publisher Inc., New York.

[3] Ying L A, 1998, *Theory of Finite Element*, Klower, Amsterdan.

[4] Jiang L S, 1984, *Finite Element Foundation*, High Education Press, Beijing, in Chinese.

[5] Guo L H and Fan T Y, 2007, Solvability on boundary-value problems of elasticity of three-dimensional quasicrystals, *Appl. Math. Mech.*, **28**(8), 1061–1070.

[6] Fikera G, 1974, *Existence Theorems of Elasticity Theory*, World Press, Moscow (in Russian).

[7] Kondrat'eb W A and Oleinik O A, 1988, *Korn Inequality of Boundary Value Problems for Systems of Theory of Elasticity in Boundless Region*, UMN Press, Moscow (in Russian).

[8] Kondrat'eb W A and Oleinik O A, 1988, Boundary-value problems for the systems of elasticity in unbounded domains, Korn's inequalities, *Russ. Math. Surv.*, **43**(5), 65–119.

Chapter 14

Nonlinear Behaviour of Quasicrystals

From Chapters 4–13, we mainly discussed the elasticity and phonon–phason dynamics of quasicrystals, which belong to the linear regime both physically and mathematically. Their mathematical treatment is relatively easy though the calculations are quite complex.

This chapter is to give a simple description of the deformation and fracture of quasicrystals with nonlinear behaviour, considering the great difficulty in this topic. For the conventional engineering materials including crystalline material, the nonlinear behaviour means mainly the plasticity. In the study on the classical plasticity, there are two different theories: one is the macroscopic plasticity theory, which is based on some assumptions concluded from certain experimental data, and the other is the so-called crystal plasticity theory, which is based on the mechanism of motion of dislocation, in some extent, can be seen as a "microscopic" theory. The difficulty for quasicrystal plasticity lies in the lack of both enough macro- and micro-data. At present, the macroscopic experiments have not, as yet, been properly undertaken. Though there is some work on the mechanism in microscopy of the plasticity, the data are very limited. This leads to the constitutive law of quasicrystals being essentially unknown. Due to this reason, the systematic mathematical analysis of the deformation and fracture of the material is not available so far.

In spite of these difficulties, the study on the plasticity of quasicrystals has aroused a great deal of attention among researchers [1–8]. But the analytic quantitative work may be at an infant stage.

Considering the readers' interest and the development level, it is beneficial to give a brief discussion on some simple problems of nonlinear behaviour of the material with some simple models and by extending results in the study of the linear regime. Of course, these discussions are not complete, which may provide some hints for further development of the area.

This chapter is structured as follows. First, we discuss some experimental results on the nonlinear deformation behaviour of quasicrystals and then describe a possible plastic constitutive relation of the material; in view of the difficulty of setting up the equations, we turn to introduce nonlinear elastic constitutive equations of quasicrystals which are available at present though not equivalent to the plastic constitutive equations. Sections 14.4 and 14.5 may be seen as applications of the macro-constitutive law in which some nonlinear solutions on quasicrystals are offered. In Section 14.6, another version of the study based on the dislocation model or "microscopic model" is exhibited; it achieves the same results as those given in Sections 14.4 and 14.5.

14.1 Macroscopic Behaviour of Plastic Deformation of Quasicrystals

At medium and low temperature, quasicrystals exhibit brittleness, but they present plasticity–ductility at high temperature. In addition, near the high-stress concentration zone, e.g., around the dislocation core or crack tip, plastic flow appears also. It was observed in experiments that plastic deformation of quasicrystals is induced by the motion of dislocations in the material. This reveals the important connection between plasticity and the structural defects in quasicrystals. To some extent, there are similarities between the phenomena of crystals and quasicrystals. But the latter presents salient structural features, fundamentally differing from those of conventional crystals. The plasticity of quasicrystals must be studied in a quite different way from that for studying the classical plasticity.

The fundamental step in studying the plasticity of quasicrystals is, of course, experimental observation.

As in the previous discussion, this chapter is mainly dealt with icosahedral and decagonal quasicrystals.

Recently, there are many experimental data for Al–Mn–Pd icosahedral quasicrystals, which are the most intensively studied and possess a very high brittle–ductile transition temperature. At a strain rate of 10^{-5}/s, the ductile range sets in at about 690°C corresponding to a homologous temperature (i.e., the absolute temperature scaled by the absolute melting temperature) of about 0.82. At lower strain rates of 10^{-6}/s and below, limited ductility can be observed down to temperatures of about 480°C. Figure 14.1 shows the stress–strain curves of the quasicrystal in the high-temperature range 730–800°C at a strain rate of 10^{-5}/s.

As another class of icosahedral quasicrystals, icosahedral Zn–Mg–Dy, the stress–strain curves measured by the experiments are similar to those given in Fig. 14.1.

Feuerbacher and Urban [9], Guyot and Canova [10], Feuerbacher *et al.* [11], etc. discussed the constitutive equation based on the experimental data on dislocation density and dislocation velocity, e.g., the plastic strain rate and the applied stress in a power-law form

$$\dot{\varepsilon}_p = B(\sigma/\hat{\sigma})^m \tag{14.1-1}$$

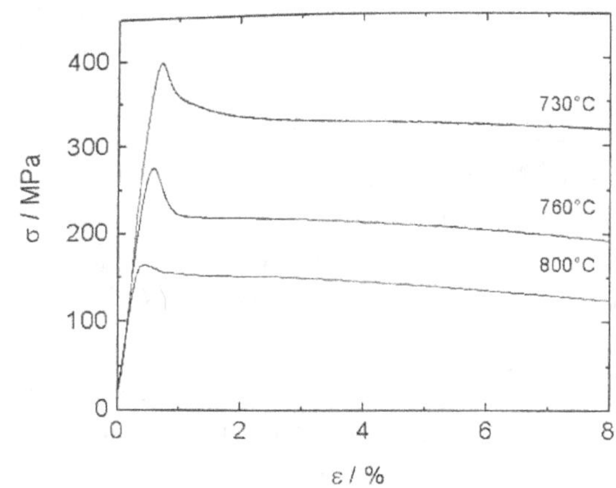

Fig. 14.1. Stress–strain curves of icosahedral Al–Mn–Pd icosahedral single quasicrystals at strain rate of 10^{-5}/s [9].

in which B and m are temperature-dependent parameters and $\hat{\sigma}$ is the internal variable that can be regarded as a reference stress representing the current microstructural state of the material and is used to accommodate the model to the description of different materials or hardening mechanism [12]. Combining relevant information, formula (14.1-1) can be used to well predict experimental curves, e.g., recorded by Fig. 14.1. It is needed to point out that, though there are some similar formulas to (14.1-1) in form in the classical plasticity, they are quite different substantively. For example, the current parameters B, m and $\hat{\sigma}$ are different from those appearing in relevant formulas in the classical plasticity. Those parameters in the classical plasticity were measured from a pure macroscopic approach rather than a dislocation model; some details about the latter can be found in Refs. [9–12], which, from the angle of methodology, are different from those adopted in the classical plasticity. Unfortunately, there is lack of comprehensive macroscopic experimental data (e.g., the data arising from multi-axial loading conditions) for quasicrystal plasticity so far.

Decagonal quasicrystals are an interesting topic for plastic deformation studies since the phase possesses quasiperiodic as well as periodic lattice directions. Therefore, the influences of periodicity and quasiperiodicity on the plastic deformation can be directly revealed on one material by investigating the plastic properties in different deformation geometries — an anisotropic behaviour is expected.

In Ref. [9], the experimental data for stress–strain curves for Al–Ni–Co decagonal quasicrystals are depicted in Fig. 14.2 (note: the results obtained for the basic cobalt (b-Co) modification of decagonal Al–Ni–Co); it presents plastic properties different from those of icosahedral quasicrystals. The reason for this is the distinguishing nature of quasicrystalline lattice between the two phases. It is different from the icosahedral phase; the decagonal phase possesses quasiperiodic as well as periodic lattice directions. Therefore, the influences of periodicity and quasiperiodicity on the plastic deformation lead to an anisotropic behaviour. To reveal this feature, Fig. 14.2 gives the data to those orientations as $A_{\|}$, $A_{\perp}$ and A_{45}, respectively. It is obvious that the effect of anisotropy is important.

The experimental data are worth while for theoretical study.

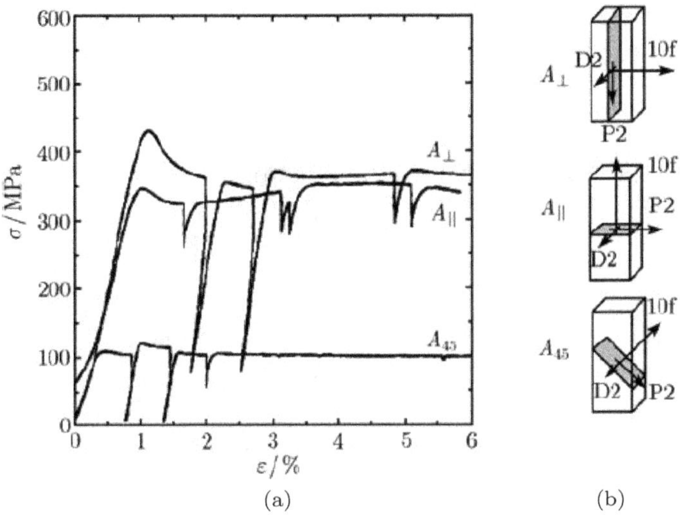

Fig. 14.2. (a) Stress–strain curves of decagonal b-Co decagonal Al–Ni–Co single quasicrystals for different specimen orientations at strain rate of 10^{-5}/s, (b) definition of the specimen orientations $A_\perp$, $A_\parallel$ and A_{45}[9].

14.2 Possible Scheme of Plastic Constitutive Equations

The formula (14.1-1) was calibrated from experimental data with the help of dislocation model assists in studying plastic constitutive equations of quasicrystals.

From the point of view macroscopically, the result is obtained for "uniaxial loading" condition as, at present, there is lack of data under multi-axial loading conditions. We assume that, if sufficient experimental data are available on yield surface/loading surface for some quasicrystals, then we can obtain an equation for yield surface such as

$$\Phi = \sigma_{eff} - Y = 0 \tag{14.2-1}$$

where $\sigma_{eff} = \sigma_e + f(H_{ij})$ denotes a generalized effective stress in which σ_e represents the classical effective stress of phonon stresses σ_{ij} and $f(H_{ij})$ the part coming from phason stresses H_{ij}. If

$$Y = \sigma_Y = \text{const} \tag{14.2-2}$$

in which σ_Y is the uniaxial yield limit of the material, then (14.2-1) represents the initial yield surface. On the other hand, if

$$Y = Y(h) \tag{14.2-3}$$

where h is related to deformation history, then (14.2-1) describes the evolution law of material deformation. When we have the yield surface/loading surface like (14.2-1), one can construct the following plastic constitutive equation as

$$\begin{aligned} \dot{\varepsilon}_{ij} &= \frac{1}{H(\sigma_{eff})}\dot{\sigma}_{eff}\frac{\partial\Phi}{\partial\sigma_{ij}} \\ \dot{w}_{ij} &= \frac{1}{H(\sigma_{eff})}\dot{\sigma}_{eff}\frac{\partial\Phi}{\partial H_{ij}} \end{aligned} \tag{14.2-4}$$

provided the flow rule of the isotropic hardening is taken, where Φ is the above-mentioned yield surface/loading surface function. The dot over physical quantities denotes the variation rate of the quantities, and $H(\sigma_{eff})$ the hardening modulus of the material can be calibrated by a test of a simple stress–strain state, e.g., given in (14.1-1). Then, adding the elastic constitutive relationship to (14.2-4), the elastic–plastic constitutive equation will be set up. The elastic constitutive law was fully discussed in previous chapters.

The expected constitutive law shown in Eq. (14.2-4), which are incremental plastic equations, can describe the effect of deformation history including the loading/unloading state within the deformation process. This may be a complete constitutive equation. There is a possible and relative simpler constitutive law belonging to total plasticity theory or deformation plasticity theory. That is, if defining the effective stress σ_{eff} and effective strain ε_{eff}, in which the former is introduced above and the latter has a similar definition and consists of phonon strains ε_{ij} as well as phason strains w_{ij}, then between the strains and stresses there are relations

$$\begin{aligned} \varepsilon_{ij} - \frac{1}{3}\varepsilon_{kk}\delta_{ij} &= \frac{3\varepsilon_{eff}}{2\sigma_{eff}}\left(\sigma_{ij} - \frac{1}{3}\sigma_{kk}\delta_{ij}\right) \\ w_{ij} - \frac{1}{3}w_{kk}\delta_{ij} &= \frac{3\varepsilon_{eff}}{2\sigma_{eff}}\left(H_{ij} - \frac{1}{3}H_{kk}\delta_{ij}\right) \end{aligned} \tag{14.2-5}$$

in which

$$\varepsilon_{kk} = \varepsilon_{xx} + \varepsilon_{yy} + \varepsilon_{zz}, \quad w_{kk} = w_{xx} + w_{yy} + w_{zz},$$

$$\sigma_{kk} = \sigma_{xx} + \sigma_{yy} + \sigma_{zz}$$

$H_{kk} = H_{xx} + H_{yy} + H_{zz}$ and we assume that

$$\varepsilon_{eff} = \begin{cases} \varepsilon_{eff}^{(e)}, & \sigma_{eff} < \sigma_0 \\ A(\sigma_{eff})^n, & \sigma_{eff} > \sigma_0 \end{cases} \tag{14.2-6}$$

where σ_0, A and n are material constants of the quasicrystals, which can be measured through a uniaxial test, and $\varepsilon_{eff}^{(e)}$ represents the quantity at the elastic deformation stage and σ_0 the uniaxial tensile yield stress.

In contrast to (14.2-4), Eq. (14.2-5) cannot describe deformation history, as they are substantively nonlinear elastic constitutive equations rather than plastic constitutive equations related to quasicrystalline materials. However, they can describe plastic deformation in the case of proportional loading and no unloading. It is evident that either (14.2-4) or (14.2-5) belong to a supposed incremental plastic constitutive law or total plastic constitutive law for quasicrystals. One cannot say whether they are correct or not due to lack of enough experimental data.

If one has the constitutive equations (14.2-4) or (14.2-5), then by coupling the equations of deformation geometry

$$\varepsilon_{ij} = \frac{1}{2}\left(\frac{\partial u_i}{\partial x_j} + \frac{\partial u_j}{\partial x_i}\right), \quad w_{ij} = \frac{\partial w_i}{\partial x_j} \tag{14.2-7}$$

and the equilibrium equations

$$\frac{\partial \sigma_{ij}}{\partial x_j} = 0, \quad \frac{\partial H_{ij}}{\partial x_j} = 0 \tag{14.2-8}$$

the basic framework of the theory of macro-plasticity, in the sense of incremental or total deformation, of quasicrystals can be set up.

At present, there is lack of such data, so Eqs. (14.2-1), (14.2-2) and (14.2-4) have not been set up yet. For the same reason, Eq. (14.2-5) have not been set up either. This is the major difficulty of macro-plasticity theory currently. It is evident that the possible theory is nonlinear, because the material parameters are not constants any more, and the mechanical behaviour is dependent on the history of the deformation process in general. The solution is of course more difficult than that for elastic deformation.

Due to the relative simplicity of Eq. (14.2-5), for some simple configurations, e.g., the anti-plane elasticity of one-dimensional hexagonal, three-dimensional cubic and three-dimensional icosahedral quasicrystals, one can make some probes for plastic analysis by using the constitutive equations.

14.3 Nonlinear Elasticity and Its Formulation

As pointed out in the previous section, due to lack of constitutive equations of plasticity of quasicrystals up to now, the analysis of plastic deformation for the material is not available. For this reason, we can perform some simpler nonlinear elastic analysis first, though the mechanisms of nonlinear elastic deformation and plastic deformation are quite different. Furthermore, there are some differences between the following nonlinear elasticity and the total plasticity introduced in Eq. (14.2-5). We here do not constrain the concrete form of the relationship between stresses and strains. The results obtained in the following may provide some useful hints for further plastic analysis.

Consider the following nonlinear elastic constitutive relations such as, i.e., define the free energy (or strain energy density)

$$F(\varepsilon_{ij}, w_{ij}) = \int_0^{\varepsilon_{ij}} \sigma_{ij} d\varepsilon_{ij} + \int_0^{w_{ij}} H_{ij} dw_{ij} \tag{14.3-1}$$

then there is

$$\sigma_{ij} = \frac{\partial F}{\partial \varepsilon_{ij}}, \quad H_{ij} = \frac{\partial F}{\partial w_{ij}} \tag{14.3-2}$$

So, formulas (14.3-1) and (14.3-2) can be seen as constitutive law for linear as well as nonlinear elasticity of quasicrystals, which, in general, cannot describe plastic deformation. If there is proportional loading and no unloading, the relationship can give an appropriate description of the plastic deformation.

In the above formulas, ε_{ij} and w_{ij} are phonon and phason strain tensors given by

$$\varepsilon_{ij} = \frac{1}{2}\left(\frac{\partial u_i}{\partial x_j} + \frac{\partial u_j}{\partial x_i}\right), \quad w_{ij} = \frac{\partial w_i}{\partial x_j} \tag{14.3-3}$$

where u_i and w_i denote the phonon and phason displacement vectors and σ_{ij} and H_{ij} the phonon and phason stress tensors, respectively, which satisfy the equilibrium equations (if the body forces and generalized body forces are omitted)

$$\frac{\partial \sigma_{ij}}{\partial x_j} = 0, \quad \frac{\partial H_{ij}}{\partial x_j} = 0 \tag{14.3-4}$$

Equations (14.3-1)–(14.3-4) are basic equations describing the nonlinear elastic deformation of a quasicrystal. In Section 14.5, we will give some applications of formulation (14.3-1)–(14.3-4), and we can show that it constitutes the basis of those nonlinear analyses given in Section 14.4.

14.4 Nonlinear Solutions Based on Some Simple Models

In this section, we give some nonlinear solutions based on some simple models which substantively are extended from the framework of Eshelby's work [13]; the classical Eshelby hypothesis was set up for linear elasticity, which can be extended into nonlinear elasticity even total plasticity (or deformation plasticity). Fan and Fan [14] carried out the investigation.

14.4.1 *Generalized Dugdale–Barenblatt model for anti-plane elasticity for some quasicrystals*

The crack problem of nonlinear deformation of quasicrystals is very interesting. For solving this problem, one of the available models is the generalized Dugdale–Barenblatt model (or generalized DB model for simplicity) [15, 16] that originated from the classical nonlinear fracture theory which was used for conventional structural materials including crystalline material.

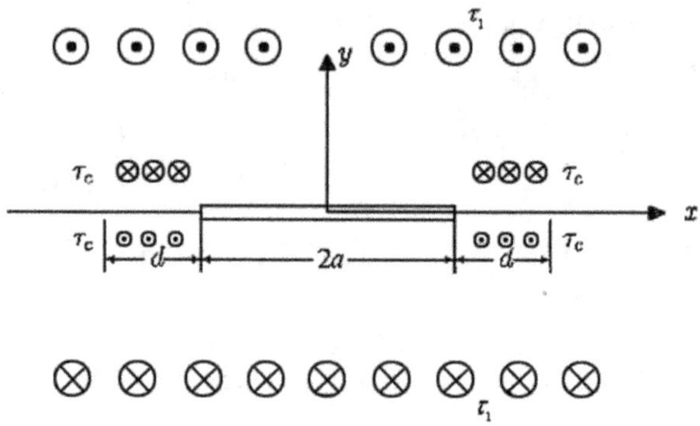

Fig. 14.3. Atomic cohesive force zone near the crack tip for anti-plane elasticity for some quasicrystals.

The linear solutions for a crack for anti-plane for some quasicrystals have been discussed in Chapters 8 and 9. We now discuss only the nonlinear solution.

Assume that there is an atomic cohesive force zone at the crack tip with a length d shown in Fig. 14.3, the value of which is unknown at the moment to be determined. In the continuum theory of quasicrystals, the atomic cohesive force zone is the plastic zone macroscopically and the distribution of atomic cohesive force must be evaluated by experimental observation. Due to lack of data on the new material, if we assume that the atomic cohesive is a constant τ_c, the shear yield limit (or shear yield strength) of the material, within the zone, then the problem is simplified. In this version, the nonlinear boundary value problem is linearized already. So, the previous formulation exhibited, e.g., in Chapter 8 can be used to solve the present problem. The most effective method in solving the problem is the complex variable function method, but the formulation and calculation are quite complicated and lengthy, more complex than those given in previous chapters. The details are omitted; we here list only the final results:

$$
\begin{aligned}
F_1(\varsigma) &= \frac{2i}{C_{44}}\left(\frac{2\theta_1}{\pi\varsigma^2}\tau_c - \tau_1\right)\frac{\varsigma^2}{\varsigma^2-1} + \frac{1}{2\pi i}\frac{2i}{C_{44}}\tau_c \ln\frac{e^{2i\theta_1}-\varsigma^2}{e^{-2i\theta_1}-\varsigma^2}\\
F_2(\varsigma) &= 0
\end{aligned}
\tag{14.4-1}
$$

in which

$$F_1(\zeta) = \varphi'(\zeta) + \frac{R_3}{C_{44}}\psi'(\zeta)$$

$$F_2(\zeta) = \psi'(\zeta) + \frac{R_3}{K_2}\varphi'(\zeta)$$

and

$$\phi(\zeta) = \phi_1(t) = \phi_1(\omega(\zeta)), \quad \psi(\zeta) = \psi_1(t) = \psi_1(\omega(\zeta))$$

where

$$t = \omega(\zeta) = \frac{a+d}{2}(\zeta + \zeta^{-1})$$

represents the conformal mapping from the t-plane (xy-plane) onto the ζ-plane; under the mapping the region of the xy-plane is transformed onto the interior of the unit circle at the ζ-plane, and $\phi'(\zeta)$ and $\psi'(\zeta)$ are derivatives of the functions of the new complex variable ζ, and θ_1 is the angle of point at the unit circle corresponding to the crack tip (i.e., $y = 0$, $x = a$, and there is the relation $\cos\theta_1 = a/(a+d)$ between the corresponding points at the t-plane and ζ-plane).

From the solution, the size of plastic zone d is determined as

$$d = a\left[\sec\left(\frac{\pi\tau_1}{2\tau_c}\right) - 1\right] \tag{14.4-2}$$

i.e., $\theta_1 = \frac{\pi\tau_1}{2\tau_c}$, and the crack tip opening displacement is

$$\delta_{III} = \frac{4K_2\tau_c a}{\pi(C_{44}K_2 - R_3^2)}\left[\ln\sec\left(\frac{\pi\tau_1}{2\tau_c}\right)\right] \tag{14.4-3}$$

It is obvious that the results are very simple and explicit, and the effects of phason to plastic deformation and fracture are explored. If taking the opening displacement as a nonlinear fracture parameter, then a fracture criterion for one-dimensional hexagonal quasicrystal

$$\delta_{III} = \delta_{IIIc} \tag{14.4-4}$$

is suggested in which δ_{IIIc} is the critical value of the crack tip opening displacement, measured by the experiment, a material constant.

In the low-stress level case, i.e., $\tau_1/\tau_c \ll 1$, in which

$$\sec\left(\frac{\pi\tau_1}{2\tau_c}\right) = 1 + \frac{1}{2}\left(\frac{\pi\tau_1}{2\tau_c}\right)^2 + \cdots$$

if we remain the first two terms and substitute them into (14.4-2)

$$d = \frac{\pi}{8}\left(\frac{K_{III}^{\parallel}}{\tau_c}\right)^2 \ll a$$

in which $K_{III}^{\parallel} = \sqrt{\pi a}\tau_1$ (refer to Section 8.2 of Chapter 8). The result is very close to the plastic size based on the "small scale yield modification" of classical fracture theory. Furthermore,

$$\ln\sec\left(\frac{\pi\tau_1}{2\tau_c}\right) = \frac{1}{2}\left(\frac{\pi\tau_1}{2\tau_c}\right)^2 + \frac{1}{12}\left(\frac{\pi\tau_1}{2\tau_c}\right)^4 + \cdots$$

If we take only the first two terms, then substitute them to (14.4-3) so

$$\delta_{III} = \frac{G_{III}}{\tau_c} \tag{14.4-5}$$

in which $G_{III} = \frac{K_2 K_{III}^{\parallel 2}}{C_{44}K_2 - R_3^2}$ represents the energy release rate in the linear elastic case (refer to Section 8.2 of Chapter 8). This shows that the parameter δ_{III} is equivalent to the energy release rate G_{III} in the linear elastic case or "small-scale yielding" plasticity.

The above results are given for one-dimensional hexagonal quasicrystals, which hold for three-dimensional icosahedral quasicrystals if the material constants C_{44}, K_2 and R_3 are replaced by $\mu, K_1 - K_2$ and R, which hold for three-dimensional cubic quasicrystals too if the constants are replaced by C_{44}, K_{44} and R_{44}, respectively.

14.4.2 *Generalized Dugdale–Barenblatt model for plane elasticity of two-dimensional point groups 5m, 10mm and 5, $\bar{5}$, 10, $\overline{10}$ quasicrystals*

The linear analysis for a crack in plane elasticity in point groups $5m$ and $10mm$ as well as in point groups $5, \bar{5}$ and $10, \overline{10}$ of two-dimensional quasicrystals is given in Sections 8.3 and 8.4 of Chapter 8 in terms of the Fourier method and complex analysis method, respectively, where we pointed out that if taking $R_1 = R$, $R_2 = 0$, then the solution of point groups $5, \bar{5}$ and $10, \overline{10}$ will exactly reduces to that on point groups $5m$ and $10mm$. In the following, we discuss the nonlinear solution only based on the complex analysis method.

For two-dimensional quasicrystals, a similar generalized Dugdale–Barenblatt model to that given in Section 14.4.1 is suggested as follows; refer to Fig. 14.4. That is, there is the so-called atomic cohesive force zone, or the plastic zone macroscopically, with length d, temporarily unknown and to be determined. If we suppose that the distribution of the atomic cohesive force distribution and magnitude of the atomic cohesive force $\sigma_c = \sigma_c(x)$ are known, furthermore σ_c = constant within the zone, then the nonlinear problem is linearized already, in which σ_c represents the macro-tensile yield strength of the quasicrystal. Therefore, the formulation introduced above can be used to solve the present problem. The powerful method for the problem is the complex variable function-conformal mapping method, which is displayed briefly in Section 14.4.1. But the calculation is much more complicated and lengthier than those listed in the previous subsection. The details of complex variable function-conformal mapping method are given in Section 11.3.10 of Chapter 11 and Section AI.3 of Major Appendix; here we list the final plastic solution only as follows.

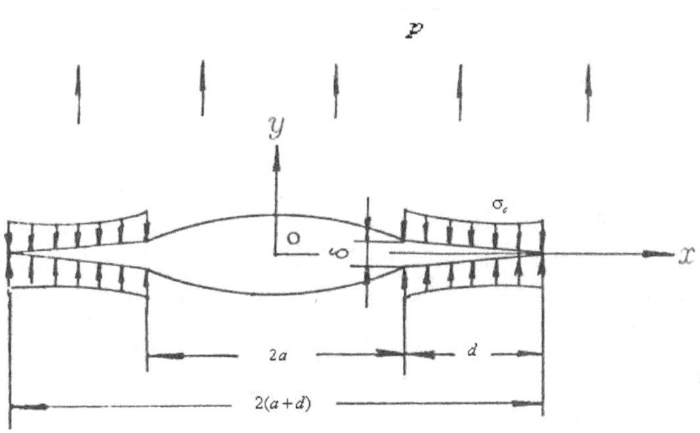

Fig. 14.4. The generalized DB model for two-dimensional decagonal quasicrystal.

The size of the plastic zone is determined as

$$d = a\left[\sec\left(\frac{\pi p}{2\sigma_c}\right) - 1\right] \tag{14.4-6}$$

which is similar to that given in (14.4-2), but the meaning is different. And the crack tip opening displacement is for point groups $5m$ and $10mm$ quasicrystals

$$\delta_I = \frac{2\sigma_c a}{\pi}\left[\frac{1}{L+M} + \frac{K_1}{MK_1 - R^2}\right]\left[\ln\sec\left(\frac{\pi p}{2\sigma_c}\right)\right] \tag{14.4-7$_1$}$$

and for point group $10, \overline{10}$

$$\delta_I = \frac{2\sigma_c a}{\pi}\left[\frac{1}{L+M} + \frac{K_1}{MK_1 - (R_1^2 + R_2^2)}\right]\left[\ln\sec\left(\frac{\pi p}{2\sigma_c}\right)\right] \tag{14.4-7$_2$}$$

which is also similar to that obtained by (14.4-3) in which $L = C_{12}$, $M = (C_{11} - C_{12})/2$ are phonon elastic constants, K_1 the phason elastic constant and R the phonon–phason coupling constant, respectively. It is very interesting that the solution (14.4-7) exactly covers the solutions of crystals and conventional structural materials, as $K_1 = R = 0$, e.g., for classical conventional materials

$$\delta_I = CTOD = \begin{cases} \dfrac{(1+\kappa)\sigma_c a}{\pi\mu}\ln\sec\left(\dfrac{\pi}{2}\dfrac{p}{\sigma_c}\right) \\ \quad = \dfrac{4(1-\nu)\sigma_c a}{\pi\mu}\ln\sec\left(\dfrac{\pi}{2}\dfrac{p}{\sigma_c}\right), \quad \textit{plane strain state} \\ \dfrac{(1+\kappa')\sigma_c a}{\pi\mu}\ln\sec\left(\dfrac{\pi}{2}\dfrac{p}{\sigma_c}\right) \\ \quad = \dfrac{4\sigma_s a}{(1+\nu)\pi\mu}\ln\sec\left(\dfrac{\pi}{2}\dfrac{p}{\sigma_c}\right), \quad \textit{plane stress state} \end{cases}$$

where $\kappa = 3 - 4\nu$ for plane strain and $\kappa' = \frac{3-\nu}{1+\nu}$ for plane stress, and ν is Poisson's ratio of the materials. So, the solutions of crystals and conventional structural materials are the specific cases of our solution of quasicrystals. Figure 14.5 shows the normalized crack tip opening displacement δ_I/a versus normalized stress p/σ_c and gives the comparison between quasicrystals and crystals as well as the

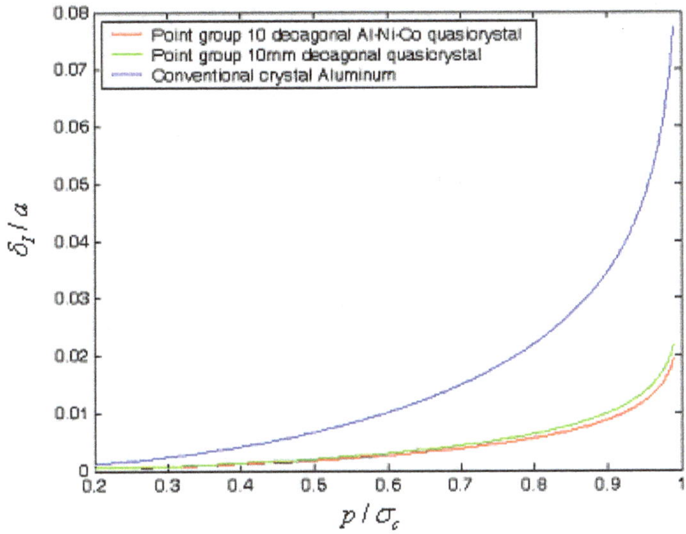

Fig. 14.5. Normalized crack tip opening displacement versus normalized stress and comparison [27].

comparison between point groups $10mm$ and $10, \overline{10}$ quasicrystals. The details of the solution including point group $10, \overline{10}$ are given in Section AI.3 of Major Appendix.

Based on the parameter δ_I given in (14.4-7), we can suggest the nonlinear fracture criterion for pentagonal and/or decagonal quasicrystals for the Mode I loading as

$$\delta_I = \delta_{Ic} \tag{14.4-8}$$

in which δ_{Ic} denotes the critical value of the crack opening displacement, measured by the experiment, a material constant.

In linear elastic case, i.e., $p/\sigma_c \ll 1$, through a similar analysis like in the previous, the crack tip opening displacement reduces to

$$\delta_I = \frac{G_I}{\sigma_c} \tag{14.4-9}$$

where G_I is defined by (8.3-19) of Chapter 8; this gives the simple connection between these two parameters, so the criterion (14.4-8) can reduce to the energy release rate criterion for linear elastic deformation of the quasicrystals discussed in Chapter 8.

14.4.3 *Generalized Dugdale–Barenblatt model for plane elasticity of three-dimensional icosahedral quasicrystals*

The schematic figure of the model for icosahedral quasicrystals is similar to Fig. 14.4; the plastic zone is the same as given in (14.4-6), and the crack tip opening displacement is as follows:

$$\delta_I = \lim_{x\to l} 2u_y(x,0) = \lim_{\phi\to\phi_2} 2u_y(x,0) = 2\left(\frac{1}{\lambda+\mu} + \frac{c_4}{c_2}\right) \cdot \frac{\sigma_s a}{\pi} \cdot \ln\sec\left(\frac{\pi}{2}\frac{p}{\sigma_s}\right) \tag{14.4-10}$$

in which

$$\begin{aligned} c_2 &= \mu(K_1 - K_2) - R^2 - \frac{(\mu K_2 - R^2)^2}{\mu K_1 - 2R^2} \\ c_4 &= c_1 R + \frac{1}{2}c_2\left(K_1 + \frac{\mu K_1 - 2R^2}{\lambda+\mu}\right) \end{aligned} \tag{14.4-11}$$

$$\text{with } c_1 = \frac{R(2K_2 - K_1)(\mu K_1 + \mu K_2 - 3R^2)}{2(\mu K_1 - 2R^2)}.$$

The variation of the normalized crack tip opening displacement versus normalized applies stress is shown in Fig. 14.6.

14.5 Nonlinear Analysis Based on the Generalized Eshelby Theory

In the previous section, we obtained some nonlinear solutions for one-, two- and three-dimensional quasicrystals by using some simple physical models. We will show that those solutions have some inherent relations to the generalized Eshelby theory originated from the classical reference [13] for crystals. Fan and Fan [14] did some work to extend the classical Eshelby model for crystals to that for quasicrystals.

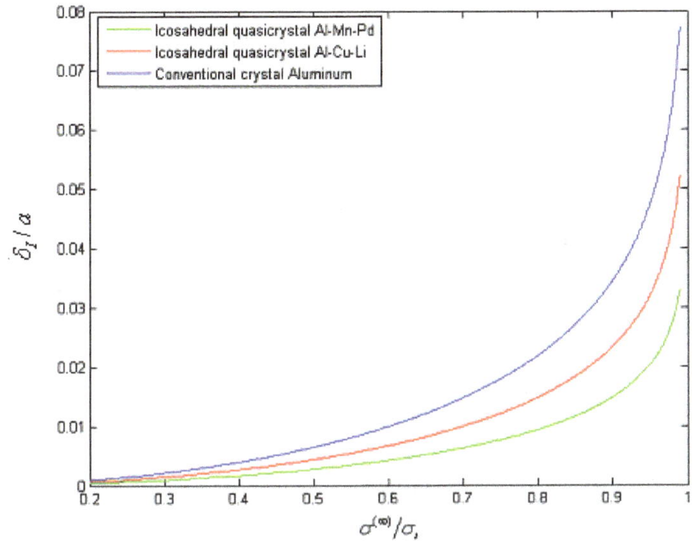

Fig. 14.6. Normalized crack tip opening displacement of icosahedral quasicrystals versus normalized applied stress [27].

14.5.1 *Generalized Eshelby energy–momentum tensor and generalized Eshelby integral*

Fan and Fan [14] (can also refer to Refs. [17, 18]) defined the generalized Eshelby energy–momentum tensor as

$$G = Fn_1 - \sigma_{ij}n_j\frac{\partial u_i}{\partial x_1} - H_{ij}n_j\frac{\partial w_j}{\partial x_1} \tag{14.5-1}$$

where F is defined in (14.3-1) and n_i the unit vector of outward normal at any point of an arc in a quasicrystal, at which there are

$$\sigma_{ij}n_j = T_i, \quad H_{ij}n_j = h_i$$

where T_i denotes the traction vector and h_i the generalized traction vector, as shown in Fig. 14.7.

Furthermore, we define an integral such as

$$E = \int_\Gamma Gd\Gamma \tag{14.5-2}$$

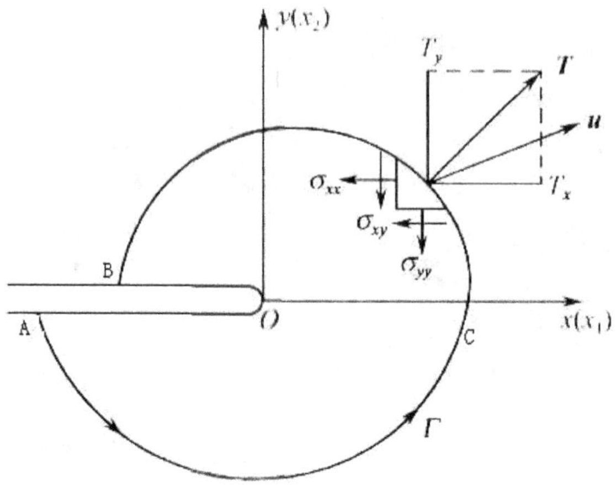

Fig. 14.7. The path of the generalized Eshelby integral.

in which Γ is an integration path enclosing a crack tip in the material. To memorize Eshelby, the integral is named as generalized Eshelby integral. The integral exhibits path independency. In Section 14.8, the proof on the path independency of the integral is given.

Because the integral presents this important characteristic, it has some applications in the fracture analysis of quasicrystals.

Another important characteristic is that the value of the integral is equal to the energy release rate G_{III} (under Mode III loading), or G_{II} (under Mode II loading), or G_I (under Mode I loading) of one-, two- and three-dimensional quasicrystals when the material is in a linear elastic deformation state. The mathematical proof of this is provided in Section 14.8.

The features of the generalized Eshelby integral (14.5-2) exhibited above imply that it may be significant for the analysis of the nonlinear fracture of quasicrystals. In the following, an application for this purpose is discussed.

14.5.2 *Relation between crack tip opening displacement and the generalized Eshelby integral*

By using the path independency of E-integral (14.5-2), we take the integration path Γ as given by the red colour and shown in Fig. 14.8

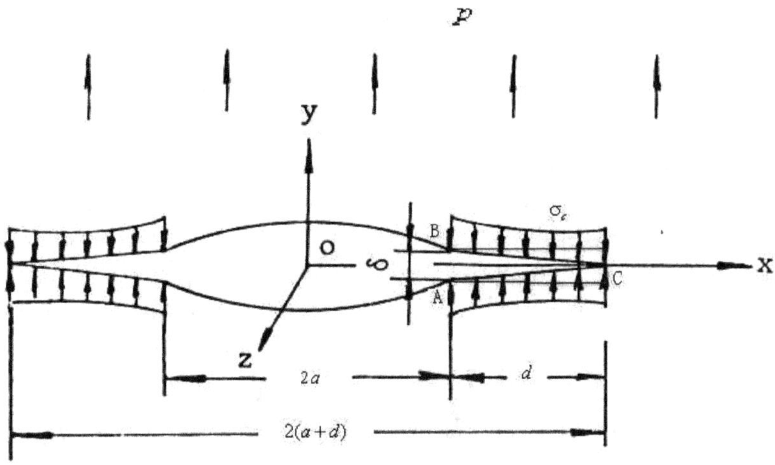

Fig. 14.8. The integration path for evaluating crack tip opening displacement.

and let the path close to the surface of the plastic zone as near as possible. Now, the integration path is $\int_{ACB}$, and along the segments AB and BC, i.e.,

$$dy = 0, \quad T_1 = 0, \quad T_2 = \sigma_c, \quad h_1 = 0, \quad h_2 = H_c$$

so that

$$\begin{aligned} E = \int_{\Gamma} G d\Gamma = \int_{ACB} G d\Gamma = \int_{A}^{B} \left(\sigma_c \frac{\partial u_y}{\partial x} dx + H_c \frac{\partial w_y}{\partial x} dx \right) \\ = \sigma_c[(u_y)_B - (u_y)_A] + H_c[(w_y)_B - (w_y)_A] \\ \approx \sigma_c[(u_y)_B - (u_y)_A] = \sigma_c \delta_I \qquad (14.5\text{-}3) \end{aligned}$$

This proves the relation between the generalized Eshelby integral and crack tip opening displacement; it shows that the E-integral presents an equivalency to the crack tip opening displacement, where σ_c is the atomic cohesive force (or the plastic yield strength in a macroscopic sense).

In general, σ_c can be understood as the cohesive force of a phonon field; similarly, we can introduce H_c as the "generalized atomic cohesive force" of the phason field of a quasicrystal material. In microscopy, the quantity H_c is meaningful, but it has not been measured macroscopically at present, the effects of which are omitted in formula (14.5-3) for this reason.

For Mode II and Mode III cracks, the proof is similar.

Since by generalized BCS model (refer to the following section) and generalized DB model one can obtain the same result on the crack tip opening displacement, which is also a direct result of the generalized Eshelby integral for quasicrystals, we realize that the generalized Eshelby energy–momentum tensor theory is the uniform physical basis of the two models.

14.5.3 *Some further interpretation on application of E-integral to the nonlinear fracture analysis of quasicrystals*

The above subsections and the appendices exhibit that the generalized Eshelby integral can be as a uniform basis of the generalized BCS model as well as the generalized DB model for the nonlinear fracture analysis of some one-, two- and three-dimensional quasicrystals. We suggest taking the E-integral as a fracture parameter and

$$E = E_c \tag{14.5-4}$$

as fracture criterion, where E_c is the critical value of the integral, a material constant, which can be measured through some conventional specimens; the discussion is given in Section 14.8.

Though the measurement of the critical value of the generalized Eshelby integral may be easier, the evaluation of the values of the E-integral for the plastic deformation of quasicrystals is very difficult. Therefore, the implementation of the criterion (14.5-4) is not so convenient in practice. Instead, people can use the fracture criterion of the crack tip opening displacement considering the equivalency between the crack tip opening displacement $\delta(\delta_I, \delta_{II}, \delta_{III})$ and E-integral. Elasto-plastic crack solutions for some one-, two- and three-dimensional quasicrystals have been found based on the generalized Dugdale–Barenblatt model, refer to Refs. [14, 18, 27], in which the size of the plastic zone and the crack tip opening displacement are determined, and with these data, an equivalent plastic fracture criterion is suggested in the previous section as

$$\delta = \delta_c \tag{14.5-5}$$

in which δ_c represents the critical value of the crack tip opening displacement, a material constant. The evaluation of the crack tip opening displacement has been exactly completed for large plates with a central crack and narrow plastic zone and introduced in Section 14.4 or refer to Refs. [14, 18, 27], and some approximate solution for other configurations may be obtained by the complex variable function method (an outline of the method was interpreted in Refs. [14, 18, 27]) and other methods (mainly the approximate methods and numerical methods). The shortcoming of this procedure lies in some difficulties of the determination of δ_c, which can be obtained through relation

$$\delta_c = \frac{E_c}{\sigma_c} \tag{14.5-6}$$

and Section 14.8 shows the measurement of E_c be easier. So, the collaboration of criterions (14.5-4) and (14.5-5) makes the nonlinear fracture analysis for quasicrystals available. Some details are given in Appendix of this chapter (i.e., Section 14.8).

14.6 Nonlinear Analysis Based on the Dislocation Model

We pointed out at the beginning of this chapter that in the study on the classical plasticity there are two different theories: one is the macroscopic plasticity theory and the other is the so-called crystal plasticity theory; to some extent, the latter can be seen as a "microscopic" theory which is based on the mechanism of motion of dislocation. We discussed quite more dislocation solutions in Chapter 7 for one- and two-dimensional quasicrystals and in Chapter 9 for three-dimensional quasicrystals; those results are helpful to explore the plastic deformation and fracture of the material. This analysis need not rely on nonlinear macroscopic constitutive law and it reduces a great difficulty in mathematical treatment. With some results of dislocation solutions here, we focus on the discussion concerning plastic flow around the crack tip for some one-, two- and three-dimensional quasicrystals, respectively.

14.6.1 *Screw dislocation pile-up for hexagonal or icosahedral or cubic quasicrystals*

For the topic, the basis is the generalized BCS model developed by Fan and co-workers [8], which is stated as follows.

Assume there is a Griffith crack with length $2l$ along the direction of z-axis in a one- or three-dimensional quasicrystal, subjected to a shear stress $\tau^{(\infty)}$ at infinity; see Fig. 14.9. Around the crack tip, there is a dislocation pile-up with size d which is unknown so far to be determined. Within the zone of dislocation pile-up, the material presents plastic flow and the stress σ_{yz} is equal to the critic shear stress τ_c, the atomic cohesive force, or the so-called flow limit of the material macroscopically. For simplicity, the external applied stress at the infinity can be removed, instead, it is applied at the crack surface. The latter is equivalent with the former from the point of

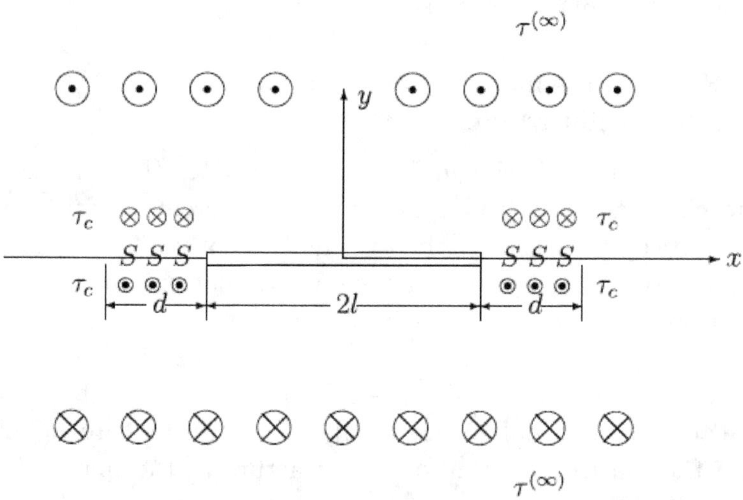

Fig. 14.9. Schematic picture of a crew dislocation pile-up coupled with a crack in anti-plane elasticity of quasicrystals.

view of the fracture behaviour for any studied systems. This is the generalized BCS model for quasicrystals.

The origin of the model was given by Bilby *et al.* [20, 21] for crystals. Reference [8] developed the model for the study of some one- and two-dimensional quasicrystals. The model can be formulated by the following boundary conditions for anti-plane elasticity:

$$\begin{aligned}
&(x^2+y^2)^{1/2} \to \infty: \quad \sigma_{ij}=0, \quad H_{ij}=0 \\
&y=0, \quad |x|<l: \quad \sigma_{yz}=-\tau^{(\infty)}, H_{yz}=0 \\
&y=0, \quad l<|x|<l+d: \quad \sigma_{yz}=-\tau^{(\infty)}+\tau_c=0, \quad H_{yz}=0
\end{aligned} \tag{14.6-1}$$

For some detailed statement for the formalism, refer to Ref. [18].

The boundary conditions (14.6-1) have shown that the nonlinear (plastic deformation) problem is linearized mathematically. So, it has been transformed into a linear problem (or an "equivalent" elasticity problem) described by the final governing equations

$$\nabla^2 u_z = 0, \quad \nabla^2 w_z = 0 \tag{14.6-2}$$

The boundary value problems (14.6-1) and (14.6-2) are the problem solved in Section 14.4, where the complex variable function method-conformal mapping method was used, but there is a simpler procedure to solve it if we need only some plastic deformation parameters around the crack tip and if we have a dislocation solution, the latter is given by Section 7.1 of Chapter 7. For this purpose, we first introduce a dislocation density function $f(\xi)$, then problems (14.6-1) and (14.6-2) can be transformed into the following singular integral equation problem such as

$$\int_L \frac{f(\xi)d\xi}{\xi - x} = \frac{\tau(x)}{A} \tag{14.6-3}$$

in which ξ denotes the dislocation source point coordinate, x denotes the field point coordinate at the real axis and L represents interval $(l, l+d)$. From the dislocation solution given in Section 7.1 of Chapter 7, the constant A is defined by

$$A = -\frac{(C_{44}K_2 - R^2)b_3^{\|}}{2\pi K_2} \tag{14.6-4}$$

and the shear stress distribution is

$$\tau(x) = \begin{cases} -\tau^{(\infty)}, & |x| < l \\ -\tau^{(\infty)} + \tau_c, & l < |x| < l+d \end{cases}$$

which comes from the boundary condition (14.6-1).

In terms of the Muskhelishvili [26] theory, Fan *et al.* [8] solved the integral equations (14.6-3) under condition (14.6-4), and the solution is

$$\begin{aligned} f(x) &= -\frac{1}{\pi^2 A}\sqrt{\frac{x+(l+d)}{x-(l+d)}}\int_L \sqrt{\frac{\xi-(l+d)}{\xi+(L+d)}}\tau(\xi)\frac{d\xi}{\xi - x} \\ &= -\frac{1}{\pi^2 A}\sqrt{\frac{x+(l+d)}{x-(l+d)}}\left\{i\left[2\tau_c\cos^{-1}\left(\frac{l}{l+d}\right) - \tau^{(\infty)}\pi\right]\right\} \\ &\quad + \frac{\tau_c}{\pi^2 A}\left[\cosh^{-1}\left|\frac{(l+d)^2 - lx}{(l+d)(l-x)}\right| - \cosh^{-1}\left|\frac{(l+d)^2 + lx}{(l+d)(l+x)}\right|\right] \end{aligned} \tag{14.6-5}$$

Because $f(x)$ should be a real number, the factor multiplying the imaginary number i must be zero in the first term on the right-hand side of (14.6-5); this leads to

$$2\tau_c\cos^{-1}\left(\frac{l}{l+d}\right) - \tau^{(\infty)}\pi = 0$$

i.e.,

$$d = l\left[\sec\left(\frac{\pi\tau^{(\infty)}}{2\tau_c}\right) - 1\right] \tag{14.6-6}$$

This is the same as (14.4-2) if l is replaced by a.

From solution (14.6-5), we evaluate the number of dislocations $N(x)$ such as

$$N(x) = \int_0^x f(\xi)d\xi \tag{14.6-7}$$

Substituting (16.6-5) (coupled with (14.6-6)) into (14.6-7), we can get $N(l+d)$ and $N(l)$, so the amount of dislocation motion is

$$\begin{aligned} \delta_{III} &= b_3^{\parallel}[N(l+d) - N(l)] = \frac{2b_3^{\parallel} l\tau_c}{\pi^2 A}\left(\ln\frac{l+d}{l}\right) \\ &= \frac{4K_2\tau_c l}{\pi(C_{44}K_2 - R^2)} \ln\sec\left(\frac{\pi\tau^{(\infty)}}{2\tau_c}\right) \end{aligned} \tag{14.6-8}$$

This is the same as (14.4-3) apart from the difference in some notations.

14.6.2 *Edge dislocation pile-up for pentagonal or decagonal two-dimensional quasicrystals*

For plane elasticity, the corresponding problem is the "edge" dislocation pile-up problem shown in Fig. 14.10; here we consider the pile-up in the pentagonal or decagonal two-dimensional quasicrystals. There are the boundary conditions

$$\begin{aligned} &(x^2+y^2)^{1/2} \to \infty: \quad \sigma_{ij} = 0, \quad H_{ij} = 0 \\ &y = 0, \quad |x| < l: \quad \sigma_{yx} = -\tau^{(\infty)}, \quad H_{yx} = 0, \quad \sigma_{yz} = 0, \quad H_{yz} = 0 \\ &y = 0, \quad l < |x| < l+d: \quad \sigma_{yx} = -\tau^{(\infty)} + \tau_c, \quad H_{yx} = 0, \\ &\sigma_{yz} = 0, \quad H_{yz} = 0 \end{aligned} \tag{14.6-9}$$

and the governing equation

$$\nabla^2\nabla^2\nabla^2\nabla^2 F(x,y) = 0 \tag{14.6-10}$$

Instead of solving the boundary value problems (14.6-9) and (14.6-10), one can determine a dislocation density which satisfies Eq. (14.6-3) if the relevant quantity A is known beforehand.

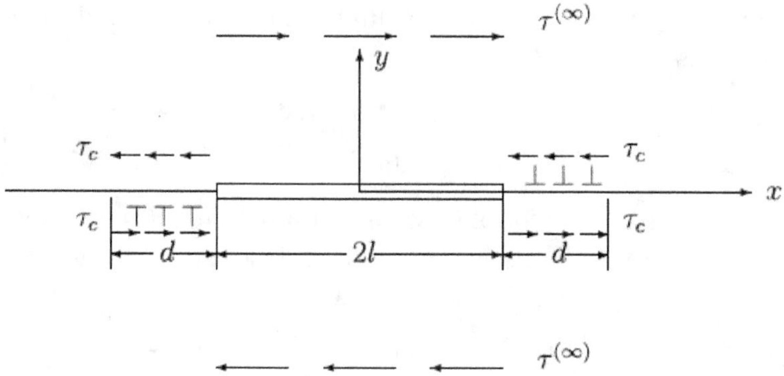

Fig. 14.10. Schematic picture of an "edge" dislocation pile-up coupled with a crack in plane elasticity of quasicrystals.

In the singular integral equation, the shear stress distribution function $\tau(x)$ is the same as the previous subsection, but here A is replaced by

$$A = \frac{b_1^{\|}}{\pi} \frac{(L+M)(MK_1 - R^2)}{(L+M)K_1 + (MK_1 - R^2)} \tag{14.6-11}$$

which has been found in the solution of Section 7.2 of Chapter 7 for point groups of $5m$ and $10mm$ quasicrystals.

According to the condition concerning factor multiplying the imaginary i to be zero in (14.6-5), the plastic zone size is determined by (14.6-6); this is similar to (14.4-6), but there are some differences in physical meaning.

With a similar calculation on the amount of dislocation motion, we can obtain the slip of the crack tip such as

$$\delta_{II} = \frac{2b_1^{\|} l}{\pi^2 A} \ln \sec\left(\frac{2\tau^{(\infty)}}{\pi\tau_c}\right) = \frac{2\tau_c l}{\pi}\left(\frac{1}{M+K_1} + \frac{K_1}{MK_1 - R^2}\right)$$
$$\times \ln \sec\left(\frac{2\tau^{(\infty)}}{\pi\tau_c}\right) \tag{14.6-12}$$

It is evident that the result of (14.6-12) is similar to that of (14.4-7), but the physical meaning is different. According to the macroscopic fracture mechanics, the crack tip slip corresponds to the Mode II crack tip displacement (i.e., the crack tip sliding displacement) δ_{II}. But for the Mode I crack tip opening displacement (14.4-7), there is no physical basis of the dislocation model, though the similar mathematical treatment can be easily given, which is omitted here.

The dislocation model provides a powerful support to the generalized cohesive force model and generalized Eshelby model presented in the previous sections, even if there is difficulty physically, we are unable to derive δ_I directly from the model and result in a slight faulty aspect of the procedure.

14.6.3 *Edge dislocation pile-up for three-dimensional icosahedral quasicrystals*

For three-dimensional icosahedral quasicrystals, the relevant dislocation pile-up problems can be formulated as

$$
\begin{aligned}
&(x^2+y^2)^{1/2}\to\infty: \quad \sigma_{ij}=0, \quad H_{ij}=0\\
&y=0, \quad |x|<l: \quad \sigma_{yx}=-\tau^{(\infty)}, \quad H_{yx}=0, \quad \sigma_{yz}=0,\\
&H_{yz}=0, \sigma_{yy}=0, \quad H_{yy}=0,\\
&y=0, \quad l<|x|<l+d: \quad \sigma_{yx}=-\tau^{(\infty)}+\tau_c, \quad H_{yx}=0,\\
&\sigma_{yz}=0, \quad H_{yz}=0, \quad \sigma_{yy}=0, \quad H_{yy}=0,\\
&\nabla^2\nabla^2\nabla^2\nabla^2\nabla^2\nabla^2 F(x,y)=0
\end{aligned}
$$

The solution through the dislocation density method can be offered similar to the above. Because the formulas are too lengthy (please refer to Section 9.6 of Chapter 9 for the dislocation solution that is the basis of the present method), the discussion is omitted.

14.7 Conclusion and Discussion

This chapter discussed the deformation and fracture exhibiting nonlinear behaviour of quasicrystals; this is a topic which has not been well studied so far. The discussion in this chapter shows that the nonlinear elastic constitutive equation possesses some meaning for the investigation due to lack of plastic constitutive equation at present. The generalized Eshelby principle based on the nonlinear elastic constitutive law plays an important role in revealing the fracture behaviour of quasicrystals under nonlinear deformation which may be a basis of the generalized DB model to some extent. We realize that the generalized Eshelby principle and generalized DB model are based on the macro-nonlinear continuous mechanics of quasicrystals. The generalized BCS model developed in Section 14.6 emphasizes the dislocation mechanism for the plastic deformation, which is a "microscopic" model and is different from those of the generalized DB hypothesis as well as the generalized Eshelby principle at physical basis and methodology. The attractiveness is that results derived from these quite different hypotheses are in exact agreement with the study.

The reader can find that even if there are lack of plastic constitutive equations for quasicrystals so far, the nonlinear analysis explored in this chapter (including the text and appendices) with considerable achievements and meaningful description encourages us to furthermore explore the physical nature and mathematical solution in the field.

14.8 Appendix: Some Mathematical Details

14.8.1 *Proof on path independency of E-integral*

Consider a single-connected region D with a closed boundary C. From the Green formula, the first, second and third terms on the left-hand side of the integral (14.5-2) will be

$$\int_C F n_1 d\Gamma = \int_C F dx_2 = \iint_D \frac{\partial F}{\partial x_1} dx_1 dx_2$$

$$\int_C \left(\sigma_{ij} n_j \frac{\partial u_i}{\partial x_1} + H_{ij} n_j \frac{\partial w_i}{\partial x_1}\right) d\Gamma$$

$$= \iint_D \frac{\partial}{\partial x_j}\left(\sigma_{ij}\frac{\partial u_i}{\partial x_1} + H_{ij}\frac{\partial w_i}{\partial x_1}\right) dx_1 dx_2$$

So

$$\int_C G d\Gamma = \iint_D \left[\frac{\partial F}{\partial x_1} - \frac{\partial}{\partial x_j}\left(\sigma_{ij}\frac{\partial u_i}{\partial x_1} + H_{ij}\frac{\partial w_i}{\partial x_1}\right)\right] dx_1 dx_2 \quad (14.8\text{-}1)$$

Meantime, (14.3-2) provides that

$$\begin{aligned}\frac{\partial F}{\partial x_1} &= \frac{\partial F}{\partial \varepsilon_{ij}}\frac{\partial \varepsilon_{ij}}{\partial x_1} + \frac{\partial F}{\partial w_{ij}}\frac{\partial w_{ij}}{\partial x_1} = \sigma_{ij}\frac{\partial \varepsilon_{ij}}{\partial x_1} + H_{ij}\frac{\partial w_{ij}}{\partial x_1}\\ &= \sigma_{ij}\frac{\partial}{\partial x_1}\left[\frac{1}{2}\left(\frac{\partial u_i}{\partial x_j} + \frac{\partial u_j}{\partial x_i}\right)\right] + H_{ij}\frac{\partial}{\partial x_1}\left(\frac{\partial w_i}{\partial x_j}\right)\\ &= \sigma_{ij}\frac{\partial}{\partial x_1}\left(\frac{\partial u_i}{\partial x_j}\right) + H_{ij}\frac{\partial}{\partial x_1}\left(\frac{\partial w_i}{\partial x_j}\right)\end{aligned}$$

Here the symmetry of tensor ε_{ij} (given by the first equation of (14.3-3)) is used. The above result can be rewritten as

$$\begin{aligned}\frac{\partial F}{\partial x_1} &= \frac{\partial}{\partial x_j}\left(\sigma_{ij}\frac{\partial u_i}{\partial x_1}\right) - \frac{\partial \sigma_{ij}}{\partial x_j}\frac{\partial u_i}{\partial x_1} + \frac{\partial}{\partial x_j}\left(H_{ij}\frac{\partial w_i}{\partial x_1}\right) - \frac{\partial H_{ij}}{\partial x_j}\frac{\partial w_i}{\partial x_1}\\ &= \frac{\partial}{\partial x_j}\left(\sigma_{ij}\frac{\partial u_i}{\partial x_1} + H_{ij}\frac{\partial w_i}{\partial x_1}\right)\end{aligned} \tag{14.8-2}$$

In the last step in driving (14.8-2), the equilibrium equations (14.3-4) are used.

Substituting (14.8-2) into (14.8-1), we have

$$\int_C G d\Gamma = 0$$

If taking $C = \Gamma + BB' - \Gamma' + A'A$ in which Γ' is a path similar to Γ, these two curves intersect the upper face of the crack at points B and B', and the lower face at points A and A', respectively. Due to

$$dx_2 = dy = 0, \quad T_i = 0, \quad h_i = 0$$

at the segments BB' and $A'A$, there are

$$\int_{BB'+A'A} G d\Gamma = 0, \quad \int_{\Gamma} G d\Gamma = \int_{\Gamma'} G d\Gamma$$

Because Γ and Γ' are arbitrary paths satisfying the requirements of E-integral definition, the last equality proves its path independency.

14.8.2 *Proof on the equivalency of E-integral to energy release rate under linear elastic case for quasicrystals*

In the text of this chapter, we argue that the E-integral is equivalent to the energy release rate when the quasicrystals are in the linear elastic deformation state; the mathematical proof about this is given here. The proof can be done for one-, two- and three-dimensional quasicrystals and for Mode I, Mode II and Mode III cracks. For simplicity, here we only discuss Mode III crack for one-dimensional hexagonal quasicrystals, but it is valid for other quasicrystals as well.

At linear elastic case, the generalized Hooke's law

$$\begin{aligned} \sigma_{ij} &= C_{ijkl}\varepsilon_{kl} + R_{ijkl}w_{kl} \\ H_{ij} &= K_{ijkl}w_{kl} + R_{klij}\varepsilon_{kl} \end{aligned}$$

reduces to

$$\begin{aligned} \sigma_{yz} &= \sigma_{zy} = 2C_{44}\varepsilon_{yz} + R_3 w_{zy} \\ \sigma_{zx} &= \sigma_{xz} = 2C_{44}\varepsilon_{zx} + R_3 w_{zx} \\ H_{zy} &= K_2 w_{zy} + 2R_3\varepsilon_{zy} \\ H_{zx} &= K_2 w_{zx} + 2R_3\varepsilon_{zx} \end{aligned}$$

For the phonon–phason coupling anti-plane shearing (or longitudinal shearing) state in the elasticity of one-dimensional hexagonal quasicrystals, one finds the free energy density (or strain energy density)

$$F = C_{44}(\varepsilon_{xz}^2+\varepsilon_{yz}^2)+\frac{1}{2}K_2(w_{zx}^2+w_{zy}^2)+R_3(\varepsilon_{zx}w_{zx}+\varepsilon_{yz}w_{zy}) \quad (14.8\text{-}3)$$

Substituting the crack solution given in Chapter 8 into (14.8-3), we have

$$\begin{aligned} F = \frac{1}{4(C_{44}K_2 - R_3^2)}&\{C_{44}[K_2 K_{III}^{\parallel} - R_3 K_{III}^{\perp}]^2 \\ &+ 2K_2[C_{44}K_{III}^{\perp} - R_3 K_{III}^{\parallel}]^2 \\ &+ 2R_3[K_2 K_{III}^{\parallel} - R_3 K_{III}^{\perp}][C_{44}K_{III}^{\perp} - R_3 K_{III}^{\parallel}]\} \quad (14.8\text{-}4) \end{aligned}$$

where $K_{III}^{\parallel} = \sqrt{\pi a}\tau_1$ and $K_{III}^{\perp} = \sqrt{\pi a}\tau_2$ are stress intensity factors associated with phonon and phason fields, respectively.

Integrating the quantity given in (14.8-4) along a path enclosing the crack tip, one can obtain the first term of the E-integral. Because of the path independency of the integral, we can take a half-circle with the crack tip as its origin with radius r so that

$$\int_{\pi}^{-\pi} F dy = \int_{\pi}^{-\pi} F r \cos\theta d\theta = 0 \tag{14.8-5}$$

where (r, θ) denote the crack tip coordinates.

Now, we calculate the second and third terms of the integral.

According to the definition of Section 14.3, it is known that between the traction, generalized traction and the phonon and phason stresses, there are

$$T_z = \sigma_{zx} n_x + \sigma_{zy} n_y = \sigma_{zx} \cos\theta + \sigma_{zy} \sin\theta$$

$$h_z = H_{zx} n_x + H_{zy} n_y = H_{zx} \cos\theta + H_{zy} \sin\theta$$

By substituting the crack solution given in Section 8.1 of Chapter 8 into the above formulae, we find

$$T_z = \frac{1}{\sqrt{2\pi r}} K_{III}^{\parallel} \sin\frac{1}{2}\theta, \quad h_z = \frac{1}{\sqrt{2\pi r}} K_{III}^{\perp} \sin\frac{1}{2}\theta \tag{14.8-6}$$

Based on the crack solution given in Section 8.1 of Chapter 8, there are

$$\begin{aligned} u_z &= \frac{K_2\tau_1 - R_3\tau_2}{C_{44}K_2 - R_3^2}\sqrt{2ar}\sin\frac{1}{2}\theta \\ w_z &= \frac{C_{44}\tau_2 - R_3\tau_1}{C_{44}K_2 - R_3^2}\sqrt{2ar}\sin\frac{1}{2}\theta \end{aligned} \tag{14.8-7}$$

In addition,

$$\frac{\partial}{\partial x} = \frac{\partial}{\partial r}\frac{\partial r}{\partial x} + \frac{\partial}{\partial \theta}\frac{\partial \theta}{\partial x} = \cos\theta\frac{\partial}{\partial r} - \sin\theta\frac{\partial}{\partial \theta}$$

Substituting these results and relations into the second and third terms of the integral, we obtain

$$-\int_{\pi}^{-\pi}\left(\sigma_{ij}n_j\frac{\partial u_i}{\partial x_1}+H_{ij}n_j\frac{\partial w_i}{\partial x_1}\right)d\Gamma$$

$$=\frac{K_2(K_{III}^{\parallel})^2+C_{44}(K_{III}^{\perp})^2-2R_3K_{III}^{\parallel}K_{III}^{\perp}}{C_{44}K_2-R_3^2} \qquad (14.8\text{-}8)$$

From (14.8-5) and (14.8-8), we find that

$$E_{III}=\frac{K_2(K_{III}^{\parallel})^2+C_{44}(K_{III}^{\perp})^2-2R_3K_{III}^{\parallel}K_{III}^{\perp}}{C_{44}K_2-R_3^2}=G_{III} \qquad (14.8\text{-}9)$$

which is just (8.1-25) given in Chapter 8.

Similarly,

$$E_I=G_I \qquad (14.8\text{-}10)$$

and

$$E_{II}=G_{II} \qquad (14.8\text{-}11)$$

where the suffixes mean the Mode I, Mode II and Mode III of the cracks. The above demonstration indicates that the generalized Eshelby integral is equivalent to the energy release rate of the crack in the linear elasticity of quasicrystals.

Under the plastic deformation of quasicrystals, in general, the generalized Eshelby integral does not represent the energy release rate; the reason of this is that the unloading can appear due to crack extension, and stress–strain relations described in (14.3-1) and (14.3-2) do not remain one-to-one correspondence; the physical background of the E-integral will be broken down. But if we define the potential energy per unit thickness for a plane (i.e., a two-dimensional) region Ω occupied by a quasicrystal and denote the boundary of this by Γ

$$V=\int_{\Omega}Fdxdy-\int_{\Gamma}(T_iu_i+h_iw_i)d\Gamma \qquad (14.8\text{-}12)$$

then the total potential energy is

$$\Pi=BV \qquad (14.8\text{-}13)$$

where B denotes the thickness of the specimen, and there is the relation

$$E = -\frac{\delta\Pi}{\delta A} = -\frac{1}{B}\frac{\delta\Pi}{\delta a} \tag{14.8-14}$$

where the E-integral can be expressed only by the "difference quotient" rather than the "differential quotient". The difference quotient is not energy release rate, and only represents the ratio between the individual differences in the energy over the difference of the crack length for the same configuration specimen but with a different initial crack length. Considering the situation, the physical meaning of (14.8-14) is left for further discussion. Even if there is the faulty aspect, (14.8-14) provides useful applications for calibrating the measurement data of the integral in experiments. The relevant discussion will be done in the following subsection.

14.8.3 *On the evaluation of the critic value of E-integral*

The E-integral provides not only the theoretical basis for generalized BCS model as well as generalized D-B model but also an effective tool for measuring the material constants E_c and δ_c. We mentioned that the measurement of δ_c is difficult, but the measurement of E_c is easier. Due to the connection between these two quantities, the value of the former can be obtained from that of the latter.

At low and conventional temperatures, quasicrystals present brittle behaviour, the fracture belongs to linear elastic fracture and the measurement of fracture toughness can be carried out by indentation technique, see, e.g., Mong *et al.* [19]. At high temperatures, the materials present large plastic deformation, and the measurement can be done by bending the experiment, in which the three-point bending specimen, see Fig. 14.11, is one of the most useful specimens. There are two procedures: either by multi-specimens testing or by single specimen testing.

For simplicity, we here discuss the procedure by single specimen testing only. For the three-point bending specimen shown in Fig. 14.11, based on formula (14.8-14), the value of E-integral is

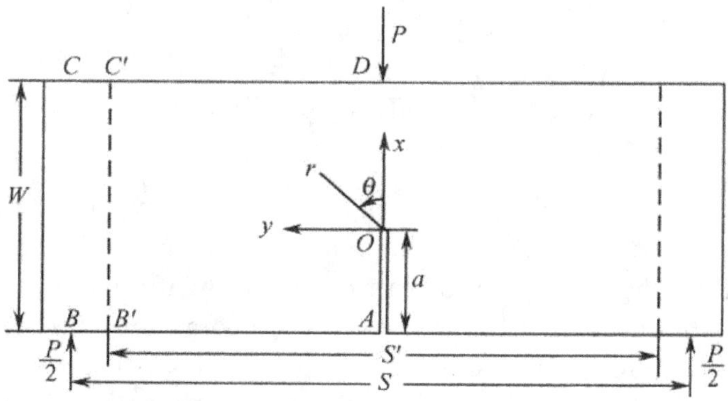

Fig. 14.11. Three-point bending specimen.

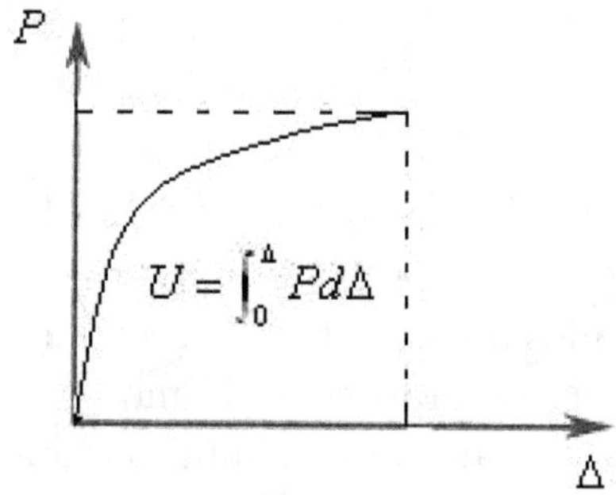

Fig. 14.12. The depiction of the specimen deformation energy.

approximately

$$E = \frac{2U}{B(W-a)} \tag{14.8-15}$$

where W represents the width, B the thickness of the specimen and U the area under the $P-\Delta$ curve (shown in Fig. 14.12), i.e.,

$$U = \int Pd\Delta \tag{14.8-16}$$

in which P is the load (force per unit thickness) at the loading point and Δ the displacement of the same point. When one observes the initiation of the crack growth, the value of the E-integral is marked as the fracture toughness of the quasicrystal.

If the phason field is absent, the material is reduced to a conventional structural material including crystals. In this case, the E-integral is reduced to a conventional Eshelby integral or J-integral; the latter was introduced by Rice [22] and Cherepanov [23] after 11 years compared with the Eshelby early publication [13], but the work of Rice has been systematically developed, especially Begley and Landes [24, 25] put forward the experimental study on J-integral, which further promotes the development of nonlinear fracture theory and its applications for conventional engineering materials; these experiences would be helpful for the experimental study of nonlinear fracture of quasicrystalline materials.

References

[1] Calliard D, 2006, Dislocation mechanism and plasticity of quasicrystals: TEM observations in icosahedral Al-Pd-Mn, *Materials Sci. Forum*, **509**(1), 49–56.

[2] Geyer B, Bartisch M, Feuerbacher M *et al.*, 2000, Plastic deformation of icosahedral Al-Pd-Mn single quasicrystals, I. Experimental results, *Phil. Mag. A*, **80**(7), 1151–1164.

[3] Messerschmidt U, Bartisch M, Geyer B *et al.*, 2000, Plastic deformation of icosahedral Al-Pd-Mn single quasicrystals, II. Interpretation of experimental results, *Phil. Mag. A*, **80**(7), 1165–1181.

[4] Urban K and Wollgarten M, 1994, Dislocation and plasticity of quasicrstals, *Materials Sci. Forum*, **150–151**(2), 315–322.

[5] Wollgarten M *et al.*, 1995, In-situ observation of dislocation motion in icosahedral Al-Pd-Mn single quasicrystals, *Phil. Mag. Lett.*, **71**(1), 99–105.

[6] Feuerbacher M, Bartsch B, Grushk B *et al.*, 1997, Plastic deformation of decagonal Al-Ni-Co quasicrystals, *Phil. Mag. Lett.*, **76**(4), 396–375.

[7] Messerschmidt U, Bartsch M, Feuerbacher M, *et al.*, Friction mechanism ofdislocation motion in icosahedral Al-Pd-Mn single quasicrystals, *Phil. Mag. A*, **79**(11), 2123–2135.

[8] Fan T Y, Trebin H-R, Messerschmidt U and Mai Y W, 2004, Plastic flow coupled with a Griffith crack in some one- and two-dimensional quasicrystals, *J. Phys.: Condens. Matter*, **16**(47), 5229–5240.

[9] Feuerbacher M and Urban K, 2003, Platic behaviour of quasicrystalline materials, in *Quasicrystals*, ed. by Trebin H R, Wiley Press, Berlin.

[10] Guyot P and Canova G, 1999, The plasticity of icosahedral quasicrystals, *Phil. Mag. A*, **79**(11), 2815–2822.

[11] Feuerbacher M, Schall P, Estrin Y *et al.*, 2001, A constitutive model for quasicrystal plasticity, *Phil. Mag. Lett.*, **81**(7), 473–482.

[12] Estrin Y, 1996, in: *Unified Constitutive Laws of Plastic Deformation*, ed. by Krausz A and Krausz K, Academic Press, New York.

[13] Eshelby J D, 1956, The continuum theory of dislocations in crystals, *Solid State Physics*, Vol. 3, ed. by Seits F *et al.*, New York: Academic Press.

[14] Fan T Y and Fan L, 2008, Plastic fracture of quasicrystals, *Phil. Mag.*, **88**(4), 523–535.

[15] Dugdale D S, 1960, Yielding of steel sheets containing slits, *J. Mech. Phys. Solids*, **32**(2), 105–108.

[16] Barenblatt G I, 1962, The mathematical theory of equilibrium of crack in brittle fracture, *Adv. Appl. Mech.*, **7**, 55–129.

[17] Fan T Y and Mai Y W, 2004, Elasticity theory, fracture mechanics and some relevant thermal properties of quasicrystalline materials, *Appl. Mech. Rev.*, **57**(5), 325–344.

[18] Fan T Y and Fan L, 2011, Relation between Eshelby integral and generalized BCS and generalized DB models for some one- and two-dimensional quasicrystals, *Chin. Phys. B*, **20**(4), 036102.

[19] Meng X M, Tong B Y and Wu Y K, 1994, Mechanical properties of quasicrystals $Al_{65}Cu_{20}Co_{15}$, *Acta Metall. Sin.*, **30**(1), 61–64 (in Chinese).

[20] Bilby B A, Cottrell A H and Swinden K H, 1963, The spread of plastic yield from a notch, *Proc. R. Soc. A*, **272**(2), 304–314.

[21] Bilby B A, Cottrell A H, Smith E *et al.*, 1964, Plastic yealding from sharp notches, *ibid*, **279**(1), 1–9.

[22] Rice J R, 1968, A path independent integral and approximate analysis of strain concentration by notches and cracks, *J. Appl. Mech.*, **35**(4), 379–386.

[23] Cherepanov G P, 1969, On crack propagation in solids, *Int. J. Solids Struct.*, **5**(8), 863–871.

[24] Begley G T and Landes J D, 1972, The J-integral as a fracture criterion, *Fracture Toughness*, ASTM STP 514, American Society for Testing and Materials, Philadelphia, pp. 1–20.

[25] Landes J D and Begley G T, 1972, The effect of specimen geometry on J_{IC}, *Fracture Toughness*, ASTM STP 514, American Society for Testing and Materials, Philadelphia, pp. 24–29.

[26] Muskhelishvili N I, 1956, *Singular Integral Equations*, English translation by Radok J R M, Noordhoff, Groningen.

[27] Li W and Fan T Y, 2009, Study on plastic analysis of crack problem of three-dimensional quasicrystalline materials, *Phil. Mag.*, **89**(31), 2823–2831.

Chapter 15

Fracture Theory of Solid Quasicrystals

Quasicrystals are brittle, and the study on the fracture behaviour of the material is significant. In the previous chapters, many of the crack problems are investigated, and the exact analytic, approximate and numerical solutions are constructed; this provides a basis for discussing the fracture theory. The exact solutions of the crack problem of quasicrystals present particular importance, which reveal the essential nature describing the fracture behaviour of this kind of material. The first exact analytic solution on a Griffith crack in decagonal quasicrystals was given by Li and Fan [1] in 1999. Afterwards, Fan and coworkers [1–3] developed the idea of linear elastic fracture mechanics of quasicrystals based on the common feature of crack solutions obtained in that date. Trebin *et al.* [4] discussed the relevant topic from another point of view. Recently, Fan and Guo [5], Zhu and Fan [6] and Li and Fan [7] observed the analytic solutions of cracks in three-dimensional icosahedral quasicrystals, Fan and Fan [8, 9] and Li and Fan [15] found the plastic analytic solutions of cracks, and Zhu and Fan [10] and Wang *et al.* [11] carried out the numerical analysis on dynamic crack of quasicrystals, respectively. And the measurement of fracture toughness of the material is also reported [12]. The quite fruit information mentioned above suggests to be needed to give a summary on the fracture study of quasicrystals. This chapter undertakes a task and may put forward some pointers for the fracture mechanics of quasicrystals.

15.1 Linear Fracture Theory of Quasicrystals

The descriptions on crack solutions in different chapters reveal some common nature of stress field and displacement field around the crack tip. For all exact solutions of cracks of one-, two- and three-dimensional quasicrystals in the linear elasticity regime, the resulting expressions of stress and displacement components hold with respect to arbitrary variables x and y. In addition, we found quite a lot of solutions of cracks of the dynamic state and nonlinear behaviour of quasicrystals; some common features are also exhibited. At first, we analyse the linear elastic solutions. As indicated in Section 8.1 of Chapter 8, within the framework of the linear elasticity and infinite-sharp crack model, stress near crack tip, i.e.,

$$\mathrm{r1/a} \ll 1 \tag{15.1-1}$$

stresses for phonons as well as for phasons appear singular in the order of $r_1^{-1/2}$(r1 $\to$ 0), and the other terms can be ignored when compared to this term. Although stress singularity is implausible, it is the result of an idealized mathematical model. Quite a few researchers indicated its severe weakness in theory and the paradox of its methodology [1,2]. However, prior to the actual establishment of a more reasonable fracture theory, we still continue using this theory in expectable near future.

If temporarily accepting this theory, we focus on the field variables near the crack tip. If in (8.3-14) only keeping the term in order of $(r_1/a)^{-1/2}$, we have

$$\sigma_{xx} = \frac{K_I^{\|}}{\sqrt{2\pi r_1}}\cos\frac{1}{2}\theta_1\left(1-\sin\frac{1}{2}\theta_1\sin\frac{3}{2}\theta_1\right)$$

$$\sigma_{yy} = \frac{K_I^{\|}}{\sqrt{2\pi r_1}}\cos\frac{1}{2}\theta_1\left(1+\sin\frac{1}{2}\theta_1\sin\frac{3}{2}\theta_1\right)$$

$$\sigma_{xy} = \sigma_{yx} = \frac{K_I^{\|}}{\sqrt{2\pi r_1}}\cos\frac{1}{2}\theta_1\cos\frac{3}{2}\theta_1$$

$$H_{xx} = -\frac{d_{21}K_I^{\|}}{\sqrt{2\pi r_1}}\sin\theta_1\left(2\sin\frac{3}{2}\theta_1+\frac{3}{2}\sin\theta_1\cos\frac{5}{2}\theta_1\right)$$

$$H_{yy} = \frac{d_{21}K_I^{\|}}{\sqrt{2\pi r_1}}\frac{3}{2}\sin^2\theta_1\cos\frac{5}{2}\theta_1$$

$$H_{xy} = -\frac{d_{21}K_I^{\|}}{\sqrt{2\pi r_1}}\frac{3}{2}\sin^2\theta_1\sin\frac{5}{2}\theta_1$$

$$H_{yx} = \frac{d_{21}K_I^{\|}}{\sqrt{2\pi r_1}}\sin\theta_1\left(2\cos\frac{3}{2}\theta_1 - \frac{3}{2}\sin\theta_1\sin\frac{5}{2}\theta_1\right) \quad (15.1\text{-}2)$$

where $d_{21} = R(K_1 - K_2)/4(MK_1 - R^2)$ given in Section 8.3 of Chapter 8 and

$$K_I^{\|} = \lim_{x\to a^+}\sqrt{2\pi(x-a)}\sigma_{yy}(x,0) = \sqrt{\pi a}p \quad (15.1\text{-}3)$$

in which

$$\sigma_{yy}(x,0) = \begin{cases} p\left(\dfrac{x}{\sqrt{x^2-a^2}} - 1\right) & |x| > a \\ -p & |x| < a \end{cases} \quad (15.1\text{-}4)$$

It is one of the results of Sections 8.3 and 8.4 of Chapter 8, and (15.1-3) represents a physical parameter describing the fracture behaviour of quasicrystals under Mode I loading conditions.

The physical meaning of the generalized surface forces $hi = H_{ij}n_j$ is clear but have not been measured so far, and we do not consider hi at the physical boundary (simply assume zero). Therefore, we only obtain the $K_I^{\|}$, but the stress intensity factor for the phason field still exits if we do not assume the generalized tractions hi to be zero. Can we use the stress intensity factor $K_I^{\|}$ in the parallel space (physical space) as the physical parameter to control the crack stability (instability) in quasicrystals? This only depends upon tests.

It can be found that $K_I^{\|}$ given in (15.1-3) is not directly related to the material constants of quasicrystals in the case for any self-equilibrium applied stress state (this is similar to that for conventional fracture mechanics for structural materials). Nevertheless, it does not mean that it cannot be used as the physical parameter to govern the crack stability/instability in quasicrystals. Further study is needed in this topic.

But the displacement field concerning the crack is strongly related with material constants (this is also similar to that of conventional fracture mechanics for structural materials), and we must distinguish the results for different quasicrystal systems.

For point groups $5m$ and $10mm$, it can be found from (8.3-17) of Chapter 8 that

$$u_y(x,0) = \begin{cases} 0 & |x| > a \\ \dfrac{p}{2}\left(\dfrac{K_1}{MK_1 - R^2} + \dfrac{1}{L+M}\right)\sqrt{a^2 - x^2} & |x| < a \end{cases} \tag{15.1-5}$$

$$wy(x,0) = 0, |x|<\infty$$

The strain energy of the system due to the existence of crack is

$$\begin{aligned} W_I &= 2\int_0^a (\sigma_{yy}(x,0) \oplus H_{yy}(x,0)(u_y(x,0) \oplus w_y(x,0))dx \\ &= 2\int_0^a \sigma_{yy}(x,0)u_y(x,0)dx \\ &= \frac{\pi a^2 p^2}{4}\left(\frac{1}{L+M} + \frac{K_1}{MK_1 - R^2}\right) \end{aligned} \tag{15.1-6}$$

which is called crack strain energy with the suffix "I" to indicate the Mode I crack.

It can be found from the solution in Sections 8.3 and 8.4 of Chapter 8 that under the assumption of the generalized surface traction $hi = H_{ij}n_j$ being free, the crack strain energy is still relevant to both the elastic constant K_1 of the phason field and the phonon–phason coupling elasticity constant R apart from that relevant to the phonon elastic constants

$$L = C_{12}, \quad M = (C_{11} - C_{12})/2$$

Similar to Section 8.1 of Chapter 8, we define the strain energy release rate (crack growth force) for point groups $5m$ and $10mm$

$$\begin{aligned} G_I &= \frac{1}{2}\frac{\partial W_1}{\partial a} = \frac{\pi a p^2}{4}\left(\frac{1}{L+M} + \frac{K_1}{MK_1 - R^2}\right) \\ &= \frac{1}{4}\left(\frac{1}{L+M} + \frac{K_1}{MK_1 - R}\right)(K_I^{\parallel})^2 \end{aligned} \tag{15.1-7}$$

for point groups $5, \overline{5}$ and $10, \overline{10}$ quasicrystals

$$\begin{aligned} G_I &= \frac{L(K_1 + K_2) + 2(R_1^2 + R_2^2)}{8(L + M)c}(K_I^{\|})^2 \\ c &= M(K_1 + K_2) - 2(R_1^2 + R_2^2) \end{aligned} \tag{15.1-8}$$

and for icosahedral quasicrystals

$$\begin{aligned} G_I^{\|} &= \frac{1}{2}\left(\frac{1}{\mu + \lambda} + \frac{c_7}{c_3}\right)(K_I^{\|})^2 \\ c_1 &= \frac{R(2K_2 - K_1)(\mu K_1 + \mu K_2 - 3R^2)}{2(\mu K_1 - 2R^2)} \\ c_3 &= \mu(K_1 - K_2) - R^2 - \frac{(\mu K_2 - R^2)^2}{\mu K_1 - 2R^2} \\ c_7 &= \frac{c_3 K_1 + 2c_1 R}{\mu K_1 - 2R^2} \end{aligned} \tag{15.1-9}$$

which also indicate that the strain energy release rate relating is not only the phonon elasticity constants $\mathrm{L} = \mathrm{C}_{12}$, $\mathrm{M} = (\mathrm{C}_{11} - \mathrm{C}_{12})/2$ but also the phason elasticity constant K_1, K_2 and phonon–phason coupling constants R_1 and R_2.

In the above relations, for point groups $5m$ and $10mm$, due to $L + M > 0$, $MK_1 - R^2 > 0$, $M + L > 0$ and crack energy W_I and crack energy release rate G_I are all positive, which are meaningful in physical sense, for point groups $5, \overline{5}$ and $10, \overline{10}$, icosahedral and other quasicrystals too.

Considering the obvious physical meaning of G_I, we recommend

$$G_I = G_{IC} \tag{15.1-6}$$

as the crack initiation criterion, where G_{IC} is the critical value, a material constant determined experimentally. With the availability of explicit expression G_I, the measurement of G_{IC} is convenient, which is to be discussed in the following section. The above results have been documented in Chapters 8 and 9 and the relevant references.

With these common features of cracks in quasicrystalline materials, the fundamental of fracture theory for the material can be set up.

15.2 Crack Extension Force Expressions of Standard Quasicrystal Samples and Related Testing Strategy for Determining Critical Value G_{IC}

Mong *et al.* [12] measured the fracture toughness for $Al_{65}Cu_{20}Co_{15}$ decagonal quasicrystals by using non-standard specimen because in that time there was lack of expressions on stress intensity factors and energy release rate. In addition, the measurement of fracture toughness used non-standard approach, e.g. the indentation approach.

During the characterization of the mechanical properties of quasicrystals, similar to conventional structural materials, standard samples are expected to use, such as cracked samples. Here, we recommend the three-point bending specimen shown in Fig. C1 in Appendix C of Chapter 14 the and compact tension specimen shown in Fig. 15.1 for determining G_{IC}. The corresponding G_I expressions are obtained by extending the formula (15.1-7) and others.

15.2.1 *Characterization of G_I and G_{IC} of three-point bending quasicrystal samples*

Due to $K_I^{\|}$ independent of material constants, according to fracture mechanics, the stress intensity factor corresponding to the three-point bending specimen as shown in Fig. C1 of Appendix C of Chapter 14 is

$$K_I^{\|} = \frac{PS}{BW^{3/2}}\left[29\left(\frac{a}{W}\right)^{1/2} - 4.6\left(\frac{a}{W}\right)^{3/2} + 21.8\left(\frac{a}{W}\right)^{5/2} - 37.6\left(\frac{a}{W}\right)^{7/2} + 38.7\left(\frac{a}{W}\right)^{9/2}\right] \tag{15.2-1}$$

Therefore, the extension of (15.1-17) leads to

$$G_I = \frac{1}{4}\left(\frac{1}{L+M} + \frac{K_1}{MK_1 - R^2}\right)\frac{PS}{BW^{3/2}}\left[29\left(\frac{a}{W}\right)^{1/2} - 4.6\left(\frac{a}{W}\right)^{3/2} + 21.8\left(\frac{a}{W}\right)^{5/2} - 37.6\left(\frac{a}{W}\right)^{7/2} + 38.7\left(\frac{a}{W}\right)^{9/2}\right] \tag{15.2-2}$$

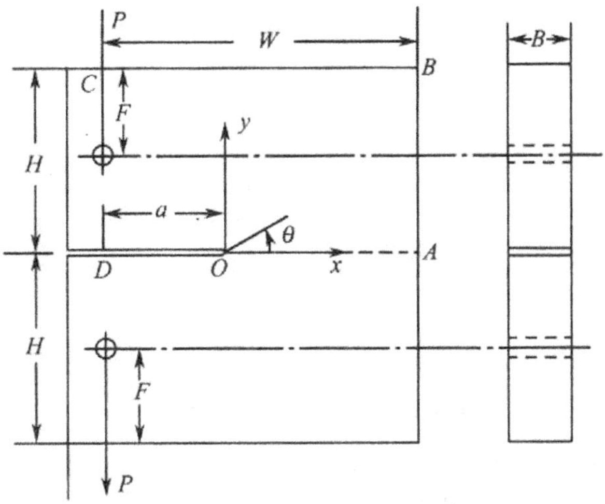

Fig. 15.1. Compact tension specimen for measuring fracture toughness of quasicrystalline material.

for point groups $5m$ and $10mm$, where S is the sample span, B is the sample thickness, W is the sample width, a is the crack length plus the size of the machined notch, and P is the external force (per unit length). Finally, the G_{IC} value can be determined by measuring the critical external force PC. For other quasicrystal systems, there are similar results.

15.2.2 *Characterization of G_I and G_{IC} of compact tension quasicrystal sample*

It can be found in fracture mechanics that the stress intensity factor

$$K_I^{\parallel} = \frac{PS}{BW^{3/2}}\left[29.6\left(\frac{a}{W}\right)^{1/2} - 185.5\left(\frac{a}{W}\right)^{3/2} + 655.7\left(\frac{a}{W}\right)^{5/2} - 1017.0\left(\frac{a}{W}\right)^{7/2} + 638.9\left(\frac{a}{W}\right)^{9/2}\right] \tag{15.2-3}$$

Therefore, for point groups $5m$ and $10mm$ quasicrystals,

$$G_I = \frac{1}{4}\left(\frac{1}{L+M} + \frac{K_1}{MK_1 - R^2}\right)\frac{P}{BW^{3/2}}\left[29.6\left(\frac{a}{W}\right)^{1/2}\right.$$

$$- 185.5\left(\frac{a}{W}\right)^{3/2} + 655.7\left(\frac{a}{W}\right)^{5/2} - 1017.0\left(\frac{a}{W}\right)^{7/2}$$
$$+ 638.9\left(\frac{a}{W}\right)^{9/2}\Bigg] \tag{15.2-4}$$

where B, W, a and P have the same meaning above. The G_{IC} value can be determined by measuring the critical external force P_C. For other quasicrystal systems, there are similar results.

15.3 Nonlinear Fracture Mechanics

In the regime for nonlinear deformation of quasicrystals, the stress intensity factor and energy release rate cannot be used as a fracture parameter, and we must carry out the elasto-plastic analysis, and the exact solutions to the crack problems have not been obtained except for a few of the special cases. Fortunately, this difficult topic has been discussed in Chapter 14 in detail, now it is needed to list some key points only.

Instead of the stress intensity factor and energy release rate, the crack tip opening displacement or Eshelby integral may be a parameter characterizing the mechanical behaviour of the crack tip under nonlinear deformation of quasicrystals. These quantities are strongly related to material constants, so the discussion must be done for distinguishing quasicrystal systems.

For one-dimensional hexagonal quasicrystals, we have obtained the crack tip sliding displacement for Mode III crack as

$$\delta_{\mathrm{III}} = \frac{4K_2\tau_c a}{\pi(\mathrm{C}_{44}\mathrm{K}_2 - R_3^2)}\left[\mathrm{lnsec}\left(\frac{\pi\tau_1}{2\tau_c}\right)\right] \tag{15.3-1}$$

and for two-dimensional quasicrystals with point groups $5m$ and $10mm$, the crack tip opening displacement for Mode I crack is

$$\delta_I = \frac{2\sigma_c a}{\pi}\left[\frac{1}{L+M} + \frac{K_1}{MK_1 - R^2}\right]\left[\ln\sec\left(\frac{\pi p}{2\sigma_c}\right)\right] \tag{15.3-2}$$

For icosahedral quasicrystal, the crack tip opening displacement is given by (14.4-10) in Chapter 14 such as

$$\delta_I = \lim_{x\to l} 2u_y(x,0) = \lim_{\phi\to\phi_2} 2u_y(x,0) = 2\left(\frac{1}{\lambda+\mu}+\frac{c_4}{c_2}\right) \cdot\frac{\sigma_s a}{\pi}\cdot\ln\sec\left(\frac{\pi}{2}\frac{p}{\sigma_s}\right)$$

in which the parameters were shown by (14.4-11)

$$c_2 = \mu(K_1-K_2) - R^2 - \frac{(\mu K_2 - R^2)^2}{\mu K_1 - 2R^2},$$

$$c_4 = c_1 R + \frac{1}{2}c_2\left(K_1 + \frac{\mu K_1 - 2R^2}{\lambda+\mu}\right)$$

with $$c_1 = \frac{R(2K_2-K_1)(\mu K_1+\mu K_2 - 3R^2)}{2(\mu K_1 - 2R^2)}.$$

Figure 15.2 shows the crack tip opening displacement versus applied stress.

And the size of the plastic zone around the crack tip is

$$d = a\left[\sec\left(\frac{\pi\tau_1}{2\tau_c}\right)-1\right]$$

for one-dimensional hexagonal quasicrystals and

$$d = a\left[\sec\left(\frac{\pi p}{2\sigma_c}\right)-1\right]$$

for two-dimensional quasicrystals with point groups $5m$ and $10mm$.

And we have the fracture criterion for Mode I crack as

$$\delta_I = \delta_{Ic} \tag{15.3-3}$$

For Mode II and Mode III cracks, there are similar criteria, which have been discussed in Chapter 14.

As pointed out in Chapter 14, the Eshelby integral can also be as a fracture parameter, and based on which one can set up a fracture criterion, the full discussion can be found there.

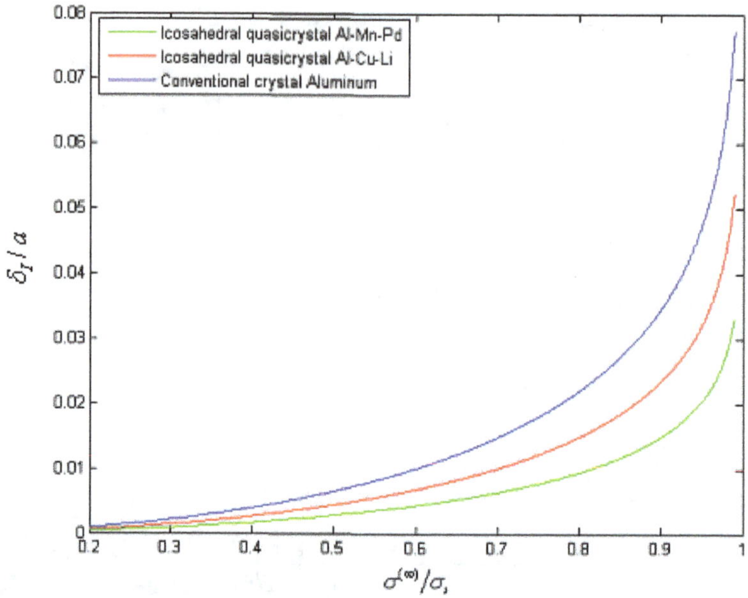

Fig. 15.2. Crack tip opening displacement of icosahedral quasicrystals versus applied stress.

The experimental measurement of the nonlinear fracture toughness of quasicrystals has been introduced in Appendix C of Chapter 14. We don't mention any more.

15.4 Dynamic Fracture

As we have known from Chapter 10, the study on the dynamics for quasicrystals presents a difficult situation, so the study for the dynamic fracture was given by numerical solutions only.

We here list some among them in Chapter 10 which might be the beneficial data for us.

By taking the so-called elasto-/hydro-dynamic equation system for quasicrystals, the dynamic crack initiation problem can be solved by the finite difference method. In linear case, the crack dynamic initiation can be described by the dynamic stress intensity factor; for some samples the results are listed in Chapter 10, which are sensitive

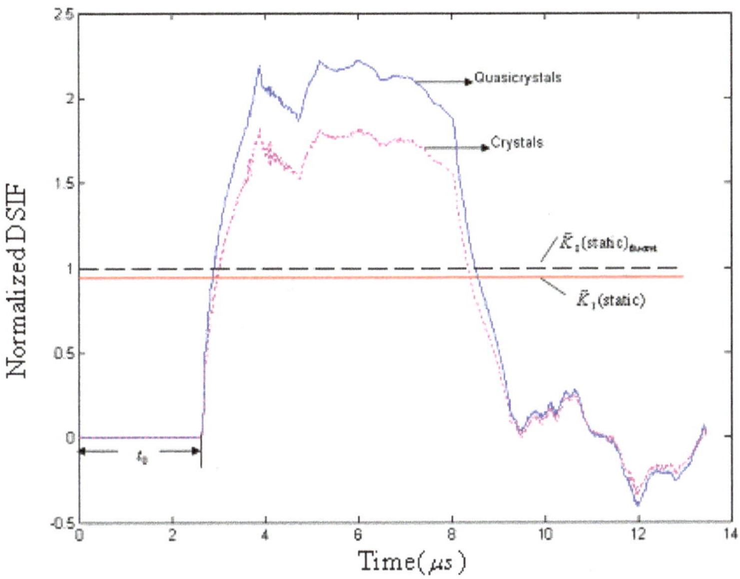

Fig. 15.3. Normalized dynamic stress intensity factor of central stationary crack specimen under Heaviside impact loading (for decagonal Al–Ni–Co quasicrystals).

functions of loading type, loading rate and sample geometry including crack geometry. One of the results can be seen in Fig. 15.3.

With the results, we can propose the fracture criterion for dynamic crack initiation

$$K_I(t) = K_{Id}(\dot{\sigma}) \tag{15.4-1}$$

in which $K_I(t)$ is the dynamic stress intensity factor evaluated by different approaches and $K_{Id}(\dot{\sigma})$ represents the dynamic fracture toughness for the initiation of the crack growth of the material, which is measured by a test, a material constant, but is the function of loading rate $\dot{\sigma}$. While for fast crack propagation/crack arrest problems we have results, e.g., shown in Fig. 15.4, for central crack specimen, there is a fracture criterion such as

$$K_I(t) \le K_{ID}(V) \tag{15.4-2}$$

where $K_I(t)$ is also the dynamic stress intensity factor, a computational quantity, while $K_{ID}(V)$ denotes the fracture toughness for the

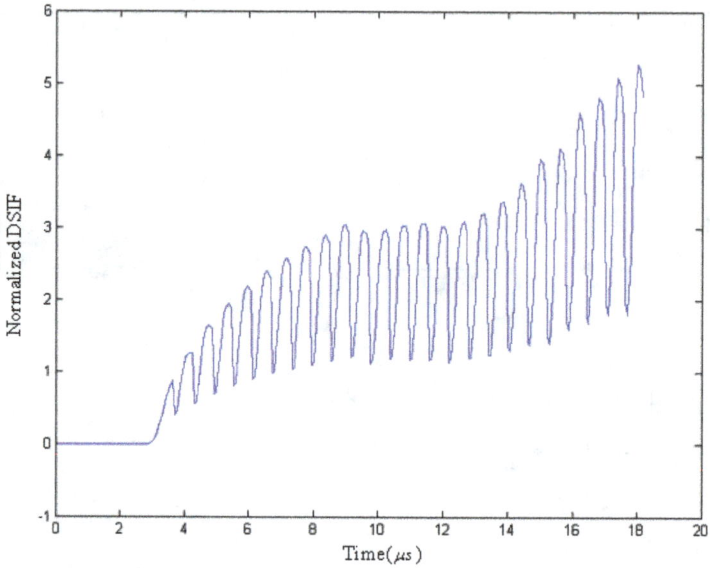

Fig. 15.4. Normalized dynamic stress intensity factor of propagating crack with constant crack speed of central crack specimen (for icosahedral Al–Pd–Mn quasicrystals).

fast propagating crack, which must be measured by a test, a material constant, but is the function of crack speed $V = da/dt$. In (15.4-2), the equality sign represents the crack propagation condition, and the inequality sign marks the crack arrest condition.

15.5 Measurement of Fracture Toughness and Relevant Mechanical Parameters of Quasicrystalline Material

Reference [12] reported the measurement of fracture toughness of two-dimensional decagonal quasicrystal $Al_{65}Cu_{20}Co_{15}$ as well as three-dimensional icosahedral Al–Li–Cu quasicrystal; the authors used the indentation approach.

15.5.1 *Fracture toughness*

The material is locally pressed, the crack around the indentation will appear as the compressible stress reaches a certain value; this describes the ability of the fracture of the material along the direction of the compressible stress. When the crack length $2a$ is greater than the 2.5 times of the length $2c$ of the diagonal of the indentation, then the fracture toughness can be evaluated by

$$K_{IC} = 0.203\ HV\sqrt{c}(^3\sqrt{(c/a)}) \tag{15.5-1}$$

in which HV denotes the hardness of the material.

Their results of the measurement for decagonal Al–Ni–Co quasicrystal is

$$K_{IC} = 1.0\text{–}1.2\ \text{MPa}\sqrt{m} \tag{15.5-2}$$

with $HV = 11.0 - 11.5$ GPa, and for icosahedral Al–Li–Cu quasicrystal is

$$K_{IC} = 0.94\ \text{MPa}\sqrt{m} \tag{15.5-3}$$

in which $HV = 4.10$ GPa.

The values of fracture toughness for general alloys for black metals measured by Ma *et al.* [13] are much greater than the above data, and those for aluminum, alloys and other coloured metals are also, e.g., for aluminum, it is 33 MPa$\sqrt{m}$, refer to Fan [14]. So, one finds that quasicrystals are very brittle.

The author of this book minds the indirect measurement for fracture toughness of quasicrystals through indentation probably is not so exact; because formula (15.5-1) is empirical, the exact measurement should be using the stress intensity factor formula. Due to the high brittleness of the material, maybe it is easy by taking the indentation approach.

15.5.2 *Tension strength*

The tensile strength σ_c is measured through the formula

$$\sigma_c = 0.187P/a^2 \tag{15.5-4}$$

Meng *et al.* [12] obtained

$$\sigma_c = 450\ \text{MPa} \tag{15.5-5}$$

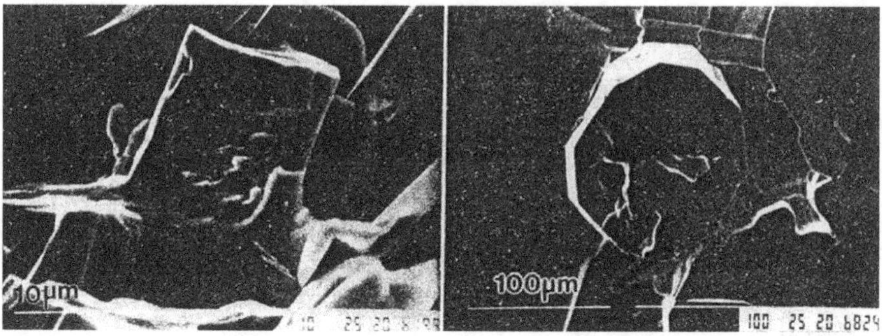

(a) (b)

Fig. 15.5. SEM morphology of grain interior hole (a) prismatic grain with perfect basic plane, (b) grain with prefect prismatic plane.

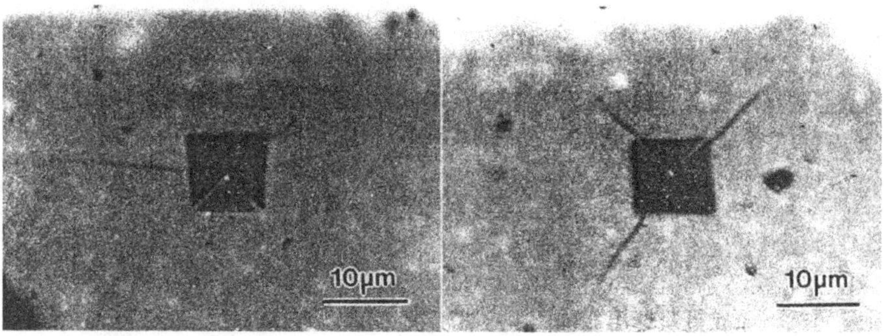

(a) (b)

Fig. 15.6. Morphology of cracks near impression under $100\,g$ load (a) before annealing, (b) after 850°C, 36 h annealing.

before annealing and

$$\sigma_c = 550 \text{ MPa} \tag{15.5-6}$$

after annealing for decagonal Al–Ni–Co quasicrystals.

Figure 15.5 shows the SEM morphology of the grain interior containing a large hole, Fig. 15.6 shows the diagram of an indentation crack under applied load $100\,g$ before annealing and after annealing, and Fig. 15.7 shows the SEM fractograph and fracture feature for decagonal $Al_{65}Cu_{20}Co_{15}$ quasicrystals.

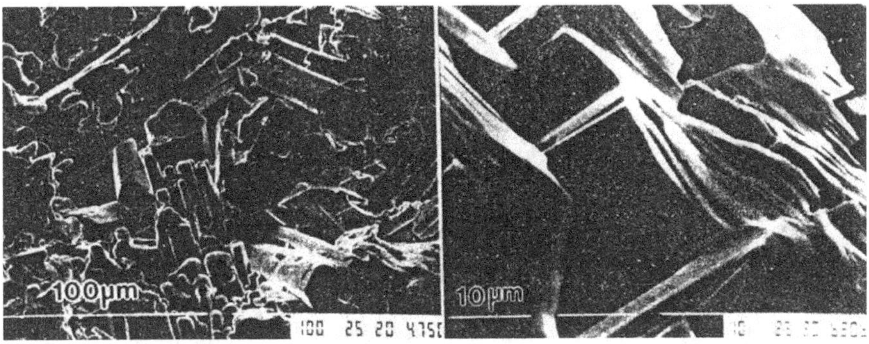

(a) (b)

Fig. 15.7. (a) SEM fractography, fracture feature of decagonal $Al_{65}Cu_{20}Co_{15}$ quasicrystal grain without plastic deformation, (b) magnification.

Fan, Tang and Chen *et al.* gave a summarizing on linear, nolinear and dynamic fracture theories of quasicrystals in Ref. [16].

References

[1] Li X F, The defect problems and their analytic solutions in elasticity of quasicrystals, Dissertation, Beijing Institute of Technology, Beijing 1999; Li X F, Fan T Y and Sun Y F, 1999, A decagonal quasicrystal with a Griffith crack, *Phil. Mag. A*, **79**(8), 1943–1952.

[2] Fan T Y, 2000, Mathematical theory of elasticity and defects of quasicrystals, *Advances of Mechanics*, **30**(2), 161–174 (in Chinese).

[3] Fan T Y and Mai Y W, 2004, Theory of elasticity, fracture mechanics and some relevant thermal properties of quasicrystalline materials, *Appl. Mech. Rev.*, **57**(5), 235–244.

[4] Rudhart C, Gumsch and Trebin H R, 2003, Crack propagation in quasicrystals, *Quasicrystals*, ed by Trebin H-R, Wiely Press, Berlin.

[5] Fan T Y and Guo L H, 2005, The final governing equation of plane elasticity of icosahedral quasicrystals, *Phys. Lett. A*, **341**(5), 235–239.

[6] Zhu A Y and Fan T Y, 2007, Elastic analysis of Mode II Griffith crack in an icosahedral quasicrystal, *Chin. Phys.*, **16**(4), 1111–1118.

[7] Li L H and Fan T Y, 2008, Complex variable method for plane elasticity of icosahedral quasicrystals and elliptic notch problem, *Science in China, G*, **51**(6), 1–8.

[8] Fan T Y and Fan L, 2008, Plastic fracture of quasicrystals, *Phil. Mag.*, **88**(4), 323–335.

[9] Fan T Y and Fan L, 2011, The relation between generalized Eshelby energy-momentum tensor and generalized BCS and DB models for some one- and two-dimensional and quasicrystals, *Chin. Phys. B*, **20**(4), 036102.

[10] Zhu A Y and Fan T Y, 2008, Dynamic crack propagation of decagonal $Al - Ni - Co$ decagonal quasicrystals, *J. Phys.: Condens. Matter*, **20**(29), 295217.

[11] Wang X F, Fan T Y and Zhu A Y, 2009, Dynamic behaviour of the icosahedral Al-Pd-Mn quasicrystals with a Griffith crack, *Chin. Phys. B*, **18**(2), 709–714.

[12] Meng X M, Tong B Y and Wu Y K, 1994, Mechanical behaviour of $Al_{65}Cu_{20}Co_{15}$ quasicrystal, *Acta Metallurgy Sinica*, **30**(1), 61–64 (in Chinese).

[13] Ma D L ed., 1994, *Handbook for Parameters of Fracture Mechanics Behaviour of General Black Metals*, Industrial Press, Beijing (in Chinese).

[14] Fan T Y, 2003, *Foundation of Fracture Theory*, Science Press, Beijing (in Chinese).

[15] Li W and Fan T Y, 2009, Study on plastic analysis of crack problem decagonal Al-Ni-Co quasicrystals of point group 10, $\overline{10}$, *Int. Mod. Phys. Lett.* B, **23**(16), 1989–1999.

[16] Fan T Y, Tang Z Y and Chen W Q, 2012, Theory of linear, nonlinear and dynamic fracture of quasicrystalline materials, *Eng. Fract. Mech.*, **82**(2), 185–194.

Chapter 16

Hydrodynamics of Solid Quasicrystals

Generalized hydrodynamics is one of the important branches on the study of solid quasicrystals, refer to Lubensky *et al.* [1, 2]. Though the discussion on phonon–phason dynamics in Chapter 10, suggested by Rochal and Norman [3] and Fan *et al.* [4], is concerned somewhat with generalized hydrodynamics, the description there is too simple and too simplified, and some nature of hydrodynamics of quasicrystals have not been detailed touched. We here intend to give a detailed introduction on the hydrodynamics of solid quasicrystals of Lubensky *et al.* The theory is concerned with many aspects of physics and mathematics, which are listed in Appendix III of Major Appendix.

Before the discovery of quasicrystals, the hydrodynamics of solids (crystals) has been developed, see, e.g., the work of Martin *et al.* [5], and Fleming and Cohen [6], which is related to the viscosity of a solid. Considering the viscosity, the number of field variables and the number of field equations are enlarged. The nature is connected to symmetry breaking. At first, we introduce some basic concepts of the viscosity of a solid, which are beneficial to understand the hydrodynamics of quasicrystals.

16.1 Viscosity of Solid

The elasticity discussed in the first 13 chapters is reversible, and the plasticity studied in Chapter 14 is one of the irreversible deformations. The irreversibility of plasticity lies in the existence of dissipation. The viscosity of a solid is another kind of dissipation.

To study the viscosity of a solid, one can take a method similar to that in fluid dynamics [7]. Introducing the velocity of mass point $\mathrm{V} = (V_x, V_y, V_z)$ and the tensor of deformation velocity

$$\dot{\xi}_{ij} = \frac{1}{2}\left(\frac{\partial V_i}{\partial x_j} + \frac{\partial V_j}{\partial x_i}\right) \tag{16.1-1}$$

then viscosity stress tensor is defined by [7]

$$\sigma'_{ij} = 2\eta\left(\dot{\xi}_{ij} - \frac{1}{3}\dot{\xi}_{kk}\delta_{ij}\right) + \zeta\dot{\xi}_{kk}\,\delta_{ij} \tag{16.1-2}$$

here only the isotropic viscosity is considered, and denote η the bulk viscosity constant ζ the shear one. Equation (16.1-2) is the constitutive law of viscosity of an isotropic solid.

The general constitutive law of viscosity of icosahedral quasicrystals is [1]

$$\sigma'_{ij} = \eta_{ijkl}\dot{\xi}_{kl} \tag{16.1-3}$$

in which η_{ijkl} represents the viscosity coefficient tensor of the viscosity of a solid.

A description of the viscosity of a solid can also be introduced by a dissipation function R such as

$$R = \frac{1}{2}\eta_{ijkl}\dot{\xi}_{ij}\dot{\xi}_{kl} \tag{16.1-4}$$

so that we have

$$\sigma'_{ij} = \frac{\partial R}{\partial \dot{\xi}_{ij}} \tag{16.1-5}$$

which is similar to the relation between the stress tensor and density of strain energy in elasticity the form. For the icosahedral quasicrystals, the viscosity coefficients can be expressed as

$$\eta_{ijkl} = \zeta\delta_{ij}\delta_{kl} + \eta\left(\delta_{ik}\delta_{jl} + \delta_{il}\delta_{jk} - \frac{2}{3}\delta_{ij}\delta_{kl}\right) \tag{16.1-6}$$

For simplicity, in the following, we consider only the simplest case, i.e., $\sigma'_{ij} = 2\eta\left(\dot{\xi}_{ij} - \frac{1}{3}\dot{\xi}_{kk}\delta_{ij}\right)$; only one viscosity constant η (the bulk or longitudinal viscosity constant) is used. For a detailed description about the viscosity of a solid physically, refer to Ref. [1].

16.2 Generalized Hydrodynamics of Solid Quasicrystals

Considering both the elasticity and viscosity of solid quasicrystals leads to a generalized hydrodynamics that is developed by Lubensky *et al.* [1, 2]; they derived the governing equations of the hydrodynamics by using Poisson bracket method. Here, there are four systems of equations: mass conservation equation, momentum conservation equations, equations of motion of phonons due to symmetry breaking and phason dissipation equations if the energy conservation equation does not consider. The mass conservation equation is

$$\frac{\partial \rho(\mathrm{r},t)}{\partial t} = -\nabla_i(\mathrm{r})(\rho V_i) \tag{16.2-1}$$

and the momentum conservation law is

$$\begin{aligned}\frac{\partial g_i(\mathrm{r},t)}{\partial t} &= -\nabla_k(\mathrm{r})(V_k g_i) + \nabla_j(\mathrm{r})\left(\eta_{ijkl}\nabla_k(\mathrm{r})V_l\right)\\ &\quad -\left(\delta_{ij} - \nabla_i(\mathrm{r})u_j\right)\frac{\delta H}{\delta u_j} + \left(\nabla_i(\mathrm{r})\mathrm{w}_j\right)\frac{\delta H}{\delta w_j}\\ &\quad -\rho\nabla_i(\mathrm{r})\frac{\delta H}{\delta \rho}g_j = \rho V_j\end{aligned} \tag{16.2-2}$$

In the meantime, one has the equations of motion of phonons due to symmetry breaking

$$\frac{\partial u_i(\mathrm{r},t)}{\partial t} = -V_j\nabla_j(\mathrm{r})u_i - \Gamma_u\frac{\delta H}{\delta u_i(\mathrm{r},t)} + V_i \tag{16.2-3}$$

and phason dissipation equations

$$\frac{\partial w_i(\mathrm{r},t)}{\partial t} = -V_j\nabla_j(\mathrm{r})w_i - \Gamma_w\frac{\delta H}{\delta w_i(\mathrm{r},t)} \tag{16.2-4}$$

A detailed discussions on the equations can be found in Major Appendix III, in which the Hamiltonian H is defined by

$$\begin{cases} H = \int\left[\dfrac{g^2}{2\rho} + \dfrac{1}{2}A\left(\dfrac{\delta\rho}{\rho_0}\right)^2 + B\left(\dfrac{\delta\rho}{\rho_0}\right)\nabla\cdot u\right]d^d\mathrm{r} + F_u + F_w + F_{uw}\\ g = \rho\mathrm{V}\end{cases} \tag{16.2-5}$$

where the integral (16.2-5) describes the contributions of the momentum and variation of mass density, the last three terms of (16.2-5) denote the contributions of phonon, phason and phonon–phason coupling, and A, B are new constants of materials describing the effect due to the variation of mass density, respectively, which are named Lubensky–Ramaswamy–Toner (LRT) constants, and the superscript of the volume element of integral represents dimension. Equations (16.2-1)–(16.2-4) are the equations of motion of hydrodynamics for solid quasicrystals, and the field variables include mass density ρ, velocity V_i (or momentum ρV_i), phonon displacement u_i and phason displacement w_i. In order to write the Hamiltonian H (16.2-5), one must give constitutive law of quasicrystals, in which the elastic constitutive equations are discussed in detail in the first 13 chapters for different quasicrystal systems, and A and B the material constants due the mass density relative variation $\frac{\delta\rho}{\rho_0}$; we named A and B as the LRT constants, where $\delta\rho = \rho - \rho_0$, and ρ_0 is the initial mass density.

The above listed equations are derived by Lubensky *et al.* [1] in 1985, but there were no details of the derivation which are supplemented in Major Appendix. Some arguments on the equations can be found in Refs. [3, 8, 9]. The amount of the filed equations including deformation geometry and constitutive law is 52 and belongs to differential-variational equations, and the solving of the initial-boundary value problems of these equations is very difficult.

16.3 Simplification of Plane Field Equations in Two-dimensional 5- and 10-fold Symmetrical Solid Quasicrystals

The equations of Lubensky *et al.* are derived for icosahedral quasicrystals but can be used for other solid quasicrystals observed so far. We here discuss point group $5, \overline{5}$ pentagonal and point group $10, \overline{10}$ decagonal quasicrystals, and only for plane field, i.e., the motion of the medium is in xy-plane if the z-axis is taken as 5- or 10-fold symmetry axis. In this case, the governing equations are the same for both point group $5, \overline{5}$ pentagonal and point $10, \overline{10}$ decagonal quasicrystals. All field variables depend only on variables x, y and t, and those of z-components are vanished. If terms $(\nabla_i(\mathrm{r})u_j)\frac{\delta H}{\delta u_j}$, $(\nabla_i(\mathrm{r})\mathrm{w}_j)\frac{\delta H}{\delta w_j}$ were omitted in Eq. (16.2-2), the governing equations (16.2-1)–(16.2-4) would reduce to (16.3-1), i.e.,

$$\left.\begin{aligned}
&\frac{\partial\rho}{\partial t}+\nabla\cdot(\rho\mathbf{V})=0\\
&\frac{\partial(\rho V_x)}{\partial t}+\frac{\partial(V_x\rho V_x)}{\partial x}+\frac{\partial(V_y\rho V_x)}{\partial y}=\eta\nabla^2V_x+\frac{1}{3}\eta\frac{\partial}{\partial x}\nabla\cdot\mathbf{V}+M\nabla^2u_x\\
&\quad+(L+M-B)\frac{\partial}{\partial x}\nabla\cdot\mathbf{u}+R_1\left(\frac{\partial^2w_x}{\partial x^2}+2\frac{\partial^2w_y}{\partial x\partial y}-\frac{\partial^2w_x}{\partial y^2}\right)\\
&\quad-R_2\left(\frac{\partial^2w_y}{\partial x^2}-2\frac{\partial^2w_x}{\partial x\partial y}-\frac{\partial^2w_y}{\partial y^2}\right)-(\mathrm{A}-\mathrm{B})\frac{1}{\rho_0}\frac{\partial\delta\rho}{\partial x}\\
&\frac{\partial(\rho V_y)}{\partial t}\frac{\partial(V_x\rho V_y)}{\partial x}+\frac{\partial(V_y\rho V_y)}{\partial y}=\eta\nabla^2V_y+\frac{1}{3}\eta\frac{\partial}{\partial y}\nabla\cdot\mathbf{V}+M\nabla^2u_y\\
&\quad+(L+M-B)\frac{\partial}{\partial y}\nabla\cdot\mathbf{u}+R_1\left(\frac{\partial^2w_y}{\partial x^2}-2\frac{\partial^2w_x}{\partial x\partial y}-\frac{\partial^2w_y}{\partial y^2}\right)\\
&\quad+R_2\left(\frac{\partial^2w_x}{\partial x^2}+2\frac{\partial^2w_y}{\partial x\partial y}-\frac{\partial^2w_x}{\partial y^2}\right)-(\mathrm{A}-\mathrm{B})\frac{1}{\rho_0}\frac{\partial\delta\rho}{\partial y}\\
&\frac{\partial u_x}{\partial t}+V_x\frac{\partial u_x}{\partial x}+V_y\frac{\partial u_x}{\partial y}=V_x+\Gamma_{\mathbf{u}}\left[M\nabla^2u_x+(L+M)\frac{\partial}{\partial x}\nabla\cdot\mathbf{u}\right.\\
&\quad\left.+R_1\left(\frac{\partial^2w_x}{\partial x^2}+2\frac{\partial^2w_y}{\partial x\partial y}-\frac{\partial w_x}{\partial y^2}\right)-R_2\left(\frac{\partial^2w_y}{\partial x^2}-2\frac{\partial^2w_x}{\partial x\partial y}-\frac{\partial^2w_y}{\partial y^2}\right)\right]\\
&\frac{\partial u_y}{\partial t}+V_x\frac{\partial u_y}{\partial x}=V_y\frac{\partial u_y}{\partial y}=V_y+\Gamma_{\mathbf{u}}\left[M\nabla^2u_y+(L+M)\frac{\partial}{\partial y}\nabla\cdot\mathbf{u}\right.\\
&\quad\left.+R_1\left(\frac{\partial^2w_y}{\partial x^2}-2\frac{\partial^2w_x}{\partial x\partial y}-\frac{\partial^2w_y}{\partial y^2}\right)+R_2\left(\frac{\partial^2w_x}{\partial x^2}+2\frac{\partial^2w_y}{\partial x\partial y}-\frac{\partial^2w_x}{\partial y^2}\right)\right]\\
&\frac{\partial w_x}{\partial t}+V_x\frac{\partial w_x}{\partial x}+V_y\frac{\partial w_x}{\partial y}=\Gamma_{\mathrm{w}}\left[K_1\nabla^2w_x+R_1\left(\frac{\partial^2u_x}{\partial x^2}-2\frac{\partial^2u_y}{\partial x\partial y}-\frac{\partial^2u_x}{\partial y^2}\right)\right.\\
&\quad\left.+R_2\left(\frac{\partial^2u_y}{\partial x^2}+2\frac{\partial^2u_x}{\partial x\partial y}-\frac{\partial^2u_y}{\partial y^2}\right)\right]\\
&\frac{\partial w_y}{\partial t}+V_x\frac{\partial w_y}{\partial x}+V_y\frac{\partial w_y}{\partial y}=\Gamma_{\mathrm{w}}\left[K_1\nabla^2w_y+R_1\left(\frac{\partial^2u_y}{\partial x^2}+2\frac{\partial^2u_x}{\partial x\partial y}-\frac{\partial^2u_y}{\partial y^2}\right)\right.\\
&\quad\left.-R_2\left(\frac{\partial^2u_x}{\partial x^2}-2\frac{\partial^2u_y}{\partial x\partial y}-\frac{\partial^2u_x}{\partial y^2}\right)\right]
\end{aligned}\right\}\tag{16.3-1}$$

in which $\nabla\cdot = \mathbf{i}\frac{\partial}{\partial x}+\mathbf{j}\frac{\partial}{\partial y}, \mathbf{V} = \mathbf{i}V_x+\mathbf{j}V_y, \mathbf{u} = \mathbf{i}u_x+\mathbf{j}u_y, \quad L = \mathrm{C}_{12}, \mathrm{M} = (\mathrm{C}_{11} - \mathrm{C}_{12})/2$.

16.4 Numerical Solution

The deformation, motion, dynamic behaviour and structure including the structure reconstruction of solid quasicrystals are often needed to solve equations such as (16.3-1), and the solving must be under appropriate initial and boundary conditions. To solve Eqs. (16.3-1) at physical time–space domain, the corresponding initial and boundary conditions must be given simultaneously. The more general discussion on the initial and boundary conditions of hydrodynamics of solid quasicrystals here has not been undertaken. We take a specimen shown in Fig. 16.1 and list the relevant conditions for the specimen, in which the computational points A_1 and A_2 and their coordinates are given as follows:

We assume that the specimen at initial time is in rest, so there are initial conditions

$$t = 0 : u_x = u_y = 0, w_x = w_y = 0, \mathrm{V}_x = \mathrm{V}_y = 0, \rho = \rho_0 \qquad (16.4\text{-}1)$$

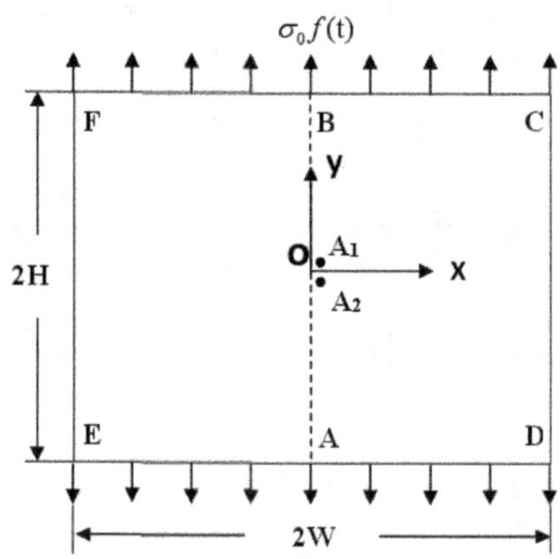

Fig. 16.1. Specimen of quasicrystal under impact loading $\mathrm{A}_1(10^{-4}m, 10^{-4}m)$ and $\mathrm{A}_2(10^{-4}m, -10^{-4}m)$.

At the surface of the specimen, there are the following boundary conditions:

$$y = \pm H, |x| < W : \sigma_{yy} = \sigma_0 f(t), \sigma_{yx} = 0, H_{yy} = H_{yx} = 0,$$
$$\mathrm{V}_x = \mathrm{V}_y = 0$$
$$x = \pm W, |y| < H : \sigma_{xx} = \sigma_{xy} = 0, H_{xx} = H_{xy} = 0,$$
$$\mathrm{V}_x = \mathrm{V}_y = 0 \tag{16.4-2}$$

in which σ_0 is a constant with the stress dimension and, $f(t)$ is the function of time, and we here take it as the Heaviside function, i.e., $f(t) = H(t)$, of course, this loading is an idealized case. The boundary conditions designed belong to the simplest one for both numerical computation and further experimental measurement, in which the non-homogeneous boundary condition is only for phonon stress, and the boundary conditions for phason field and viscosity flow field are zero. As an example, we calculate a decagonal Al–Ni–Co quasicrystal, with the experimentally measured material data $\rho_0 = 4.1868 \times 10^{-3}\mathrm{g/mm}^3$, viscosity coefficient $\eta = 1\mathrm{cm}^2/s$ approximately [1] (note that here η in fact is corresponding to η/ρ; refer to Ref. [1]) and elastic moduli $C_{11} = 2.3430$, $C_{12} = \mathrm{L} = 0.5741$ $(10^{12}\mathrm{dyn/cm}^2)$, which are obtained by resonant ultrasound spectroscopy; we have also chosen $K_1 = 1.22$ and $K_2 = 0.24$ $(10^{12}\mathrm{dyn/cm}^2)$ estimated by Monte Carlo simulation, $\mathit{\Gamma}_u = 4.8 \times 10^{-8}$, $\mathit{\Gamma}_w = 4.8 \times 10^{-10}$ $(\mathrm{cm}^3\mu\mathrm{s/g})$, refer to Chapter 10, A $\sim 10^{10}\mathrm{dyn/cm}^2$ and B $\sim 10^{10}\mathrm{dyn/cm}^2$ by an estimation. The coupling constants R_1,R_2 have not been measured so far. In computation, we take $R_1/M = 0.001$ and $R_2/M = 0.002$, where $M = (C_{11} - C_{12})/2$, $\sigma_0 = 5\mathrm{MPa}$, $H = 20\mathrm{mm}$ and $W = 10\mathrm{mm}$.

The specimen shown in Fig. 16.1 is the simplest specimen for exploring the hydrodynamic behaviour of the quasicrystal through computational analysis. The upper and lower surfaces of the specimen are subjected impact external stress $\sigma_{yy} = \sigma_0\mathrm{H(t)}$, where the loading function is taken to be the Heaviside function

$$H(t) = \begin{cases} 0, & t = 0 \\ 1, & t > 0 \end{cases}$$

Of course, this is an extreme case (at $t = 0$, the loading rate is infinite).

Equations (16.3-1) with initial and boundary value conditions (16.4-1) and (16.4-2) are named initial-boundary value problems (16.3-1)–(16.4-1), (16.4-2), in solving partial differential equations (16.3-1) under the initial-boundary value conditions (16.4-1)–(16.4-2). We solve the initial-boundary value problem by a finite difference method. The numerical solutions with highly stability are obtained.

The impact loading at the upper and lower surfaces of the specimen is the disturbance source, if the initial propagating rate of the disturbance source is denoted by v_0, which can be evaluated by $v_0 = \frac{H}{t_0}$, where t_0 the time of propagation of the disturbance source to the central transverse section, our computational point is very close to the location. The computation for time t_0 is very easy which can directly check our computation.

In Fig. 16.2 the variation of mass density versus time is illustrated, which explicitly explores the wave propagation character of the physical process under the given boundary conditions of the specimen. Before the arriving of the wave emanated from the upper surface in the computational point A_1 whose distance to upper surface is almost equal to H (completely similar to the wave emanated from lower surface arriving in the computational point A_2), the mass density maintains its initial value ρ_0, the "propagation time" is equal to $t_0 = 2.545\,\mu s$, which is evaluated by $t_0 = \frac{H}{v_0}$ and $v_0 = \frac{H}{t_0} =$ 7.8585 km/s which is very close to the value of the longitudinal wave speed $c_L = \sqrt{(L + 2M + A - 2B)/\rho_0} = 7.4807$ km/s (the formula of the longitudinal wave, refer to Lubensky *et al.* [1]). This is a direct check to the computation physically, though there is a smaller difference between the values of v_0 and c_L. It is understandable because the longitudinal wave speed formula is a linear elastic result, and the present case is a nonlinear medium, the disturbance source propagating speed v_0 from the upper surface to the computational point is not exact by equal to the linear longitudinal wave rate, but these two values are very close to each other. This shows that the mathematical formulation of the initial-boundary value

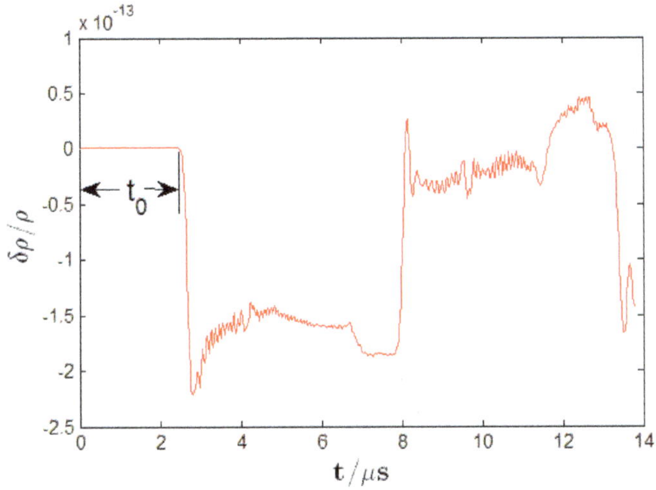

Fig. 16.2. Variation of mass density of the computational point A_1 (or A_2) of specimen versus time.

problem (16.3-1)–(16.4-1), (16.4-2) is correct physically, the solving method is effect, and the solution presents satisfactory precise.

The phenomenon indicates that for the complex system of wave propagation and diffusion coupling, the phonon wave propagation plays a dominating role. At the moment that the wave arrives at the central, because the dynamic load is an impacting force which produces a tensile wave, the value of mass density at central section decreases steeply to the minimum.

After the longitudinal waves emanated from upper and lower surfaces confluent with each other at the central section, and the arrival of succeeded transverse wave and the reflected waves by lateral surfaces of the specimen, the material at the location is compressible, the value of $\delta\rho/\rho_0$ will be increasing and then grows into a maximum value 2.527×10^{-14} at time $8.149\,\mu s$ indicating ρ beyond the initial value ρ_0 of mass density.

Figure 16.3 gives the time variation of velocities (the velocities of mass point to the coordinates) at point A_1 in x and y directions, respectively, in which the wave behaviour is still exhibited; before the "propagation time" t_0, the values of velocities are zero. The order

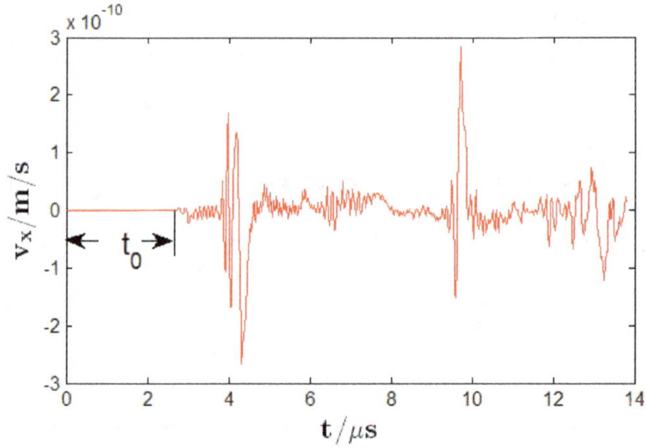

(a)

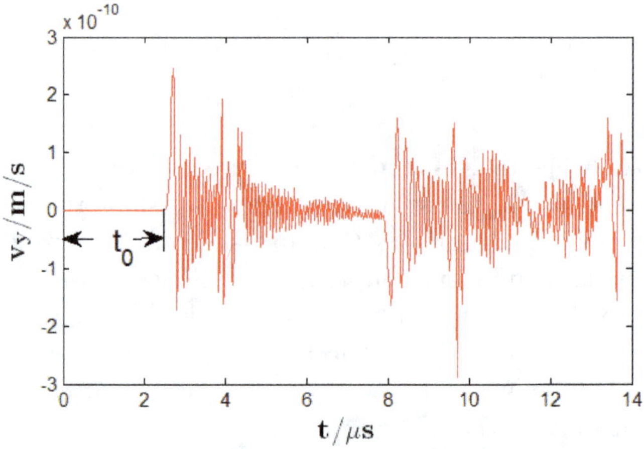

(b)

Fig. 16.3. Velocity fields at the computational point A_1 of specimen versus time: (a) V_x and (b) V_y.

of their magnitude is 10^{-10}m/s for the material; this quantitatively explores the hydrodynamic character of the motion pointed out by Lubensky *et al.* Though one knows the motion should be very slow, beforehand we did not know the realistic order of magnitude of the rate, and naturally there was no possibility to know the exact values of the rate, in that time the analysis was only qualitative.

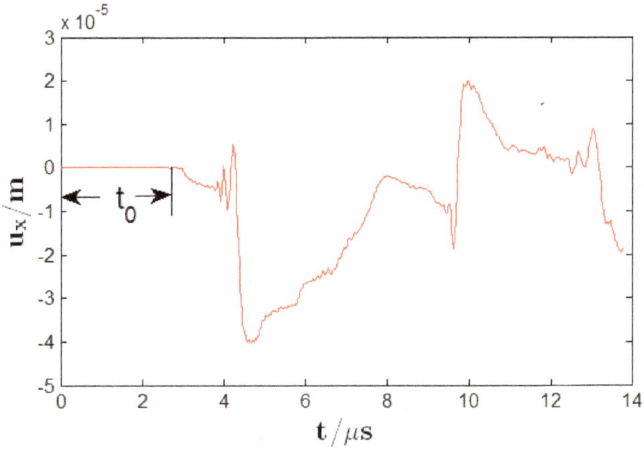

(a)

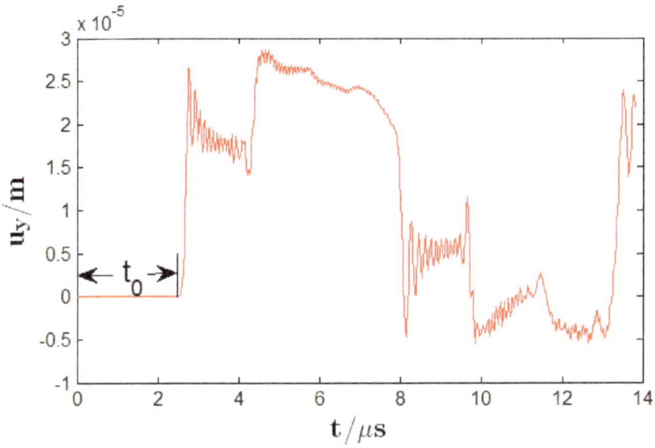

(b)

Fig. 16.4. Phonon displacement fields at the computational point A_1 versus time: (a) u_x and (b) u_y.

Figure 16.4 illustrates the time variation of phophon dispacements at point A_1 in directions x and y, respectively. The results show the wave propagation behaviour of the field variables, which is evident. Of course, the performance of u_x and u_y is different from each other, which is resulted from the boundary conditions and specimen configuration.

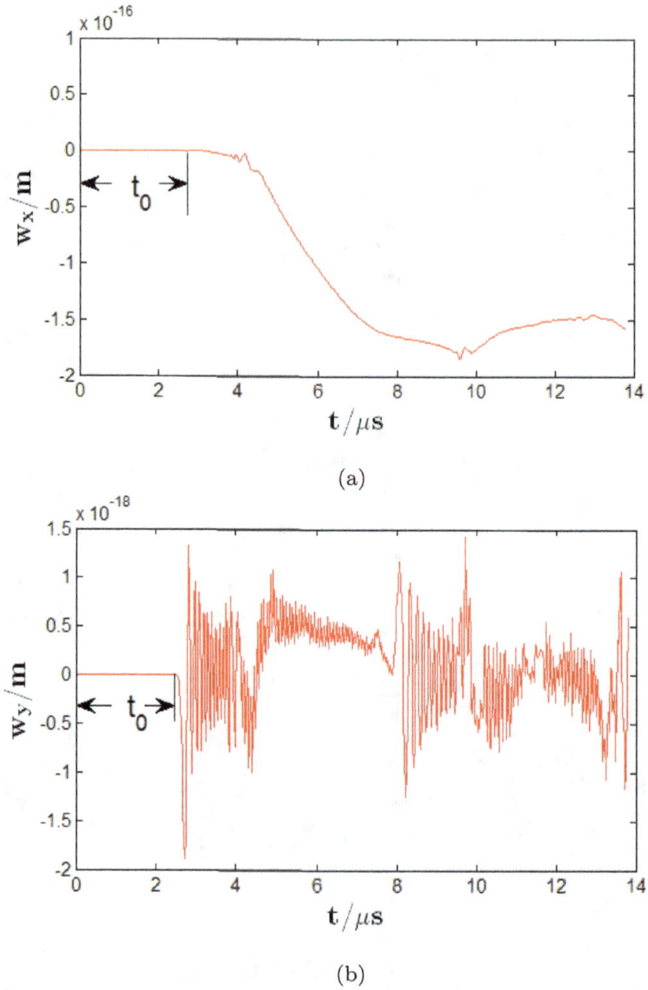

Fig. 16.5. Phason displacement fields at the computational point A_1 of specimen versus time: (a) w_x and (b) w_y.

The results on phason displacements w_x and w_y are ploted in Fig. 16.5, and their order of magnitude is very smaller than that of phonons, which is understandable because the external disturbance source comes from the phonon stress only, and the corresponding phason boundary conditions as homogeneous (the zero boundary conditions). If we take the phonon stress boundary

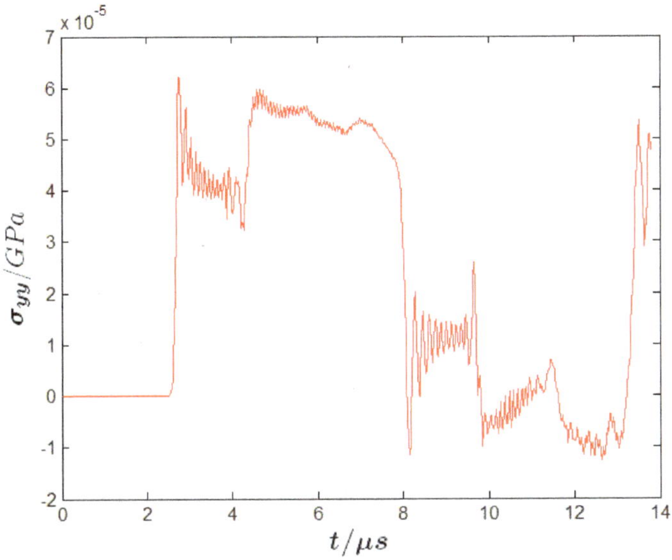

Fig. 16.6. Elastic normal stress at the computational point A_1 (or A_2) of specimen versus time.

conditions as homogeneous, and phason stress boundary conditions as non-homogeneous, then the order of the magnitude of solutions of phasons is much greater than that of the present results. In addition, the phason displacement field represents diffusion rather than wave propagation in physical nature, but the profile of the figures of w_x and w_y does not exhibit pure diffusion feature which is exhibited for the solutions of a simple diffusion equation due to the strong coupling of wave equations and the controlling of the mixed boundary conditions.

Figures 16.6–16.8 show the time variation of phonon, phason and viscosity normal stresses at the computational point; it is evident that phonon stress is greater by two order of magnitude than phason one, and the phonon and phason stresses are much greater than viscosity one. The outline of phason stress describes certain diffusive behaviour and so is evident different from that of the phonon one, while the outline of the phonon and viscosity ones exhibits a typical wave propagation character and presents a strong oscillation.

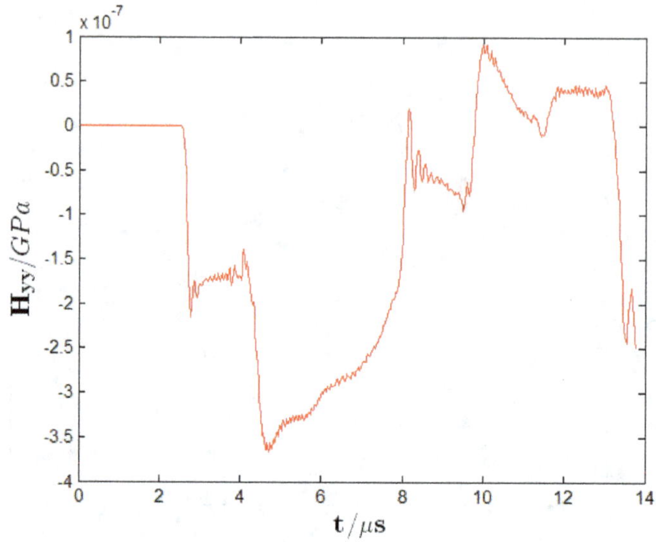

Fig. 16.7. Phason normal stress at the computational point A_1 (or A_2) of specimen versus time.

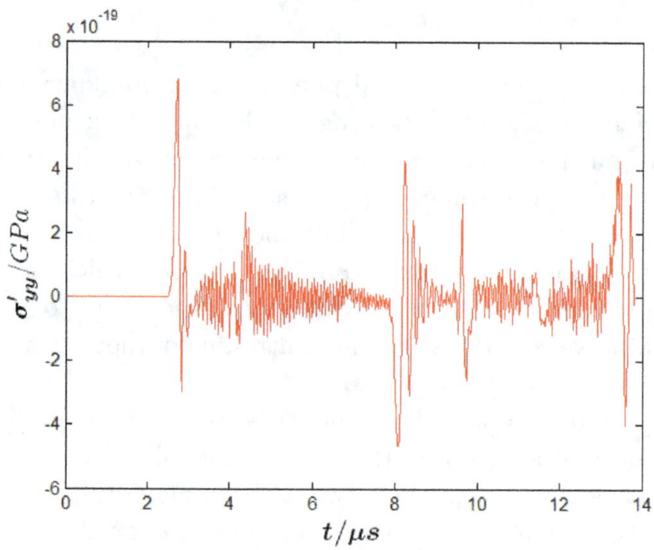

Fig. 16.8. Viscosity normal stress at the computational point A_1 (or A_2) of specimen versus time.

16.5 Conclusion and Discussion

This chapter gives a preliminary introduction on the hydrodynamics of solid quasicrystals of Lubensky *et al.* The numerical computation verifies the solvability and high accuracy of equations of Lubensky *et al.* [1], and the integrity, strictness and thinking profundity of their theory are proved in the detailed derivation in Appendix III in Major Appendix. The numerical computation verifies the correctness of the formulation developed here too. However, it is carried out through only a simple specimen, it describes only one of the dynamic processes of solid quasicrystals, the results are limited for a local computational point and a short time duration, and it conveys limited information. Nevertheless, the numerical solving and computer implementation for such a complex substantive system are significant, and the obtained stable and physically meaningful results are also significant.

The computational results reveal the phonon and phonon–phason coupling, and dominate the deformation and motion of the matter, which is undoubted for most quasicrystals. The computational results show that the variation of mass density is small, and the viscosity stresses are small for the present computed specimen; these do not mean the common response for all dynamic processes of the matter. Even if the weakness of the effects of compressibility and viscosity at the computed specimen, they may present significance in other dynamic processes. We know similar phenomena appear often in physics, for example, the gravitation due to curved space–time is very weak in general, but it is important for some special cases, for example, the gravitational waves observed recently [11].

So far, the solutions in time-spatial domain of hydrodynamics of Lubensky *et al.* are very few; our understanding on the topic is very limited, developing the computation is significant.

The work of hydrodynamics study may help us enhance the understanding of the deformation and motion of solid quasicrystals. Of course, the analysis of hydrodynamics here is a preliminary attempt, it needs further development on both the aspects in the theory and computation.

The theory of hydrodynamics of solid quasicrystals has been extended and developed by Fan and his coworkers to the study of generalized dynamics of soft-matter quasicrystals by introducing the fluid phonon and equation of state, and they have got some approximate solutions, which is a very complicated problem and beyond the scope of this chapter, hence the details are omitted here.

References

[1] Lubensky T C, Ramaswamy S and Toner J, 1985, Hydrodynamics of icosahedral quasicrystals, *Phys. Rev. B*, 32(11), 7444–7452.
[2] Lubensky T C, 1988, Symmetry, elasticity and hydrodynamics of quasistructures, in *Introduction to Quasicrystals*, ed. by Jaric V M, Boston, Academic Press, pp. 199–280.
[3] Rochal S B and Norman V L, 2002, Minimal model of the phonon-phason dynamicsof quasicrystals, *Phys. Rev. B*, 66, 144–204.
[4] Fan TY, Wang X F, Li W and Zhu A Y, 2009, Elasto-hydrodynamics of quasicrystals, *Phil. Mag*, **89**(6), 501–512.
[5] Martin P C, Parodi O and Pershan P S, United hydrodynamic theory for crystals, liquid crystals and normal fluids, *Phys. Rev. A*, **6**(6), 2401–2420.
[6] Fleming P D and Cohen C, 1976, Hydrodynamics of solids, *Phys. Rev. B*, **13**(2), 500–516.
[7] Landau L D and Lifshitz E M, 1988, *Theory of Elasticity*, Oxford, Pergamon Press.
[8] Khannanov S K, 2002, Dynamics of elastic and phason field in quasicrystals, *Phys Met Metallogr. (USSR)*, **93**, 397–403.
[9] Coddens G, 2006, On the problem of the relation between phason elasticity and phason dynamics in quasicrystals, *Eur. Phys. J.B*, **54**(1), 37–65.
[10] Cheng H, Fan T Y and Wei H, 2016, Time-space solutions for hydrodynamics of pentagonal/decagonal quasicrystals, *Appl. Math. Mech.*, **37**(10), 1393–1404.
[11] Abbott B P *et al.*, 2016, Observation of gravitation waves from a binary hole merger, *Phys. Rev. Lett.*, **116**, 061102.

Chapter 17

Two-dimensional Quasicrystals of the Second Kind

In Chapters 6–8, 10 and 11, we discussed the first kind of two-dimensional quasicrystals, their elasticity, dynamics and solutions, where we mentioned apart from the first kind of two-dimensional quasicrystals, there is another kind of one, i.e., the second kind of two-dimensional quasicrystals, although the study of this kind of quasicrystals in solid has not been developed yet.

An 18-fold symmetry quasicrystal was first reported in soft matter [1]. It belongs to the second kind of two-dimensional quasicrystals, which implies that there is the structure of the second kind of two-dimensional quasicrystals in solid possibly. Hu *et al.* [2] and Yang *et al.* [3] predicted the existence of this kind of solid quasicrystal from the point of view of group theory, although in that time Hu and coworkers did not give the name "two-dimensional quasicrystals of second kind". The discovery of this kind of quasicrystal by experiments will be an event in coming time. Fan and coworkers developed the theory of this kind of quasicrystal. In this chapter, we give an introduction to this possible kind of two-dimensional quasicrystal in solid and in static state.

17.1 The Point Groups of the Second Kind of Two-dimensional Quasicrystals

The symmetries on two-dimensional quasicrystals of the second kind have not well been studied, but Tang and Fan [4] put forward

Table 17.1. Point groups of second kind of two-dimensional quasicrystals.

Axis	Plus m_h	Plus m_v	Plus 2_h	Plus multi-operators
7	$\overline{14}$	$7m$	72	$\overline{14}m2$
$\overline{7}$		$\overline{7}m$	$\overline{7}m$	
14	$14/m$	$14mm$	1422	$14/mmm$
$\overline{14}$		$\overline{14}m2$	$\overline{14}m2$	
9	$\overline{18}$	$9m$	92	$\overline{18}m2$
$\overline{9}$		$\overline{9}m$	$\overline{9}m$	
18	$18/m$	$18mm$	1822	$18/mmm$
$\overline{18}$		$\overline{18}m2$	$\overline{18}m2$	

the point group classification and the group representation on the structure.

Based on the Schoenflies method, the point groups of 7-, 14-, 9- and 18-fold symmetry quasicrystals are listed in Table 17.1.

In Chapter 3, we discussed the two-dimensional quasicrystals of the first kind having 57 point groups, and if considering the ones of the second kind, then the two-dimensional quasicrystals have 83 point groups.

The theory of group representation including the character tables of the second kind of two-dimensional quasicrystals given in Ref. [3] is quite complicated, and the introduction about the derivation needs a very large volume of space, which is not able to be conducted here. The key results concerning this are the determination of quadratic invariants of strain tensors of phonons, first and second phasons and their couplings (i.e., the determination of all independent non-zero components of physical modulus tensors of the material) so the constitutive equations, which are presented in the subsequent sections. But the profound analysis was given by Ma [5] in his famous monograph on the theory of groups.

17.2 Six-dimensional Embedding Space

Different from the first kind of two-dimensional quasicrystals, it requires a six-dimensional embedding space to describe its quasilattice of the second kind of two-dimensional quasicrystals, which includes possible 7-, 9-, 14- and 18-fold symmetry quasicrystals in solid, not yet reported by experiments, but the 18-fold symmetry quasicrystals have been observed in soft matter.

The six-dimensional embedding space was suggested by Hu *et al.* [2] and Yang *et al.* [3], which consists of parallel space $E_{\|}^2$ and two perpendicular spaces $E_{\perp 1}^2$ and $E_{\perp 2}^2$, i.e., we recall because of the importance (and readers are not so familiar)

$$E^6 = E_{\|}^2 \oplus E_{\perp 1}^2 \oplus E_{\perp 2}^2 \tag{17.2-1}$$

where $E_{\perp 1}^2$ is defined as the first perpendicular space, and $E_{\perp 2}^2$ the second one. Based on this concept, the Landau–Anderson expansion may be extended as

$$\rho(\mathbf{r}) = \sum_{\mathbf{G} \in L_R} \rho_{\mathbf{G}} \exp\{i\mathbf{G} \cdot \mathbf{r}\} = \sum_{\mathbf{G} \in L_R} |\rho_{\mathbf{G}}| \exp\{-i\Phi_{\mathbf{G}} + i\mathbf{G} \cdot \mathbf{r}\} \tag{17.2-2}$$

with the extended phase (phase angle) term

$$\Phi_n = \mathbf{G}_n^{\|} \cdot \mathbf{u} + \mathbf{G}_n^{\perp 1} \cdot \mathbf{v} + \mathbf{G}_n^{\perp 2} \cdot \mathbf{w} \tag{17.2-3}$$

where $\mathbf{G}_n^{\|}$ represents reciprocal lattice vector in parallel space $E_{\|}^2$, $\mathbf{G}_n^{\perp 1}$ and $\mathbf{G}_n^{\perp 2}$ the reciprocal lattice vectors in the first $E_{\perp 1}^2$ and second $E_{\perp 2}^2$ perpendicular spaces, $\mathbf{u}$ the phonon displacement field in parallel space and $\mathbf{v}$ and $\mathbf{w}$ the first and second phason displacement fields in the two perpendicular spaces, respectively.

In the case for the second kind of quasicrystals, the theory for the first kind of quasicrystals introduced in Chapter 6 cannot be used to describe the structure from the symmetry point of view. Since Hu *et al.* in 1994 [2] and Yang *et al.* in 1995 [3] proposed to adopt the six-dimensional embedding space hypothesis and extended the Landau–Anderson theory, all 7-, 9-, 14-, and 18-fold symmetries of the second kind of quasicrystals can be described with Eqs. (17.2-2) and (17.2-3).

If only considering the $x-y$ quasiperiodic plane with the convention from Hu *et al.* [2] and Yang *et al.* [3], the displacements are defined as $u = (u_x, u_y)$, $v = (v_x, v_y)$, $w = (w_x, w_y)$ and the corresponding strain fields can be derived as

$$\varepsilon_{ij} = \frac{1}{2}\left(\frac{\partial u_i}{\partial x_j} + \frac{\partial u_j}{\partial x_i}\right), \quad v_{ij} = \frac{\partial v_i}{\partial x_j}, \quad w_{ij} = \frac{\partial w_i}{\partial x_j} \tag{17.2-4}$$

Then, the corresponding constitutive law is

$$\begin{aligned} \sigma_{ij} &= \frac{\partial f_{def}}{\partial \varepsilon_{ij}} = C_{ijkl}\varepsilon_{kl} + r_{ijkl}v_{kl} + R_{ijkl}w_{kl} \\ \tau_{ij} &= \frac{\partial f_{def}}{\partial v_{ij}} = T_{ijkl}v_{kl} + r_{klij}\varepsilon_{kl} + G_{ijkl}w_{kl} \\ H_{ij} &= \frac{\partial f_{def}}{\partial w_{ij}} = K_{ijkl}w_{kl} + R_{klij}\varepsilon_{kl} + G_{klij}v_{kl} \end{aligned} \tag{17.2-5}$$

the meanings of σ_{ij}, C_{ijkl} are as those defined in the previous section, r_{ijkl} the phonon–first phason coupling (i.e., the $u{-}v$ coupling) elastic constant tensor, R_{ijkl} the phonon–second phason coupling (i.e., the $u{-}w$ coupling) elastic constant tensor, τ_{ij} the stress tensor associated with phason strain tensor v_{ij}, T_{ijkl} the phason elastic constants corresponding to $\tau_{ij}{-}v_{ij}$, the meanings of H_{ij} and K_{ijkl} are the same discussed beforehand, corresponding to field w, but it is the second phason field now, and G_{ijkl} the elastic constants of coupling (i.e., the $v{-}w$ coupling) between first–second phason fields. According to group representation theory [2, 3], the independent non-zero elastic constants of phonons, phasons and coupling elastic constants between phonon–phason and first phason–second phason are determined.

17.3 Possible 18-fold Symmetry Solid Quasicrystals and Their Elasticity

The quasicrystals of 18-fold symmetry in colloids were observed in 2011 [1] for the first time; their generalized dynamics have been discussed in monograph [10]. The diffraction pattern of the matter is shown in Chapter 18, which is the only one experimental result in colloid. Hu *et al.* [2] predicted the existence of the 18-fold symmetry structure 17 years before the experimental observation. They gave a point group 18 mm to describe the possible solid quasicrystals, but the observed quasicrystals are in soft matter rather than in solid up to now. Although the result is different than their prediction, it is still meaningful. In the 1990s, people guessed that there might be soft-matter quasicrystals, but they could not estimate the observation of 18-fold symmetry ones. In addition, does the point group 18 mm correspond to the structure shown by the diffraction pattern mentioned above? The question needs to be replied. According to the group theory, there might be other point groups with 18-fold symmetry excepted point group 18 mm, refer to Chapter 18, but we here discuss only the quasicrystals of point group 18 mm for simplicity.

The quasicrystals of 18-fold symmetry are two-dimensional quasicrystals. If we consider the plane field and assume the z-direction is the direction of the 18-fold rotation axis, the displacements are $\mathbf{u} = (u_x, u_y)$, $\mathbf{v} = (v_x, v_y)$, $\mathbf{w} = (w_x, w_y)$, which are two-dimensional displacement fields. The corresponding strain fields are

$$\varepsilon_{ij} = \frac{1}{2}\left(\frac{\partial u_i}{\partial x_j} + \frac{\partial u_j}{\partial x_i}\right), \quad v_{ij} = \frac{\partial v_i}{\partial x_j}, \quad w_{ij} = \frac{\partial w_i}{\partial x_j} \tag{17.3-1}$$

and the corresponding generalized Hooke's law is

$$\left.\begin{aligned} \sigma_{ij} &= \frac{\partial f_{def}}{\partial \varepsilon_{ij}} = C_{ijkl}\varepsilon_{kl} + r_{ijkl}v_{kl} + R_{ijkl}w_{kl} \\ \tau_{ij} &= \frac{\partial f_{def}}{\partial v_{ij}} = T_{ijkl}v_{kl} + r_{klij}\varepsilon_{kl} + G_{ijkl}w_{kl} \\ H_{ij} &= \frac{\partial f_{def}}{\partial w_{ij}} = K_{ijkl}w_{kl} + R_{klij}\varepsilon_{kl} + G_{klij}v_{kl} \end{aligned}\right\} \tag{17.3-2}$$

in which $f_{def} = f_{def}(\mathbf{u}, \mathbf{v}, \mathbf{w})$ denotes the elastic deformation energy density (or strain energy density) of the system, which will be given in (17.3-11), the meanings of σ_{ij} and C_{ijkl} are as those defined in the previous chapters, r_{ijkl} the phonon–first phason coupling (i.e., the u–v coupling) elastic constants, R_{ijkl} the phonon–second phason coupling (i.e., the u–w coupling) elastic constants, τ_{ij} the stress tensor associated with phason strain tensor v_{ij}, T_{ijkl} the phason elastic constants corresponding to τ_{ij}–v_{ij} and the meanings of H_{ij} and K_{ijkl} are the same discussed beforehand, corresponding to field $\mathbf{w}$, but it is the second phason field now, and G_{ijkl} the elastic constants of coupling (i.e., the v–w coupling) between first and second phason fields.

According to group representation theory, the independent nonzero elastic constants of phonons are only two, L and M, i.e.,

$$C_{ijkl} = L\delta_{ij}\delta_{kl} + M(\delta_{ik}\delta_{jl} + \delta_{il}\delta_{jk}) \quad (i,j,k,l = 1,2) \quad (17.3\text{-}3)$$
$$L = C_{12}, M = (C_{11} - C_{12})/2 = C_{66}$$

This is identical to those in the plane field of solid quasicrystals of 5-, 8-, 10- and 12-fold symmetries, and this is shown in Table 17.2. where $x = x_1, y = x_2$.

The phonons are decoupled with the first phasons, so

$$r_{ijkl} = 0(i,j,k,l = 1,2) \quad (17.3\text{-}4)$$

The phonons are decoupled with the second phasons too, and

$$R_{ijkl} = 0(i,j,k,l = 1,2) \quad (17.3\text{-}5)$$

In addition, there are elastic constants of second phasons T_{ijkl}

$$T_{ijkl} = T_1\delta_{ik}\delta_{jl} + T_2(\delta_{ij}\delta_{kl} - \delta_{il}\delta_{jk})(i,j,k,l = 1,2)$$
$$T_{1111} = T_{2222} = T_{2121} = T_1$$
$$T_{1122} = T_{2211} = -T_{2112} = -T_{1221} = T_2 \quad (17.3\text{-}6)$$

which are given in Table 17.3.

Table 17.2. Phonon elastic constants of 18-fold symmetry quasicrystals.

	11	22	12	21
11	C_{11}	C_{12}	0	0
22	C_{12}	C_{11}	0	0
12	0	0	C_{66}	C_{66}
21	0	0	C_{66}	C_{66}

Table 17.3. Elastic constants of second phasos.

	11	22	12	21
11	T_1	T_2	0	0
22	T_2	T_1	0	0
12	0	0	T_1	$-T_2$
21	0	0	$-T_2$	T_1

Table 17.4. The coupling elastic constants between first and second phasons.

$v_{ij}\backslash w_{ij}$	11	22	12	21
11	G	G	0	0
22	$-G$	$-G$	0	0
12	0	0	$-G$	G
21	0	0	$-G$	G

The coupling elastic constants between first and second phasons

$$G_{ijkl} = G(\delta_{i1} - \delta_{i2})(\delta_{ij}\delta_{kl} - \delta_{ik}\delta_{jl} + \delta_{il}\delta_{jk})(i, j, k, l = 1, 2) \quad (17.3\text{-}7)$$

are listed in Table 17.4.

17.4 Equation System of Elasticity of Quasicrystals of 18-fold Symmetry with Point Group 18 mm

In the statics of quasicrystals with 18-fold symmetry apart from elastic displacement fields u $= (u_x, u_y)$, v $= (v_x, v_y)$, w $= (w_x, w_y)$, the elastic constitutive law has been listed in (17.2-5), as mentioned above.

The equations of elastic statics in the two-dimensional case can be easily derived if we consider only for point group 18 mm.

Denote the elastic energy as

$$F_{el} = F_u + F_v + F_w + F_{uv} + F_{uw} + F_{vw}, \quad (17.4\text{-}1)$$

which represents free energies of phonons, first phasons, second phasons, phonon–phason couplings and first–second phason coupling, respectively,

$$\begin{aligned} F_u &= \int \frac{1}{2} C_{ijkl}\varepsilon_{ij}\varepsilon_{kl} d^d\mathbf{r} \\ F_v &= \int \frac{1}{2} T_{ijkl} v_{ij} v_{kl} d^d\mathbf{r} \\ F_w &= \int \frac{1}{2} K_{ijkl} w_{ij} w_{kl} d^d\mathbf{r} \\ F_{uv} &= \int (r_{ijkl}\varepsilon_{ij} v_{kl} + r_{klij} v_{ij}\varepsilon_{kl}) d^d\mathbf{r} \end{aligned} \quad (17.4\text{-}2)$$

$$F_{uw} = \int (R_{ijkl}\varepsilon_{ij}w_{kl} + R_{klij}w_{ij}\varepsilon_{kl})d^d\mathbf{r}$$

$$F_{vw} = \int (G_{ijkl}v_{ij}w_{kl} + G_{klij}w_{ij}v_{kl})d^d\mathbf{r}$$

for the present case, then F_u, F_w, F_{uw} in Chapter 5 are replaced by $F_u, F_v, F_w, F_{uv}, F_{uw}, F_{vw}$, respectively.

The deformation energy density for defining the constitutive equation in (17.4-1) is given for the following:

$$\begin{aligned}
f_{def}(u,v,w) = f_{def}(\varepsilon_{ij}, v_{ij}, w_{ij}) &= \frac{1}{2}L(\nabla\cdot u)^2 + M\varepsilon_{ij}\varepsilon_{ij}\\
&+T_1[(v_{11}+v_{22})^2+(v_{21}-v_{12})^2]+T_2[(v_{11}-v_{22})^2\\
&+(v_{21}+v_{12})^2]+K_1[(w_{11}+w_{22})^2+(w_{21}-w_{12})^2]\\
&+T_2[(w_{11}-w_{22})^2+(w_{21}+w_{12})^2]\\
&+G[(v_{11}-v_{22})(w_{11}-w_{22})+(v_{21}+v_{12})(w_{21}+w_{12})]\\
&=F_u+F_v+F_w+F_{vw}, (x=x_1, y=x_2, i=1,2, j=1,2)
\end{aligned} \tag{17.4-3}$$

and then the elastic constitutive equations, so the whole constitutive equations are determined at last [4]

$$\left.\begin{aligned}
\sigma_{xx} &= (L+2M)\varepsilon_{xx} + L\varepsilon_{yy}\\
\sigma_{yy} &= L\varepsilon_{xx} + (L+2M)\varepsilon_{yy}\\
\sigma_{xy} &= \sigma_{yx} = 2M\varepsilon_{xy}\\
\tau_{xx} &= T_1v_{xx} + T_2v_{yy} + G(w_{xx} - w_{yy})\\
\tau_{yy} &= T_2v_{xx} + T_1v_{yy} + G(w_{xx} - w_{yy})\\
\tau_{xy} &= T_1v_{xy} - T_2v_{yx} - G(w_{yx} + w_{xy})\\
\tau_{yx} &= -T_2v_{xy} + T_1v_{yx} + G(w_{yx} + w_{xy})\\
H_{xx} &= K_1w_{xx} + K_2w_{yy} + G(v_{xx} + v_{yy})\\
H_{yy} &= K_2w_{xx} + K_1w_{yy} - G(v_{xx} + v_{yy})\\
H_{xy} &= K_1w_{xy} - K_2w_{yx} - G(v_{xy} + v_{yx})\\
H_{yx} &= K_1w_{yx} - K_2w_{xy} + G(v_{xy} - v_{yx})
\end{aligned}\right\} \tag{17.4-4}$$

The stress equilibrium equations are

$$\left.\begin{aligned}\frac{\partial\sigma_{xx}}{\partial x}+\frac{\partial\sigma_{xy}}{\partial y}=0,\ \frac{\partial\sigma_{yx}}{\partial x}+\frac{\partial\sigma_{yy}}{\partial y}=0,\\ \frac{\partial\tau_{xx}}{\partial x}+\frac{\partial\tau_{xy}}{\partial y}=0,\ \frac{\partial\tau_{yx}}{\partial x}+\frac{\partial\tau_{yy}}{\partial y}=0,\\ \frac{\partial H_{xx}}{\partial x}+\frac{\partial H_{xy}}{\partial y}=0,\ \frac{\partial H_{yx}}{\partial x}+\frac{\partial H_{yy}}{\partial y}=0,\end{aligned}\right\} \tag{17.4-5}$$

Substituting (17.4-4) into (17.4-4), the final governing equations of elastic equilibrium can be obtained

$$\left.\begin{aligned}&M\nabla^2 u_x+(L+M)\frac{\partial}{\partial x}\nabla\cdot\mathbf{u}=0\\ &M\nabla^2 u_y+(L+M)\frac{\partial}{\partial y}\nabla\cdot\mathbf{u}=0\\ &T_1\nabla^2 v_x+G\left(\frac{\partial^2 w_x}{\partial x^2}-\frac{\partial^2 w_x}{\partial y^2}\right)-2G\frac{\partial^2 w_y}{\partial x\partial y}=0\\ &T_1\nabla^2 v_y+2G\frac{\partial^2 w_x}{\partial x\partial y}+G\left(\frac{\partial^2 w_y}{\partial x^2}-\frac{\partial^2 w_y}{\partial y^2}\right)=0\\ &K_1\nabla^2 w_y-2G\frac{\partial^2 v_x}{\partial x\partial y}+G\left(\frac{\partial^2 v_y}{\partial x^2}-\frac{\partial^2 v_y}{\partial y^2}\right)=0\\ &K_1\nabla^2 w_y-2G\frac{\partial^2 v_x}{\partial x\partial y}+G\left(\frac{\partial^2 v_y}{\partial x^2}-\frac{\partial^2 v_y}{\partial y^2}\right)=0\end{aligned}\right\} \tag{17.4-6}$$

In Eq. (17.4-6), there are six equations with six field variables, i.e., $u_x, u_y, \mathrm{v}_x, \mathrm{v}_y w_x, w_x$, the amount of field variables is six, and the amount of field equations is six too, which is quite different from that for the first kind of two-dimensional quasicrystals in plane problem, discussed in Chapter 6.

17.5 Possible 7-fold Symmetry Solid Quasicrystals and Their Elasticity

Apart from 18-fold symmetrical quasicrystals, the possible 7-, 9- and 14-fold symmetrical quasicrystals are the second kind of two-dimensional quasicrystals and are different from those of 5-, 8-,

10- and 12-fold symmetry quasicrystals. For this reason, we must introduce the so-called six-dimensional embedding space that we have been introduced in Section 17.2.

According to the hypothesis of the six-dimensional embedding space, there are phonon displacement field **u** in parallel space and the first and second phason displacement fields **v** and **w** in the two perpendicular spaces, respectively.

The quasicrystals of seven-fold symmetry have not been observed that far yet, so there is no diffraction pattern of the matter like that was shown in figures in Ref. [1] or Chapter 18. Hu *et al.* [2] predicted the existence of 7-fold symmetry structure. They gave a point group 7 m to describe the possible solid quasicrystals, but the solid quasicrystals have not been observed too. Although there is a lack of experimental results, the prediction of Hu *et al.* is meaningful. Based on the point group 7 m, we can discuss the phonon and phason constitutive laws. We should point out that except for point group 7 m there might be other point groups of 7-fold symmetry quasicrystals, but the discussion here is only for point group 7 m quasicrystals.

If we assume the z-direction is the direction of 7-fold rotation axis, the displacements are $u = (u_x, u_y)$, $v = (v_x, v_y)$, $w = (w_x, w_y)$, which are two-dimensional displacement fields. The corresponding strain fields are

$$\varepsilon_{ij} = \frac{1}{2}\left(\frac{\partial u_i}{\partial x_j} + \frac{\partial u_j}{\partial x_i}\right), \quad v_{ij} = \frac{\partial v_i}{\partial x_j}, \; w_{ij} = \frac{\partial w_i}{\partial x_j}$$

Based on the group representation theory, there are two couplings: between phonons and second phasons, and first and second phasons. Accordingly, the corresponding free-energy densities are defined by

$$f_u = \frac{1}{2}L\varepsilon_{ii}\varepsilon_{ii} + M\varepsilon_{ij}\varepsilon_{ij}$$

$$f_v = \frac{1}{2}T_1 v_{ij}v_{ij} + T_2(v_{xx}v_{yy} - v_{xy}v_{xy})$$

$$
\begin{aligned}
f_w &= \frac{1}{2}K_1 w_{ij}w_{ij} + K_2(w_{xx}w_{yy} - w_{xy}w_{xy}) \\
f_{uw} &= R[(\varepsilon_{xx} - \varepsilon_{yy})(w_{xx} + w_{yy}) + 2\varepsilon_{xy}(w_{yx} - w_{xy})] \\
f_{vw} &= G[(v_{11} - v_{22})(w_{11} - w_{22}) + (v_{21} + v_{12})(w_{21} + w_{12})]
\end{aligned}
\tag{17.5-1}
$$

the total deformation energy density $f_{def} = f_u + f_v + f_w + f_{uw} + f_{vw}$ can be obtained. The generalized Hooke's law is defined by

$$
\begin{aligned}
\sigma_{ij} &= \frac{\partial f_{def}}{\partial \varepsilon_{ij}} = C_{ijkl}\varepsilon_{kl} + r_{ijkl}v_{kl} + R_{ijkl}w_{kl} \\
\tau_{ij} &= \frac{\partial f_{def}}{\partial v_{ij}} = T_{ijkl}v_{kl} + r_{klij}\varepsilon_{kl} + G_{ijkl}w_{kl} \\
H_{ij} &= \frac{\partial f_{def}}{\partial w_{ij}} = K_{ijkl}w_{kl} + R_{klij}\varepsilon_{kl} + G_{klij}v_{kl}
\end{aligned}
$$

$$
\left.
\begin{aligned}
\sigma_{xx} &= (L+2M)\varepsilon_{xx} + L\varepsilon_{yy} + R(w_{xx} + w_{yy}) \\
\sigma_{yy} &= L\varepsilon_{xx} + (L+2M)\varepsilon_{yy} - R(w_{xx} + w_{yy}) \\
\sigma_{xy} &= \sigma_{yx} = 2M\varepsilon_{xy} + R(w_{yx} - w_{xy}) \\
\tau_{xx} &= T_1 v_{xx} + T_2 v_{yy} + G(w_{xx} - w_{yy}) \\
\tau_{yy} &= T_2 v_{xx} + T_1 v_{yy} - G(w_{xx} - w_{yy}) \\
\tau_{xy} &= T_1 v_{xy} - T_2 v_{yx} + G(w_{yx} + w_{xy}) \\
\tau_{yx} &= -T_2 v_{xy} + T_1 v_{yx} + G(w_{yx} + w_{xy}) \\
H_{xx} &= K_1 w_{xx} + K_2 w_{yy} + R(\varepsilon_{xx} - \varepsilon_{yy}) + G(v_{xx} - v_{yy}) \\
H_{yy} &= K_2 w_{xx} + K_1 w_{yy} + R(\varepsilon_{xx} - \varepsilon_{yy}) - G(v_{xx} - v_{yy}) \\
H_{xy} &= K_1 w_{xy} - K_2 w_{yx} - 2R\varepsilon_{xy} + G(v_{xy} + v_{yx}) \\
H_{yx} &= K_1 w_{yx} - K_2 w_{xy} + 2R\varepsilon_{xy} + G(v_{xy} + v_{yx})
\end{aligned}
\right\}
\tag{17.5-2}
$$

According to a similar derivation in Section 17.4, we can obtain the equation system of the possible 7-fold symmetry soft-matter quasicrystals as follows:

$$\left.\begin{aligned}
&M\nabla^2 u_x + (L+M)\frac{\partial}{\partial x}\nabla\cdot\mathbf{u} + R\left(\frac{\partial^2 w_x}{\partial x^2} + 2\frac{\partial^2 w_y}{\partial x\partial y} - \frac{\partial w_x}{\partial y^2}\right)\\
&M\nabla^2 u_y + (L+M)\frac{\partial}{\partial y}\nabla\cdot\mathbf{u} + R\left(\frac{\partial^2 w_y}{\partial x^2} - 2\frac{\partial^2 w_x}{\partial x\partial y} - \frac{\partial^2 w_y}{\partial y^2}\right) = 0\\
&T_1\nabla^2 v_x + G\nabla^2 w_x = 0\\
&T_1\nabla^2 v_y + G\nabla^2 w_y = 0\\
&K_1\nabla^2 w_x + R\left(\frac{\partial^2 u_x}{\partial x^2} - 2\frac{\partial^2 u_y}{\partial x\partial y} - \frac{\partial^2 u_x}{\partial y^2}\right) + G\nabla^2 v_x = 0\\
&\quad K_1\nabla^2 w_y + R\left(\frac{\partial^2 u_y}{\partial x^2} + 2\frac{\partial^2 u_x}{\partial x\partial y} - \frac{\partial^2 u_y}{\partial y^2}\right) + G\nabla^2 v_y = 0
\end{aligned}\right\} \tag{17.5-3}$$

in which $\nabla = \mathbf{i}\frac{\partial}{\partial x} + \mathbf{j}\frac{\partial}{\partial y}$, $\nabla 2 = \frac{\partial^2}{\partial x^2} + \frac{\partial^2}{\partial y^2}$ and $L = C_{12}, M = (C_{11} - C_{12})/2$ are phonon elastic constants, T_1, K_1 the elastic constants of the first and second phasons, R, G the coupling elastic constants between the phonons and second phasons and between first and second phasons, η the fluid viscosity (for simplicity, only the scalar version of η_{ijkl} is considered), Γ_u, Γ_v and Γ_w the phonon, first phason and second phason dissipation coefficients and A and B the material constants due to the variation of mass density, respectively.

Equations (17.5-3) consist of six field variables, i.e., phonon field $u = (u_x, u_y)$, first phason field $v = (v_x, v_y)$ and second phason field, respectively. The amount of equations is also six. At the present case, Eqs. (17.5-3) are consistent mathematically and solvable.

Because of both the coupling between phonons and second phasons and between first and second phasons, these kinds of quasicrystals are the very interesting ones which will be observed in the near future.

17.6 The Possible 9-fold Symmetrical Quasicrystals with Point Group $9m$

The quasicrystals of 9-fold symmetry has not been observed that far yet, so there is no a diffraction pattern of the matter like that was given for 18-fold symmetry quasicrystals (but in soft matter, refer to Ref. [1] or Chapter 18). Hu *et al.* [2] predicted the existence of 9-fold symmetry structures in solids. They gave a point group $9m$ to describe the possible solid quasicrystals, but the solid quasicrystals have not been observed too. Although there is a lack of experimental results, the prediction of Hu *et al.* is meaningful. We should point out that except for point group $9m$ there might be other point groups of 9-fold symmetry quasicrystals, please refer to Chapter 2, but the discussion here is only on point group $9m$ quasicrystals for simplicity.

Based on the point group $9m$, we can discuss the phonon and phason constitutive laws.

According to the theory of group representation, the possible 9-fold symmetrical quasicrystals of soft matter have similar constitutive equations to those of 18-fold symmetrical quasicrystals. The concrete results are as follows:

$$\begin{cases} f_{def}(u,v,w) = f_{def}(\varepsilon_{ij}, v_{ij}, w_{ij}) = f_u + f_v + f_w + f_{vw} \\ f_u = \frac{1}{2}L(\nabla \cdot u)^2 + M\varepsilon_{ij}\varepsilon_{ij} \\ f_v = T_1[(v_{xx}+v_{yy})^2 + (v_{yx}-v_{xy})^2] + T_2[(v_{xx}-v_{yy})^2 + (v_{yx}+v_{xy})^2] \\ f_w = K_1[(w_{xx}+w_{yy})^2 + (w_{yx}-w_{xy})^2] + K_2[(w_{xx}-w_{yy})^2 + (w_{yx}+w_{xy})^2] \\ f_{vw} = G[(v_{xx}-v_{yy})(w_{xx}+w_{yy}) + (v_{yx}+v_{xy})(w_{yx}-w_{xy})] \\ (x = x_1, y = x_2, i = 1,2, j = 1,2) \end{cases} \tag{17.6-1}$$

for 9-fold symmetry quasicrystals with point group $9m$, and substituting the elastic deformation energy density (17.6-1) into the general relationship

$$\sigma_{ij} = \frac{\partial f_{def}}{\partial \varepsilon_{ij}} = C_{ijkl}\varepsilon_{kl} + r_{ijkl}v_{kl} + R_{ijkl}w_{kl}$$

$$\tau_{ij} = \frac{\partial f_{def}}{\partial v_{ij}} = T_{ijkl}v_{kl} + r_{klij}\varepsilon_{kl} + G_{ijkl}w_{kl}$$

$$H_{ij} = \frac{\partial f_{def}}{\partial w_{ij}} = K_{ijkl}w_{kl} + R_{klij}\varepsilon_{kl} + G_{klij}v_{kl}$$

one can obtain the elastic constitutive law

$$\left.\begin{aligned}
\sigma_{xx} &= (L+2M)\varepsilon_{xx} + L\varepsilon_{yy} \\
\sigma_{yy} &= L\varepsilon_{xx} + (L+2M)\varepsilon_{yy} \\
\sigma_{xy} &= \sigma_{yx} = 2M\varepsilon_{xy} \\
\tau_{xx} &= T_1 v_{xx} + T_2 v_{yy} + G(w_{xx} + w_{yy}) \\
\tau_{yy} &= T_2 v_{xx} + T_1 v_{yy} - G(w_{xx} + w_{yy}) \\
\tau_{xy} &= T_1 v_{xy} - T_2 v_{yx} + G(w_{yx} - w_{xy}) \\
\tau_{yx} &= -T_2 v_{xy} + T_1 v_{yx} + G(w_{yx} - w_{xy}) \\
H_{xx} &= K_1 w_{xx} + K_2 w_{yy} + G(v_{xx} - v_{yy}) \\
H_{yy} &= K_2 w_{xx} + K_1 w_{yy} + G(v_{xx} - v_{yy}) \\
H_{xy} &= K_1 w_{xy} - K_2 w_{yx} - G(v_{xy} + v_{yx}) \\
H_{yx} &= K_1 w_{yx} - K_2 w_{xy} + G(v_{xy} + v_{yx})
\end{aligned}\right\} \tag{17.6-2}$$

for 9-fold symmetry quasicrystals and collaborating the fluid constitutive law.

With these relations and by using equilibrium equations (17.4-5), we can obtain the equations of motion plus the equation of the state for 9-fold symmetry quasicrystals in soft matter as follows:

$$\left.\begin{aligned}
&M\nabla^2 u_x + (L+M)\frac{\partial}{\partial x}\nabla\cdot\mathbf{u} = 0 \\
&M\nabla^2 u_y + (L+M)\frac{\partial}{\partial y}\nabla\cdot\mathbf{u} = 0 \\
&T_1\nabla^2 v_x + G\left(\frac{\partial^2 w_x}{\partial x^2} - \frac{\partial^2 w_x}{\partial y^2}\right) + 2G\frac{\partial^2 w_y}{\partial x\partial y} = 0 \\
&T_1\nabla^2 v_y + 2G\frac{\partial^2 w_x}{\partial x\partial y} + G\left(\frac{\partial^2 w_y}{\partial x^2} - \frac{\partial^2 w_y}{\partial y^2}\right) = 0 \\
&K_1\nabla^2 w_x + G\left(\frac{\partial^2 v_x}{\partial x^2} - \frac{\partial^2 v_x}{\partial y^2}\right) - 2G\frac{\partial^2 v_y}{\partial x\partial y} = 0 \\
&K_1\nabla^2 w_y - 2G\frac{\partial^2 v_x}{\partial x\partial y} + G\left(\frac{\partial^2 v_y}{\partial x^2} - \frac{\partial^2 v_y}{\partial y^2}\right) = 0
\end{aligned}\right\} \tag{17.6-3}$$

which are similar to those of (17.4-5), in which $\nabla = i\frac{\partial}{\partial x} + j\frac{\partial}{\partial y}$, $\nabla^2 = \frac{\partial^2}{\partial x^2} + \frac{\partial^2}{\partial y^2}$, and $L = C_{12}$, $M = (C_{11}–C_{12})/2$ are phonon elastic constants, T_1, K_1 the elastic constants of the first and second phasons, $R=0$, G the coupling elastic constant between first and second phasons, Eqs. (17.6-3) consist of six field variables, i.e., phonon field $u=(u_x,u_y)$, first phason field $v=(v_x,v_y)$ and second phason field $w=(w_x,w_y)$. In the present case, Eqs. (17.6-3) are consistent mathematically and solvable.

17.7 Possible 14-fold Symmetry Solid Quasicrystals and Their Elasticity: The Possible 14-fold Symmetrical Quasicrystals with Point Group 14*mm*

The quasicrystals of 14-fold symmetry have not been observed that far yet, so there is no a diffraction pattern of the matter like that was shown in Ref. [1] for soft matter or in Chapter 18. Hu *et al.* [3] predicted the existence of 14-fold symmetry structure. They gave a point group 14*mm* to describe the possible solid quasicrystals, but the solid quasicrystals have not been observed so far. Although there is lack of experimental results, the prediction of Hu *et al.* is meaningful. We should point out that apart from the point group 14*mm*, there might be other point groups of 14-fold symmetry quasicrystals according to the group theory, refer to Chapter 18 or Ref. [6], but the discussion here is only for point group 14*mm* quasicrystals.

Based on the point group 14*mm*, we can discuss the phonon and phason constitutive laws.

According to the theory of group representation, the possible 14-fold symmetrical quasicrystals of soft matter have similar constitutive equations to those of 7-fold symmetrical quasicrystals, but there are some distinctions, and the elastic deformation energy density is

$$f_{def}(u,v,w)=f_{def}(\varepsilon_{ij},v_{ij},w_{ij})=f_u+f_v+f_w+f_{uw}+f_{vw}$$

$$f_u=\frac{1}{2}L\varepsilon_{ii}\varepsilon_{ii}+M\varepsilon_{ij}\varepsilon_{ij}$$

$$f_v=\frac{1}{2}T_1v_{ij}v_{ij}+T_2(v_{xx}v_{yy}-v_{xy}v_{xy})$$

$$
\begin{aligned}
f_w &= \frac{1}{2}K_1 w_{ij}w_{ij} + K_2(w_{xx}w_{yy} - w_{xy}w_{xy}) \\
f_{uw} &= R[(\varepsilon_{xx} - \varepsilon_{yy})(w_{xx} + w_{yy}) + 2\varepsilon_{xy}(w_{yx} - w_{xy})] \\
f_{vw} &= G[(v_{11} + v_{22})(w_{11} - w_{22}) + (v_{21} - v_{12})(w_{21} + w_{12})]
\end{aligned} \tag{17.7-1}
$$

for 14-fold symmetry quasicrystals, respectively. With the deformation energy density and the definition of the generalized Hooke's law, the concrete relation between stresses and strains can be obtained, and adding the relations between fluid stresses and deformation rate components, we have

$$
\left.
\begin{aligned}
\sigma_{xx} &= (L + 2M)\varepsilon_{xx} + L\varepsilon_{yy} + R(w_{xx} + w_{yy}) \\
\sigma_{yy} &= L\varepsilon_{xx} + (L + 2M)\varepsilon_{yy} - R(w_{xx} + w_{yy}) \\
\sigma_{xy} &= \sigma_{yx} = 2M\varepsilon_{xy} + R(w_{yx} - w_{xy}) \\
\tau_{xx} &= T_1 v_{xx} + T_2 v_{yy} + G(w_{xx} - w_{yy}) \\
\tau_{yy} &= T_2 v_{xx} + T_1 v_{yy} + G(w_{xx} - w_{yy}) \\
\tau_{xy} &= T_1 v_{xy} - T_2 v_{yx} - G(w_{yx} + w_{xy}) \\
\tau_{yx} &= -T_2 v_{xy} + T_1 v_{yx} + G(w_{yx} + w_{xy}) \\
H_{xx} &= K_1 w_{xx} + K_2 w_{yy} + R(\varepsilon_{xx} - \varepsilon_{yy}) + G(v_{xx} + v_{yy}) \\
H_{yy} &= K_2 w_{xx} + K_1 w_{yy} + R(\varepsilon_{xx} - \varepsilon_{yy}) - G(v_{xx} + v_{yy}) \\
H_{xy} &= K_1 w_{xy} - K_2 w_{yx} - 2R\varepsilon_{xy} + G(v_{yx} - v_{xy}) \\
H_{yx} &= K_1 w_{yx} - K_2 w_{xy} + 2R\varepsilon_{xy} + G(v_{yx} - v_{xy})
\end{aligned}
\right\} \tag{17.7-2}
$$

Completely similar, utilizing the procedure mentioned above, the equations of motion of 14-fold symmetry quasicrystals, we have the final governing equations as follows:

$$\left.\begin{aligned}
&M\nabla^2 u_x + (L+M)\frac{\partial}{\partial x}\nabla\cdot\mathbf{u} + R\left(\frac{\partial^2 w_x}{\partial x^2} + 2\frac{\partial^2 w_y}{\partial x\partial y} - \frac{\partial w_x}{\partial y^2}\right) = 0\\
&M\nabla^2 u_y + (L+M)\frac{\partial}{\partial y}\nabla\cdot\mathbf{u} + R\left(\frac{\partial^2 w_y}{\partial x^2} - 2\frac{\partial^2 w_x}{\partial x\partial y} - \frac{\partial^2 w_y}{\partial y^2}\right)\\
&T_1\nabla^2 v_x + G\left(\frac{\partial^2 w_x}{\partial x^2} - \frac{\partial^2 w_x}{\partial y^2}\right) - 2G\frac{\partial^2 w_y}{\partial x\partial y} = 0\\
&T_1\nabla^2 v_y + 2G\frac{\partial^2 w_x}{\partial x\partial y} + G\left(\frac{\partial^2 w_y}{\partial x^2} - \frac{\partial^2 w_y}{\partial y^2}\right) = 0\\
&K_1\nabla^2 w_x + R\left(\frac{\partial^2 u_x}{\partial x^2} - 2\frac{\partial^2 u_y}{\partial x\partial y} - \frac{\partial^2 u_x}{\partial y^2}\right) + G\left(\frac{\partial^2 v_x}{\partial x^2} - \frac{\partial^2 v_x}{\partial y^2}\right) + 2G\frac{\partial^2 v_y}{\partial x\partial y} = 0\\
&K_1\nabla^2 w_y + R\left(\frac{\partial^2 u_y}{\partial x^2} + 2\frac{\partial^2 u_x}{\partial x\partial y} - \frac{\partial^2 u_y}{\partial y^2}\right) - 2G\frac{\partial^2 v_x}{\partial x\partial y} - G\left(\frac{\partial^2 v_y}{\partial x^2} - \frac{\partial^2 v_y}{\partial y^2}\right) = 0
\end{aligned}\right\} \tag{17.7-3}$$

in which $\boldsymbol{V} = \mathbf{i}V_x + \mathbf{j}V_y$, $\boldsymbol{u} = \mathbf{i}u_x + \mathbf{j}u_y$, $\nabla = \mathbf{i}\frac{\partial}{\partial x} + \mathbf{j}\frac{\partial}{\partial y}$, $\nabla^2 = \frac{\partial^2}{\partial x^2} + \frac{\partial^2}{\partial y^2}$, and $L = C_{12}$, $M = (C_{11} - C_{12})/2$ are phonon elastic constants, T_1, K_1 the elastic constants of the first and second phasons, and R, G the coupling elastic constants between the phonons and second phasons and between first and second phasons. Equations (17.7-3) consist of six field variables, i.e., phonon field $\boldsymbol{u} = (u_x, u_y)$, first phason field $\boldsymbol{v} = (v_x, v_y)$ and second phason field $\boldsymbol{w} = (w_x, w_y)$, respectively. The amount of equations is also six, so the problem can be solved if the boundary conditions are appropriately given.

17.8 Solution of Phason Field of Dislocation of Possible 9-fold Symmetry Solid Quasicrystals

Here, we first consider a special mathematical solution of Eqs. (17.6-4) in static case and discuss the dislocation in a 9-fold symmetry quasicrystal of soft matter. For a dislocation in the second kind of two-dimensional quasicrystals, the Burgers vector can be expressed by $\mathbf{b} = \mathbf{b}^{\|} \oplus \mathbf{b}_1^{\perp} \oplus \mathbf{b}_2^{\perp} = \left(b_1^{\|}, b_2^{\|}, b_{11}^{\perp}, b_{12}^{\perp}, b_{21}^{\perp}, b_{22}^{\perp}\right)$, where

$$\int_{\Gamma} du_j = b_j, \int_{\Gamma} dv_j = b_{1j}^{\perp}, \int_{\Gamma} dw_j = b_{2j}^{\perp}, j = 1, 2. \tag{17.8-1}$$

In the above, the first two components are in the physical space or parallel space, and the last four components are in the perpendicular space, in which the first and second phason fields are existed, respectively.

The equation system (17.6-4) offers a basis and possibility to solve various solutions for the matter distribution, deformation, motion and reconstruction of the new phase. As a simplified model, we first consider a static dislocation problem, i.e., the fluid effects can be omitted for the time being. In the case, the above equations reduce to

$$M\nabla^2 u_x + (L + M)\frac{\partial}{\partial x}\nabla \cdot \mathbf{u} = 0 \tag{17.8-2}$$

$$M\nabla^2 u_y + (L + M)\frac{\partial}{\partial y}\nabla \cdot \mathbf{u} = 0 \tag{17.8-3}$$

and

$$T_1\nabla^2 v_x + G\left(\frac{\partial^2 w_x}{\partial x^2} - \frac{\partial^2 w_x}{\partial y^2}\right) + 2G\frac{\partial^2 w_y}{\partial x \partial y} = 0 \tag{17.8-4}$$

$$T_1\nabla^2 v_y + 2G\frac{\partial^2 w_x}{\partial x \partial y} + G\left(\frac{\partial^2 w_y}{\partial x^2} - \frac{\partial^2 w_y}{\partial y^2}\right) = 0 \tag{17.8-5}$$

$$K_1\nabla^2 w_x + G\left(\frac{\partial^2 v_x}{\partial x^2} - \frac{\partial^2 v_x}{\partial y^2}\right) - 2G\frac{\partial^2 v_y}{\partial x \partial y} = 0 \tag{17.8-6}$$

$$K_1\nabla^2 w_y - 2G\frac{\partial^2 v_x}{\partial x \partial y} + G\left(\frac{\partial^2 v_y}{\partial x^2} - \frac{\partial^2 v_y}{\partial y^2}\right) = 0 \tag{17.8-7}$$

From the above, for the 9-fold symmetry quasicrystals, the displacement fields in the parallel space and perpendicular space are decoupled. For those in the parallel space, we can directly write the dislocation solution according to the classical one (the phonon one, i.e., the solution (7.3-1), which holds for the present case) and so, in what follows, we neglect the part given in (7.3-1) and only solve the phason part.

To obtain the dislocation-induced phason field, we introduce two biharmonic functions f_1 and f_2 and choose

$$v_x = -G\left(\frac{\partial^2 f_1}{\partial x^2} - \frac{\partial^2 f_1}{\partial y^2}\right) - 2G\frac{\partial^2 f_2}{\partial x \partial y} \tag{17.8-8}$$

$$v_y = -2G\frac{\partial^2 f_1}{\partial x \partial y} - G\left(\frac{\partial^2 f_2}{\partial x^2} - \frac{\partial^2 f_2}{\partial y^2}\right) \tag{17.8-9}$$

$$w_x = T_1\nabla^2 f_1, w_y = T_1\nabla^2 f_2 \tag{17.8-10}$$

where

$$\nabla^2\nabla^2 f_j = 0, \quad j = 1, 2 \tag{17.8-11}$$

It is easily found that the coupled partial differential equations are all automatically satisfied. Of course, it is mentioned that other general representations are given and omitted.

17.9 Conclusion and Discussion

In this chapter, we only give a simple discussion on the two-dimensional quasicrystals of the second kind because the solid quasicrystals have not been discovered yet. Although there is the above difficulty, in Refs. [7–11], some experts of quasicrystal study were working for a possible quasicrystal in solid with 7-fold symmetry, etc., but we have not found enough experimental data. This shows that the observation on two-dimensional quasicrystals in solids of the second kind is carried out still. So, the discussions in this chapter on quasicrystals in solids of the second kind is meaningful, and the observation of 18-fold symmetry quasicrystals in soft matter has been successful by Fischer *et al.* [1] and promotes us to do probe for studying this kind of quasicrystal in soft matter, and we give a simple discussion on the theory of groups which proves the elasticity on this kind of quasicrystal in solids, and they should exist, hence the discovery of the phase in solids is possible.

References

[1] Fischer S, Exner A, Zielske K, Perlich J, Deloudi S, Steuer W, Linder P and Foestor S, 2011, Colloidal quasicrystals with 12-fold and 18-fold diffraction symmetry, *Proc. Natl. Acad. Sci.*, **108**(5), 1810–1814.

[2] Hu C Z, Ding D H, Yang W G and Wang R H, 1994, Possible two-dimensional quasicrystals structures with six-dimensional embedding space, *Phys. Rev. B.*, **49**(14), 9423–9427.

[3] Yang W G, Wang R H, Ding DH and Hu C Z, 1995, Group-theoretical derivations of quadratic elastic invariants of two-dimensional quasicrystals of rank five and rank seven, *J. Phys. Condens. Matter.*, **7**, 7099–7112.

[4] Tang Z Y and Fan T Y, 2016, Point groups and group representation theory for second kind of two-dimensional quasicrystals, unpublished work.

[5] Ma Z Q, 2019, *Groups Theory for Physicists*, 2nd edition, World Scientific, Singapore.

[6] Li X F and Fan T Y, 2016, Dislocations in the second kind two-dimensional quasicrystals of soft matter, *Physica. B.*, **52**, 175–180.

[7] Newkome G. R ed., *Comprehensive Heterocyclic Chemistry II. Vol. 9: Seven Membered and Larger Rings and all Fused Derivatives*, (Elsevier Science (Pergamon), Oxford, 1996).

[8] Janner A, 2010, Strongly correlated structure of axial-symmetric proteins. II. Pentagonal, heptagonal, octagonal, nonagonal and ondecagonal symmetries, *Acta. Crystallogr. D. Biol.Crystallogr.*, **61**(3), 256–268.

[9] Steurer W, 2007, Boron-based quasicrystals with sevenfold symmetry, *Phil. Mag.*, **87**(18), 2707–2712.

[10] Socolar J E S, 1990, Weak matching rules for quasicrystals, *Commun. Math. Phys.*, **129**(3), 599–619.

[11] Orsini-Rosenberg H and Steurer W, 2011, Ab initio investigations on the stability of seven-fold approximants, *Phil. Mag.*, **91**(19–21), 2567–2578.

Chapter 18

Identification and Phase Transitions of Soft-matter Quasicrystals

The discussions in Chapters 3–17 are mainly concerned with solid quasicrystals (QC), e.g., their structures and properties. In this chapter, the focus is on the identification and phase transition of various supramolecular quasicrystals in different soft matter systems. These quasicrystals exhibit unique symmetries, such as 10-fold, 12-fold or 18-fold. The 12-fold symmetry quasicrystals (dodecagonal quasicrystal, DDQC) have been observed in many types of soft matter, while the 18-fold symmetry quasicrystals have only been observed in colloids in 2011. The soft-matter 10-fold symmetry quasicrystals (decagonal quasicrystal, DQC) were observed most recently and have been reported in 2022. These experimental results have close ties to chemistry and are discussed in detail in Section 18.1.

From a physics perspective, the soft-matter 10-fold DQCs discovered in giant molecules are more important than the 12- and 18-fold ones. Moreover, a unique phase transition from 10-fold DQC to 12-fold DDQC was identified. The mechanism of this peculiar phase transition demonstrates a fundamental difference between soft quasicrystals and hard atomic ones. This distinction, as well as the transition from 12-fold DDQC to its crystalline approximant, Frank–Kasper σ(FKσ) structure, is discussed in Section 18.2, which delves deeper into the phase transitions of supramolecular quasicrystals in various soft-matter systems.

18.1 Identification of Supramolecular Quasicrystals in Soft Matter

18.1.1 *Basic notions and discovery of soft quasicrystals*

It is well known that quasicrystals have a long-range order but no translational periodicity. Therefore, sharp diffraction spots can be observed but are unable to be described by 230 space groups in both real and reciprocal spaces. There are three types of quasicrystals: one-, two- and three-dimensional quasicrystals. In two-dimensional quasicrystals, the direction perpendicular to the quasiperiodic layers is periodic, resulting in a true layer structure within which no transitional order exists. Typical two-dimensional quasicrystals of the first kind include pentagonal, octagonal, decagonal and dodecagonal polygonal quasicrystals with 5-, 8-, 10 and 12-fold symmetry, respectively. The typical two-dimensional quasicrystals of the second kind include 7-, 9-, 14- and 18-fold symmetry, respectively. The three-dimensional quasicrystals have no periodic directions, represented by icosahedral quasicrystals. The icosahedral symmetry extends along three dimensions. Three independent vectors in traditional crystallography are not enough to index the diffraction peaks in quasicrystals, instead, at least five linearly independent vectors are necessary [1]. The necessary n vectors span independently in n-dimensional space. In other words, the quasicrystals in three-dimensional space can be constructed as periodic crystals in n-dimensional space. The real structures of quasicrystals in three-dimensional physical space can be obtained by appropriate projection/section techniques preserving the symmetries from n-dimensional space [2]. In fact, five indices are needed for two-dimensional polygonal quasicrystals and six indices for three-dimensional icosahedral quasicrystals.

Before the discovery of alloy quasicrystals, Roger Penrose created a set of prototiles to tile a plane quasi-periodically according to certain matching rules while preserving 5-fold rotational symmetry, which became the starting motivation of much work on theory of quasicrystals. This famous third type of Penrose tiling (P3) is composed of two types of rhombi tiles with different angles. An equivalent

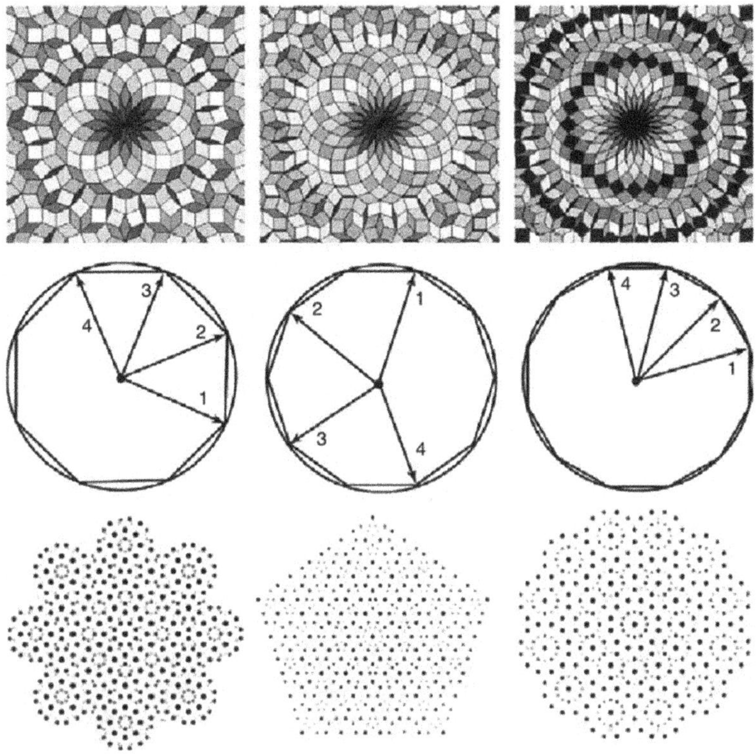

(a) (b) (c)

Fig. 18.1. Two-dimensional tilings (top), projection of the unit vectors of the polygonal reciprocal lattices in a four-dimensional space (middle) and simulated diffraction patterns (bottom), all along the highest symmetry zone axis for (a) octagonal, (b) decagonal and (c) dodecagonal quasicrystals [3].

Penrose tiling can also be constructed in three-dimension, resulting in icosahedral symmetry. Besides five-fold Penrose tiling, it is also possible to construct two-dimensional tiling with other rotational symmetries including 8-fold, 10-fold, or 12-fold, illustrated in Figs. 18.1(a)–18.1(c) for some examples [3]. Their corresponding reciprocal space lattice vectors can be obtained through projection from higher-dimensional spaces and showed in Fig. 18.1 middle array. We can see that four vectors are projected on the plane to describe related rotational symmetry. Their simulated diffractions exhibit

8-fold, 10-fold and 12-fold symmetries, respectively. The fifth vector projected is perpendicular to the plane.

After the first report by Zeng *et al.* in 2004 [4], the quasicrystalline phase study in soft matters experienced a rapid development. Especially, the DDQC with 12-fold rotational symmetry has been frequently observed in diverse soft-matter systems, such as block copolymers [5], colloidal particles [6] and giant molecules [7, 8]. The DDQC phase belongs to two-dimensional quasicrystals and has a strong tie to periodic Frank–Kasper (FK) phases. The formation of the FK and quasicrystalline phases results from the subtle competitions between thermodynamic driving forces, i.e., both enthalpic and entropic origins, and molecular shapes, which might open new possibilities to design materials with different properties.

At this moment, the physic model in two-dimensional quasicrystals is treated as a combination of a framework and the decoration with atoms [9]. The framework is just the tiling pattern on a two-dimensional plane which has been discussed in Fig. 18.1. The dodecagonal tiling pattern proposed in Fig. 18.1(c) is mainly composed of rhombus with the same edge length. The rhombus structure is relatively rare in real quasicrystals, and triangles and squares occupy the most space. The ideally quasiperiodic dodecagonal quasicrystal structure is still not constructed yet, but many methods have been developed to create tiling pattern with 12-fold using only triangles and squares. The deterministic dodecagonal quasicrystal with triangle-square tiling is presented in Fig. 18.2(a) generated by the projection method [10]. The 12-fold symmetry of the tiles results from the long-range orientations of the edges restricted in 12 directions with intervals of 30° and equal frequency. Three various tiling methods of $3^2 \cdot 4 \cdot 3 \cdot 4, 3^3 \cdot 4^2$ and 3^6 can be identified, among which 3^6 has two different orientations rotated by 30° and causes also two orientations for dodecagons consisting of 12 triangles and 6 squares. According to the self-similarity principle, the second generation of dodecagons in a larger scale can be constructed by connecting centres of the smallest dodecagons, illustrated by the red lines in Fig. 18.2(a). They have $2+\sqrt{3}$ times larger edge length than that of original ones. Continuing this process we can obtain higher

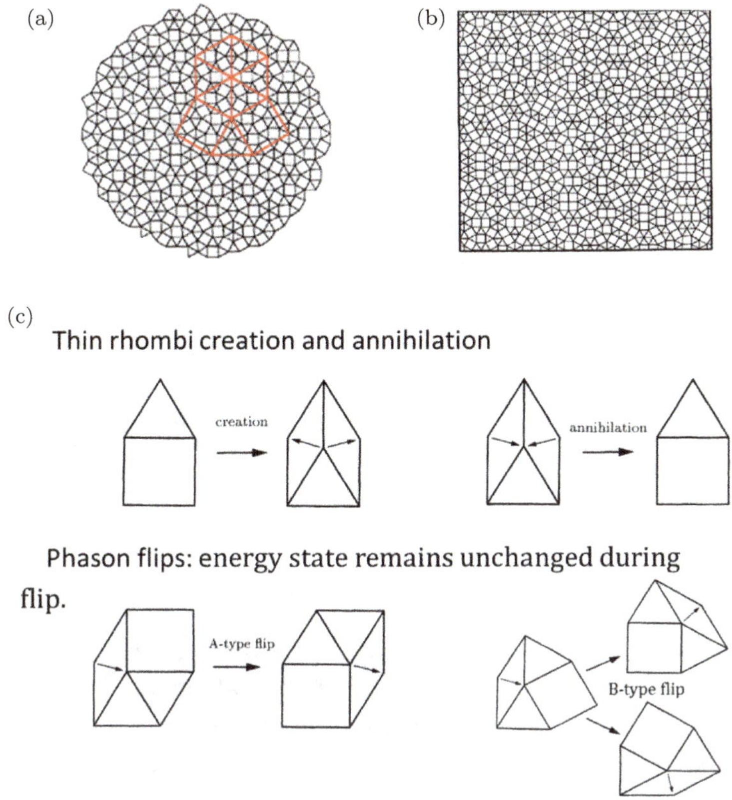

Fig. 18.2. (a) Dodecagonal quasilattice generated by the projection method [10]. (b) Dodecagonal quasicrystal with random tiling of triangles and squares. (c) Scheme of phason flips in random tiling dodecagonal quasicrystal [11].

generations. Considering the corresponding areas of an equilateral triangle and square, it is amazing to note that squares and triangles each occupy exactly half of the total area [11].

Besides the deterministic quasicrystal tiling, various random tilings can be constructed using the same number of triangles and squares, Fig. 18.2(b) as one example. In details, a square-triangle tiling contains exclusively equilateral squares and triangles. The random tilings belong to an ensemble of tilings satisfying certain boundary conditions, which require a long-range bond orientational order but exhibit 12-fold diffraction symmetry. This ensemble thus has many probabilities for random tilings [11]. A proposed mechanism of random tiling formation is the special local atom sites relocation, phason flips, which are related with phase strain (Fig. 18.2(c)). A pair of rhombi shapes can be created accompanying the change of energy states in the local sites. Each rhombus then follows A-type, B-type or bounce phason flips through the square-triangle tiling, in a way analogous to the operation of a zip fastener. This pair of two rhombi would sooner or later annihilate back into a square and triangle, acting the role as a zipper of squares and triangles in their whole life and creating randomness. Why do the quasicrystals always have random tilings? A reasonable answer is the minimized free energy in random tiling caused by the maximized density of entropy. This entropy arises from coordinated atomic relocations, where each such relocation is equivalent to a rearrangement of tiles. In the condition of either infinite temperature or zero Hamiltonian, the maximally random tiling would occur and the phason strain vanishes [12].

The real space structures of DDQC are built from the combination of the random tiling framework and the decoration units, resulting in a layered structure with a 12-fold symmetry along the perpendicular direction. To get a better understanding of real-space packing structure of DDQC, the readers are referred to a beautiful work about binary nanoparticle blends [6]. A quasicrystalline superlattice form in a binary nanocrystal blends 6.8 nm $CoFe_2O_4$ and 12.0 nm Fe3O4. A three-dimensional structure model through electron tomography and direct imaging of surface topography was clearly constructed. Although the nanocrystals are inorganic, this DDQC superlattice shares very similar structure features as those of DDQC phases observed in soft matters.

18.1.2 *12-fold dodecagonal quasicrystals in soft matters*

In 2004, Zeng *et al.* reported the first identified DDQC structure [4]. The single domain grown of this dendrimer (Fig. 18.3(a)) revealed the distinctive 12-fold symmetry SAXS pattern (Fig. 18.3(c)). When the sample is shoot by the incident beam perpendicular to the 12-fold

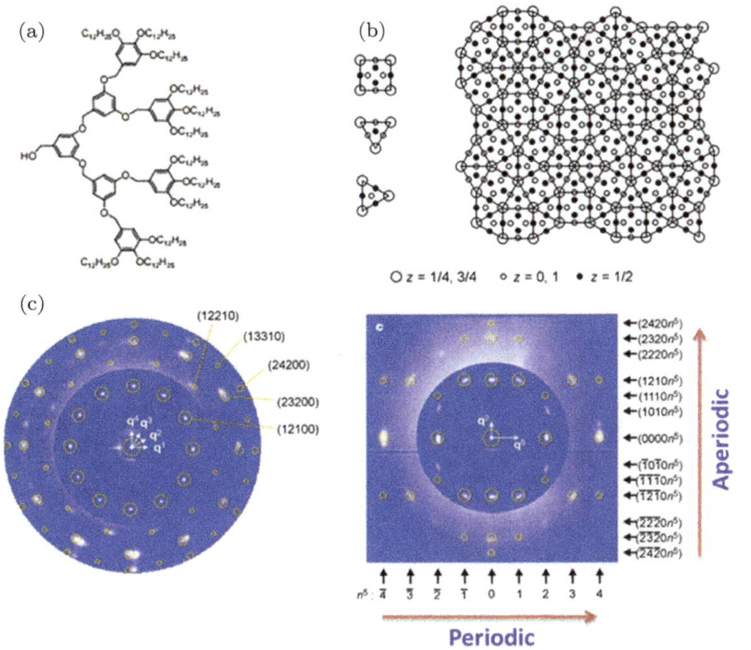

Fig. 18.3. (a) Chemical structure of the dendrimer resulting dodecagonal quasicrystal structure. (b) Three basic decorated tiles to generate the model of the dodecagonal quasicrystal. (c) Single-domain diffraction pattern along the 12-fold axis (on the left) and diffraction pattern shoot perpendicular to the 12-fold axis [4].

axis, a different pattern was acquired with periodicity along one direction and this pattern can be repeated every 30° rotation along the 12-fold axis (Fig. 18.3(c)). The full quasicrystal structure is constructed by decoration with three basic tiling units: the square and two triangles (Fig. 18.3(b)). The simulated diffraction pattern of dodecagonal quasicrystals marked by open circles matched rather well with the experimental one. Note that this quasiperiodic structure exists in the scaled-up micellar phases, representing an unprecedented mode of organization in soft matter.

The dodecagonal quasicrystal phase was later revealed by Bates *et al.* in diblock or tetrablock copolymers [5], thanks to the almost unlimited structural flexibility of block copolymers by tuning

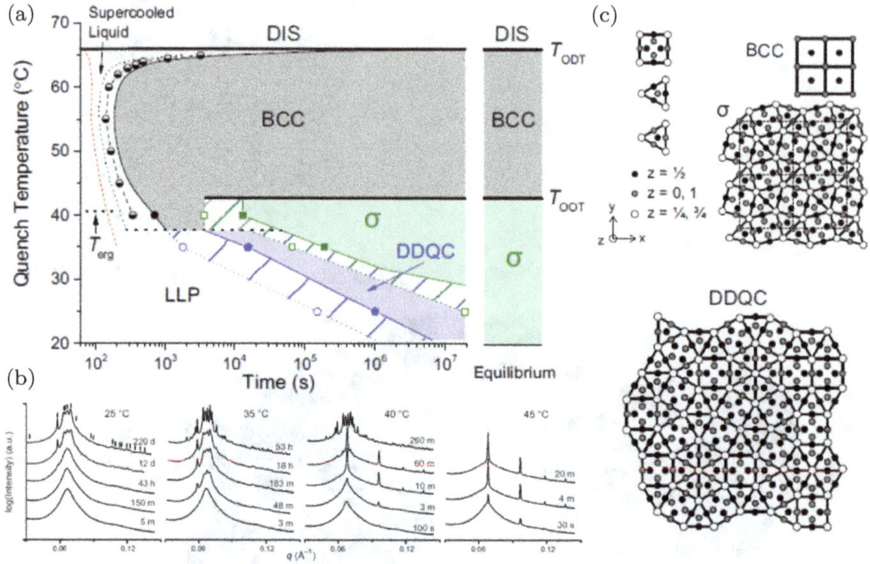

Fig. 18.4. (a) Time–temperature–transformation diagram illustrating the BCC-forming regions (shaded grey). Dotted lines, dashed lines and half-filled circles represent the achievement of 5%, 50% and 95% of the maximum measured value of G′, respectively. Open and filled symbols in the complex phase region indicate the first evidence of nucleation and completion of ordering in SAXS, respectively. Lines connecting these symbols are guides to the eye. The red-dashed line approximates the times during DMS experiments after the initiation of a quench when the sample equilibrated at the target temperature. (b) Synchrotron SAXS measurements of IL-5.1 ordering kinetics. Selected time-resolved SAXS data were collected at different times after rapidly quenching from the disordered state above TODT to 25°C, 35°C, 40°C and 45°C. (c) View of the BCC, σ and DDQC phases as decorated tilings of the x–y plane [5].

chemical compositions, molecular weights and molecular topology. In a sphere (micelle) forming poly(isoprene-*b*-lactide) (IL) diblock copolymer melt, the supercooled disordered state evolves into a metastable DDQC phase that transforms with time to the FK σ phase, where the lifetime of the DDQC phase is strongly temperature dependent (Fig. 18.4). Besides, the DDQC phase can evolve only from the supercooled disordered state. Directly cooling from the high temperature BCC phase leads to the FK σ phase, without passing through the dodecagonal quasicrystal phase [5]. This systematic work strongly suggests that the dodecagonal quasicrystal structure is a metastable intermediate state before the formation of the equilibrium FK σ phase in many soft-matter systems. Exceptionally, as in metals or alloys, the structure evolution is mediated by the exchange of charge, and the transition in diblock copolymers is mediated by the exchange of mass via the redistribution of diblock copolymer molecules among deformed micelles.

More recently, Cheng *et al.* observed the dodecagonal quasicrystal phases in developed giant molecular systems, including giant surfactants [7] or giant shape amphiphiles [8]. For example, in a giant surfactant series which has one molecular nanoparticle head and four polystyrene tails, a unique sequence of HEX $\rightarrow$ A15 $\rightarrow$ FKσ $\rightarrow$ DDQC $\rightarrow$ BCC phases was determined with increasing v_f^{PS} [7] (Fig. 18.5). The characteristic structure features of dodecagonal quasicrystal were clearly observed in both reciprocal and real spaces.

Interestingly, a non-periodic random tiling DDQC pattern was discovered in ABC star-branched three-component polymer copolymer systems, including [13, 14]. The strong segregation between three distinct polymers would drive the junction points aligned into one-dimensional length, resulting in self-organized two-dimensional tilings (Fig. 18.6). Note that the dodecagonal quasicrystal pattern constructed from ABC-type star polymers is distinct from the dodecagonal quasicrystal we discussed in dendrimers, giant surfactants, diblock copolymers and other soft systems, since it's purely two-dimensional tiling by columnar phase domains. Thus, no spherical building blocks are involved in these two-dimensional tilings. These results are similarly reported in another class of thermotropic columnar liquid crystal phases, which often have the T-shaped bolaamphiphiles and facial amphiphiles [15, 16]. Generally, the aromatic rod-like mesogen cores lie perpendicular to the column axis, constructing the cells of columns which are composed of flexible side groups. The aromatic rod walls take a polygonal tiling in two dimensions.

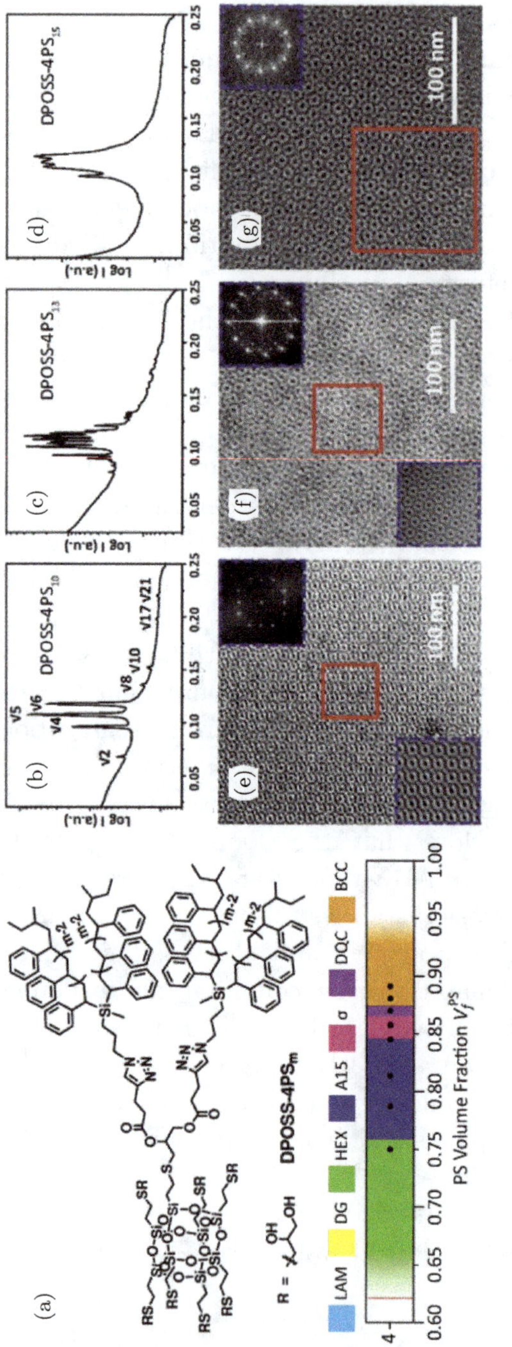

Fig. 18.5. (a) Chemical structures and phase diagram of giant surfactant DPOSS-4PS$_m$. (b, e), (c, f) and (d, g) SAXS and TEM results of FK A15 superlattice by DPOSS-4PS$_{10}$, FK σ superlattice by DPOSS-4PS$_{13}$ and DDQC (d, g) by DPOSS-4PS$_{15}$ [7].

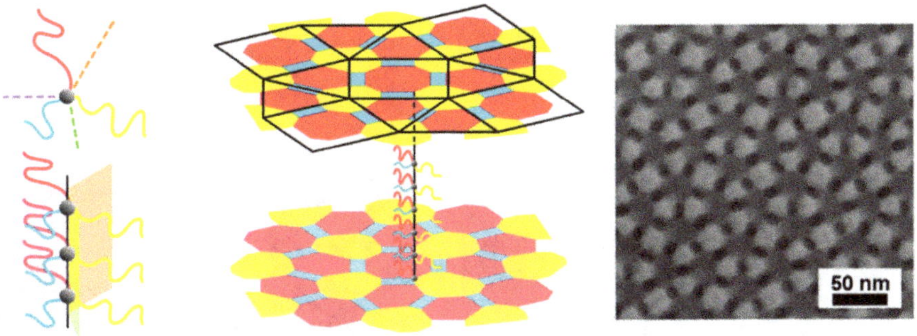

(a) (b) (c)

Fig. 18.6. Schematic drawings of the intramolecular phase separation of ABC star-shaped terpolymer chains in bulk (a) and their nanodomain assembly (b), with the typical bright-field transmission electron micrograph shown in (c) [14].

18.1.3 *10-fold axial decagonal DQC in soft matter*

It is interesting that the quasicrystal with 12-fold is dominated in soft matters, while there are only a few reports on supramolecular quasicrystals adopting other symmetries including, especially the 10-fold ones, which are commonly identified in the metallic counterparts.

Here, it is worth mentioning that Chen *et al.* discovered large-area 10-fold quasicrystalline superlattices from anisotropic colloidal particles (Fig. 18.7) [17]. They first synthesized wurtzite-truncated tetrahedron-shaped quantum dots and functionalized with organic ligands and then conducted the self-assembly by a using a liquid/air interface method. As result, a large area of 10-fold DQC structures emerges at the interface between the air and solvent. Electron tomography further indicates that the resultant DQC structures adopt an axial periodicity with six layers. It is further suggested that this unique order arises from a flexible polygon tiling rule. However, due to the bulk feature of the colloidal systems, it is not clear about the stability and the phase transition behaviour of this thin-film state DQC structure.

Recently, Cheng *et al.* have found an unprecedent axial 10-fold DQC supramolecular quasicrystal in a pair of binary giant molecular blends: OP8 and OP14 — two giant molecules with similar chemical

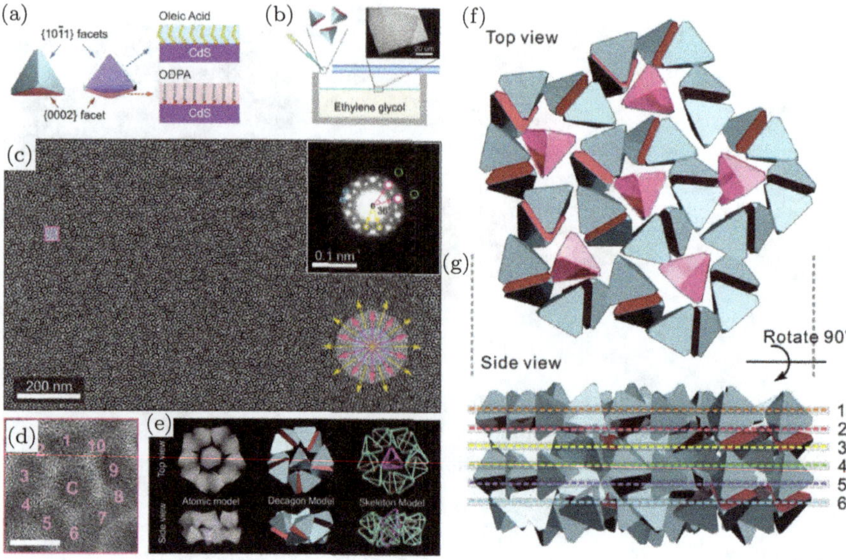

Fig. 18.7. (a) Anisotropic colloidal particles exhibit three facets coated with aliphatic oleic acid (blue) and one facet coated with hydrophilic octadecylphosphonic acid (red). (b) Liquid/air interface mediated self-assembling procedures. (c) TEM image of 10-fold DQC and the corresponding selected area electron diffraction (SA-ED) pattern (inset: pink, orange and green cycles indicate 10-basis vectors; blue cycles indicate 20-basis vectors). (d) TEM of a decagon-derivative unit (scale bar, 10 nm). (e) Computer-generated models (top and side views) of a decagon-derivative unit. (f and g) A computer-generated model of a double-decker QC-SL with six interconnected decagon-derivative units shown from a top view (f) and a side view (g) [17].

structures but containing different numbers of OPOSS cages at their peripheries [18]. When annealed separately, OP8 generates a larger FK σ phase composed of larger double-shell mesoatoms. On the other hand, OP14 self-assembles into a smaller FK σ phase with typical single-shell packings. Surprisingly, when mixed well together, the binary blends of these two giant molecules generate a DQC with an unprecedent phase transition sequence during thermal annealing (i.e., DQC $\rightarrow$ DDQC $\rightarrow$ FK σ). With comprehensive characterization, this soft DQC was confirmed to adopt an axial arrangement, that is, the packings within each layer are quasiperiodic, while the arrangement among layers is regular, adopting an A–B–A–C manner along the axial direction that is perpendicular to the quasiperiodic two-dimensional plane. In addition, the quasiperiodic patterns within each layer are found to arise from the random combinations of two characteristic tiling elements: a deformed hexagon and rhombus. Moreover, further annealing at higher temperatures leads the assembly an unprecedented phase transition from DQC to DDQC; when annealed for even longer time, the structure finally evolves to its stable crystalline FK σ structures. This transition sequence is unique, and we will elaborate the detailed mechanism in the following sections.

18.1.4 *Soft quasicrystals with other symmetry*

Foerster *et al.* reported the quasicrystal phase of PI_{30}-b-PEO_{120} micelles self-assembled in concentrated aqueous solution [19] (Fig. 18.8). In the concentration region between 13% and 18%, an

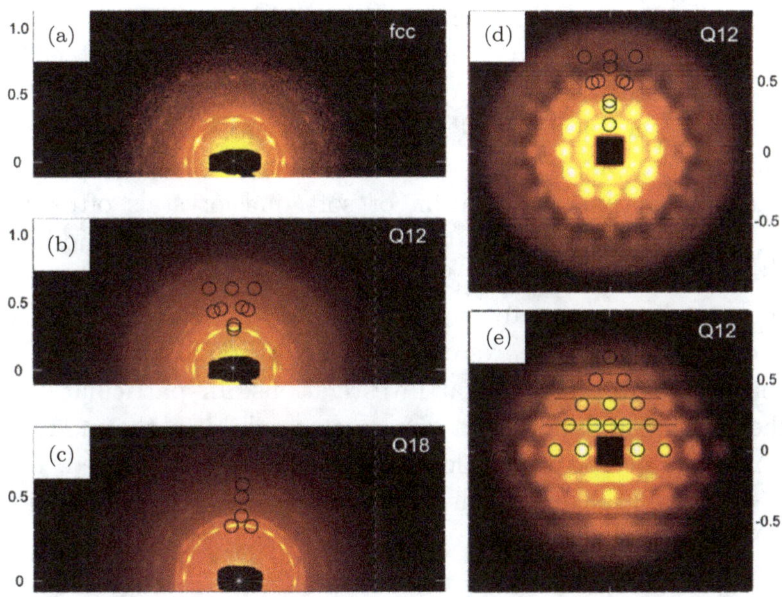

Fig. 18.8. Synchrotron SAXS patterns of FCC phase (a) and the quasicrystalline phases Q12 (b) and Q18 (c). The SANS patterns have also been recorded for Q12 phase, parallel (d) and normal (e) to the 12-fold rotation axis [19].

intriguing phase sequence occurred with decreasing temperature: Q18 $<$ 15°C $<$ Q12 $<$ 25°C $<$ FCC. Here, the Q18 and Q12 represent the quasicrystal phase with 18-fold and 12-fold symmetries. Note that quasicrystal phases are spontaneously formed from single component block copolymer micelles via self-assembly in concentrated solutions. This lyotropic block copolymer colloidal system provides the only example of a quasicrystal with 18-fold symmetry.

18.2 Phase Transitions and Formation Mechanisms of Soft Quasicrystals

The discovery of quasicrystals in soft matter enlarges the scope of the quasicrystal study. A distinct feature of quasicrystal in soft matter is that the constructing spherical motifs are assembled from multiple, identical molecular constituents. These spherical motifs scaling from several to tens of nanometers spontaneously form volume partition asymmetry during self-assembly. The thermodynamic environments of a quasicrystal state in soft matter could be different from the metallic alloys, regarding the introduced enthalpic and entropic factors of constitutional molecules and supramolecular motifs. More importantly, the hierarchical structures of spherical motifs greatly complicate the phase behaviour of soft-matter self-assembly. The emergence and phase transition of soft quasicrystals often result from changes not only in the packing of spherical motifs but also in their compositions. As a result, the underlying dynamics of the quasicrystal state in soft matter are expected to differ significantly and require systematic study through both theoretical calculations and experimental investigations. In particular, nearly all the quasicrystals found in soft matter exhibit a thermodynamically metastable feature, at least experimentally confirmed. In this section, we will delve into the dynamics of the phase transition of soft QC.

18.2.1 *Phase transitions of 12-fold DDQC*

In contrast to most metallic QC, the formation of soft DDQC is always accompanied by the formation of its crystalline approximants, namely, the FK σ phases. Moreover, the DDQC phase is inherently metastable and tends to evolve into the more stable FK σ structure. This unique phase transition behaviour has been observed in various homogeneous soft-matter systems, such as block copolymers, dendrimers and precise giant molecules.

The first reported observation of this unusual phase transition from DDQC to FK σ was made by Zeng *et al.* in 2004 [4]. By heating a dendrimer from room temperature, the cone-shaped molecules aggregate into supramolecular spherical motifs, resulting in the formation of a unique axial DDQC structure. This structure remains stable up to 71°C, after which it transitions into the FK σ with $P4_2/mnm$ symmetry (Fig. 18.9).

A similar phase transition was identified in the poly(isoprene-b-lactide) (IL) diblock copolymer melt and studied in detail by Bates *et al.* [5] (Fig. 18.4). Upon quenching from the disordered melt state to different temperatures, the overcooled block copolymer melts exhibit temperature-dependent phase behaviours. Samples quenched to 25°C and 35°C both show a typical scattering pattern of DDQC structures, which are metastable and evolve into the FK σ phase

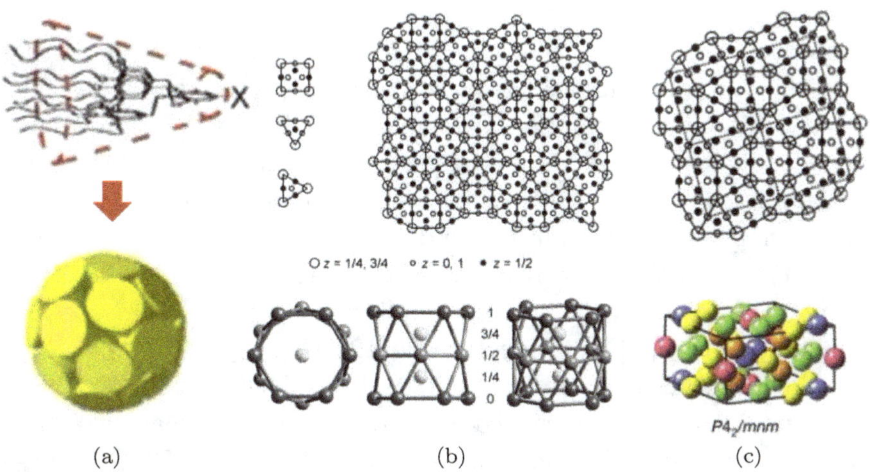

Fig. 18.9. (a) Structural illustration of the dendrimer and its formation of supramolecular spherical motifs. (b) Structure of soft DDQC. (c) Structure of FK σ with $P4_2/mnm$ symmetry [4].

after a longer period of time (Fig. 18.4(b)). In contrast, the sample quenched to 45°C gradually grows into ordered body-centred cubic structures. Remarkably, the DDQC phase evolves only from the supercooled disordered state. Cooling to temperatures below T_{OOT} (i.e., the order–order phase transition temperature between FK σ and BCC) from the BCC phase (grown at $T_{\mathrm{OOT}} < T < T_{\mathrm{ODT}}$) leads to the direct formation of the FK σ phase without passing through the DDQC phase. Despite the fact that the DDQC state only forms from a supercooled disordered state at a sufficiently high cooling rate, it is noteworthy that this quasicrystalline state can exist well above the glass transition temperature (T_g) of either block. This suggests that the behaviour of this soft solid is not solely driven by the conventional particle core or matrix vitrification. Furthermore, the high energy required to move a core block through the corona region results in the phase transition from DDQC to FK σ being primarily driven by the re-arrangement of spherical motifs rather than single-chain diffusion across domains.

A more quantitative study about the transition dynamics from DDQC to FK σ was studied by Cheng *et al.* in the assembly of giant dendrons (Fig. 18.10) [8]. Similar formation of DDQC structures was identified after quenching the disordered giant dendrons to different temperatures. Alike the block copolymer counterpart, DDQC structures only grow in samples quenched below certain temperatures. The detailed kinetics information was documented with the time-resolve small angle X-ray scattering (SAXS). A typical Avrami treatment was applied to analyse the grow of each phase at different conditions. As the slope of the Avrami plot indicates the growth dimensionality (n) of the ordered structures, the DDQC structures grown at 110°C adopts a higher growth dimensionality (n_{DDQC}, 110°C $=$ 1.28) compared with the FK σ structures at 125°C ($n_{\mathrm{FK}\sigma}$, 125°C = 1.15). In addition, an intriguing transition from DDQC to FK σ was captured in samples quenched at 118°C, whereas clear two-stage process was observed. The sample initially forms the DDQC phase and, after a relatively long isothermal time of approximately 80 minutes, completes a transition from the DQC phase to the FK σ phase. The growth of the DDQC process shows a comparable

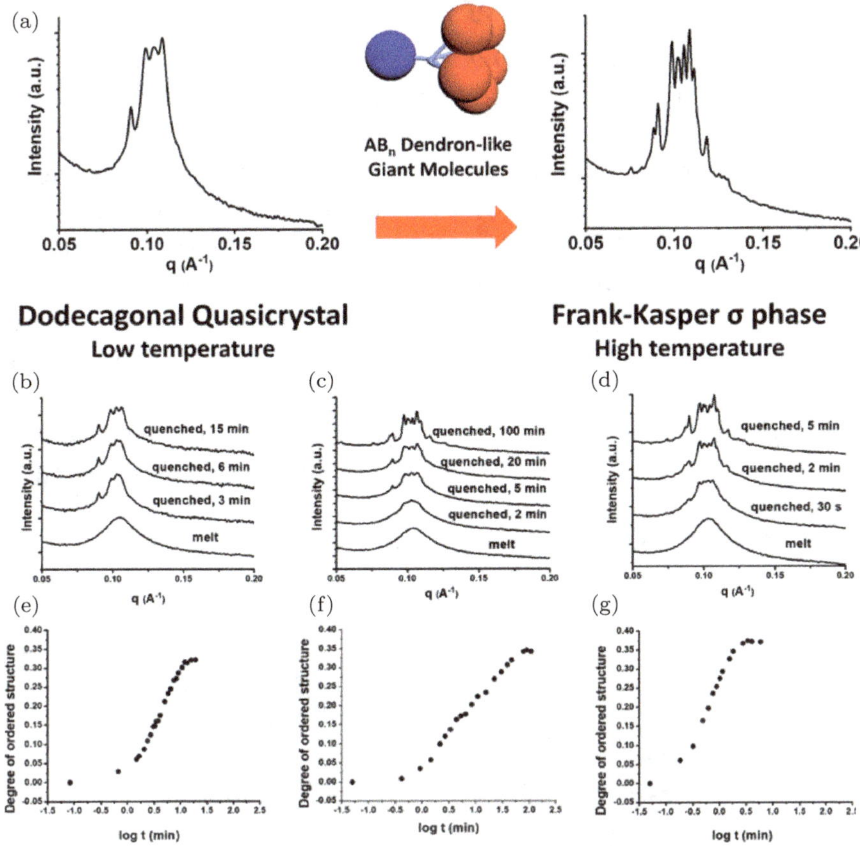

Fig. 18.10. (a) Structural illustration and phase transitions of DPOSS-MPOSS5. (b–d) *In situ* SAXS results at different isothermal times. Samples were initially disordered at 150°C and then quenched to different isothermal temperatures (with a cooling rate of 100°C/min): (b) 110°C, (c) 118°C, and (d) 125°C. (e–g) Degree of structural order versus isothermal time, based on *in situ* SAXS results (degree of structural order is calculated as $S_{\text{diffraction}}/S_{\text{pattern}}$ — the integrated area of all the feature diffraction peaks versus the one of the corresponding SAXS pattern) [8].

growth dimensionality ($n_{\text{DDQC}} = 1.10$), while the transition process adopts a much smaller dimensionality ($n_{\text{DDQC}} \rightarrow \text{FK}\sigma = 0.31$), likely due to the fact that growing the σ phase is at the expense of the DQC phase, leading to an overall slower growth rate and a smaller n value.

18.2.2 *Phase transitions of DQC by giant molecules*

In the previously mentioned DQC by giant molecules, combined with molecular dynamic simulations, Cheng *et al.* further revealed the mechanism of the phase transitions. OP8 and OP14 have similar chemical structures but contain different numbers of OPOSS cages at periphery (Fig. 18.11(a)) [18]. When annealed separately, OP8 generates a larger FK $\sigma(P4_2/mnm)$ phase composed of a larger double-shell mesoatoms (i.e., double-shell secondary supramolecular spherical motifs, (Fig. 18.11(b)), further confirmed by molecular dynamic simulation; on the other hand, OP14 self-assembles into a smaller FK σ phase via the typical single-shell packings (Fig. 18.11(c)). Moreover, the molecular dynamic simulation suggests that both FK σ phases arise due to the coexistence of multiple mesoatomic species, i.e., the [8 × OP8] and [10 × OP8] in the FK σ from OP8 (Fig. 18.11(b)) and the [3 × OP14] and [4 × OP14] from OP14 (Fig. 18.11(c)).

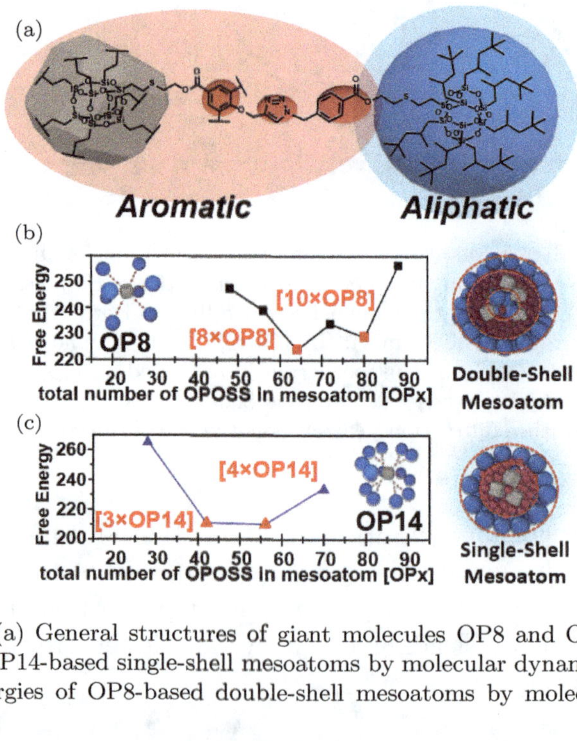

Fig. 18.11. (a) General structures of giant molecules OP8 and OP14. (b) Free energies of OP14-based single-shell mesoatoms by molecular dynamic simulation. (c) Free energies of OP8-based double-shell mesoatoms by molecular dynamic simulation.

In their binary blends, the molecular packings have been greatly complicated: besides the coexistence of multiple mesoatomic species within the same packing modes, both the single-shell and double-shell mesoatoms can also occur and coexist depending on the local distribution of OP8 and OP14. As a result, the well-mixed binary mixture generates a DQC phase with unprecedent phase transition sequences during thermal annealing (i.e., DQC $\rightarrow$ DDQC $\rightarrow$ FKσ). At the beginning of assembling, a temporal generation of the large double-shell mesoatoms greatly broaden the overall mesoatomic volume distribution, giving rise to a metastable decagonal quasicrystal phase. Later, the larger double-shell mesoatoms decompose into smaller single-shell ones, and the mesoatomic volume distributions gradually unify, affording 12-fold dodecagonal quasicrystal phases. Further arrangement of these single-shell mesoatoms renders dodecagonal quasicrystal assembly into its stable crystalline approximant — the FK σ structures Figs. 18.12(d) and 18.12(e).

The above-mentioned cases have provided plausible answers to the interesting question of why metallic quasicrystals prefer 10-fold decagonal quasicrystalline ordering and the condensed soft supramolecular quasicrystals predominantly exhibit 12-fold ordering. The formation of DQC usually requires a wider volume distribution of the constructing mesoatoms, which is hard to achieve for the "single-component" supramolecular assembly. However, for atomic alloy, a similar criterion can be easily fulfiled by multi-element metals [20]. Therefore, decagonal quasicrystal structures are seldom seen in soft matter but widely exist in metal alloys. On the contrary, 12-fold DDQC structures possess a relatively narrower volume distribution (nearly identical with FK σ structures). For the metallic atoms, they can easily nucleate into the stable FK σ (or other crystalline approximants) crystals instead of being trapped in metastable dodecagonal quasicrystal state [9]. For supramolecular assemblies in soft matter, since rearranging the mesoatoms needs either higher kinetic energy or longer time, less stable local configurations can preserve at more general conditions [21]. Thus, DDQC structures are more frequently observed in condensed soft matters and less in metal alloys.

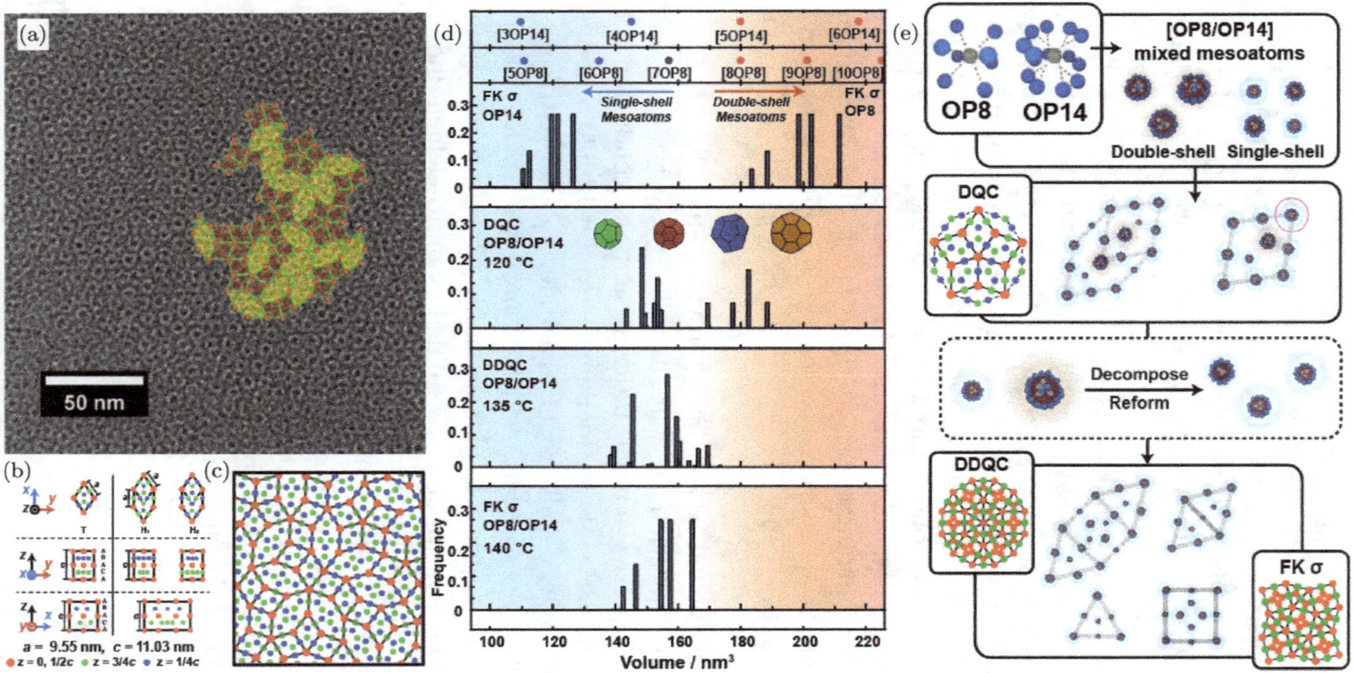

Fig. 18.12. Self-assembly of binary giant molecular blends at mass ratio mOP8/mOP8 = 1:2. (a) TEM result of decagonal quasicrystal superlattice. (b and c) Mg–Zn–Y axial decagonal quasicrystal structural model by the giant molecular assembly. (d) Mesoatomic volume distribution of the FK σ by unblended OP8 and OP14 (top); DQC, DDQC, FK σ by the binary blends. (e) Schematic illustration for the mesotomic mechanism during the phase transition sequence DQC $\rightarrow$ DDQC $\rightarrow$ FKσ [18].

18.2.3 *Simulation of QCphase formation in soft matters*

Note that in all the experimental results of soft matter or even in most simulations, quasicrystals mostly occur as aperiodic only in two dimensions. The three-dimensional icosahedral quasicrystals so far have only been observed in complicated metal alloys via

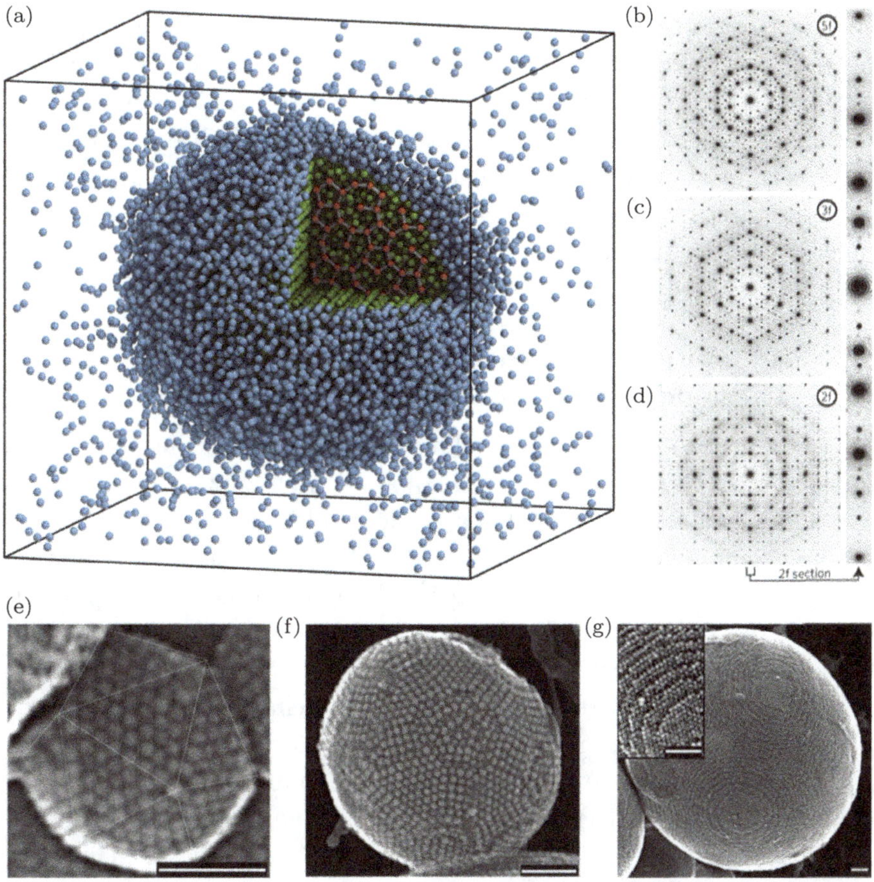

Fig. 18.13. Three-dimensional icosahedral quasicrystal phase. (a) Computational self-assembly of a one-component icosahedral quasicrystal. Diffraction patterns along five fold (b), three fold (c) and two fold (d) axes exhibit Bragg peaks and weak diffuse scattering [22]. (e–g) Entropy-driven experimental formation of large icosahedral colloidal clusters by spherical confinement [23].

the cooperation of several different atomic elements. A very recent intriguing molecular dynamics simulation result by Glotzer *et al.* shows that an icosahedral quasicrystal can be obtained from one-component system of uniform particles interacting via a tunable, isotropic pair Figs. 18.13(a)–18.13(d) [22]. Almost at the same time, another group reported that the entropy and spherical confinement suffice for the formation of icosahedral clusters consisting of up to 100,000 particles [23]. Specifically, tens of thousands of nanoparticles (below 10 nm size) compressed under spherical confinement spontaneously crystallize into icosahedral clusters with obvious icosahedral symmetry morphology Figs. 18.13(c)–18.13(g). This formation of dynamics and structure assembly process of icosahedral quasicrystal in this finding may help direct the right design in soft matter or nanoparticle systems to achieve three-dimensional icosahedral quasicrystal structures.

18.3 A Columnar 12-fold Quasicrystal with Thermodynamic Stability

Very recently, Zeng et alreported a two-dimensional columnar liquid quasicrystal (CLQC) with dodecagonal symmetry [24]. Unlike previous dodecagonal quasicrystals based on random tiling (Fig. 18.2(a)), a honeycomb structure based on a strictly quasiperiodic tessellation of tiles is observed. This soft matter dodecagonal quasicrystal phase is constructed from a specific T-shaped polyphilic molecule with three non-miscible chemical affinities (Fig. 18.14(a)): an aromatic rigid part (in yellow), two terminal flexible alkyl chains (in red) and one polar flexible side chain (in grey) [25]. The nano-segregation between the three non-miscible blocks drives the assembly of the molecular columns (Fig. 18.14(b)). Two different columnar phases are observed for this system: a periodic $p4gm^{\mathrm{L}}$ phase with a square symmetry around 70°C (Fig. 18.14(c)), and a CLQC with a dodecagonal symmetry around 75°C (Fig. 18.14(d)). These two phases have very similar diffraction pattern, with CLQC exhibiting notably fewer split peaks (Fig. 18.14(e)). The q^2 ratio of the first four observed peaks in CLQC phase is 1:2:3:3.72, very similar to the expected q^2 ratio

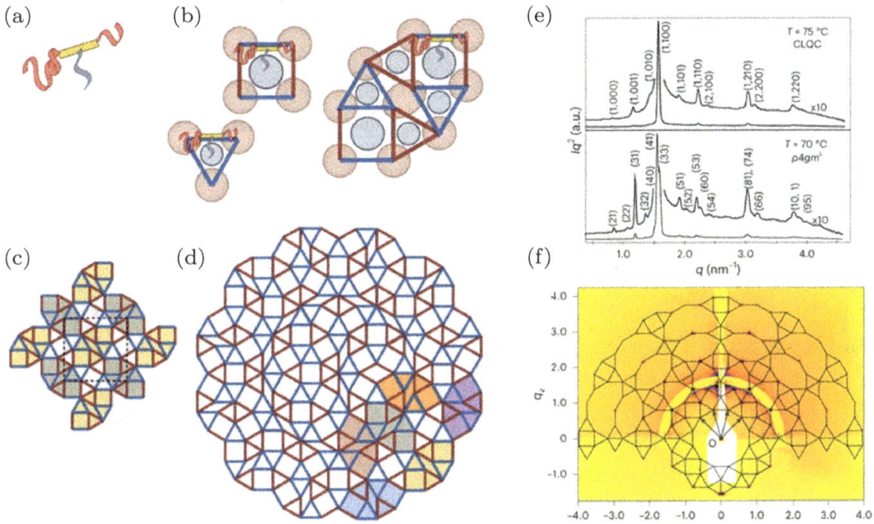

Fig. 18.14. (a) T-shaped polyphilic molecule containing three non-miscible chemical affinities: an aromatic rigid part (in yellow), two terminal flexible alkyl chains (in red) and one polar flexible side chain (in grey). (b) Dodecagonal clusters made up of triangular, square and trapezoidal cells. A red area is filled by alkyl chains, a grey area by polar chains, and the aromatic cores are located in-between red and grey areas. (c), (d), Two different arrangements of the columns found for the same T-shaped molecule depending on the temperature: a periodic columnar liquid crystalline phase with a square symmetry (c) and a columnar liquid quasicrystalline phase with a dodecagonal symmetry (d) [25]. (e) Indexed powder X-ray diffractograms of the CLQC and $p4gm^L$ phases, collected at 75 and 70°C, respectively. (f) GIXRD pattern of the CLQC phase recorded at 75°C and mapped into the q space, with the reciprocal lattice overlaid on top [24].

of 1:2:3:3.732 $(1:2:3:2+3-\sqrt{2}+3)$ between the (1000), (1001), (1010) and (1100) peaks of a 2D quasicrystal with a 12-fold rotational symmetry, with 4 instead of 2 reciprocal lattice basis vectors (Fig. 18.14(f)). Surprisingly, the CLQC phase was observed in both heating and cooling after melting into isotropic liquid, indicating the thermodynamic stability of this soft mattery quasicrystal phase. A continuous phase transition of $p_4gm^L \rightarrow$ CLQC $\rightarrow$ P6/mmm were taken from the SAXS data, but without measurable enthalpy.

The CLQC structure consists of dodecagonal clusters made up of triangular, square and trapezoidal cells that are optimal for local packing. To maximize the presence of such dodecagonal clusters, the system abandons periodicity but adopts a quasiperiodic structure that follows strict packing rules. The stability of random-tiling dodecagonal quasicrystals is often attributed to the entropy of disordering when strict tiling rules are broken, at the sacrifice of the long-range positional order. However, Zeng et al demonstrate that the driving force for quasiperiodicity in this new CLQC could be attributed primarily to optimization of the packing [24], that is, a lowering of the system energy rather than an increase in the randomness and/or entropy, as in the case of previously found soft quasicrystals.The quasicrystal stability may rest on energy minimization alone, or with only minimal entropic intervention.

18.4 Conclusion and Discussion

Soft-matter quasicrystals are formed mainly from supramolecular assembly. Apart for the direct assembly in colloidal systems, most soft quasicrystals are formed *via* a hierarchical self-assembling procedure. In this regard, packings on molecular level direct the formation to spherical supramolecular motifs of different features; in addition, the arrangement of these secondary spherical motifs further determine the final structure of the supramolecular assembly. The large dimensions of these secondary motifs (1–10 nm) greatly disfavour the relaxations of these structural units, even triggering the jamming of the system, rendering the assembly into a cooperative glassy state [26, 27]. These hierarchical procedures wildly exist in abroad kinds of soft matter and their assemblies in not only bulk states but also solutions. As a consequence, this lagging dynamics creates a wide window for the existence of many metastable states, so many metastable spherical packing structures can be captured, such as the broadly observed DDQC structures, even its seldom appears in their atomic counterparts. In fact, the metastability is perhaps the most important feature of soft matter; the metastable states play an important role in the phase transitions in many kinds of soft matter; for a more detailed discussion, refer to monograph [28].

The discovery of quasicrystalline structures in soft matter is inspiring. It not only reinforces the scale invariance principles but more importantly opens up new possibilities of structure engineering in soft materials. Nevertheless, concerning the emergence of various quasicrystalline structures in soft matter, there still remain many challenges to be addressed, such as the construction of soft quasicrystals of other types, e.g., the three-dimensional icosahedral quasicrystals. Besides, a more profound challenge is to establish a physical model to describe not only the formation of quasicrystalline state but also other phases in metastable states in general.

Before Zeng *et al.* [24], Fan and his group studied a series of generalized dynamics of soft-matter quasicrystals with thermodynamic stability [29], in which they introduced the fluid phonon as well as equations of state of soft matter, and carried out a series solutions, which were compared with those of the classical fluid dynamics. The comparisons show the dynamics is significant. Due to limitations of space, the details of the dynamics have not been introduced in this chapter yet.

References

[1] Abe E, Yan Y and Pennycook J, 2004, Quasicrystals as cluster Aggregates, *Nat. Mater.* **3**, 759–767.

[2] Yamamoto A, 1996, Crystallography of quasiperiodic crystals, *Acta Crystallogr. Sect. A Found. Crystallogr.*, **52**, 509–560.

[3] Graef M. E. M. M. D, 2012, *Structure of Materials: An Introduction to Crystallography, Diffraction and Symmetry*, Cambridge University Press, Cambridge.

[4] Zeng X, Ungar G, Liu Y, Percec V, Dulcey A E and Hobbs J K, 2004, Supramolecular dendritic liquid quasicrystals, *Nature*, **428**, 157–160.

[5] Gillard T M, Lee S and Bates F S, 2016, Dodecagonal quasicrystalline order in a diblock copolymer melt, *Proc. Natl. Acad. Sci.*, **113**, 5167–5172.

[6] Ye X, Chen J, Eric Irrgang M, Engel M, Dong A, Glotzer S C and Murray C. B, 2017, Quasicrystalline nanocrystal superlattice with partial matching rules, *Nat. Mater.*, **16**, 214–219.

[7] Yue K, Huang M, Marson R L, He J, Huang J, Zhou Z, Wang J, Liu C, Yan X, Wu K, Guo Z, Liu H, Zhang W, Ni P., Wesdemiotis C, Zhang W-B, Glotzer S C and Cheng S Z D, 2016, Geometry induced

sequence of nanoscale Frank–Kasper and quasicrystal mesophases in giant surfactants, *Proc. Natl. Acad. Sci.*, **113**, 14195–14200.

[8] Feng X, Liu G, Guo D, Lang K, Zhang R, Huang J, Su Z, Li Y, Huang M, Li T and Cheng S Z D, 2019, Transition kinetics of self-assembled supramolecular dodecagonal quasicrystal and Frank-Kasper σ phases in ABn dendron-like giant molecules, *ACS Macro Lett.*, **8**, 875–881.

[9] Ishimasa T, 2011, Dodecagonal quasicrystals still in progress, *Isr. J. Chem.*, **51**, 1216–1225.

[10] Baake M, Klitzing R and Schlottmann M, 1992, Fractally shaped acceptance domains of quasiperiodic square-triangle tilings with dedecagonal symmetry, *Phys. A: Stat. Mech. Appl.*, **191**, 554–558.

[11] Oxborrow M and Henley C L, 1993, Random square-triangle tilings: A model for twelvefold-symmetric quasicrystals, *Phys. Rev. B*, **48**, 6966–6998.

[12] Leung P W, Henley C L and Chester G V, 1989, Dodecagonal order in a two-dimensional Lennard-Jones system, *Phys. Rev. B*, **39**, 446–458.

[13] Asai Y, Takano A and Matsushita Y, 2015, Creation of cylindrical morphologies with extremely large oblong unit lattices from ABC block terpolymer blends, *Macromolecules*, **48**, 1538–1542.

[14] Hayashida K, Dotera T, Takano A and Matsushita Y, 2007, Polymeric quasicrystal: Mesoscopic quasicrystalline tiling in ABC star polymers, *Phys. Rev. Lett.*, **98**, 1–4.

[15] Ungar G, Zeng X and Frank–Kasper, 2005, Quasicrystalline and related phases in liquid crystals, *Soft Matter*, **1**, 95.

[16] Zeng X, Kieffer R, Glettner B, Nürnberger C, Liu F, Pelz K, Prehm M, Baumeister U, Hahn H, Lang H, Gehring G A, Weber C H M, Hobbs J K, Tschierske C and Ungar G, 2011, Complex multicolor tilings and critical phenomena in tetraphilic liquid crystals, *Science*, **331**, 1302–1306.

[17] Nagaoka Y, Zhu H, Eggert D and Chen O, 2018, Single-component quasicrystalline nanocrystal superlattices through flexible polygon tiling rule, *Science*, **362**, 1396–1400.

[18] Liu Y, Liu T, Yan X.-Y, Guo Q.-Y, Lei H, Huang Z, Zhang R, Wang Y, Wang J, Liu F, Bian F.-G, Meijer E. W, Aida T, Huang M and Cheng S Z D, 2022, Expanding quasiperiodicity in soft matter: Supramolecular decagonal quasicrystals by binary giant molecule blends, *Proc. Natl. Acad. Sci.*, **119**, 21613–21621.

[19] Fischer S, Exner A, Zielske K, Perlich J, Deloudi S, Steurer W, Lindner P and Förster S, 2011, Colloidal quasicrystals with 12-fold and 18-fold diffraction symmetry, *Proc. Natl. Acad. Sci.*, **108**, 1810–1814.

[20] Steurer W, 2004, Twenty years of structure research on quasicrystals. Part I. Pentagonal, octagonal, decagonal and dodecagonal quasicrystals, *Zeitschrift für Krist. - Cryst. Mater.*, **219**, 391–446.

[21] Dzugutov M, 1996, A universal scaling law for atomic diffusion in condensed matter, *Nature*, **381**, 137–139.

[22] Engel M, Damasceno P F, Phillips C L and Glotzer S C, 2015, Computational self-assembly of a one-component icosahedral quasicrystal, *Nat. Mater.*, **14**, 109–116.

[23] de Nijs B, Dussi S, Smallenburg F, Meeldijk J D, Groenendijk D J, Filion L, Imhof A, Van Blaaderen A and Dijkstra M, 2015, Entropy-driven formation of large icosahedral colloidal clusters by spherical confinement, *Nat. Mater.*, **14**, 56–60.

[24] Zeng X, Glettner B, Baumeister U, Chen B, Ungar G, Liu F, Tschierske C, 2023, A columnar liquid quasicrystal with ahoneycomb structure that consists oftriangular, square and trapezoidal cells, *Nat. Chem.*, **15**, 625–632.

[25] Impéror-Clerc M, 2023, A liquid quasicrystal of columns, *Nat. Chem.*, **15**, 591–592.

[26] Liu Y, Liu G, Zhang W, Du C, Wesdemiotis C and Cheng S Z D, 2019, Cooperative soft-cluster Glass in giant molecular clusters, *Macromolecules*, **52**, 4341–4348.

[27] Liu Y, Yan X, Guo Q, Lei H, Liu X, Li X, Wu Y, Zhang W, Liu G and Cheng S Z D, 2022, Recent progress on the rheology of giant molecules, *Macromol. Chem. Phys.*, **224**, 2200357.

[28] Cheng S Z D, 2008, *Phase Transitions in Polymers — The Role of Metastable States*, Elsevier Science Publishers, B. V. North Holland.

[29] Fan T Y, Yang W G, Cheng H and Sun X H, 2022, *Generalized Dynamics of Soft-Matter Quasicrystals and Its Applications*, Springer, Heidelberg.

Chapter 19

Photonic Band Gap and Application of Two-dimensional Photonic Quasicrystals

Starting from Chapter 7, we have given a series of applications on the theory in different areas. However, in this chapter, we will discuss a kind of special application of quasicrystals, which present an important role in device physics.

The two-dimensional quasicrystals are discussed in Chapters 6, 7, 8, 17 and 18, whose higher folds of rotational symmetry are dependent upon their symmetry structure, or say, their group structures; for this, refer to Chapters 6 and 17 or the monograph of group theory [1]. Photonic quasicrystals (PQCs) have a bigger structural size than the conventional quasicrystals investigated in the solid physics. Furthermore, more complex PQCs with multi-fold rotational symmetry can be formed by using a multi-beam interfering method. PQCs have abundant photonic bandgap and can be applied in micro-cavity lasers, sensors, filters and so on.

In this chapter, we construct two-dimensional PQCs with different rotational symmetries by using multi-beam holographic interference. Their photonic band-gap (PBG) properties are investigated by finite element method (FEM). The results show that 10-fold PQCs formed by five-beam interference and 12-fold PQCs are easier to form PBG. At the same time, the 10-fold PQCs were also prepared in the SU8 photo-resist film by using single-prism holographic interfering lithography. On the other hand, the application of 10-fold PQCs was also investigated in the sensor of liquid refractive index.

This will give guidance for selecting an appropriate PQC structure to form a photonic integrated device.

19.1 Introduction

Compared with periodic photonic crystals (PCs), PQCs have many advantages, such as easy appearance of a complete PBG [2–4], higher PBG isotropy [5,6] and richer defect modes [7]. Because of abundant PBG and localization properties, PQCs have been used for designing waveguides [5,8], sensors [9], laser microcavities [10], filters [3,11], etc. This shows that PQC has great potential in the design of small-size and multi-functional optical integrated circuits.

Research of PQC band-gap is the basis of application. Searching for wider complete PBG characteristics [12] and developing more application fields of PQC have been the goal pursued by scientists. Therefore, various PQC structures are designed and implemented by different methods. Among all these methods, multiple-beam holographic interfering method is considered to be the simplest method for fabricating PQC structures with large areas and without defects [13, 14]. In this chapter, the multi-beam interfering method is used to construct different PQC structures and the FEM method (for this method, refer to Chapter 12) is utilized to investigate their PBG characteristics and compare the ability to form band gaps. At the same time, 10-fold PQCs were prepared by using single-prism holographic lithography and its application in refractive index sensor will be investigated.

19.2 Design and Formation of Holographic PQCs

Multi-beam interference system is shown in Fig. 19.1. By using this method, two-dimensional PQC structures can be obtained [13]. The interfering patterns in the medium are expressed as follows:

$$I(r) = \sum_{j,l} E_j E_l * \exp\left\{ -i\left[(\vec{k_j} - \vec{k_l}) \cdot \vec{r} \right] \right\} \tag{19.2-1}$$

where E_j and $\vec{k_j}$ are the amplitude and wave vector of beam j, respectively. $\vec{r} = (x, y, z)$ is the position vector. The wave vectors of

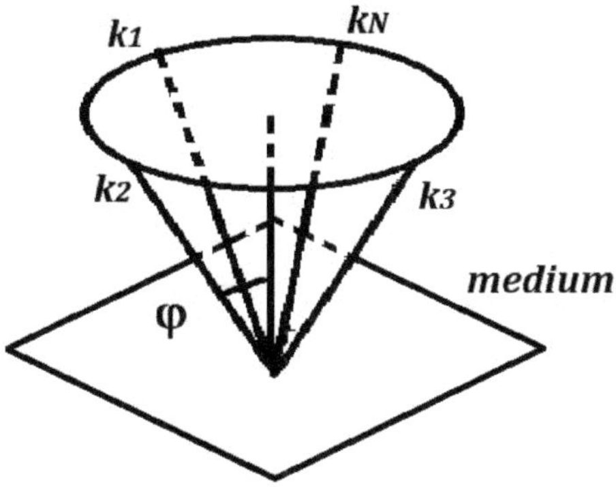

Fig. 19.1. Formation of multi-beam interfering system.

the mth beam are

$$\overrightarrow{k_m} = \frac{2\pi n_w}{\lambda}\left(\cos\frac{2(m-1)\pi}{N}\sin\varphi, \sin\frac{2(m-1)\pi}{N}\sin\varphi, \cos\varphi\right) \tag{19.2-2}$$

where $m = 1 \rightarrow N$ is the number of beams, n_w represents the refractive index of the medium and φ is the angle between each beam and central axis. We can get the interferograms by substituting $\varphi = 29°$, $n_w = 1.5$, $\lambda = 355\,\text{nm}$ into Eq. (19.2-1) and Eq. (19.2-2). The two-dimension different PQCs are shown in Fig. 19.2. Among these figures, Figs. 19.2(b) and 19.2(e)–19.2(h) are formed by interfering 5, 7, 9, 11 and 13 beams, while Figs. 19.2(a), 19.2(c) and 19.2(d) are obtained by 8, 10, 12 beams interference. All these structures possess even-fold rotational symmetry. From Figs. 19.2(b) and 19.2(c), we can see that 10-fold PQC structures are different by the interference of five beams and ten beams, with the increase in beams, and a sparser structure appears in Fig. 19.2(c).

19.3 Band Gap of 8-fold PQCs

In Fig. 19.3(a), the $8\,\mu\text{m} \times 8\,\mu\text{m}$ central region in Fig. 19.2(a) is selected for calculating the PBG of 8-fold PQCs. A more detailed setting is shown in Fig. 19.3(b). Blue spots stand for medium rods with dielectric constant $\varepsilon_1 = 14$ and radius $r = 0.12\,\mu\text{m}$, and

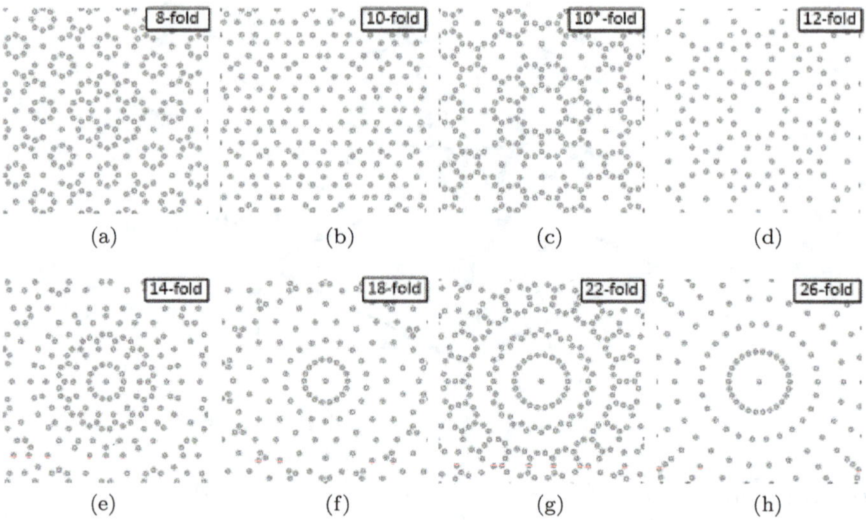

Fig. 19.2. Different two-dimensional PQC structures by interference of (a) 8 beams, (b) 5 beams, (c) 10 beams, (d) 12 beams, (e) 7 beams, (f) 9 beams, (g) 11 beams and (h) 13 beams.

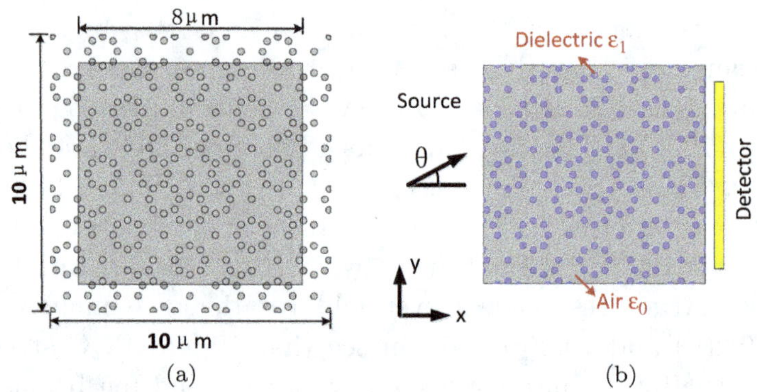

Fig. 19.3. (a) The selected eightfold PQC and (b) the modelling setup.

the grey background is air with $\varepsilon_0 = 1$. Light is incident on the sample at an angle θ to the x-axis and is detected on the right side. θ is in the range from $0°$ to $22.5°$. PQCs' band gap properties are investigated by calculating the transmittance and reflection spectra of the sample [15].

When a TE polarized wave is incident on the sample, the change of transmission and reflection with the normalized frequency a/λ is calculated at different incident angels θ, as shown in Fig. 19.4. The lattice constant is chosen as $a = 1\,\mu$m and λ is the wavelength. From the figure, we can see that the biggest PBGs happen in the frequency range from 0.476 to 0.523. The electric field distributions at the frequency $a/\lambda = 0.507$ and $a/\lambda = 0.730$ are also calculated and shown in Fig. 19.5. They are corresponding to the frequency in the

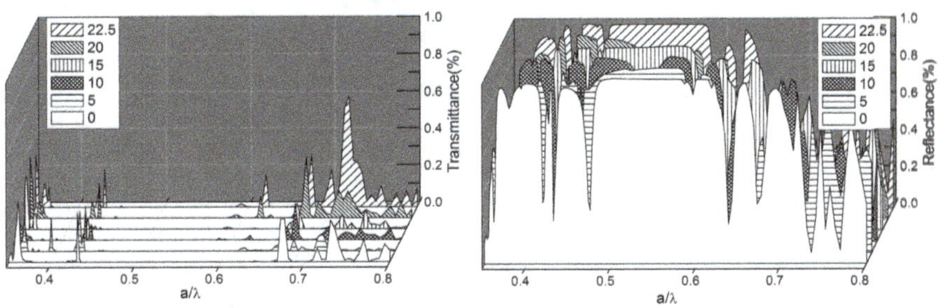

(a) (b)

Fig. 19.4. (a) Transmission spectrum and (b) reflection spectrum of eightfold PQC for different incidence angles.

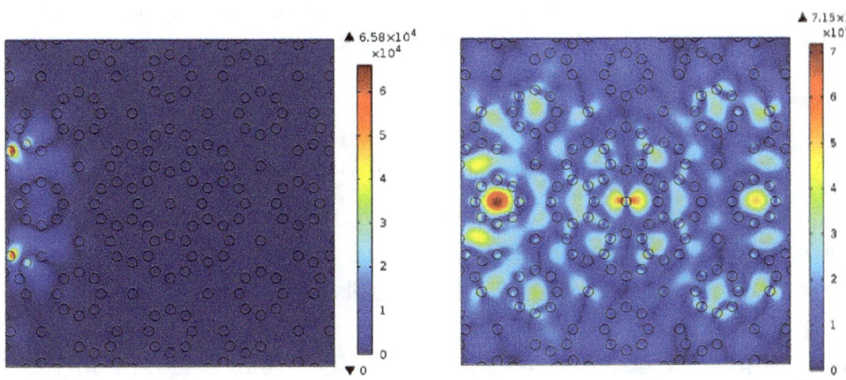

(a) (b)

Fig. 19.5. The electric field distribution (a) $a/\lambda = 0.507$ in the band gap and (b) $a/\lambda = 0.730$ outside the band gap.

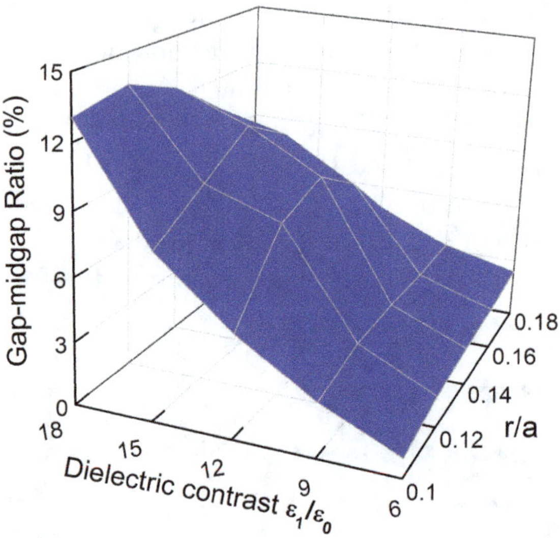

Fig. 19.6. Relative PBG versus parameters $\varepsilon_1/\varepsilon_0$ and r/a of eightfold PQC.

band gap and outside the band gap. We can see that the light in the band gap is forbidden to propagate in the PQC, as in Fig. 19.5(a).

Then, the maximum band gap of an 8-fold quasicrystal was further investigated when the dielectric contrast $\varepsilon_1/\varepsilon_0$ and fill factor r/a changed. The result is shown in Fig. 19.6. The relative PBG is expressed by the gap–midgap ratio, which is the ratio of the band gap width to central band. From the figure, we can see that the maximum PBG happens at $r/a = 0.12$ and $\varepsilon_1/\varepsilon_0 = 18$.

19.4 Band Gap of Multi-fold Complex PQCs

In the same way, the relative band gap $\Delta\omega/\omega_c$ is optimized for all the PQCs formed by interference, where $\Delta\omega$ and ω_c are the width and central frequency of the PBG, respectively. The results are shown in Fig. 19.7. The figure reveals that five-beam-interference 10-fold PQC and 12-fold PQC are easier to generate PBG than others, and PQC above 12-fold symmetry has smaller PBG. This is consistent with the viewpoint in Ref. [16].

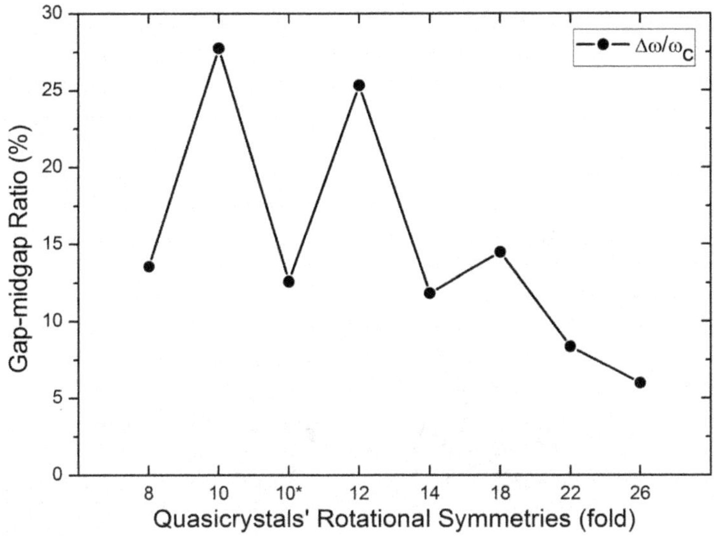

Fig. 19.7. Comparison of relative band gap for different PQCs.

19.5 Fabrication of 10-fold Holographic PQCs

19.5.1 *Material and writing system*

The sample preparation includes four steps: clean of glass substrates, spin coating of SU8 photo-resist film, holographic lithography of the film and post processing of the exposed film. The main material used in the sample preparation is SU-8 photoresist, a high contrast, epoxy-based photo-resist resin from MICRO CHEM Cop., which can produce a photosensitive reaction under ultraviolet light and record interference light field. After the SU8 film is irradiated by the interfering laser, Propylene Glycol Monomethyl Ether Acetate (PGMEA, C6H12O3) is used to develop the exposed SU8 film to obtain PQC structures [17].

The writing system is shown in Fig. 19.8. A 355 nm laser beam from a triple-frequency Nd:YAG laser (YG980, Quantel Co. France) was expanded and collimated and then split into six umbrella-like beams by using a top-cut prism invented by our group. The cutting

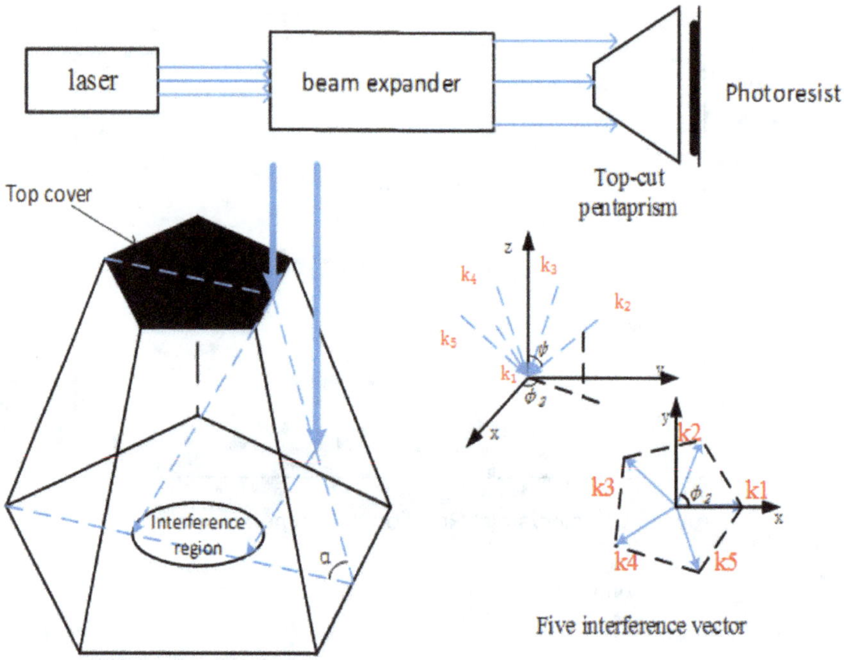

Fig. 19.8. The diagram of the interfering system.

angle of the prism is $\alpha = 54.7°$. The central beam was covered and the other beams irradiate the photo-resist medium. The two-dimensional interference patterns were recorded in the films.

19.5.2 *Experimental results*

Scanning electronic microscopy (SEM S-4800, Hitachi Limited, Japan) and Scanning Probe Microscopy (Multimode 8 SPM, Bruker) are utilized to observe the two-dimensional PQC micro-structures. Before SEM measurement, the 1 cm $\times$ 1 cm sample needs to be coated with a thin gold film. For SPM, the film topography was obtained by using the tapping mode.

SEM micrographs of two samples with different exposure conditions are measured and shown in Figs. 19.9 and 19.10, respectively. Figure 19.9 indicates the patterns with different magnifications when the film is irradiated by five beams with the same intensity

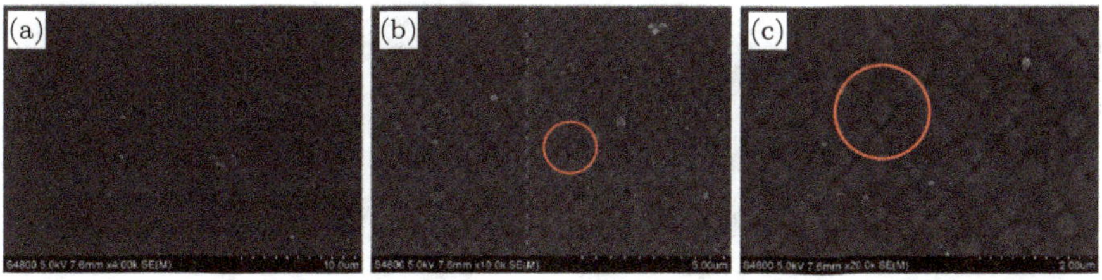

Fig. 19.9. SEM micrographs of PQC prepared with the same beam intensity.

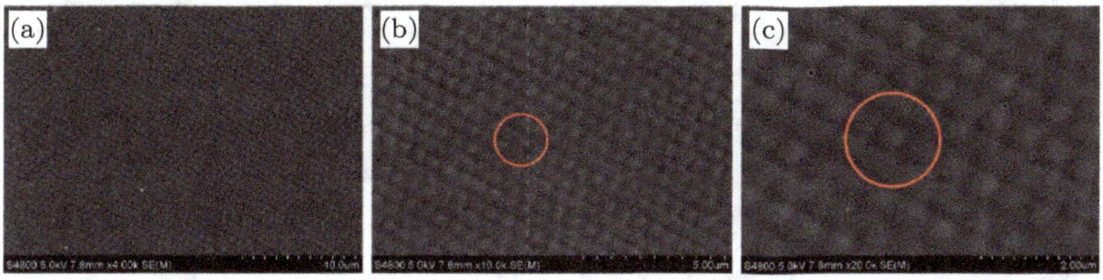

Fig. 19.10. SEM micrograph of PQC structure with two beams weaker.

40 mJ/cm^2. From the figure, we can see a 10-fold rotational symmetrical structure, as circled in red. While in Fig. 19.10, because the intensity of the two beams is changed as 10 mJ/cm^2, the symmetry is broken and the structure is not perfect.

At the same time, we also provide SPM photographs to observe the PQC structures more clearly. Figures 19.11(a) and 19.11(b) are photographs with different magnifications of the same sample. The exposure intensity and time of the sample are 20 mJ/cm^2 and 4s, respectively. Figure 19.11(b) is the magnification of the marked area in Fig. 19.11(a). From this figure, we can observe the 10-fold rotational symmetry clearly. In Fig. 19.11(b), the red circle stands for a unit cell, whose diameter is about 1.6 μm. The gap between two adjacent dots is about 490 nm. To reveal the rotational symmetry more clearly, a 532-nm Nd:YVO_4 laser is incident on the sample perpendicularly to observe the diffraction pattern, which is shown in Fig. 19.11(c). Figure 19.11(d) shows the calculated intensity distributions. From this figure, we can see the same symmetrical

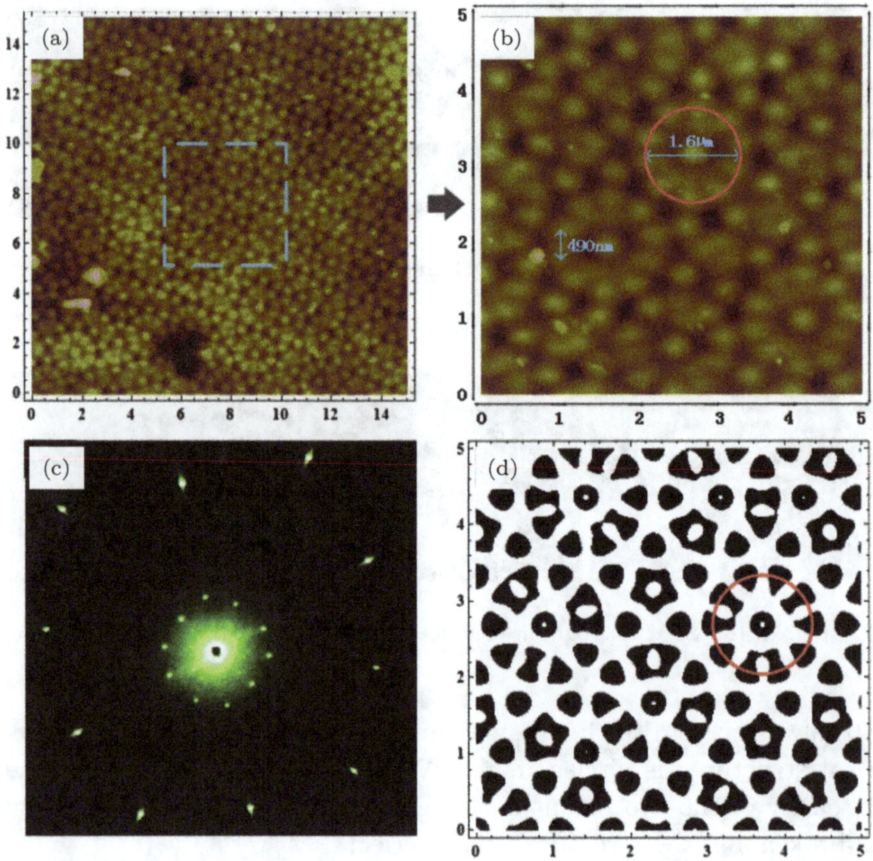

Fig. 19.11. SPM micrographs (a) $14\,\mu\text{m} \times 14\,\mu\text{m}$, (b) $5\,\mu\text{m} \times 5\,\mu\text{m}$, (c) diffraction patterns and (d) the calculated intensity distribution.

structure as the experimental result. The sizes of the unit cell and adjacent dots are $1.62\,\mu\text{m}$ and 500 nm, respectively. These two values are a little bigger than the experimental results. This is caused by the shrinkage of the photo-resist film in the developing and drying process.

19.6 PQC Sensor of Liquid Refractive Index

Ultra small optical sensors have considerable interest because their small volumes may produce abnormally high sensitivity and because of their suitability for integration on micro-fluid for "lab on a chip" applications [18]. PCs have good application prospects in the field of microsensors because of their good band gap and defect modes. However, in all PC sensors, the common characteristics are that we have to introduce the defects such as microcavities (point defects) or waveguide (line defects) structures or to destroy the symmetry to obtain sharp resonance peaks. Introduction of defects will largely increase the difficulty in preparing PC sensors. For PQC, no defects need to be introduced. It is more special than periodic PC in localized states and has more ideal photonic band-gaps and more abundant resonance peaks, so it can be used for designing various sensors [19].

19.6.1 *Design and optimization of the sensor*

The sensor is designed and shown in Fig. 19.12. Figure 19.12(a) shows the schematic configuration of the sensor. The sensing unit in Fig. 19.12(b) is based on the prepared PQC, comprising of silicon pillars arranged in an air background. In this model, the silicon rods are perpendicular to the XY plane, and their radius can be arbitrarily adjusted within the range of 0–150 nm. Figure 19.12(c) shows the three-dimensional view of the PQC. The structure expands in the X–Y direction and stretches in the Z direction. Figure 19.12(d) indicates the light propagating geometry and characteristic parameters of the structure. And the simulation was carried out by using FEM method.

In the calculation, the refractive index of silicon rods is set as the interpolation function of the wavelength (the data come from https://refractiveindex.info/). The minimum lattice constant is kept unchanged as $a = 600\,\text{nm}$, and the length of the PQC sensing unit is $l = 18\,\mu\text{m}$. By controlling the wavelength scanning and intensity detection, the transmission spectra can be obtained in the range of 1000–2000 nm. The resonance peak with a high Q-value and in a large band gap is selected as the sensing wavelength. The sensor is constituted by changing the refractive index of the background and measuring the location of the resonant wavelength. By altering the the radius r of Si pillars and the width d of the sensing unit, different types of sensors are designed and optimized.

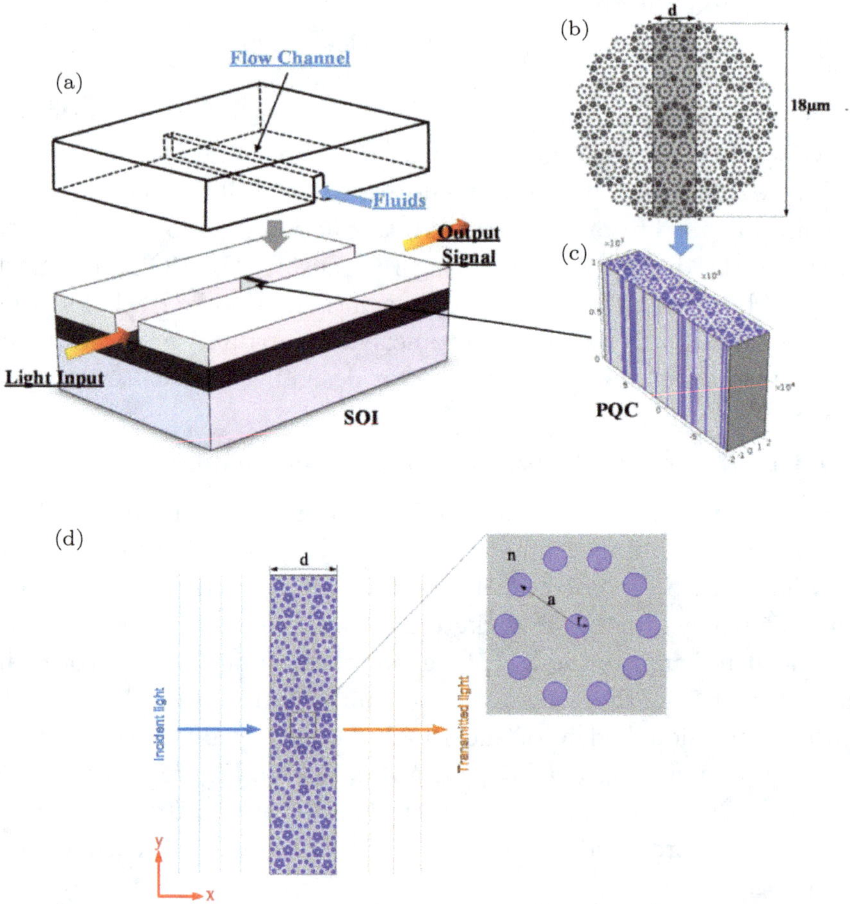

Fig. 19.12. Formation of PQC sensor (a) Schematic configuration of the device. (b) 10-fold PQC structure. (c) Three-dimensional view of the PQC. (d) Simulation geometry.

First, the transmission spectrum of the PQC was analysed and shown in Fig. 19.13 when the thickness of the PQC is set as $d = 4\,\mu\text{m}$, the radius of Si rods $r = 100\,\text{nm}$ and the refractive index $n = 1$ and $n = 1.333$ of air background and the filled water. Figure 19.13(a) is the transmission spectra with the incident light wavelength ranging from 1000 nm to 1800 nm by 0.1 nm step size. Analysis shows that in the range of 1500–1600 nm (shown in the red box), there is only one

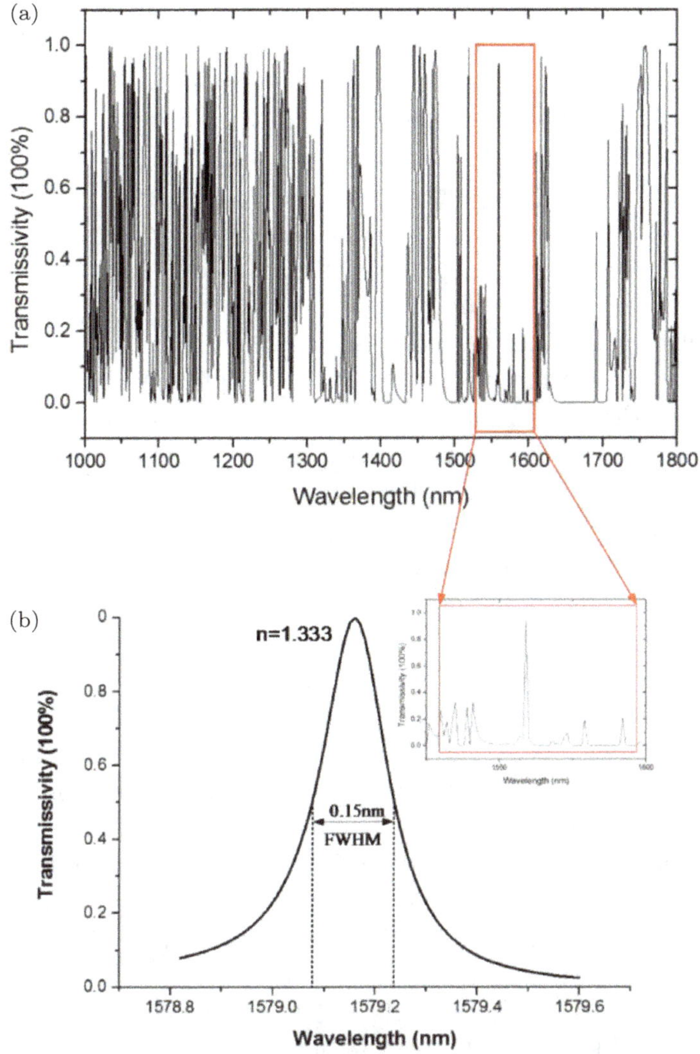

Fig. 19.13. (a) The transmission spectrum of the PQC with the liquid refractive index 1.333. (b) Magnification of the transmission spectrum.

resonance peak with a high Q factor at 1579.1 nm (Q = 10527.3, FWHM = 0.15 nm). The peak exists in a relatively wider band gap from 1570.6 nm to 1591.6 nm which has fewer interference peaks and is more suitable for sensing, as shown in Fig. 19.13(b). So, the detecting range of this sensing structure will be set from 1570.6 nm to 1591.6 nm.

In the same way, the resonant or sensing wavelength can be obtained as 1923.8 nm and 1170.1 nm for another two kinds of PQC sensing units with rod radii of $r = 50$ nm and $r = 150$ nm, respectively. While keeping r unchanged, when the width of the sensing unit is valued as $d = 8\,\mu$m and $d = 12\,\mu$m, the resonance wavelengths are 1628.1 nm and 1596 nm, respectively. These data will be referred in the following calculation.

Selecting the width of the sensing PQC unit as $d = 4\,\mu$m, three types of sensors were designed for three different PQC structures with the rod radius $r = 50$ nm, 100 nm and 150 nm, respectively. The relationship between the resonant peak location and refractive index of the background filling liquid is obtained and shown in Fig. 19.14. Figures 19.14(a)–19.14(c) are corresponding to three different structures. From the figure, a good linear relationship can be observed from the fitting line and the peak is shifted towards the long wavelength with the increase in the refractive index. For the three PQC sensors, the corresponding sensitivity is 992.73 nm/RIU (refractive index unit), 216.91 nm/RIU and 60 nm/RIU, respectively. The sensitivity is decreased with the increase in the rod radius. That is to say, the larger the quantity of the filling liquid, the more sensitive the sensor will be. This is reasonable. However, the fabrication of PQC structures with the rod radius 50 nm is more difficult than that of the other structures. After a trade-off between the sensitivity and fabrication difficulty, the PQC structure with the rod radius $r = 100$ nm is chosen as the sensing unit.

Keeping the radius of the rod $r = 100$ nm unchanged and changing the width of PQC as $d = 8\,\mu$m and $12\,\mu$m, another two kinds of sensing structures are designed and the sensing characteristics are shown in Fig. 19.15. The slope of the fitting line shows that the sensitivity of the sensor is 216.91 nm/RIU, 284.18 nm/RIU and 234.55 nm/RIU for the PQC sensing unit with the width of $d = 4\,\mu$m, $8\,\mu$m and $12\,\mu$m, respectively. The corresponding sensing unit and the transmission spectra are also indicated in each figure. Considering the

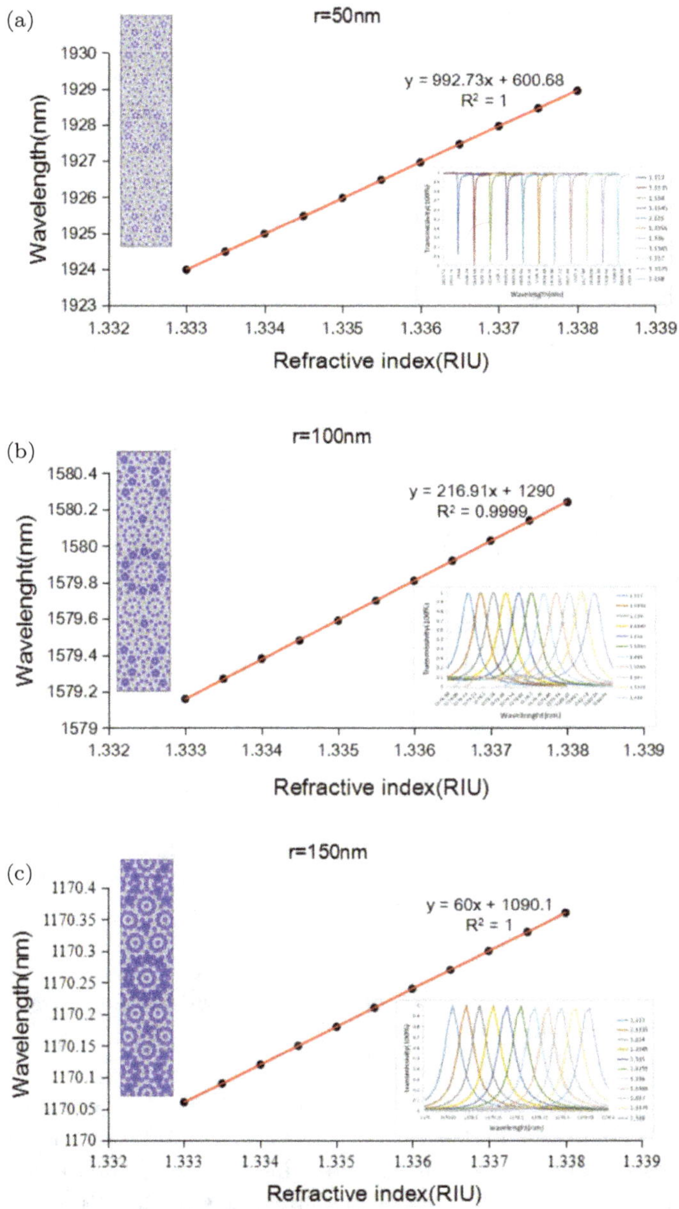

Fig. 19.14. Sensing characteristics of three different types of PQC sensors with the rod radius of (a) $r = 50\,\text{nm}$, (b) $r = 100\,\text{nm}$, (c) $r = 150\,\text{nm}$.

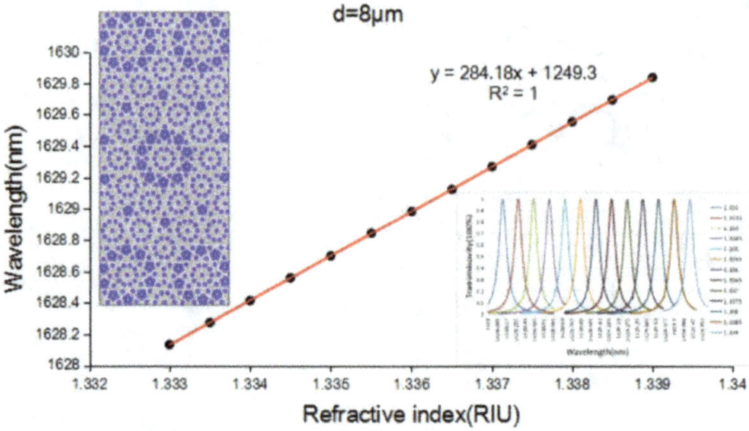

(a)

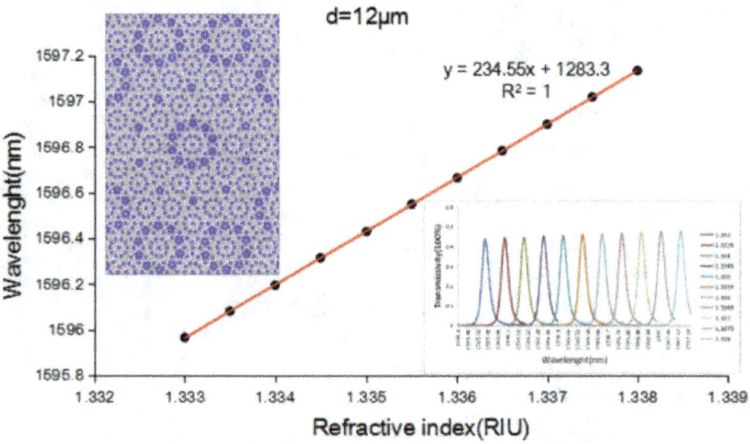

(b)

Fig. 19.15. Sensing characteristics of PQC units with different widths of (a) $d = 8\,\mu$m, (b) $d = 12\,\mu$m.

measurement error, there is little difference among the sensitivity. Therefore, it can be concluded that the width of the sensing PQC unit only affects the position and width of the PQC band gap but does not affect the sensor sensitivity.

19.6.2 *Performance analysis of the sensor*

The measurement range and accuracy of the sensor are also investigated. The selected sensor has parameters of $d = 4\,\mu$m and $r = 100\,$nm. Figure 19.16(a) shows the transmission spectrum of refractive index from 1.27 to 1.42 with 0.01 step size. The different line colour represents the spectra of PQC filled with different liquids. It can be found that the position of the resonance peak shifts equally to the long wavelength with the increase in the refractive index of the filling liquids. The corresponding sensing characteristics are shown in Fig. 19.16(b). The black dots represent the calculated results and the red line shows the fitting curve. From the figure, we can see that a monotonic linear relationship exists between the peak positions and refractive index. The sensitivity of the sensor can be obtained as 221 nm/RIU.

To further analyse the resolution of the sensor, we reduce the step size of the refractive index from 0.01 to 0.0001. Figure 19.17 was obtained in the range 1.333–1.3339 with 0.0001 increment of each step. Figure 19.17(a) shows the shift of the transmission spectra. Transmission peaks can still be distinguished. The corresponding sensing characteristics are shown in Fig. 19.17(b), which also show good linearity. We also find that with the 0.0001 increase in the

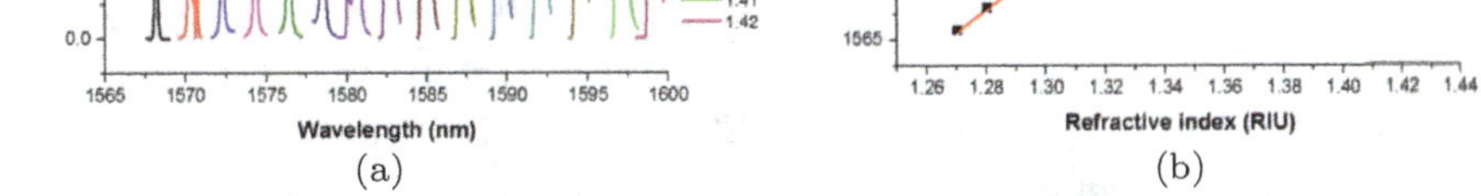

Fig. 19.16. (a) Transmission spectrum of PQC with different refractive index from 1.28 to 1.42 with 0.01 step size, (b) sensing characteristic curve.

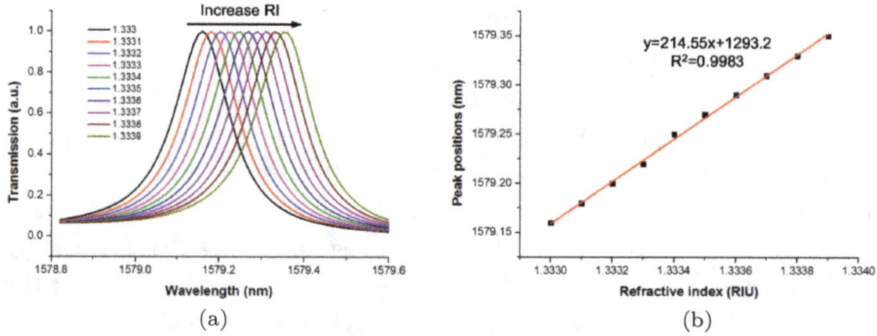

Fig. 19.17. The limiting resolution of the sensor: (a) the transmission spectra with the step size 0.0001 of refractive index, (b) sensing characteristic curve.

refractive index, the peak wavelength shifts about 0.02 nm, which is the detection limit of our spectrometer, so the sensing accuracy of this sensor is 10^{-4} RIU. At the same time, the measurement range is from 1.2731 to 1.4185 as observed from Fig. 19.16(b).

19.7 Conclusions

The multi-beam holographic interfering method has been used to form various PQC structures with different rotational symmetries. By using the FEM method to calculate the transmission and reflection characteristics, we compare and optimize their PBGs. On the other hand, 10-fold PQCs were prepared by using single-prism holographic interference lithography. This will be helpful for the design and fabrication of PQC-based photonic integrated circuits.

Furthermore, using the natural defect mode of the PQC, a microfluidic device has been designed. Under the condition of 0.02 nm wavelength accuracy, the resolution can reach up to 10^{-4} RIU while the measurement range of the refractive index is from 1.273 to 1.418. A refractive index sensitivity of 221.8 nm/RIU was also achieved, which would be very suitable for the detection of biomolecules.

At the same time, in order to help readers better understand the relationship between photonic quasicrystals and liquid crystals, the characteristics of cholesteric liquid crystals (CLCs) and the relevance between CLC and PC, refer to [20–24].

References

[1] Ma Z Q, 2019, *Group Theory for Physicists*, 2nd edition, World Scientific Press, Singapore.

[2] Rechtsman M C, Jeong H C, Chaikin P M, Torquato S and Steinhardt P J, 2008, Optimized structures for photonic quasicrystals, *Phys. Rev. Lett.*, **101**, 073902.

[3] Romero-Vivas J, Chigrin D N, Lavrinenko A V and Torres C S, 2005, Resonant add-drop filter based on a photonic quasicrystal, *Opt. Express*, **13**, 826–835.

[4] Hase M, Miyazaki H, Egashira M, Shinya N, Kojima K M and Uchida S I, 2002, Isotropic photonic band-gap and anisotropic structures in transmission spectra of two-dimensional five-fold and eight-fold symmetric quasi-periodic photonic crystals, *Phys. Rev. B*, **66**, 214205.

[5] Jin C, Cheng B, Man B, Li Z, Zhang Z, Ban S and Sun B, 1999, Band-gap and wave guiding effect in a quasiperiodic photonic crystal, *Appl. Phys. Lett.*, **75**, 1848–1850.

[6] Yin J, Huang X, Liu S and Hu S, 2007, Photonic band-gap properties of 8-fold symmetric photonic quasicrystals, *Opt. Commun.*, **269**, 385–388.

[7] Mnaymneh K and Gauthier R C, 2007, Mode localization and band-gap formation in defect-free photonic quasicrystals, *Opt. Express*, **14**, 5089–5099.

[8] Rose P, Zito G, Di Gennaro E, Abbate G and Andreone A, 2012, Control of the light transmission through a quasiperiodic waveguide, *Opt. Express*, **20**, 26056–26061.

[9] Wang S, Sun X, Wang C, Peng G, Qi Y and Wang X, 2017, Liquid refractive index sensor based on a 2D 10-fold photonic quasicrystal, *J. Phys. D Appl. Phys.*, **50**, 365102.

[10] Ren J, Sun X and Wang S, 2018, A low threshold nanocavity in a two-dimensional 12-fold photonic quasicrystal, *Opt. Laser Technol.* **101**, 42–48.

[11] Ren J, Sun X and Wang S, 2018, A narrow-band filter based on 2D 8-fold photonic quasicrystal, *Superlattice Microstruct.*, **116**, 221–226.

[12] Florescu M, Torquato S and Steinhardt P J, 2009, Complete band-gaps in two-dimensional photonic quasicrystals, *Phys. Rev. B*, **80**, 145112.

[13] Sun X, Wang S, Liu W and Jiang L, 2016, A simple configuration for fabrication of 2D and 3D photonic quasicrystals with complex structures, *Opt. Commun.*, **369**, 138–144.

[14] Sun X, Liu W, Wang G and Tao X, 2012, Optics design of a top-cut prism interferometer for holographic photonic quasicrystals, *Opt.Commun.*, **285**, 4593–4598.

[15] Xi X Y and Sun X H, 2019, Photonic band-gap properties of twodimensional photonic quasicrystals with multiple complex structures, *Superlattice Microst.*, **129**, 247–251.

[16] Sellers S, Man W, Sahba S and Florescu M, 2017, Local self-uniformity in photonic networks, *Nat. Commun.*, **8**, 14439.

[17] Wang S, Sun XH, Li WY, Liu W, Jiang L and Han J, 2016, Fabrication of photonic quasicrystalline structures in the sub-micrometer scale, *Superlattice Microst.*, **93**, 122–127.

[18] Pal S *et al.*, 2013, Selective virus detection in complex samplematrices with photonic crystal optical cavities, *Biosens. Bioelectron.* **44**, 229–234.

[19] Wang S, Sun X H, Wang C, Peng G D, Qi Y L and Wang X S, 2017, Liquid refractive index sensor based on a 2D 10-fold photonic quasicrystal, *J. Phys. D: Appl. Phys.*, **50**(1–6), 365102.

[20] Dunmur D. and Toriyama K., 1999, *Physical Properties of Liquid Crystals*, ed. by Demus D *et al.*, Wiley-VCH, Weinheim, pp. 124–128.

[21] Mitova M, Nouvet E and Dessaud N, 2004, Polymer-stabilized cholesteric liquid crystals as switchable photonic broad band-gaps, *Eur. Phys. J. E*, **14**, 413–419.

[22] Hrozhyk U A, Serak S V, Tabiryan N V, White T J and Bunning T J, 2011, Nonlinear optical properties of fast, photo-switchable cholesteric liquid crystal band-gaps, *Opt. Mater. Expr.*, **1**, 943–952.

[23] Hwang J, Ha N Y, Chang H J, Byoungchoo P and Wu J W, 2004, Enhanced optical nonlinearity near the photonic band-gap edges of a cholesteric liquid crystal, *Opt. Lett.*, **29**, 2644–2646.

[24] Costello M J, Meiboom S and Sammon M, 1984, Electron microscopy of a cholesteric liquid crystal and its blue phase, *Phys. Rev. A*, **29**, 2957–2959.

Chapter 20

Concluding Remarks and Perspectives

In this chapter, we would give a conclusion on the previous text.

Elasticity is one of the important branches of natural science since the time of Hooke, Bernoulli, Euler, Navier, Cauchy, Saint-Venant, among the others. It is emphasized by great theoretical physicists including Sommerfeld [1], Landau and Lifshitz [2] and presented in their famous theoretical physics courses. However, those they discussed belong to the classical elasticity or crystal elasticity in scope of continuum mechanics, which presents important meaning for engineering as well as science including modern physics. For example, one application of the theory combining Planck's quantization by Debye [3] and Born [4] to specific heat of crystals at low temperatures (refer to [1] for the detail) leads to phonon concept and the appearance of lattice dynamics, which is an important basis of modern solid state physics.

Elasticity of quasicrystals is a branch of condensed matter physics rather than classical mechanics. Landau and Anderson did great contributions to condensed matter physics, and the symmetry-breaking principle sets up the paradigm of the branch of physics. The basis of the elasticity is phonon, phason and phonon–phason coupling, which are originated from Landau elementary excitation principle. The concept is a quantum mechanics description of the collective excitation of massive atoms, indicating that the elasticity of quasicrystals is beyond classical mechanics and essentially connected with quantization.

Apart from quantization, the symmetry and phase are also the basis of the elasticity of quasicrystals. The quantization, symmetry and phase belong to the main themes of 20th-century physics. Anderson [5] revealed the nature of phonon from the phase of Landau's order parameter in three-dimensional reciprocal space. Based on Landau–Anderson's theory, the pioneering work of Bak [6] created the elasticity study of quasicrystals, in which the phason elementary excitation was introduced first by him for studying the atom local rearrangement in the quasicrystal lattice, and Lubensky *et al.* [7] found the hydrodynamics of quasicrystals and further explained the nature of phason. Rocal and Lorman [8] and Fan *et al.* [9] suggested phonon–phason dynamics of quasicrystals. Fan and his group also developed the defect dynamics of the matter; they gave the systematic analytic and numerical solutions. We call the above hydrodynamics, phonon–phason dynamics and defect dynamics as generalized dynamics of quasicrystals, which are as important as that of elasticity in the discussion of this book.

If limited only to the intuitive idea of classical mechanics, the phason cannot be understood, although the phason concept was drawn from incommensurate phase physics at the beginning. These discussions may help us to deepen our understanding of elasticity and generalized dynamics of quasicrystals. Of course, this inevitably increases the complexity and difficulty of the study.

Based on the above concepts and principles, the fundamentals of the theory of elasticity and generalized dynamics of solid quasicrystals have been developed, and quite fruitful achievements are obtained, which are partly introduced in the first 16 chapters of this book.

Although the quasicrystal elasticity and generalized dynamics are different from classical elasticity and elastodynamics in nature, there are some connections in methodology between these two disciplines. The development of the theory of quasicrystal elasticity was learned from the crystal elasticity [10–12], as well as from the crystal plasticity [13–16], crystal dislocation [17, 18] and classical crack theories [19, 20]. Besides, there are much inherent connections between quasicrystal elasticity and group theory [21–23] (in which Hu *et al.* [22] developed the group representation theory and determined all independent and non-zero elements of physical property tensors of quasicrystal systems discovered in that time and set up the basis of the constitutive law of the quasicrystal elasticity), partial differential equations [12, 24] (in which a quite lot of new types of the equations have been derived for the elasticity of quasicrystals), Fourier analysis [18, 25], complex analysis [27], integral equations [19,20,26] and computational mathematics [28,29]. Without these new partial differential equations and the help of these powerful mathematics tools, the development of the elasticity of quasicrystals is impossible. These problems are also introduced in the text of the first 16 chapters and Major Appendix, while some among them are quite detailed. The obtained exact analytic solutions are explored in the physics-mathematics compatibility and consistency and shown the power of analytic mathematics for the elasticity of solid quasicrystals. These results contributed to the physics as well as applied mathematics. However, the development level of the generalized dynamics of quasicrystals cannot reach the same level as just mentioned for the elasticity, or say there is a gap between them because the dynamic problems are more complicated, and corresponding mathematics has not been fully developed yet so far.

Considering some developments of quasicrystal study in recent years, we gave some new contents in the present edition of our book.

In Chapter 17, we supplemented a theory of elasticity of the second-kind two-dimensional solid quasicrystals of 7-, 9-, 14- and 18-fold symmetry, which were predicted by Hu *et al.* [30] based on group theory. Although the possible solid quasicrystals have not been observed so far yet, the 18-fold symmetry soft-matter quasicrystal was reported in 2011 by Fischer *et al.* [31] in colloids, after 17 years of the prediction of Hu *et al.* Tang and Fan [32] gave the point groups for this kind of quasicrystal. Ma [23] made a systematic and profound analysis of related group theory.

In this new edition, we paid effort to concern the newly developed branch. For instance, Chapter 18 was preliminary to introduce the soft-matter quasicrystal formation, metastability and phase transitions [33, 34]. However, the metastability is a complex thermodynamic as well as dynamic problem [35], in which the dynamic problem is named molecular dynamic mechanics and apparently belongs to one of the generalized dynamics. As previously mentioned, the development level of generalized dynamics in solid quasicrystals is lag behind that of elasticity, while the study of soft-matter quasicrystals is more difficult than that of solid quasicrystals. One of the reasons lies in the complexity in the thermodynamics of soft-matter quasicrystals, for example, the metastability of the soft matter. According to the analysis of some theories, the soft-matter quasicrystals are stable in thermodynamics. However, according to the experimental results, they are in metastable state. The analysis of polymer self-consistent field theory also supports the viewpoint of metastable states. This problem is beyond the scope of this book, and we would not give further discussion. Overall, the development of the theory of soft-matter quasicrystals needs the co-operation among researchers of chemistry, physics and engineering.

However, very recently, Zeng *et al.* [36] observed the columnar 12-fold soft-matter quasicrystals with thermodynamic stability, which rewrites the situation of soft-matter quasicrystals so far and promotes the development of the branch of quasicrystal study. Before this work of Zeng *et al.* [36], Fan *et al.* [37] studied the generalized dynamics of soft-matter quasicrystals with thermodynamic stability, in which the fluid phonon concept was introduced as well as an equation of state for soft-matter. They derived the equations of motion in soft matter quasicrystals, and the computed results were compared with some results of well-known fluid dynamics, the comparison shows the dynamics is significant. Due to the limitation of space, the detail about the dynamics could not be introduced yet.

In Chapters 19, an application of photon band-gap of photonic quasicrystals was discussed, which may be useful in the device physics [38].

Although in Chapter 18 we touched in a few of meso- and micro-scale problems in the discussion of soft-matter quasicrystals, there exists an evident defect from the whole scale of the book that the microscopic structures, especially the electronic structures of quasicrystals have not been touched. One of reasons is due to lack of the theory for quasicrystal lattice similar to Bloch theorem [39] in crystal lattice so far, and the quantum mechanics cannot be directly used in quasicrystals. People pay great efforts to study the electronic structures in quasicrystals through the periodic lattice approximations of quasicrystalline materials, e.g. [40], which are also called the approximation phases of quasicrystals. However, the so-called approximation phases close to quasicrystals are in fact crystals. Although the efforts have resulted in many useful results, the fundamental difficulty has not been overcome, i.e., the direct application of the quantum mechanics to quasicrystals is not successful. This leads to the loss in quite a lot of important fields of quasicrystal study, and prevents the further development of the science branch of quasicrystals. Fortunately, researchers are putting persistent efforts for the direct application of quantum mechanics to quasicrystals. We hope these efforts would gain successes in the future and promote giant development of quasicrystal study.

References

[1] Sommerfeld A, 1952, *Vorlesungen ueber theoretische Physik, Band II: Mechanik der deformirbaren Medien*, Diederich Verlag, Wiesbaden.

[2] Landau L D and Lifshitz E M, 1986, *Theoretical Physics, Vol. VII: Theory of Elasticity*, Pergamon Press, Oxford.

[3] Debye P, 1912. Eigentuemlichkeit der spezifischen Waermen bei tiefen Temperaturen, *Arch de Genéve*, **33**(7), 256–258.

[4] Born M and von Kármán Th, 1913, Zur Theorie der spezifischen Waermen, *Physikalische Zeitschrift*, **14**(1), 15–19; Born M and Huang K, 1954, *Dynamic Theory of Crystal Lattices*, Clarendon Press, Oxford.

[5] Anderson P W, 1984, *Basic Notation of Condensed Matter Physics*, Benjamin-Cummings, Menlo Park.

[6] Bak P, 1985, Phenomenological theory of icosahedral incommensurate ("quaisiperiodic") order in Mn-Al alloys, *Phys. Rev. Lett.*, **54**(8), 1517–1519; Bak P, 1985, Symmetry, stability and elastic properties of icosahedral incommensurate crystals, *Phys. Rev. B*, **32**(9), 5764–5772.

[7] Lubensky T C, Ramaswamy S and Toner J, 1985, Hydrodynamics of icosahedral quasicrystals, *Phys. Rev. B*, **32(11)**, 7444–7452.

[8] Rochal S B and Lorman V L, 2002, Minimal model of the phonon-phason dynamics on icosahedral quasicrystals and its application for the problem of internal friction in the i-AIPdMn alloys, *Phys. Rev. B*, **66**(14), 144204.

[9] Fan T Y, Wang X F, Li W and Zhu A Y, 2009, Elasto-hydrodynamics of quasicrystals, *Phil. Mag.*, **89**(6), 501–512.

[10] De P and Pelcovits R A, 1987, Linear elasticity theory of pentagonal quasicrystals, *Phys. Rev. B*, **35**(16), 8609–8620.

[11] Ding D H, Yang W G, Hu C Z and Wang R H, 1993, Generalized elasticity theory of quasicrystals, *Phys. Rev. B*, **48**(10), 7003–7010.

[12] Li X F and Fan T Y, 1998, New method for solving elasticity problems of some planar quasicrystals and solution, *Chin. Phys. Lett.*, **15**(4), 278–280.

[13] Feurbacher M and Urban K, 2003, Plastic behaviour of quasicrystalline materials, in *Quasicrystals*, ed. by Trebin H-R, Wiley Press, Berlin.

[14] Messerschmidt U, 2009, *Dislocation Dynamics during Plastic Deformation*, Heidelberg, Springer-Verlag.

[15] Fan T Y and Fan L, 2008, Plastic fracture of quasicrystals, *Phil. Mag.*, **88**(4), 323–335.

[16] Fan T Y and Fan L, 2011, Relation between generalized Eshelby integral and generalized BCS and generalized DB models for some one- and two-dimensional quasicrystals, *Chin. Phys. B*, **20**(4), 036102.

[17] Yang S H and Ding D H, 1998, *Crystal Dislocation Theory, Vol.2*, Science Press, Beijing (in Chinese).

[18] Li X F, Duan X Y, Fan T Y and Sun Y F, 1999, Elastic field for a straight dislocation in a decagonal quasicrystal, *J. Phys. Condens. Matter,* **11**(3), 703–711.

[19] Fan T Y, Trebin H-R, Messerschmidt U, and Mai Y W, 2004, Plastic flow coupled with a Griffith crack in some one- and two-dimensional quasicrystals, *J. Phys. Condens. Matter*, **16**(27), 5229–5240.

[20] Li X F, Fan T Y and Sun Y F, 1999, A decagonal quasicrystals with a Griffith crack, *Phil. Mag. A*, **79**(8), 1943–1952.

[21] Janssen T, 1992, The symmetry operations for n-dimensional periodic and quasi-periodic structures, *Zeitschrift fuer Kristallographie,* **198**(1–2), 17–32.

[22] Hu C Z, Wang R H and Ding D H, 2000, Symmetry groups, physical property tensors, elasticity and dislocations in quasicrystals, *Rep. Prog. Phys.,* **63**(1), 1–39.

[23] Ma Z Q, 2019, *Group Theory for Physicists*, 2nd edition, World Scientific, Singapore.

[24] Fan T Y and Guo L H, 2005, Final governing equation of plane elasticity of icosahedral quasicrystals, *Phys. Lett. A*, **341**(5), 235–239.

[25] Ding D H, Wang R H, Yang W G, and Hu C Z, 1995, Elasticity theory of straight dislocation in quasicrystals, *Phil. Mag. Lett.*, **72**(5), 353–359.

[26] Zhu A Y and Fan T Y, 2009, Elastic analysis of a Griffith crack in icosahedral Al-Pd-Mn quasicrystals, *Int. J. Mod. Phys. B*, **23**(10), 1–16.

[27] Fan T Y, Tang Z Y, Li L H and Li W, 2010, The strict theory of complex variable function method of sextuple harmonic equation and applications, *J. Math. Phys.*, 51(5), 053519.

[28] Zhu A Y and Fan T Y, 2008, Dynamic crack propagation in a decagonal Al-Ni-Co quasicrystal, *J. Phys. Condens. Matter*, **20**(29), 295217.

[29] Cheng H, Fan T Y, Sun J J and Wei H, 2016, Time-space solutions for hydrodynamics of pentagonal/decagonal quasicrystals, *Appl. Math. M*ech., **37**(10), 1393–1403.

[30] Hu C Z, Yang W G, Ding D H and Wang R H, 1994, Possible two-dimensional quasicrystals structure with six-dimensional embedding space, *Phys. Rev. B*, **49**, 9423–9427.

[31] Fischer S, Exner A, Zielske K, Perlich J, Deloudi S, Steurer W, Lindner P, Foerster S and Eaborn C, 2011, Colloidal quasicrystals with 12-fold and 18-fold diffraction symmetry. *Proc. Natl. Acad. Sci.*, **108**, 1810–1814.

[32] Tang Z Y and Fan T Y, 2017, Point groups and group representation theory of the second kind of two-dimensional quasicrystals, unpublished work.

[33] Yue K, Huang M J, R Marson, He J L, Huang J H, Zhou Z, Liu C, Yan X S, Wu K, Wang J, Guo Z H, Liu H, Zhang W, Ni P H, Wesdemiotis C, Wen-Bin Zhang W B, Sharon, Glotzer S C and Cheng S Z D, 2016, Geometry induced sequence of nanoscale Frank-Kasper and quasicrystal mesophases in giant surfactants. *Proc. Nat. Ac. Sci.*, **113**, 1392–1400.

[34] Liu Y, Liu T, Yan X Y, Guo Q Y, Lei H, Huang Z, Zhang R, Wang Y, Wang J, Liu F, Bian F, Meijer E W, Aida T, Huang M and Cheng S Z D, 2022, Expanding quasiperiodicity in soft matter: Supramolecular decagonal quasicrystals by binary giant molecule blends. *Proc. Nat. Acad. Sci.*, **119**, e2115304119.

[35] Cheng S Z D, 2008, *Phase Transitions of Polymers: The Role of Metastability State*, Elsevier Science Publisher, B.V. North Holland.

[36] Zeng X B *et al.*, 2023, A columnar liquid quasicrystal with a honeycomb structure that consists of triangular, square and trapezoidal cells, *Nat. Chem.*, **15**, 625–632.

[37] Fan T Y, Yang W G, Cheng H and Sun X H, 2022, *Generalized Dynamics of Soft-Matter Quasicrystals and Its Applications*, Springer, Singapore.

[38] Wang S, Sun X H, Wang C, Peng G D, Qi Y L and Wang X S, 2017, Liquid refractive index sensor based on a 2D 10-fold photonic quasicrystal, *J. Phys. D: Appl. Phys.*, **50**(1–6), 365102.

[39] Bloch F, 1928, Ueber die Quantenmechanik der Elektronik in Kristallgittern, *Zeitschrift fuer Physik*, **52**, 555–600.

[40] Quilichini M and Janssen T, 1997, Phonon excitations in quasicrystals, *Rev. Modern Phys.*, **69**(1), 277–314.

Major Appendix: On Some Mathematical Additional Materials

Mathematical solutions for individual boundary value problems are given in the previous chapters of the text, in which crack problems are more difficult than dislocation problems. Due to the complexity of boundary conditions for the formers, we must use some special techniques, in which the complex function method collaborating conformal mapping and Fourier analysis collaborating dual integral equations are the best effective procedures. The use of the procedures has been exhibited in the relevant chapters. Though some appendixes in relevant chapters were given in the text for discussing some special problems, we list some necessary additional materials on complex variable functions and dual integral equations in two separate parts in this appendix. It is interesting, in particular, between these two parts there are some close inherent connections. Probably, these materials are not necessary for readers majoring applied mathematics but may be helpful for young scholars and post-graduate students who are not majoring applied mathematics.

Appendix I

Additional Calculations Related to Complex Analysis

AI.1 *Additional derivation of solution (8.2-19)*

In Chap. 8, we emphasized importance of complex analysis (see [1–3]) and pointed out that the Muskhelishvili's [1] method limited the comformal mapping to the rational functions. However our practice breaks the limitation, e.g. [4, 5], the calculation is handled as below.

Formula (8.2-19) is obtained from the integral of (8.2-6) in which the conformal mapping $\omega(\zeta)$ is given by (8.2-17). The difficulty of the calculation lies in the integral path which is a part of the circle rather than the whole circle. Substitution of (8.2-17) into the right-hand side of (8.2-6) yields

$$F(\zeta) = -\frac{p}{2\pi i}\int_{-1}^{1}\frac{\omega(\sigma)}{(\sigma-\zeta)^2}d\sigma = -\frac{p}{2\pi i}\int_{-1}^{1}\frac{\omega'(\sigma)}{\sigma-\zeta}d\sigma$$

where

$$\omega'(\sigma) = -\frac{4H\alpha(1-\beta)}{\pi}\frac{1-\sigma}{1+\sigma} \times \frac{(1+\sigma)^2}{[(1+\sigma)^2+\alpha(1-\sigma^2)^2][(1+\sigma)^2+\beta\alpha(1-\sigma)^2]}$$

Substituting $e^{i\varphi}$ by $(1+ix)/(1-ix)$ yields that

$$\frac{1-\sigma}{1+\sigma} = -2x, d\sigma = \frac{2i}{(1-ix)^2}dx, (1+\sigma)^2 = \frac{4}{(1-ix)^2},$$

$$\frac{1}{\sigma-\zeta} = \frac{1-ix}{1-\zeta+ix(1+\zeta)} = \frac{(1-ix)[1-\zeta-ix(1+\zeta)]}{(1-\zeta)^2+x^2(1+\zeta)^2} = \frac{1-\zeta-x^2(1+\zeta)-2ix}{(1-\zeta)^2+x^2(1+\zeta)^2}$$

So,

$$F(\zeta) = -\frac{pH\alpha(1-\beta)}{\pi^2} \times \int_{-1}^{1}\frac{ix[1-\zeta-x^2(1+\zeta)-2ix]}{(1-\alpha x^2)(1-\beta\alpha x^2)[(1-\zeta)^2+x^2(1+\zeta)^2\}}dx$$

$$= -\frac{4pH\alpha(1-\beta)}{\pi^2} \times \int_{0}^{1}\frac{x^2}{(1-\alpha x^2)(1-\beta\alpha x^2)[(1-\zeta)^2+x^2(1+\zeta)^2\}}dx$$

$$
\begin{aligned}
&= \frac{2pH}{\pi^2}\frac{\alpha(1-\beta)(1-\zeta^2)}{[\alpha(1-\zeta)^2+(1+\zeta)^2][\beta\alpha(1-\zeta)^2+(1+\zeta)^2]} \\
&\quad \times \left[\arctan\left(\frac{1+\zeta}{1-\zeta}\right) - \arctan\left(\frac{1+\zeta}{-1-\zeta}\right)\right] \\
&\quad - \frac{4pH}{\pi^2}\frac{\sqrt{\alpha}\arctan h\sqrt{\alpha}}{\alpha(1-\zeta)^2+(1+\zeta)^2} - \frac{\sqrt{\beta\alpha}\arctan h\sqrt{\alpha}}{\beta\alpha(1-\zeta)^2+(1+\zeta)^2}\Bigg]
\end{aligned} \tag{AI-1}
$$

In the last step, the evaluation is done in ***Mathematica3.0*** [6].

By considering

$$A = \ln\frac{1+\sqrt{\alpha}}{1-\sqrt{\alpha}} = 2\arctan h\sqrt{\alpha}$$

$$M = \ln\frac{1+\sqrt{\gamma\alpha}}{1-\sqrt{\gamma\alpha}} = 2\arctan h\sqrt{\gamma\alpha}$$

$$\arctan\left(\frac{1+\zeta}{-1+\zeta}\right) = -2\arctan\left(\frac{1+\zeta}{1-\zeta}\right) = \frac{i}{2}\ln\left(\frac{i-\zeta}{1-i\zeta}\right)$$

(AI.1) is just formula (8.2-19).

In the calculation, if let $L \to 0$, then the integral (8.2-7) can be obtained.

AI.2 *Additional derivation of solution (11.3-53)*

In example 3 of Section 11.3, the calculation is quite lengthy, and here, we provide some details on the evaluation. As an example, we can show the derivation on function $\Phi_4(\zeta)$, which is

$$\Phi_4(\zeta) = d_1(X+iY)\ln\zeta + B\omega(\zeta) + \Phi_4^*(\zeta)$$

in which the first two terms are known, see the text, and the single-valued analytic function $\Phi_4^*(\zeta)$ satisfies the following boundary condition:

$$\Phi_4^*(\sigma) + \overline{\Phi_3^*(\sigma)} + \frac{\omega(\sigma)}{\omega'(\sigma)}\cdot\overline{\Phi_4^{*\prime}(\sigma)} = f_0$$

where

$$f_0 = \frac{i}{32c_1}\int (T_x + iT_y)ds - (d_1 - d_2)(X + iY)\ln\sigma \\ - \frac{\omega(\sigma)}{\overline{\omega'(\sigma)}}\cdot d_1(X - iY)\cdot\sigma - 2B\omega(\sigma) - (B' - iC')\overline{\omega(\sigma)} \quad \text{(AI-2)}$$

and (referring to Fig. 11.4)

$$T_x = -pcon(n,x), T_y = -p\cos(n,y) \text{ at } \widehat{z_1Mz_2}$$

$$(T_x + iT_y)ds = \begin{cases} ipdz & \widehat{z_1Mz_2} \\ 0 & \widehat{z_2Nz_1} \end{cases} \quad \text{(AI-3)}$$

$$X + iY = \int (T_x + iT_y)ds = ip(z_1 - z_2)$$

Multiplying both sides of the above boundary equation by $\frac{1}{2\pi i}\frac{d\sigma}{\sigma-\zeta}$ and integrating around the unit circle, we have

$$\frac{1}{2\pi i}\int_\gamma \frac{\Phi_4^*(\sigma)}{\sigma-\zeta}d\sigma + \frac{1}{2\pi i}\int_\gamma \frac{\omega(\sigma)}{\overline{\omega'(\sigma)}}\frac{\overline{\Phi_4^{*\prime}(\sigma)}}{\sigma-\zeta}d\sigma + \frac{1}{2\pi i}\int_\gamma \frac{\overline{\Phi_3^*(\sigma)}}{\sigma-\zeta}d\sigma \\ = \frac{1}{2\pi i}\int_\gamma \frac{f_0}{\sigma-\zeta}d\sigma \quad \text{(a)}$$

in which according to Cauchy's integral formula (referring to formula (11.7-5)), there is

$$\frac{1}{2\pi i}\int_\gamma \frac{\Phi_4^*(\sigma)}{\sigma-\zeta}d\sigma = \Phi_4^*(\zeta)$$

and in terms of analytic extension principle and the Cauchy theorem (referring to Appendix of Chap. 11)

$$\frac{1}{2\pi i}\int_\gamma \frac{\omega(\sigma)}{\overline{\omega'(\sigma)}}\frac{\overline{\Phi_4^{*\prime}(\sigma)}}{\sigma-\zeta} = 0$$

and according to formula (11.7-9)

$$\frac{1}{2\pi i}\int_\gamma \frac{\overline{\Phi_3^*(\sigma)}}{\sigma-\zeta}d\sigma = const$$

So formula (a) reduces to

$$\Phi_4^*(\zeta) = \frac{1}{2\pi i}\int_\gamma \frac{f_0}{\sigma-\zeta}d\sigma + const$$

And substituting (A.I-2) and (A.I-3) into the right-hand side yields

$$\begin{aligned}\frac{1}{2\pi i}\int_\gamma \frac{f_0}{\sigma-\zeta}d\sigma = &\frac{pR_0}{2\pi i c_1}\int_{\sigma_1}^{\sigma_2}\left(\sigma+\frac{m}{\sigma}\right)\frac{d\sigma}{\sigma-\zeta}\\ &+\frac{pz_2}{2\pi i}\int_{\sigma_1}^{\sigma_2}\frac{d\sigma}{\sigma-\zeta}+\frac{p(z_1-z_2)}{2\pi i}\frac{1}{2\pi i}\int_\gamma \frac{\ln\sigma}{\sigma-\zeta}d\sigma\\ &-\frac{p(\overline{z_1}-\overline{z_2})}{2\pi i}\frac{1}{2\pi i}\int_\gamma \frac{\sigma^2+m}{1-m\sigma^2}\frac{d\sigma}{\sigma-\zeta}\end{aligned}$$

in which

$$\int_{\sigma_1}^{\sigma_2}\left(\sigma+\frac{m}{\sigma}\right)\frac{d\sigma}{\sigma-\zeta} = \sigma_2-\sigma_1-\frac{m}{\zeta}\ln\frac{\sigma_2}{\sigma_1}+\left(\zeta+\frac{m}{\zeta}\right)\ln\frac{\sigma_2-\zeta}{\sigma_1-\zeta}$$

$$\int_{\sigma_1}^{\sigma_2}\frac{d\sigma}{\sigma-\zeta} = \ln\frac{\sigma_1-\zeta}{\sigma_2-\zeta}$$

In accordance with Cauchy's theorem (referring to formula (AI-4)),

$$\frac{1}{2\pi i}\int_\gamma \frac{\sigma^2+m}{1-m\sigma^2}\frac{d\sigma}{\sigma-\zeta} = 0$$

because the integrand is a single-valued analytic function in the region outside the unit circle γ.

The remaining term is

$$I(\zeta) = \frac{1}{2\pi i}\int_\gamma \frac{\ln\sigma}{\sigma-\zeta}d\sigma$$

For calculating it, we consider

$$\frac{dI}{d\zeta} = \frac{1}{2\pi i}\int_\gamma \frac{\ln\sigma}{(\sigma-\zeta)^2}d\sigma = -\frac{1}{2\pi i}\int_\gamma \ln\sigma d\frac{1}{\sigma-\zeta}$$

$$= -\frac{1}{2\pi i}\left[\frac{\ln\sigma}{\sigma-\zeta}\right]_{\sigma=\exp(i\varphi_1)}^{\sigma=\exp i(\varphi_1+2\pi)} + \frac{1}{2\pi i}\int_\gamma \frac{d\sigma}{\sigma(\sigma-\zeta)}$$

$$= -\frac{1}{2\pi i}\frac{1}{\sigma_1-\zeta}\ln\frac{\exp i(\varphi_1+2\pi)}{\exp(i\varphi_1)} - \frac{1}{\zeta} = -\frac{1}{\sigma_1-\zeta} - \frac{1}{\zeta}$$

So

$$I(\zeta) = \ln(\sigma_1-\zeta) - \ln\zeta + const$$

Hence, function $\Phi_4^*(\zeta)$ is determined so the function $\Phi_4(\zeta)$ in which the constant term is omitted is

$$\Phi_4(\zeta) = \frac{1}{32c_1}\cdot\frac{p}{2\pi i}\cdot\left[-\frac{mR_0}{\zeta}\ln\frac{\sigma_2}{\sigma_1} + z\ln\frac{\sigma_2-\zeta}{\sigma_1-\zeta}\right.$$
$$\left. + z_1\ln(\sigma_1-\zeta) - z_2\ln(\sigma_2-\zeta)\right] + ip(d_1-d_2)(z_1-z_2)\ln\zeta$$

which is just the first formula of (11.3-53), where d_1 and d_2 were given by (11.3-34) in the text. The others can be similarly derived. In the derivation, the classical work of Muskhelishvili [1] is referred.

AI.3 *Details of complex analysis of solution (14.4-7) of generalized cohesive force model for plane plasticity of two-dimensional point groups 5m, 10mm and 10, $\overline{10}$ quasicrystals*

The elasticity solution (11.3-53) based on complex analysis can be used to solve the present problem.

The generalized Dugdale–Barenblatt model or generalized cohesive model for decagonal quasicrystals makes the plastic problem to be linearized, so the final governing equation is reduced to solve the equation

$$\nabla^2\nabla^2\nabla^2\nabla^2 G = 0 \qquad \text{(AI-4)}$$

under boundary conditions

$$\begin{cases} \sigma_{yy} = p, \sigma_{xx} = \sigma_{xy} = 0, & \sqrt{x^2+y^2} \to \infty \\ \quad H_{xx} = H_{yy} = H_{xy} = H_{yx} = 0 & \\ \sigma_{yy} = \sigma_{xy} = 0, H_{yy} = H_{yx} = 0 & y = 0, |x| < a \\ \sigma_{yy} = \sigma_c, \ \sigma_{xy} = 0, H_{yy} = H_{yx} = 0 & y = 0, a < |x| < a + d \end{cases} \tag{AI-5}$$

which can be decomposed into two cases, among them one is

$$\begin{cases} \sigma_{xx} = \sigma_{xy} = \sigma_{yy} = 0, & \sqrt{x^2+y^2} \to \infty \\ \quad H_{xx} = H_{yy} = H_{xy} = H_{yx} = 0 & \\ \sigma_{yy} = \sigma_{xy} = 0, H_{yy} = H_{yx} = 0 & y = 0, |x| < a \\ \sigma_{yy} = \sigma_c, \sigma_{xy} = 0, H_{yy} = H_{yx} = 0 & a < |x| < a + d \end{cases} \tag{AI-6}$$

and the other is

$$\begin{cases} \sigma_{yy} = p, \sigma_{xx} = \sigma_{xy} = 0, & \sqrt{x^2+y^2} \to \infty \\ \quad H_{xx} = H_{yy} = H_{xy} = H_{yx} = 0 & \\ \sigma_{yy} = \sigma_{xy} = 0, H_{yy} = H_{yx} = 0 & y = 0, |x| < a + d \end{cases} \tag{AI-7}$$

The solution of problems (AI-31) and (AI-33) can be obtained from (11.3-53), i.e., if put $m = 1, R_0 = (a+d)/2$, then the elliptic hole reduced to a Griffith crack with half-length $(a+d)$. In Fig. 11.3, let $z_1 = (a+d, +0), z_2 = (a, +0)$, and from (11.3-53), we can obtain a solution, similarly put $z_1 = (a+d, -0), z_2 = (a, -0)$, $z_1 = (-a-d, +0), z_2 = (-a, +0)$, $z_1 = (-a-d, -0), z_2 = (-a, -0)$, respectively, and from (11.3-53), one can find the other corresponding three solutions, by superposing which one can obtain solution.

$$\begin{cases} \Phi_4^{(1)}(\zeta) = \dfrac{1}{32c_1} \cdot \dfrac{\sigma_c (a+d)\varphi_2}{\pi} \cdot \dfrac{1}{\zeta} - \dfrac{1}{32c_1} \cdot \dfrac{\sigma_c}{2\pi i} \\ \qquad \times \left[z \left(\ln \dfrac{\sigma_2 - \zeta}{\overline{\sigma_2} - \zeta} + \ln \dfrac{\sigma_2 + \zeta}{\overline{\sigma_2} + \zeta} \right) - l \ln \dfrac{(\zeta - \sigma_2)(\zeta + \overline{\sigma_2})}{(\zeta + \sigma_2)(\zeta - \overline{\sigma_2})} \right] \\ \Phi_3^{(1)}(\zeta) = \dfrac{1}{32c_1} \cdot \dfrac{\sigma_c (a+d)\varphi_2}{\pi} \cdot \dfrac{2\zeta}{\zeta^2 - 1} \\ \qquad - \dfrac{1}{32c_1} \cdot \dfrac{\sigma_c a}{2\pi i} \ln \dfrac{(\zeta - \sigma_2)(\zeta + \overline{\sigma_2})}{(\zeta + \sigma_2)(\zeta - \overline{\sigma_2})} \end{cases}$$

(AI-8)

where $\sigma = e^{i\varphi}$ represents the value of ζ at the unit circle in the mapping plane, and $\sigma_2 = e^{i\varphi_2}$, $a = (a+d)\cos\varphi_2$.

And the solution of problems (AI-31) and (AI-34), as the solution of Griffith crack problem, is known, i.e.,

$$\begin{cases} \Phi_4^{(2)}(\zeta) = -\dfrac{1}{32c_1}\dfrac{p}{2}(a+d)\dfrac{1}{\zeta} \\ \Phi_3^{(2)}(\zeta) = -\dfrac{p}{32c_1}(a+d)\left[\dfrac{\zeta}{(\zeta^2-1)}\right] \end{cases} \tag{AI-9}$$

The superposition of (A.I-35) and (A.I-36) gives the total solution for $\Phi_4(\zeta) = \Phi_4^{(1)}(\zeta) + \Phi_4^{(2)}(\zeta)$, $\Phi_3(\zeta) = \Phi_3^{(1)}(\zeta) + \Phi_3^{(2)}(\zeta)$, for example, the first term of $\Phi_4(\zeta)$ is

$$-\frac{1}{32c_1}\frac{p}{2}(a+d)\frac{1}{\zeta} + \frac{1}{32c_1}\cdot\frac{\sigma_c(a+d)\varphi_2}{\pi}\cdot\frac{1}{\zeta} \tag{AI-10}$$

and $\Phi_2(\zeta)$ has not been listed here because it is too lengthy. So, the stresses and displacements are determined already. In addition, we know that

$$\sigma_{ij}, H_{ij} \sim \Phi'(\zeta)/\omega'(\zeta) \tag{AI-11}$$

where $\Phi(\zeta)$ means $\Phi_4(\zeta)$ or $\Phi_3(\zeta)$, and $\omega'(\zeta) \sim 1/(1-\zeta^2)$.

From Section 14.4 in the text, we know that there is no stress singularity at the generalized Dugdale–Barrenblatt crack tip; this fact and conjunct with (AI-37) and (AI-38) require that the value of formula (AI-37) must be zero, which leads to (14.4-6) in the text. Considering this, the final version of $\Phi_4(\zeta)$ is

$$\begin{aligned} \Phi_4(\zeta) = -\frac{1}{32c_1}\cdot\frac{\sigma_c}{2\pi i}\Bigg[z\left(\ln\frac{\sigma_2-\zeta}{\overline{\sigma_2}-\zeta} + \ln\frac{\sigma_2+\zeta}{\overline{\sigma_2}+\zeta}\right) \\ - a\ln\frac{(\zeta-\sigma_2)(\zeta+\overline{\sigma_2})}{(\zeta+\sigma_2)(\zeta-\overline{\sigma_2})}\Bigg] \end{aligned} \tag{AI-12}$$

And the displacement at the crack surface presents the form

$$u_y(x,0) = (128c_1c_2 - 64c_3)\mathrm{Im}(\Phi_4(\zeta))_{\zeta=\sigma} \tag{AI-13}$$

After some calculation we find that

$$u_y(x,0) = \frac{(4c_1c_2 - 2c_3)}{c_1} \cdot \frac{\sigma_c(a+d)}{2\pi} \cdot \left[\cos\varphi \ln \frac{\sin(\varphi_2 - \varphi)}{\sin(\varphi_2 + \varphi)} - \cos\varphi_2 \ln \frac{(\sin\varphi_2 - \sin\varphi)}{(\sin\varphi_2 + \sin\varphi)}\right] \quad \text{(AI-14)}$$

So, the crack tip opening displacement

$$\begin{aligned}\delta_t = CTOD &= \lim_{x\to l} 2u_y(x,0) = \lim_{\varphi\to\varphi_2} 2u_y(x,0)\\ &= \frac{(8c_1c_2 - 4c_3)\sigma_s a}{c_1\pi} \ln\sec\left(\frac{\pi}{2}\frac{\sigma^{(\infty)}}{\sigma_s}\right) \quad \text{(AI-15)}\end{aligned}$$

in which the constants c_1, c_2, c_3 are defined in Section 11.3, so the solution holds for point groups $5m, 10mm$ as well as point groups $5, \bar{5},\ 10, \overline{10}$ quasicrystals. When we assume $R_1 = R,\ R_2 = 0$ in the equation (A.I-15), δ_t will be the corresponding solution of point groups $5m$ and $10mm$ quasicrystals, i.e.,

$$\delta_t = CTOD = \frac{2\sigma_s a}{\pi}\left[\frac{1}{L+M} + \frac{K_1}{MK_1 - R^2}\right] \ln\sec\left(\frac{\pi}{2}\frac{\sigma^{(\infty)}}{\sigma_s}\right)$$

which is just (14.4-7). If we let $K_1 = R = 0,\ L = \lambda,\ M = \mu$ in above formula, then it exactly reduces to the classical Dugdale–Barenblatt solution holding for engineering material (or structural material) including crystalline material (referring to Section 14.4).

More details can be found in the article given by [7].

AI.4 *On the calculation of integral (9.2-14)*

In formulas (9.2-14), $y > 0$ results in the integrals being convergent. We let ξ to extend to complex number $\xi_1 + i\xi_2$ and take integration path at complex ξ-plane similar to Fig. 11.7, and by physical consideration $k(K_1 - K_2) > 0$, $\mu(K_1 - K_2) - R^2 > 0$, and $k = \mu^{(c)}/h$, then we can find the integrand of (9.2-14) in the interior of the region enclosing by the integration path which is analytic except for the poles

$$\xi_1^{(1)} = \frac{k(K_1 - K_2)}{R^2 - \mu(K_1 - K_2)} < 0, \quad \xi_1^{(2)} = -\frac{k(K_1 - K_2)}{R^2 - \mu(K_1 - K_2)} > 0 \tag{AI-16}$$

at real axis ξ_1 and by a generalized Jordan lemma the integral along the big half-circle is zero.

If in Fig. AI.2 put $\omega = \xi$, $\omega_1 = \xi_1$, $\omega_2 = \xi_2$, $\sqrt{k/m} = \xi_1^{(1)}$, $-\sqrt{k/m} = \xi_1^{(2)}$, according to the additional integration path at complex ξ-plane, through a similar manner for evaluating integral (AI-18), then we can obtain results (9.2-15) and (9.2-16) respectively where $\xi_1^{(1)}, \xi_1^{(2)}$ are defined by (AI-16).

AI.5 *On the calculation of integral (8.8-9)*

$$\begin{aligned}\varphi'(\zeta) &= \frac{p}{2\pi i}\int_{-1}^{1}\frac{2w}{\pi}\cdot\left(\frac{-\sigma\tan\left(\frac{\pi a}{2w}\right)}{\left[1+(1-\sigma^2)\tan^2\left(\frac{\pi a}{2w}\right)\right]\sqrt{1-\sigma^2}}\right)\cdot\frac{1}{\sigma-\zeta}d\sigma \\ &= -\frac{pw}{\pi^2 i}\tan\left(\frac{\pi a}{2w}\right)\int_{-1}^{1}\left(\frac{\sigma}{\left[1+(1-\sigma^2)\tan^2\left(\frac{\pi a}{2w}\right)\right]\sqrt{1-\sigma^2}}\right) \\ &\quad \cdot\frac{1}{\sigma-\zeta}d\sigma\end{aligned} \tag{AI-17}$$

Put $m = \tan\left(\frac{\pi a}{2w}\right)$, so $m^2 = \tan^2\left(\frac{\pi a}{2w}\right)$

We calculate

$$\begin{aligned}I(\zeta) &= \int_{-1}^{1}\left(\frac{\sigma}{\left[1+(1-\sigma^2)\tan^2\left(\frac{\pi a}{2w}\right)\right]\sqrt{1-\sigma^2}}\right)\cdot\frac{1}{\sigma-\zeta}d\sigma \\ &= \int_{-1}^{1}\left(\frac{1}{[1+(1-\sigma^2)m^2]\sqrt{1-\sigma^2}}\right)d\sigma \\ &\quad + \int_{-1}^{1}\left(\frac{\zeta}{[1+(1-\sigma^2)m^2]\sqrt{1-\sigma^2}(\sigma-\zeta)}\right)d\sigma \\ &= I_1 + I_2\end{aligned} \tag{AI-18}$$

The calculation of the first integral is easy, and we now calculate the second one. Put $\zeta = b$, $\sigma = x$ such that

$$I_2 = \int_{-1}^{1} \left(\frac{b}{[1 + (1 - x^2)m^2]\sqrt{1 - x^2}(x - b)} \right) d\sigma$$

Denote $x = \sin t$, $t \in \left[-\frac{\pi}{2}, \frac{\pi}{2}\right]$, $dx = \cos t dt$, where

$$\begin{cases} t = -\dfrac{\pi}{2}, & \text{as } x = -1 \\ t \neq -\dfrac{\pi}{2}, & \text{as } x \neq -1 \end{cases}$$

$$I_2 = \int_{-\pi/2}^{\pi/2} \frac{b}{[1 + m^2 \cos t] \cdot (\sin t - b) \cdot \cos t} \cdot \cos t d\sigma$$

$$= \int_{-\pi/2}^{\pi/2} \frac{b}{[1 + m^2 \cos^2 t](\sin t - b)} dt$$

$$= \int_{0}^{\pi/2} \frac{b}{[1 + m^2 \cos^2 t](\sin t - b)} dt$$

$$+ \int_{-\pi/2}^{0} \frac{b}{[1 + m^2 \cos^2 t](\sin t - b)} dt \quad \text{(AI-19)}$$

For the second integral of (AI-19), put

$$t = -x, \quad dt = -dx, \quad t = -\frac{\pi}{2}, \quad \text{as } x = \frac{\pi}{2}; \quad t = 0, \quad \text{as } x = 0$$

$$\int_{-\pi/2}^{0} \frac{b}{[1 + m^2 \cos^2 t](\sin t - b)} dt$$

$$= -\int_{0}^{\pi/2} \frac{b}{[1 + m^2 \cos^2 x](\sin x + b)} dx$$

So,

$$\int_{-\pi/2}^{\pi/2} \frac{b}{[1 + m^2 \cos^2 t](\sin t - b)} dt$$

$$= \int_{0}^{\pi/2} \frac{b}{[1 + m^2 \cos^2 t](\sin t - b)} dt$$

$$- \int_{0}^{\pi/2} \frac{b}{[1 + m^2 \cos^2 t](\sin t + b)} dt$$

$$
\begin{aligned}
&= \int_0^{\pi/2} \frac{2b^2}{[1+m^2\cos^2 t](\sin^2 t - b^2)} dt \\
&= A \cdot \int_0^{\pi/2} \frac{1}{[1+m^2\cos^2 t]} dt + B \cdot \int_0^{\pi/2} \frac{1}{(\sin^2 t - b^2)} dt
\end{aligned} \tag{AI-20}
$$

The constants A and B can be determined by

$$
A = \frac{2m^2b^2}{[m^2(1-b^2)+1]}
$$

$$
B = \frac{2b^2}{[m^2(1-b^2)+1]}
$$

Therefore,

$$
A \cdot \int_0^{\pi/2} \frac{1}{[1+m^2\cos^2 t]} dt = A \cdot \frac{\pi}{2} \cdot \cos\left(\frac{\pi a}{2w}\right)
$$

$$
\begin{aligned}
&B \cdot \int_0^{\pi/2} \frac{1}{(\sin^2 t - b^2)} dt \\
&\quad = B \cdot \int_0^{\pi/2} \frac{1/2b}{(\sin t - b)} dt - B \cdot \int_0^{\pi/2} \frac{1/2b}{(\sin t + b)} dt \\
&\quad = B \cdot 1/2b \left[\int_0^{\pi/2} \frac{1}{(\sin t - b)} dt - \int_0^{\pi/2} \frac{1}{(\sin t + b)} dt \right]
\end{aligned}
$$

From the integral table, one can find

$$
\begin{aligned}
&B \cdot \int_0^{\pi/2} \frac{1}{(\sin^2 t - b^2)} dt \\
&= \frac{B}{2b\sqrt{1-b^2}} \left\{ \ln\left| \frac{-b+1-\sqrt{1-b^2}}{-b+1+\sqrt{1-b^2}} \right| - \ln\left| \frac{1-\sqrt{1-b^2}}{1+\sqrt{1-b^2}} \right| \right. \\
&\quad \left. - \ln\left| \frac{b+1-\sqrt{1-b^2}}{b+1+\sqrt{1-b^2}} \right| + \ln\left| \frac{1-\sqrt{1-b^2}}{1+\sqrt{1-b^2}} \right| \right\}
\end{aligned}
$$

$$= \frac{B}{2b\sqrt{1-b^2}}\left\{\ln\left|\frac{-b+1-\sqrt{1-b^2}}{-b+1+\sqrt{1-b^2}}\right|\cdot\left|\frac{b+1+\sqrt{1-b^2}}{b+1-\sqrt{1-b^2}}\right|\right\}$$

$$= \frac{B}{2b\sqrt{1-b^2}}\left\{\ln\left|\frac{(-2b\sqrt{1-b^2})^2}{4b^2(b^2-1)}\right|\right\} = \frac{B}{2b\sqrt{1-b^2}}\ln\left|\frac{1}{b}\right|$$

Therefore,

$$\varphi'(\zeta) = -\frac{pw}{\pi i}\sin\left(\frac{\pi a}{2w}\right) - \frac{pw}{\pi i}\sin\left(\frac{\pi a}{2w}\right)\frac{\zeta^2\tan^2(\frac{\pi a}{2w})}{\left[1+(1-\zeta^2)\tan^2\left(\frac{\pi a}{2w}\right)\right]}$$
$$+\frac{pw}{\pi^2 i}\tan\left(\frac{\pi a}{2w}\right)\frac{\ln|\zeta|}{\left[1+(1-\zeta^2)\tan^2\left(\frac{\pi a}{2w}\right)\right]}\cdot\frac{\zeta}{\sqrt{1-\zeta^2}}d\sigma \tag{AI-21}$$

It is easy to find that $\varphi'(0) = -\frac{pw}{\pi i}\sin\left(\frac{\pi a}{2w}\right)$.

References

[1] Muskhelishvili N I, 1953, *Some Basic Problems of the Mathematical Theory of Elasticity*, Noordhoff Ltd, Groningen.

[2] Privalov I I, 1984, *Introduction to Complex Variable Functions Theory*, Science, Moscow (in Russian).

[3] Lavrentjev M A and Schabat B V, 1986, *Methods of Complex Variable Functions Theory*, 6th edition, National Technical-Theoretical Literature Press, Moscow (in Russian).

[4] Fan T Y, 1990, Semi-infinite crack in a strip, *Chin. Phys. Lett.*, **8**(9), 401–404.

[5] Shen D W and Fan T Y, 2003, Two collinear semi-infinite cracks in a strip, *Eng. Fract. Mech.*, **70**(8), 813–822.

[6] Wolfran St., 1996, *The Mathematica Book*, 3rd edition, by Wolfran St, Cambridge University Press, Cambridge.

[7] Fan T Y and Fan L, 2008, Relation between Eshelby integral and generalized BCS and generalized DB models*, *Chin. Phys. B.*, **20**(3), 036102.

[8] Li W, 2011, Analytic solutions of a finite width strip with a single edge crack of two-dimensional quasicrystals, *Chin. Phys. B.*, **20**(11), 116201.

Appendix II

Dual Integral Equations and Some Additional Calculations

AII.1 Dual Integral Equations

It is well known that the Fourier transform or Hankel transform is a very useful tool in solving partial differential equations which have been shown in Chapters 7–9 though the introduction is very limited. For non-harmonic and non-multi-harmonic equations, the complex potential method is not effective, so we have to use the Fourier transform, Hankel transform, Mellin transform or others. After the transform, the boundary value problems of the dislocations are reduced to some algebraic equations to solve (this is relatively simpler), while those of the cracks are concluded for solving the following dual integral equations:

$$\left.\begin{aligned} &\int_0^\infty y^\alpha f(y) J_v(xy)dy = g(x), \quad 0 < x < 1 \\ &\int_0^\infty f(y) J_v(xy)dy = 0, \qquad\qquad x > 1 \end{aligned}\right\} \tag{AII-1}$$

or

$$\left.\begin{aligned} &\int_0^\infty y^{\alpha_j} \sum_{k=1}^n a'_{jk} f_j(y) J_{v_j}(xy)dy = g_j(x), \quad 0 < x < 1 \\ &\int_0^\infty \sum_{k=1}^n a_{jk} f_j(y) J_{v_j}(xy)dy = 0, \qquad\qquad x > 1 \\ &(j = 1, 2, \ldots, n) \end{aligned}\right\} \tag{AII-2}$$

or

$$\left.\begin{aligned}&\int_0^\infty\int_0^\infty g_1(\xi_1,\xi_2,s,x_1,x_2)f(\xi_1,\xi_2,s)J_\alpha(\xi_1x_1)\\&\qquad J_\beta(\xi_2x_2)d\xi_1d\xi_2=h(x_1,x_2,s),\quad (x_1,x_2)\in\Omega_1\\&\int_0^\infty\int_0^\infty g_2(\xi_1,\xi_2,s,x_1,x_2)f(\xi_1,\xi_2,s)J_\alpha(\xi_1x_1)\\&\qquad J_\beta(\xi_2x_2)d\xi_1d\xi_2=0,\qquad\qquad\quad (x_1,x_2)\in\Omega_2\end{aligned}\right\}\tag{AII-3}$$

Among them, Eqs. (AII-1) are the simplest ones, which are discussed in the following. Equations (AII-2) are dealt with multi-unknown functions, and (AII-3) are the two-dimensional dual integral equations; these two kinds of dual integral equations are more complicated.

In Eqs. (AII-1), $f(x)$ is the unknown function to be determined, $g(x)$ is the known one, α and v are constants and $J_v(xy)$ is the first kind of Bessel function of v order. Titchmarsh [1] and Busbridge [2] gave the analytic solution of the equations. Various authors [3–11] discussed the solutions with different methods. Here, only the procedure of Refs. [1, 2] is introduced. Titchmarsh [1] gave formal solution for the case $\alpha > 0$. Busbridge extended the discussion to the case $\alpha > -2$ and gave proof for the existence of the solution. The solution is given through a complex integral as follows:

$$f(x)=\frac{1}{2\pi i}\int_{k-i\infty}^{k+i\infty}2^{s-\alpha}\frac{\Gamma\left(\frac{1}{2}+\frac{1}{2}v+\frac{1}{2}s\right)}{\Gamma\left(\frac{1}{2}+\frac{1}{2}v+\frac{1}{2}\alpha-\frac{1}{2}s\right)}\psi(s)x^{-s}ds\tag{AII-4}$$

in which $s=\sigma+i\tau$ and

$$\psi(s)=\frac{1}{2\pi i}\int_{C-i\infty}^{C+i\infty}\frac{\Gamma\left(\frac{1}{2}+\frac{1}{2}v-\frac{1}{2}\alpha+\frac{1}{2}w\right)}{\Gamma\left(\frac{1}{2}+\frac{1}{2}v+\frac{1}{2}w\right)}\cdot\frac{\overline{g}(\alpha+1-w)}{w-s}dw\tag{AII-5}$$

where $w = u + iv$ (in which v represents the imaginary part of a complex variable w; do not confuse with v — the suffix of the Bessel function, which represents the order of the Bessel function), $\sigma < u$

and

$$\overline{g}(\alpha + 1 - w) = \int_0^1 g(x)x^{\alpha - w}dx$$

In the above formulas, $\Gamma(x)$ represents the Euler gamma function. The solution (AII-4) holds for both $\alpha > 0$ and $\alpha > -2$.

For $\alpha > 0$, the solution can be expressed by real integrals as

$$f(x) = \frac{(2x)^{1-\alpha/2}}{\Gamma(\alpha/2)} \int_0^1 \mu^{1+\alpha/2} J_{\nu+\alpha/2}(\mu x)d\mu \int_0^1 g(\rho\mu)\rho^{\nu+1} \times (1 - \rho^2)^{\alpha/2-1}d\rho \tag{AII-4'}$$

and for $\alpha > -2$, which is of the form

$$f(x) = \frac{2^{-\alpha/2}x^{-\alpha}}{\Gamma(1+\alpha/2)} \left[x^{1+\alpha/2} J_{\nu+\alpha/2}(x) \int_0^1 y^{\nu+1}(1-y^2)^{\alpha/2} g(y)dy + \int_0^1 y^{\alpha+1}(1-y^2)^{\alpha/2}dy \int_0^1 (xu)^{2+\alpha/2} g(yu) J_{\nu+1+\alpha/2}(xu)du\right] \tag{AII-4''}$$

Theorem. *If* $\alpha > -2$, $-v-1 < \alpha - \frac{1}{2} < v+1$, *the Mellin transforms of* $g(x)$ *and* $f(x)$ *exist; the latter is analytic in the strip region* $-v <$ *Re* $s = \sigma < \alpha$ *and has the order* $O(|t|^{\sigma-\alpha+\varepsilon})$ $(\varepsilon > 0, t \to \infty)$, *where* $s = \sigma + i\tau$ *is the Mellin transform parameter, then Eqs. (AII-1) have one and only one solution (AII-4).*

Proof. Because the strict proof given by Busbridge [7] is very lengthy, we cannot quote all its details here, instead, only a rough outline of the proof is figured out in the following. One can find that in the proof, quite a lot of complex variable function knowledge is used; this seems that the theory on dual integral equations presents inherent connection with complex analysis. So Section AII.1 is helpful for the present discussion too.

At first, assume that $0 < \alpha < 2$, $-\nu - 1 < \alpha - \frac{1}{2} < \nu + 1$, and the Mellin transform of $f(x)$

$$\overline{f}(s) = \int_0^\infty f(x)x^{s-1}dx, \quad s = \sigma + i\tau$$

is analytic in the region $-\nu < \sigma < \alpha$, and assume that as $\varepsilon > 0$, as $t \to \infty$, it has order $O(|t|^{-\alpha+\varepsilon})$ (in fact, this is a Lemma, but we omit the proof for simplicity).

According to the definition, the Mellin transform of function $y^\alpha J_\nu(xy)$ is

$$\overline{J}_\alpha(s) \equiv \int_0^\infty [y^\alpha J_\nu(xy)]y^{s-1}dy = \frac{2^{\alpha+s-1}}{x^{\alpha+s}} \frac{\Gamma\left(\frac{1}{2}\alpha + \frac{1}{2}v + \frac{1}{2}s\right)}{\Gamma\left(1 - \frac{1}{2}\alpha + \frac{1}{2}v - \frac{1}{2}s\right)} \tag{AII-6}$$

Recall that $s = \sigma + i\tau$. By using the notation of relevant Mellin transforms, the left-hand sides of the first and second equations in (AII-1) become

$$\int_0^\infty y^\alpha f(y)J_\nu(xy)dy = \frac{1}{2\pi i}\int_{C-i\infty}^{C+i\infty} \overline{f}(s)\overline{J}_\alpha(1-s)ds$$

$$\int_0^\infty f(y)J_\nu(xy)dy = \frac{1}{2\pi i}\int_{C-i\infty}^{C+i\infty} \overline{f}(s)\overline{J}_0(1-s)ds$$

and substituting (AII-6) into the above formulas yields

$$\frac{1}{2\pi i}\int_{C-i\infty}^{C+i\infty} \frac{2^{\alpha-s}}{x^{1-s}} \frac{\Gamma\left(\frac{1}{2} + \frac{1}{2}\alpha + \frac{1}{2}v - \frac{1}{2}s\right)}{\Gamma\left(\frac{1}{2} - \frac{1}{2}\alpha + \frac{1}{2}v + \frac{1}{2}s\right)}\overline{f}(s)ds = g(x), \quad 0 < x < 1$$

$$\frac{1}{2\pi i}\int_{C-i\infty}^{C+i\infty} \frac{2^{\alpha-s}\Gamma\left(\frac{1}{2} + \frac{1}{2}v - \frac{1}{2}s\right)}{\Gamma\left(\frac{1}{2} + \frac{1}{2}v + \frac{1}{2}s\right)}\overline{f}(s)ds = 0, \quad x > 1$$

Put

$$\overline{f}(s) = \frac{2^{\alpha-s}\Gamma\left(\frac{1}{2} + \frac{1}{2}v + \frac{1}{2}s\right)}{\Gamma\left(\frac{1}{2} + \frac{1}{2}\alpha + \frac{1}{2}v - \frac{1}{2}s\right)}\psi(s) \tag{AII-7}$$

Then, the above equations reduce to

$$\left.\begin{aligned} &\frac{1}{2\pi i}\int_{C-i\infty}^{C+i\infty} \frac{\Gamma\left(\frac{1}{2} + \frac{1}{2}v + \frac{1}{2}s\right)}{\Gamma\left(\frac{1}{2} + \frac{1}{2}v - \frac{1}{2}\alpha + \frac{1}{2}s\right)}\psi(s)x^{s-1-\alpha}ds = g(x), \quad 0 < x < 1 \\ &\frac{1}{2\pi i}\int_{C-i\infty}^{C+i\infty} \frac{\Gamma\left(\frac{1}{2} + \frac{1}{2}v - \frac{1}{2}s\right)}{\Gamma\left(\frac{1}{2} + \frac{1}{2}v + \frac{1}{2}\alpha - \frac{1}{2}s\right)}\psi(s)x^{s-1}ds = 0, \qquad x > 1 \end{aligned}\right\} \tag{AII-8}$$

Multiplying $x^{\alpha-w}$ to the first one of (AII-8), where $w = u + iv$, $\sigma - u > 0$, then integrating over $(0,1)$ to x and

$$\frac{1}{2\pi i}\int_{C-i\infty}^{C+i\infty}\frac{\Gamma\left(\frac{1}{2}+\frac{1}{2}v+\frac{1}{2}s\right)}{\Gamma\left(\frac{1}{2}+\frac{1}{2}v-\frac{1}{2}\alpha+\frac{1}{2}s\right)}\psi(s)\frac{ds}{s-w} = \overline{g}(\alpha - w + 1) \tag{AII-9}$$

where $u < C$, and

$$\overline{g}(\alpha - w + 1) = \int_0^1 g(x)x^{\alpha-w}dx$$

The left-hand side of Eq. (AII-9) is analytic everywhere in the strip zone

$$-\nu < \sigma < \alpha$$

except for the simple pole $s = w$ and behaves order $O(|t|^{-\alpha+\varepsilon})$. If we move the integration path from $\sigma = C$ to $\sigma = C' < u$, see Fig. AII.1, based on Cauchy's integral formula (referring to formula (AI-5)),

$$\frac{1}{2\pi i}\int_{C'-i\infty}^{C'+i\infty}\frac{\Gamma\left(\frac{1}{2}+\frac{1}{2}v+\frac{1}{2}s\right)}{\Gamma\left(\frac{1}{2}+\frac{1}{2}v-\frac{1}{2}\alpha+\frac{1}{2}s\right)}\psi(s)\frac{ds}{s-w}$$
$$= \overline{g}(\alpha - w + 1) - \frac{\Gamma\left(\frac{1}{2}+\frac{1}{2}v+\frac{1}{2}w\right)}{\Gamma\left(\frac{1}{2}+\frac{1}{2}v-\frac{1}{2}\alpha+\frac{1}{2}w\right)}\psi(w)$$

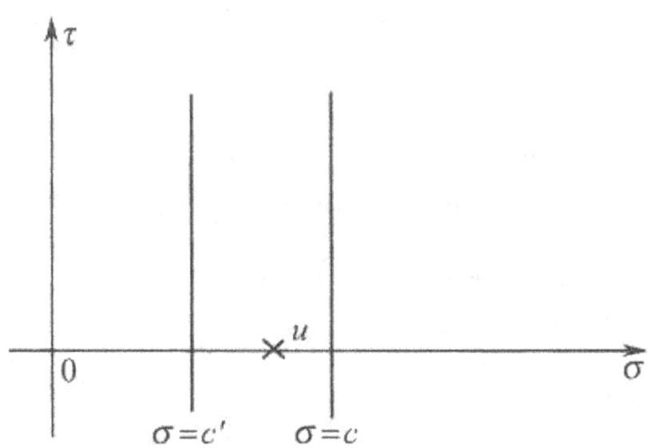

Fig. AII.1. The integration path in $s = \sigma + i\tau$-plane.

This translation of the integration line corresponds to form a closed region, and the value of the integral around the closed region is just equal to the second term of the above formula including the sign of the term. Because the left-hand side is analytic as $u > C'$, so the right-hand side. In addition,

$$\psi(w) - \frac{\Gamma\left(\frac{1}{2}+\frac{1}{2}v-\frac{1}{2}\alpha+\frac{1}{2}w\right)}{\Gamma\left(\frac{1}{2}+\frac{1}{2}v+\frac{1}{2}\alpha\right)}\overline{g}(\alpha-w+1) \tag{AII-10}$$

is analytic for the case

$$\frac{1}{2}+\frac{1}{2}\nu-\frac{1}{2}\alpha+\frac{1}{2}w \neq 0,-1,-2,\ldots$$

Integrating function (AII-10) along a big rectangle whose corners are the points

$$C-iT, C+iT, -T+iT, -T-iT \quad (T>|\nu|)$$

and one can find that the absolute values of the integrals

$$\left|\int_{C-iT}^{-T+iT}\right|, \left|\int_{-T+iT}^{-T-iT}\right|, \left|\int_{-T-iT}^{C-iT}\right|$$

have order $O(|T|^{-\alpha/2+\varepsilon})$, and the value of ε can be always taken less than $\alpha/2$, then they can tend to zero as $T \to \infty$. According to Cauchy's integral theorem (referring to formula (AI-4)),

$$\frac{1}{2\pi i}\int_{C-i\infty}^{C+i\infty}\left\{\psi(s) - \frac{\Gamma\left(\frac{1}{2}+\frac{1}{2}v-\frac{1}{2}\alpha+\frac{1}{2}w\right)}{\Gamma\left(\frac{1}{2}+\frac{1}{2}v+\frac{1}{2}s\right)}\overline{g}(\alpha-s+1)\right\}\frac{ds}{s-w} = 0, \quad (u<C) \tag{AII-11}$$

Similarly, multiplying x^{-w} to the second one of Eq. (AII-8), where $\sigma - w < 0$, then integrating to x over $(1,\infty)$

$$\frac{1}{2\pi i}\int_{C'-i\infty}^{C'+i\infty}\frac{\Gamma\left(\frac{1}{2}+\frac{1}{2}v-\frac{1}{2}s\right)}{\Gamma\left(\frac{1}{2}+\frac{1}{2}v+\frac{1}{2}\alpha-\frac{1}{2}s\right)}\psi(s)\frac{ds}{s-w} = 0, \quad (u>C')$$

Move the integration path and find that

$$\psi(w) = \frac{1}{2\pi i}\int_{C-i\infty}^{C+i\infty}\psi(s)\frac{ds}{s-w}, \quad (u<C) \tag{AII-12}$$

Comparing (AII-11) and (AII-12), we can find (AII-5), so the solution (AII-4), in which

$$\overline{g}(\alpha - s + 1) = \int_0^1 g(\xi)\xi^{\alpha-s}d\xi$$

The theorem is proved.

Furthermore, the form of the real integral of the solution can be obtained as follows. In fact,

$$\frac{1}{s-w} = \int_0^1 \eta^{s-w-1}d\eta$$

If we exchange the integration order, then (AII-5) may be rewritten as

$$\psi(w) = \int_0^1 g(\xi)\xi^{\alpha}d\xi \int_0^1 \eta^{-w-1}d\eta$$
$$\times \frac{1}{2\pi i}\int_{C-i\infty}^{C+i\infty} \frac{\Gamma\left(\frac{1}{2}+\frac{1}{2}v-\frac{1}{2}\alpha+\frac{1}{2}s\right)}{\Gamma\left(\frac{1}{2}+\frac{1}{2}v+\frac{1}{2}s\right)}\left(\frac{\xi}{\eta}\right)^{-s}ds$$

in which the integration [12, 13]

$$\frac{1}{2\pi i}\int_{C-i\infty}^{C+i\infty} \frac{\Gamma\left(\frac{1}{2}+\frac{1}{2}v-\frac{1}{2}\alpha+\frac{1}{2}s\right)}{\Gamma\left(\frac{1}{2}+\frac{1}{2}v+\frac{1}{2}s\right)}\left(\frac{\xi}{\eta}\right)^{-s}ds$$
$$= \begin{cases} \dfrac{2}{\Gamma\left(\frac{1}{2}\alpha\right)}\xi^{1+\nu-\alpha}(\eta^2-\xi^2)^{\alpha/2-1}\eta^{1-\nu}, & \eta \geq \xi \\ 0, & 0 < \eta < \xi \end{cases}$$

So,

$$\psi(w) = \frac{2}{\Gamma(\alpha/2)}\int_0^1 g(\xi)\xi^{1+\nu}d\xi \int_{\xi}^1 \eta^{-w-\nu}(\eta^2-\xi^2)^{\alpha/2-1}d\eta$$

By exchanging the integration order, we may find that

$$\psi(w) = \frac{2}{\Gamma(\alpha/2)}\int_0^1 \eta^{-w-\nu}d\eta \int_0^1 g(\xi)\xi^{1+\nu}(\eta^2-\xi^2)^{\alpha/2-1}d\xi$$
$$= \frac{2}{\Gamma(\alpha/2)}\int_0^1 \eta^{\alpha-\omega}d\eta \int_0^1 g(\xi)\xi^{1+\nu}(1-\xi^2)^{\alpha/2-1}d\xi$$

Substituting it into (AII-4) yields

$$f(x) = \frac{2}{\Gamma(\alpha/2)} \int_0^1 \eta^\alpha d\eta \int_0^1 g(\eta\zeta)\zeta^{1+\nu}(1-\zeta^2)^{\alpha/2-1} d\zeta$$
$$\times \frac{1}{2\pi i} \int_{C-i\infty}^{C+i\infty} 2^{s-\alpha}(x\eta)^{-s} \frac{\Gamma\left(\frac{1}{2}+\frac{1}{2}v+\frac{1}{2}s\right)}{\Gamma\left(\frac{1}{2}+\frac{1}{2}v+\frac{1}{2}\alpha-\frac{1}{2}s\right)} ds$$

By utilizing the inversion of the Mellin transform [12, 13],

$$\frac{1}{2\pi i} \int_{C-i\infty}^{C+i\infty} 2^{s-\alpha} \frac{\Gamma\left(\frac{1}{2}+\frac{1}{2}v+\frac{1}{2}s\right)}{\Gamma\left(\frac{1}{2}+\frac{1}{2}v+\frac{1}{2}\alpha-\frac{1}{2}s\right)} (at)^{-s} ds$$
$$= 2^{-\alpha/2}(at)^{1-\alpha/2} J_{\nu+\alpha/2}(at)$$

Then, one finds (AII.4′).

For the case $\alpha > -2$, the derivation is similar. For some details, the reader can refer to Busbridge [2].

In the following, some examples are discussed in detail, which are the dual integral equations that, appeared in Chapters 8 and 9, respectively, where only the solutions were listed without derivation details.

AII.2 Additional Derivation on the Solution of Dual Integral Equations (8.3-8) and (9.7-4)

Equations (8.3-8) in the text are

$$\left.\begin{aligned}
&\frac{2}{d_{11}} \int_0^\infty [C(\xi)\xi - 6D(\xi)] \cos(\xi x) d\xi = -p, && 0 < x < a \\
&\int_0^\infty \xi^{-1}[C(\xi)\xi - 6D(\xi)] \cos(\xi x) d\xi = 0, && x > a \\
&\frac{2}{d_{12}} \int_0^\infty D(\xi) \cos(\xi x) d\xi = 0, && 0 < x < a \\
&\int_0^\infty \xi^{-1} D(\xi) \cos(\xi x) d\xi = 0, && x > a
\end{aligned}\right\} \quad \text{(AII-13)}$$

It is evident that the second pair of dual integral equations (AII-13) has the zero solution, i.e., $D(\xi) = 0$, and we only consider the first pair in the equations, which is

$$\left.\begin{aligned} &\frac{2}{d_{11}}\int_0^\infty C(\xi)\xi\cos(\xi x)d\xi = -p, \quad 0 < x < a \\ &\int_0^\infty C(\xi)\cos(\xi x)d\xi = 0, \qquad\qquad x > a \end{aligned}\right\} \tag{AII-14}$$

and is similar to that of (9.7-4) in Chapter 9. Because of

$$\cos(\xi x) = \left(\frac{\pi\xi x}{2}\right)^{1/2} J_{-1/2}(\xi x)$$

and denoting

$$\xi^{1/2}C(\xi) = f(\xi), \quad \eta = a\xi, \quad \rho = \frac{x}{a}, \quad g(\rho) = a\left(\frac{\pi a d_{11}}{2\rho}\right)^{1/2} p$$

then (AII-14) is reduced to

$$\left.\begin{aligned} &\frac{2}{d_{11}}\int_0^\infty \eta f(\eta)J_{-1/2}(\eta\rho)d\eta = g(\rho), \quad 0 < \rho < 1 \\ &\int_0^\infty f(\eta)J_{-1/2}(\eta\rho)d\eta = 0, \qquad\qquad \rho > 1 \end{aligned}\right\} \tag{AII-14$'$}$$

which becomes one of the standard dual integral equations shown by (AII-1) with

$$\alpha = 1, \quad \nu = -1/2, \quad g(\rho) = g_0\rho^{-1/2}, \quad g_0 = const = a(\pi a d_{11})^{1/2}p$$

In this case, it is very easy to calculate the solution of dual integral equations (AII-14) (or (AII-14$'$)) by formulas (AII-4) and (AII-5), but the key step is choosing the integration path. In the previous introduction on the Titchmarsh–Busbridge solution, we mentioned that it must require that $-\nu > k > \alpha$, $-\nu < C < \alpha$ and $k < C$. In the present case, $\nu = -1/2$, $\alpha = 1$ such that $1/2 < k < 1$ and $1/2 < C < 1$. The concrete calculation is as follows. Due to

$$\overline{g}(\alpha + 1 - t) = \int_0^1 g(\rho)\rho^{\alpha - t}d\rho = g_0\int_0^1 \rho^{-1/2}\rho^{1-t}d\rho = \frac{g_0}{\frac{3}{2} - t}$$

where $t = t_1 + it_2$ represents a complex variable and requires that $t_1 < 3/2$. Substituting the relevant data and the above result into

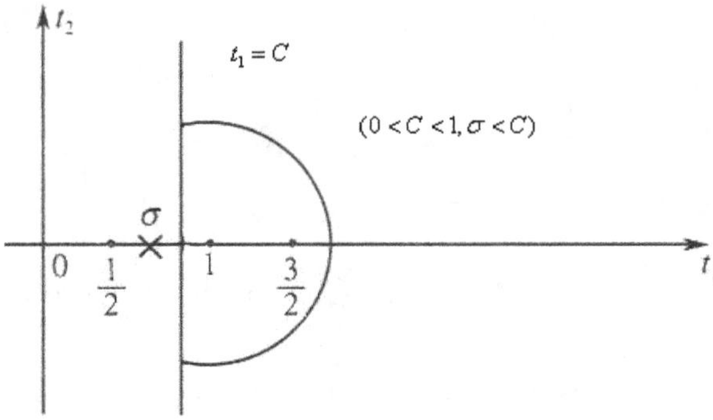

Fig. AII.2. The integration path in $t = t_1 + it_2$-plane.

(AII-5), we have

$$\psi(s) = g_0 \frac{1}{2\pi i} \int_{C-i\infty}^{C+i\infty} \frac{\Gamma\left(-\frac{1}{4} + \frac{t}{2}\right)}{\Gamma\left(\frac{1}{4} + \frac{t}{2}\right)} \frac{1}{t-s} \frac{1}{\frac{3}{2} - t} dt$$

The integration path is shown in Fig. AII.2.

In this case, the integrand has only one pole at point $t = 3/2$ of first order. According to the formula for evaluating residual (AI-15), the above integral is very easily obtained as

$$\psi(s) = g_0 \frac{\Gamma\left(\frac{1}{2}\right)}{\Gamma(1)} \frac{1}{\frac{3}{2} - s} = g_0 \frac{\sqrt{\pi}}{\frac{3}{2} - s} \tag{AII-15}$$

whereas $\Gamma(1) = 1, \Gamma\left(\frac{1}{2}\right) = \sqrt{\pi}$. Substituting the result into formula (AII-4) leads to

$$f(\eta) = g_0 \sqrt{\pi} \frac{1}{2\pi i} \int_{k-i\infty}^{k+i\infty} \frac{2^{s-\alpha} \Gamma\left(\frac{1}{4} + \frac{s}{2}\right)}{\Gamma\left(\frac{1}{4} - \frac{s}{2}\right)} \eta^{-s} ds \tag{AII-16}$$

In terms of the inversion of the Mellin transform [13],

$$\begin{aligned} &\frac{1}{2\pi i} \int_{k-i\infty}^{k+i\infty} 2^{s-\lambda} \frac{\Gamma\left(\frac{1}{2} + \frac{1}{2}\mu + \frac{1}{2}s\right)}{\Gamma\left(\frac{1}{2} + \frac{1}{2}\mu + \frac{1}{2}\lambda - \frac{1}{2}s\right)} (\beta\eta)^{-s} ds \\ &= 2^{-\lambda/2} (\beta\eta)^{1-\lambda/2} J_{\mu+\lambda/2}(\beta\eta) \end{aligned} \tag{AII-17}$$

In formula (AII-16), $\mu = -1/2, \lambda = 3, \beta = 1$, so

$$f(\eta) = g_0 \left(\frac{\pi}{2\eta}\right)^{1/2} J_1(\eta)$$

and

$$C(\xi) = \xi^{-1/2} f(\xi) = \frac{\pi a d_{11}}{2} \xi^{-1} J_1(a\xi)$$

This is just the result given by (8.3-10) and (9.7-8), the difference between them lies in a constant factor.

The calculation through (AII-4′) and (AII-4) yields the same result, so the correctness of the result is demonstrated.

AII.3 Additional Derivation on the Solution of Dual Integral Equations (9.8-8)

In Section 9.8, in the text of chapter, the dual integral equations

$$\begin{aligned} &\int_0^\infty \xi A_i(\xi) J_0(\xi r) d\xi = M_i p_0, \quad 0 < r < a \\ &\int_0^\infty A_i(\xi) J_0(\xi r) d\xi = 0, \qquad r > a \end{aligned} \tag{AII-18}$$

are solved and we obtained the solution (9.8-8). We here give the details for the derivation.

According to the standard type of the equations, here $\alpha = 1$, $\nu = 0$, $g(\rho) = g_0 = const$, and put $\rho = r/a$, so

$$\bar{g}(\alpha + 1 - t) = \int_0^1 g(\rho)\rho^{\alpha - t} d\rho = \frac{g_0}{2 - t} \quad (\text{Re}\, t = t_1 < 2)$$

$$\psi(s) = g_0 \frac{1}{2\pi i} \int_{C-i\infty}^{C+i\infty} \frac{\Gamma\left(\frac{1}{2}\right)}{\Gamma\left(\frac{1}{2} + \frac{t}{2}\right)} \frac{1}{t - s} \frac{1}{2 - t} dt = g_0 \frac{2}{\sqrt{\pi}} \frac{1}{2 - s}$$

in which $s = \sigma + i\tau$, and the integral is evaluated through the residual of pole $t = 2$ and the integration path is shown in Fig. AII.3.

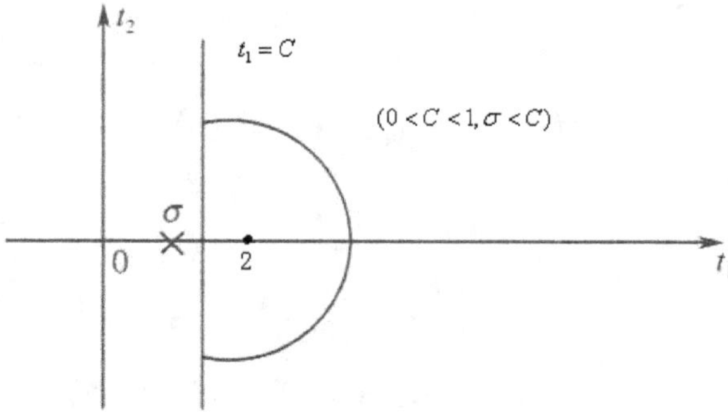

Fig. AII.3. Integration path at the $t = t_1 + it_2$-plane.

Substituting the result into (AII-4) yields

$$A_i(\xi) = f(\xi)$$

$$= g_0 \frac{2}{\sqrt{\pi}} \frac{1}{2\pi i} \int_{k-i\infty}^{k+i\infty} \frac{2^s \Gamma\left(\frac{1}{2} + \frac{s}{2}\right)}{\Gamma\left(2 - \frac{s}{2}\right)} \xi^{-s} ds = \frac{2g_0}{\sqrt{2\pi}} \xi^{-1/2} J_{3/2}(\xi)$$

which is just the solution (9.8-9), a little bit difference with that lies in that we here used the normalized coordinate $\rho = r/a$. In the last step of the calculation, the inversion of the Mellin transform (AII-17) was used.

The evaluation through formulas (AII-4′) and (AII-4″) finds the same result, this checks the correctness of the above calculation.

The above two subsections demonstrate the effect and simplicity of the complex variable function method in evaluating solutions of Titchmarsh–Busbridge dual integral equations.

The system of dual integral equations (AII-2) and its applications are discussed by Fan [14], and the two-dimensional dual integral equations (AII-3) are solved approximately by Fan and Sun [15], in which some applications are also given.

The application of integral transforms in the elasticity of quasicrystals may be effective and more widely used than that of the complex variable function method and has got much analytic solutions, see, e.g., Li [16], Zhou and Fan [17], Zhou [18] and Zhu and Fan [19, 20]; due to the limitation of the space, much results have not been quoted. The application of the method to crack problems often leads to some dual integral equations, so it is helpful for a discussion about this.

References

[1] Titchmarsh B C, 1937, *Introduction to the Fourier Integrals*, Clarenden Press, Oxford.

[2] Busbridge I W, 1938, Dual integral equations, *Math. Soc. Proc.*, **44**(2), 115–129.

[3] Weber H, 1873, Ueber die Besselschen Functionen und ihre Anwendung auf die Theorie der electrischen Stroeme, *J. fuer reihe und angewandte Mathematik*, **75**(1), 5–105.

[4] McDonald H M, 1895, The electrical distribution induced on a disk placed in any fluid of force, *Phil. Mag.*, **26**(1), 257–260.

[5] King L V, 1935, On the acoustic radiation pressure on circular disk, *Roy. Soc. London Proc. Ser A*, **153**(1), 1–16.

[6] Copson E T, 1947, On the problem of the electric disk, *Edinburg Math. Soc. Proc.*, **8**(1), 5–14.

[7] Tranter C J, 1951, On some dual integral equations, *Quar. J. Math. Ser 2*, **2**(1), 60–66.

[8] Sneddon I N, 1951, *Fourier Transforms*, McGraw-Hill, New York.

[9] Gordon A N, 1954, Dual integral equations, *London Math. Soc. J.*, **29**(5), 360–369.

[10] Noble B, 1955, On some dual integral equations, *Quar. J. Math. Ser* 2, **6**(2), 61–67.

[11] Noble B, 1963, The solutions of Bessel function dual integral equations by a multiplying factor method, *Cambridge Phil. Soc. Proc.*, **59**(4), 351–362.

[12] Watson G N, 1955, *A Treatise on the Theory of Bessel Functions*, Amazon Com, New York.

[13] Erdelyi A, 1953, *Higher Transcendental Functions*, Vol. 2, McGraw-Hill, New York.

[14] Fan T Y, 1979, Dual integral equations and system of dual integral equations and their applications in solid mechanics and fluid mechanics, *Mathematica Applicata Sinica*, **2**(3), 212–230.

[15] Fan T Y and Sun Z F, 2007, A class of two-dimensional dual integral equations and applications, *Appl. Math. Mech.*, **28**(2), 247–252.

[16] Li X F, 1999, Defects and their analytic solution in elasticity of quasicrystals, Dissertation, Beijing Institute of Technology.

[17] Zhou W M and Fan T Y, 2001, Plane elasticity problem of two-dimensional octagonal quasicrystal and crack problem, *Chin. Phys.*, **10**(8), 743–747.

[18] Zhou W M, 2000, Mathematical analysis of elasticity and defects of two- and three-dimensional quasicrystals, Dissertation, Beijing Institute of Technology, Beijing (in Chinese).

[19] Zhu A Y and Fan T Y, 2007, Elastic analysis of a mode II crack in an icosahedral quasicrystal, *Chin. Phys.*, **16**(5), 1111–1119.

[20] Zhu A Y and Fan T Y, 2009, Elastic analysis of a Griffith crack in an icosahedral Al-Pd-Mn quasicrystal, *Int. J. Modern Phys. B*, 23(10), 1–16.

Appendix III

Poisson Brackets in Condensed Matter Physics, Lie Group and Lie Algebra and Their Applications

In Chapter 16, the equations of motion of icosahedral quasicrystals were listed, and the details of the derivation of the equations have not been given in the text, which need a tool: Poisson brackets in condensed matter physics. In this appendix, we first introduce the method and then give the derivation of those equations.

AIII.1 Poisson Brackets in Condensed Matter Physics

Due to symmetry breaking or too complex, the derivation of some equations of motion of hydrodynamics of some substantive systems cannot be obtained directly by conventional conservations laws. The Poisson brackets in condensed matter physics become a useful method for the derivation, which simplifies the calculation. The method is originated from Landau and his school in former Soviet Union and Russia; refer to [1–6]. Physicists Martin *et al.* [7], Fleming and Cohen [8] in US developed the method to hydrodynamics of crystals and liquid crystals, but their derivations were still lengthy. Lubensky *et al.* [9] further developed the approach in deriving the hydrodynamic equations of quasicrystals, simplified the derivation and made it arrive in systematization.

Poisson brackets come from the classical analytic mechanics, i.e., for two mechanical quantities f, g, there is the following relation:

$$\{f, g\} = \sum_i \left(\frac{\partial f}{\partial q_i} \frac{\partial g}{\partial p_i} - \frac{\partial f}{\partial p_i} \frac{\partial g}{\partial q_i} \right) \tag{AIII-1}$$

which is the Poisson bracket, where p_i, q_i denote the canonic momentum and canonic coordinate. According to the terminology of physics, (AIII-1) is named classical Poisson bracket hereafter. Relative to the classical Poisson bracket (AIII-1), there is a quantum Poisson bracket, which is related to the commutation relation in quantum mechanics

$$[\hat{A}, \hat{B}] = \hat{A}\hat{B} - \hat{B}\hat{A} \tag{AIII-2}$$

in which $\hat{A}$, $\hat{B}$ represent two operators, e.g., $\hat{A}$ represents coordinate operator x_α, $\hat{B}$ the momentum operator p_β, then

$$[x_\alpha, p_\beta] = \mathrm{i}\hbar\delta_{\alpha\beta}, [x_\alpha, x_\beta] = 0, [p_\alpha, p_\beta] = 0 \tag{AIII-3}$$

where $\mathrm{i} = \sqrt{-1}, \hbar = h/2\pi, h$ the Planck constant and $\delta_{\alpha\beta}$ unit tensor. Equation (AIII-3) is named quantum Poisson bracket. In the quantum mechanics, mechanical quantities present operators. Equation (AIII-3) holds for any operator, in general.

There is inherent connection between the quantum Poisson bracket and classical Poisson bracket, i.e.,

$$\lim_{\hbar \to 0} \frac{\mathrm{i}\left[\hat{A}\hat{B} - \hat{B}\hat{A}\right]}{\hbar} = \{A, B\} \tag{AIII-4}$$

This is a well-known result in the quantum mechanics.

Landau [4] introduced the limit passing over (AIII-4) from the quantum Poisson bracket to the classical Poisson bracket in deriving the hydrodynamic equations of superfluids. He takes the expansion of mass density and momentum, such as

$$\hat{\rho}(\mathrm{r}) = \sum_\alpha m_\alpha \delta(\mathrm{r}_\alpha - \mathrm{r}) \tag{AIII-5}$$

$$\hat{g}_k(\mathrm{r}) = \sum_\alpha \hat{p}_k^\alpha \delta(\mathrm{r}_\alpha - \mathrm{r}) \tag{AIII-6}$$

whose quantum Poisson brackets are

$$\begin{aligned}
[\hat{\rho}(\mathrm{r}_1), \hat{\rho}(\mathrm{r}_2)] &= 0 \\
[\hat{p}_k(\mathrm{r}_1), \hat{\rho}(\mathrm{r}_2)] &= i\hbar\hat{\rho}(\mathrm{r}_1)\nabla_k(\mathrm{r}_1)\delta(\mathrm{r}_1 - \mathrm{r}_2) \qquad \text{(AIII-7)} \\
[\hat{p}_k(\mathrm{r}_1), \hat{p}_l(\mathrm{r}_2)] &= i\hbar\left(\hat{p}_l(\mathrm{r}_1)\nabla_k(\mathrm{r}_1) - \hat{p}_k(\mathrm{r}_2)\nabla_k(\mathrm{r}_2)\right)\delta(\mathrm{r}_1 - \mathrm{r}_2)
\end{aligned}$$

where $\nabla_k(\mathrm{r}_1)$ represents the derivative carrying out on coordinate r_1 and $\nabla_l(\mathrm{r}_2)$ on coordinate r_2.

By using the limit passing over (AIII-4) from the quantum Poisson to the classical Poisson bracket, from (AIII-7), one can obtain the corresponding classical Poisson brackets:

$$\begin{aligned}
\{p_k(\mathrm{r}_1), \rho(\mathrm{r}_2)\} &= \rho(\mathrm{r}_1)\nabla_k(\mathrm{r}_1)\delta(\mathrm{r}_1 - \mathrm{r}_2) \\
\{p_k(\mathrm{r}_1), p_l(\mathrm{r}_2)\} &= \left(p_l(\mathrm{r}_1)\nabla_k(\mathrm{r}_1) - p_k(\mathrm{r}_2)\nabla_k(\mathrm{r}_2)\right)\delta(\mathrm{r}_1 - \mathrm{r}_2)
\end{aligned} \qquad \text{(AIII-8)}$$

Lubensky *et al.* [9] extended the discussion to quasicrystals, and they introduced the Landau expassion to phonon field u_i and phason field w_i as follows:

$$u_k(\mathrm{r}) = \sum_\alpha u_k^\alpha \delta(\mathrm{r}_\alpha - \mathrm{r}) \qquad \text{(AIII-9)}$$

$$w_k(\mathrm{r}) = \sum_\alpha w_k^\alpha \delta(\mathrm{r}_\alpha - \mathrm{r}) \qquad \text{(AIII-10)}$$

By using the limit passing over (AIII-4) from the quantum Poisson to the classical Poisson bracket, from (AIII-9) and (AIII-10), one can find the corresponding classical Poisson brackets as follows:

$$\{u_k(\mathrm{r}_1),\ g_l(\mathrm{r}_2)\} = \left(-\delta_{kl} + \nabla_l(\mathrm{r}_1)u_k\right)\delta(\mathrm{r}_1 - \mathrm{r}_2) \qquad \text{(AIII-11)}$$

$$\{w_k(\mathrm{r}_1), g_l(\mathrm{r}_2)\} = \left(\nabla_l(\mathrm{r}_1)w_k\right)\delta(\mathrm{r}_1 - \mathrm{r}_2) \qquad \text{(AIII-12)}$$

It is evident that (AIII-12) is quite different from (AIII-11), this leads to the dissipation equations of phasons given in the subsequent discussion are quite different from those of phonons. The relevant derivations are carried out by Lubensky *et al.* [9].

AIII.2 Other Relevant Formulas

Apart from Poisson brackets, it needs some other formulas in the derivation of hydrodynamic equations of quasicrystals, which is related to the Langevin equation or generalized Langevin equation.

It is well known that the conventional Langevin equation stands for

$$\frac{\partial \Psi(\mathrm{r}, t)}{\partial t} = -\Gamma \Psi(\mathrm{r}, t) + F_s \tag{AIII-13}$$

in which $\Psi(\mathrm{r}, t)$ is a mechanical quantity, Γ represents a resistant force and F_s a stochastic force. The equation describes a stochastic process. Ginzburg and Landau extended it to the case of multi-variables

$$\frac{\partial \Psi_\alpha(\mathrm{r}, t)}{\partial t} = -\Gamma_{\alpha\beta} \frac{\delta H}{\delta \Psi_\beta(\mathrm{r}, t)} + (F_s)_\alpha \tag{AIII-14}$$

where $H = H[\Psi(\mathrm{r}, t)]$ denotes a energy functional, which can also be named Hamiltonian, $\frac{\delta H}{\delta \Psi_\beta(\mathrm{r},t)}$ represents a variation of $H = H[\Psi(\mathrm{r}, t)]$ on $\Psi_\beta(\mathrm{r}, t)$ and $\Gamma_{\alpha\beta}$ represents the elements of resistant matrix (or dissipation kinetic coefficient matrix), and the meaning of the definitions of other quantities is the same as before. Equation (AIII-14) is a kind of generalized Langevin equation, which can also be extended in a more wide sense. If the macroscopic quantity $\Psi_\alpha(\mathrm{r}, t)$ may be seen as the thermodynamic average of microscopic quantity $\Psi_\alpha^\mu(\mathrm{r}, \{q^\alpha\}, \{p^\alpha\})$, i.e.,

$$\Psi_\alpha(\mathrm{r}, t) = \langle \Psi_\alpha^\mu(\mathrm{r}, \{q^\alpha\}, \{p^\alpha\}) \rangle \tag{AIII-15}$$

in which p^α, q^α are the canonic momentum and canonic coordinate, and the microquantities obey the microscopic Liouville equation

$$\frac{\partial \Psi_\alpha^\mu}{\partial t} = \{H^\mu, \Psi_\alpha^\mu\} \tag{AIII-16}$$

where $H^\mu(\{q^\alpha\}, \{p^\alpha\})$ denotes the Hamiltonian.

In d-dimensional space, the partial derivative of macroquantity $\Psi_\alpha(\mathrm{r}, t)$ on time

$$\frac{\partial \Psi_\alpha(\mathrm{r}, t)}{\partial t}$$

consists of various terms: one among them is

$$-\int\left(\{\Psi_\beta(\mathrm{r}'),\ \Psi_\alpha(\mathrm{r})\}\frac{\delta H}{\delta \Psi_\beta(\mathrm{r}', t)}\right) d^d\mathrm{r}' \tag{AIII-17}$$

and the other is

$$\int\left(\frac{\delta\{\Psi_\beta(\mathrm{r}'), \Psi_\alpha(\mathrm{r})\}}{\delta \Psi_\beta(\mathrm{r}', t)}\right) d^d\mathrm{r}' \tag{AIII-18}$$

Combining (AIII-17) and (AIII-18), (AIII-14) is generalized as

$$\begin{aligned}\frac{\partial \Psi_\alpha(\mathrm{r}, t)}{\partial t} = &-\int\left(\{\Psi_\beta(\mathrm{r}'), \Psi_\alpha(\mathrm{r})\}\frac{\delta H}{\delta \Psi_\beta(\mathrm{r}', t)}\right) d^d\mathrm{r}' \\ &+\int\left(\frac{\delta\{\Psi_\beta(\mathrm{r}'), \Psi_\alpha(\mathrm{r})\}}{\delta \Psi_\beta(\mathrm{r}', t)}\right) d^d\mathrm{r}' - \Gamma_{\alpha\beta}\frac{\delta H}{\delta \Psi_\beta(\mathrm{r}, t)} + (F_s)_\alpha\end{aligned} \tag{AIII-19}$$

where $d^d\mathrm{r}' = dV$ represents the volume element of the integral. Based on the formulas (AIII-8), (AIII-11) and (AIII-12), Lubensky *et al.* [9] utilized Eq. (AIII-19) to derive the hydrodynamic equations of quasicrystals. This will be given in the following subsection. In the derivation, the last term in (AIII-19) is omitted.

AIII.3 Derivation of Hydrodynamic Equations of Solid Quasicrystals

The derivation of equation of mass conservation is very simple, which is the same as that in the conventional fluid, and it is omitted here.

At first, consider the derivation of the equation of motion of phonon. Put $\Psi_\alpha(\mathrm{r}, t) = u_i(\mathrm{r}, t), \Psi_\beta(\mathrm{r}', t) = g_j(\mathrm{r}', t)$ in (AIII-19) and omit the second and fourth terms on the right-hand side of the equation, then

$$\frac{\partial u_i(\mathrm{r},t)}{\partial t} = -\int \left(\left\{u_i(\mathrm{r}'), g_j(\mathrm{r})\right\} \frac{\delta H}{\delta g_j(\mathrm{r}',t)}\right) d^d\mathrm{r}' - \Gamma_{\mathbf{u}} \frac{\delta H}{\delta u_i(\mathrm{r},t)}$$

Substituting bracket (AIII-11) into the integral on the right-hand side yields

$$\begin{aligned} \frac{\partial u_i(\mathrm{r},t)}{\partial t} &= \int (-\delta_{ij} + \nabla_j(\mathrm{r})u_i)\, \delta(\mathrm{r}-\mathrm{r}') \frac{g_j(\mathrm{r}')}{\rho(\mathrm{r}')} d^d\mathrm{r}' + \Gamma_{\mathrm{u}} \frac{\delta H}{\delta u_i(\mathrm{r},t)} \\ &= -V_j \nabla_j(\mathrm{r}) u_i - \Gamma_{\mathbf{u}} \frac{\delta H}{\delta u_i(\mathrm{r},t)} + V_i \end{aligned} \tag{AIII-20}$$

where Γ_{u} denotes the phonon dissipation kinematic coefficient, and the Hamiltonian is defined by

$$\begin{aligned} H = H[\Psi(\mathrm{r},t)] &= \int \frac{\mathrm{g}^2}{2\rho} d^d\mathrm{r} + \int \left[\frac{1}{2} A \left(\frac{\delta\rho}{\rho_0}\right)^2 + B\left(\frac{\delta\rho}{\rho_0}\right) \nabla \cdot \mathrm{u}\right] \\ &\quad \times d^d\mathrm{r} + F_{el} = H_{kin} + H_{density} + F_{el} \\ F_{el} &= F_{\mathrm{u}} + F_{\mathrm{w}} + F_{\mathrm{uw}}, \mathrm{g} = \rho \mathrm{V} \end{aligned} \tag{AIII-21}$$

and A, B are the constants that describe density variation, which are named Lubensky–Ramaswamy–Toner (LRT) constants, the last terms of (AIII-21) represent elastic energies, which consist of phonons, phasons and phonon–phason coupling parts, respectively:

$$\begin{aligned} F_{\mathrm{u}} &= \int \frac{1}{2} C_{ijkl} \varepsilon_{ij} \varepsilon_{kl} d^d\mathrm{r} \\ F_{\mathrm{w}} &= \int \frac{1}{2} K_{ijkl} w_{ij} w_{kl} d^d\mathrm{r} \\ F_{\mathrm{uw}} &= \int (R_{ijkl} \varepsilon_{ij} w_{kl} + R_{klij} w_{ij} \varepsilon_{kl})\, d^d\mathrm{r} \end{aligned} \tag{AIII-22}$$

where C_{ijkl} the phonon elastic constants K_{ijkl} phason elastic constants and R_{ijkl}, R_{klij} the phonon–phason coupling elastic constants, and the strain tensors ε_{ij}, w_{ij} are defined by

$$\varepsilon_{ij} = \frac{1}{2}\left(\frac{\partial u_i}{\partial x_j} + \frac{\partial u_j}{\partial x_i}\right), \quad w_{ij} = \frac{\partial w_i}{\partial x_j} \tag{AIII-23}$$

the associated stress tensors are related through the generalized Hooke's law

$$\left.\begin{aligned} \sigma_{ij} &= \frac{\partial F}{\partial \varepsilon_{ij}} = C_{ijkl}\varepsilon_{kl} + R_{ijkl}w_{kl} \\ H_{ij} &= \frac{\partial F}{\partial w_{ij}} = K_{ijkl}w_{kl} + R_{klij}\varepsilon_{kl} \end{aligned}\right\} \tag{AIII-24}$$

Equation (AIII-20) is just Eq. (16.3-3) in the text of Chapter 16.

Now, consider the derivation of phason dissipation equations. In (AIII-19), put $\Psi_\alpha(\mathrm{r},t) = w_i(\mathrm{r},t), \Psi_\beta(\mathrm{r}',t) = g_j(\mathrm{r}',t)$, neglecting the second and fourth terms on the right-hand side and then substituting the Poisson bracket (AIII-12) into it leads to

$$\frac{\partial w_i(\mathrm{r},t)}{\partial t} = -\int \left(\{w_i(\mathrm{r}'), g_j(\mathrm{r})\}\frac{\delta H}{\delta g_j(\mathrm{r}',t)}\right) d^d\mathrm{r}' - \Gamma_{\mathrm{w}}\frac{\delta H}{\delta w_i(\mathrm{r},t)}$$

then the dissipation equation of phason is

$$\begin{aligned} \frac{\partial w_i(\mathrm{r},t)}{\partial t} &= \int (\nabla_j(\mathrm{r})w_i)\,\delta(\mathrm{r}-\mathrm{r}')\frac{g_j(\mathrm{r}')}{\rho(\mathrm{r}')}d^d\mathrm{r}' - \Gamma_{\mathrm{w}}\frac{\delta H}{\delta w_i(\mathrm{r},t)} \\ &= -V_j\nabla_j(\mathrm{r})w_i - \Gamma_{\mathrm{w}}\frac{\delta H}{\delta w_i(\mathrm{r},t)} \end{aligned} \tag{AIII-25}$$

which is just Eq. (16.3-4) in the text of Chapter 16, and Γ_{w} denotes the phason dissipation coefficient and the Hamiltonian is defined by (AIII-21).

Comparing (AIII-20) and (AIII-25), it is found that the physical meaning of phonon and phason in hydrodynamic sense is quite different from each other. According to the explanation of Lubensky *et al.* [9], the phonon represents wave propagation, while the phason represents diffusion.

Of course, the other difference between phonon and phason is that they belong to the different irreducible representations of point groups, which has been discussed in Chapter 4. The derivation of momentum equation (16.3-2) is somehow lengthy. The calculation is related to momentum $g_j = \rho V_j$, mass density ρ, phonon u_i and phsaon w_i; this means that the integrands on the right-hand side of (AIII-19) need to use simultaneously the Poisson brackets listed in (AIII-8), (AIII-11) and (AIII-12), i.e.,

$$\begin{aligned}\frac{\partial g_i(\mathrm{r},t)}{\partial t} = &-\int\left(\{g_i(\mathrm{r}),\rho(\mathrm{r}')\}\frac{\delta H}{\delta\rho(\mathrm{r}',t)}\right)d^d\mathrm{r}'\\ &-\int\left(\{g_i(\mathrm{r}),g_j(\mathrm{r}')\}\frac{\delta H}{\delta g_j(\mathrm{r}',t)}\right)d^d\mathrm{r}'\\ &-\int\left(\{g_i(\mathrm{r}),u_j(\mathrm{r}')\}\frac{\delta H}{\delta u_j(\mathrm{r}',t)}\right)d^d\mathrm{r}'\\ &-\int\left(\{g_i(\mathrm{r}),w_j(\mathrm{r}')\}\frac{\delta H}{\delta w_j(\mathrm{r}',t)}\right)d^d\mathrm{r}'\\ &\int\left(\frac{\delta\{g_i(\mathrm{r}),\Psi_\beta(\mathrm{r}')\}}{\delta\Psi_\beta(\mathrm{r}',t)}\right)d^d\mathrm{r}'+\Gamma_g\frac{\delta H}{\delta g_i(\mathrm{r},t)},\Gamma_g=\eta_{ijkl}\end{aligned} \quad \text{(AIII-26)}$$

in which the first integral on the right-hand side can be evaluated as

$$\begin{aligned}&\int\left(\{g_i(\mathrm{r}),\rho(\mathrm{r}')\}\frac{\delta H}{\delta\rho(\mathrm{r}',t)}\right)d^d\mathrm{r}'\\ &=\int\left(\rho(\mathrm{r})\nabla_i\delta(\mathrm{r}-\mathrm{r}')\frac{\delta(H_{kin}+H_{density})}{\delta\rho(\mathrm{r}',t)}\right)d^d\mathrm{r}'\\ &=\rho(\mathrm{r})\nabla_i\int\left(\delta(\mathrm{r}-\mathrm{r}')\frac{\delta(H_{kin}+H_{density})}{\delta\rho(\mathrm{r}',t)}\right)d^d\mathrm{r}'\\ &=\rho(\mathrm{r})\nabla_i\left(\frac{\delta H_{density}}{\delta\rho}\right)+\rho(\mathrm{r})\nabla_i\left(-\frac{\mathrm{g}^2}{2\rho^2}\right)\\ &=\rho(\mathrm{r})\nabla_i\left(\frac{\delta H_{density}}{\delta\rho}\right)-g_j\nabla_iV_j\end{aligned}$$

Similarly, the second to fifth integrals are evaluated, and then making some algebraic manipulations yields

$$\begin{aligned}\frac{\partial g_i(\mathrm{r},t)}{\partial t} &= -\nabla_k(\mathrm{r})(V_k g_i) + \nabla_j(\mathrm{r})\left(\eta_{ijkl}\nabla_k(\mathrm{r})V_l\right) \\ &\quad - (\delta_{ij} - \nabla_i u_j)\frac{\delta H}{\delta u_i(\mathrm{r},t)} - (\nabla_i w_j)\frac{\delta H}{\delta w_i(\mathrm{r},t)} \\ &\quad -\rho\nabla_i(\mathrm{r})\frac{\delta H}{\delta\rho(\mathrm{r},t)}, g_j = \rho V_j \end{aligned} \tag{AIII-27}$$

in which η_{ijkl} denotes the viscosity coefficient tensor of a solid, and the viscous stress tensor is

$$\sigma'_{ij} = \eta_{ijkl}\dot{\xi}_{kl} \tag{AIII-28}$$

with the deformation velocity tensor

$$\dot{\xi}_{kl} = \frac{1}{2}\left(\frac{\partial V_k}{\partial x_l} + \frac{\partial V_l}{\partial x_k}\right) \tag{AIII-29}$$

Equations (AIII-20), (AIII-25), (AIII-27) and mass density conservation equation

$$\frac{\partial \rho}{\partial t} + \nabla_k(\rho V_k) = 0 \tag{AIII-30}$$

consist of the equations of hydrodynamic equations of icosahedral quaiscrystals, which are obtained by Lubensky *et al.* [9], in which there are field variables: mass density ρ, velocities V_i (or momentums $g_i = \rho V_i$), phonon displacements u_i and phason displacements w_i.

After the publication of the work in 1985, which is cited by many authors, in the meantime, there are some arguments [11–13], in which Ref. [11] suggested some simplifications to the equations, e.g.,

$$\frac{\partial \rho}{\partial t} + \nabla_k(\rho V_k) = 0 \tag{AIII-31}$$

$$\begin{aligned}\frac{\partial g_i(\mathrm{r},t)}{\partial t} &= \nabla_j(\mathrm{r})\left(\eta_{ijkl}\nabla_k(\mathrm{r})V_l\right) - \frac{\delta H}{\delta u_i(\mathrm{r},t)} \\ &\quad - \rho\nabla_i(\mathrm{r})\frac{\delta H}{\delta\rho(\mathrm{r},t)}, g_j = \rho V_j\end{aligned} \tag{AIII-32}$$

$$\frac{\partial u_i(\mathrm{r},t)}{\partial t} = -\Gamma_{\mathbf{u}} \frac{\delta H}{\delta u_i(\mathrm{r},t)} + V_i \tag{AIII-33}$$

$$\frac{\partial w_i(\mathrm{r},t)}{\partial t} = -\Gamma_{\mathbf{w}} \frac{\delta H}{\delta w_i(\mathrm{r},t)} \tag{AIII-34}$$

AIII.4 Lie Group and Lie Algebra and Their Applications

The above derivation indicates that, the Poisson brackets (AIII-8), (AIII-11) and (AIII-12) are fundamental, which can also be derived based on the concept of Lie group. In this section, we give a brief introduction on the derivation, and in some considerations, the introduction in brief here is a supplement to the previous derivation; the readers may pay no attention to the following discussion if they are not interested in.

Just like the point groups discussed in the first 15 chapters, the Lie is a group, which satisfy the axioms, referring to the Appendix of Chapter 1. However, there is difference between point group and Lie group; the later is a kind of continuous group. The momentum operator mentioned previously is a generator of movement group, the spin operator is a generator of rotation group in spin space and the quantum Poisson brackets are connected inherent to the Lie group, so Ref. [2] suggests the concept of "group Poisson bracket".

Assume g to be an element of group G, and it has relation to the m real continuous parameters α_i, i.e.,

$$g(\alpha_i) \in G, \quad \alpha_i \in \mathbb{R}, \; i = 1, 2, \ldots, m \tag{AIII-35}$$

$\mathbb{R}$ denotes a real space.

The notion "$\cdot$" connects two elements $a(\alpha_i)$ and $b(\beta_i)$ and gives another element $c(\gamma_i) \in G$:

$$c(\gamma_i) = a(\alpha_i) \cdot b(\beta_i), \quad i = 1, 2, \ldots, m \tag{AIII-36}$$

For the continuous parameters, there is

$$\gamma_i = \varphi_i(\alpha_1, \alpha_2, \ldots, \alpha_m, \beta_1, \beta_2, \ldots, \beta_m) \tag{AIII-37}$$

If φ_i is a single-valued analytic function of $\alpha_1, \alpha_2, \ldots, \alpha_m, \beta_1, \beta_2, \ldots, \beta_m$, this group is the Lie group. The concept of a single-valued analytic function can be found in Appendix I of the Major Appendix of this book.

People usually take a parameter α_i and identical element E (the concept of identical element E referring to the Appendix of Chapter 1), then $E(\alpha_i) = 0$. The generator of the Lie group is taken to be L_i, which can be expressed by the following partial differential:

$$L_i = \mathrm{i}\frac{\partial a(\ldots, \alpha_i, \ldots)}{\partial \alpha_i}|_{\alpha_i=0} \tag{AIII-38}$$

The group element a can be expressed by the following expansion:

$$a(\ldots, \alpha_i, \ldots) = E(\ldots, 0, \ldots) + \mathrm{i}\alpha_i L_i + A(\alpha_i) \tag{AIII-39}$$

where $A(\alpha_i)$ is an infinitely small quantity related to α_i.

The infinitesimal element in the Lie group presents our important sense in this kind of group. Assume matrix $D(A)$ to be the expression matrix of element A of Lie group. The parameter α_i of infinitesimal element $A(\alpha)$ is an infinitesimal quantity. The matrix $D(A)$ can be expanded as follows:

$$D(A) = 1 - \mathrm{i}\sum_{j=1}^{N} \alpha_j I_j \tag{AIII-40}$$

and

$$I_j = \mathrm{i}\frac{\partial D(A)}{\partial \alpha_j}\bigg|_{\alpha_j = 0} \tag{AIII-41}$$

in which N I_j are named the generators of representation matrix, Lie algebra is constituted through the commutation relation between generators

$$[L_i, L_j] = \mathrm{i}C_{ij}^k L_k, i, j, k = 1, 2, \ldots, m \tag{AIII-42}$$

where C_{ij}^k is called the structure constant. The antisymmetry, linearity and Jacobi identity of Lie algebra are as follows:

$$[L_i, L_j] = -[L_j, L_i] \tag{AIII-43}$$

$$[\alpha L_i + \beta L_j, L_k] = \alpha[L_i, L_k] + \beta[L_j, L_k],\ \alpha, \beta \in \mathbb{R} \tag{AIII-44}$$

$$[L_i, [L_j, L_k]] + [L_k, [L_i, L_j]] + [L_j, [L_k, L_i]] = 0 \tag{AIII-45}$$

respectively. In classical continuum mechanics, the coordinate transformation

$$x^k \to x^k + u^k(\mathrm{r}) \tag{AIII-46}$$

is often used, which is called the translation group or movement group, or the infinitesimal movement group. It is interesting, in particular, that $u^k(\mathrm{r})$ presents evident physical meaning, and represents displacement, or phonon. Note that x^k here is a contravariant vector, while x_i is the covariant vector. We recall that the physical quantities mentioned previously present very close connection to the group algebra because the momentum operator is the generator of movement group, the spinor operator is the generator of rotation group in spinor space, etc. Some correlation between physical quantities $a, b, c, \ldots$ and elements of transformation group $A, B, C, \ldots$ of transformation group

$$\{a, b, c, \ldots\} \to \{A, B, C, \ldots\} \tag{AIII-47}$$

may be set up. The linear combination of group element A can be given by the following linear expression:

$$A = \sum_{g \in G} A(g)g, \quad A(g) \in \mathbb{R} \tag{AIII-48}$$

where $A(g)$ can be understood as the coefficients of the expansion, but the series is for discrete groups only, and it should be replaced by the integral for continuous groups, and in this case the group elements vary continuously.

Assume A can be transformed according to the following version:

$$A \to gAg^{-1} \tag{AIII-49}$$

Suppose δg is an infinitesimal transformation, if $g = 1 + \delta g$, then the linear approximation is

$$A \to A + \delta A \tag{AIII-50}$$

and

$$\delta A = [\delta g, A] \tag{AIII-51}$$

The infinitesimal transformation δg is of the form

$$\delta g = \frac{\mathrm{i}}{\hbar}\int \alpha^k(\mathrm{r})L^k(\mathrm{r})d^d\mathrm{r} \tag{AIII-52}$$

in which $\alpha^k(\mathrm{r})$ are the local infinitesimal "angular", $L^k(\mathrm{r})$ the generators of local transformation group, $\mathrm{i} = \sqrt{-1}, \hbar = h/2\pi$ and h the Planck constant.

For the movement group, take $\alpha^k(\mathrm{r}) = u^k(\mathrm{r})$, and the generator is the momentum, then there is from (AIII-51) and (AIII-52)

$$\delta A(\mathrm{r}) = \frac{\mathrm{i}}{\hbar}\int \alpha^k(\mathrm{r}')\left[L^k(\mathrm{r}'), A(\mathrm{r})\right] d^d\mathrm{r}' \tag{AIII-53}$$

This equation shows that δA is the linear functional of "angular" $\alpha^k(\mathrm{r})$ of infinitesimal local transformation, and the corresponding variation is

$$\frac{\delta A(\mathrm{r})}{\delta \alpha^k(\mathrm{r}')} = \frac{\mathrm{i}}{\hbar}\left[L^k(\mathrm{r}'), A(\mathrm{r})\right] \tag{AIII-54}$$

The limit passing over from quantum mechanics to classical mechanics is

$$\frac{\delta \hat{A}}{\delta \alpha} = \frac{\mathrm{i}}{\hbar}\left[\hat{L}, \hat{A}\right] \to \frac{\delta A}{\delta \alpha} = \{L, A\} \tag{AIII-55}$$

Recall again that $\hat{L}, \hat{A}$ represent operators in quantum mechanics and L, A the field variables in classical mechanics. So the right-hand side of (AIII-54) may be written as

$$\frac{\delta a}{\delta \alpha} = \{l, a\} \tag{AIII-56}$$

in which a can represent any field variables $a, b, c, \ldots$ of hydrodynamics and l the generator $L^k(\mathrm{r})$ corresponding to the group, so from (AIII-56),

$$\frac{\delta a(\mathrm{r})}{\delta \alpha^k(\mathrm{r}')} = \left\{l^k(\mathrm{r}'), a(\mathrm{r})\right\} \tag{AIII-57}$$

Furthermore,

$$\frac{\delta l^m(\mathrm{r})}{\delta \alpha^k(\mathrm{r}')} = \left\{l^k(\mathrm{r}'), l^m(\mathrm{r})\right\}, \{a, a\} = \{a, b\} = \{b, b\} = 0 \tag{AIII-58}$$

At the finite temperature, the Hamiltonian can be expressed by

$$H = \int \varepsilon(\mathbf{p}, \rho, s) d^d \mathrm{r}$$

$$d\varepsilon = V^k dp_k + \mu d\rho + T ds$$

where ε denotes the energy density, the others are the same before, $\mathrm{p} = (p_x, p_y, p_z), \rho$, the momentum and mass density, s the entropy, $\mathrm{V} = (V_x, V_y, V_z)$ the velocity, μ the chemical potential and T the absolute temperature, respectively, so

$$\begin{aligned} \delta p_k &= -u^l \nabla_l p_k - p_k \nabla_k u^k \\ \delta \rho &= -u^l \nabla_l \rho - \rho \nabla_k u^k \\ \delta s &= -u^l \nabla_l s - s \nabla_k u^k \end{aligned} \qquad \text{(AIII-59)}$$

From (AIII-46) and (AIII-47), we obtain

$$\begin{aligned} \{p_k(\mathrm{r}_1), \rho(\mathrm{r}_2)\} &= \rho(\mathrm{r}_1) \nabla_k(\mathrm{r}_1) \delta(\mathrm{r}_1 - \mathrm{r}_2) \\ \{p_k(\mathrm{r}_1), p_l(\mathrm{r}_2)\} &= (p_l(\mathrm{r}_1) \nabla_k(\mathrm{r}_1) - p_k(\mathrm{r}_2) \nabla_k(\mathrm{r}_2)) \, \delta(\mathrm{r}_1 - \mathrm{r}_2) \end{aligned} \qquad \text{(AIII-60)}$$

This is identical to those (AIII-8) given by using the Poisson bracket of condensed matter physics, which belong to Ref. [2].

If we use the above method to quasicrystals, there are

$$\{u_k(\mathrm{r}_1), g_l(\mathrm{r}_2)\} = (-\delta_{kl} + \nabla_l(\mathrm{r}_1) u_k) \, \delta(\mathrm{r}_1 - \mathrm{r}_2) \qquad \text{(AIII-61)}$$

$$\{w_k(\mathrm{r}_1), g_l(\mathrm{r}_2)\} = (\nabla_l(\mathrm{r}_1) w_k) \, \delta(\mathrm{r}_1 - \mathrm{r}_2) \qquad \text{(AIII-62)}$$

which are the same as those given by Lubensky *et al.* [9] who obtained directly by using the Poisson bracket method, i.e., (AIII-11) and (AIII-12).

This shows the power of the Lie group theory. Reference [4] proved that there are the above results, and applying Liouville equation, one can obtain equations of motion of many complex systems.

References

[1] Landau L D and Lifshitz M E, 1988, *Fluid Mechanics, Theory of Elasticity*, Pergamon, Oxford.

[2] Landau L D, 1941, The theory of superfluidity of heilium II, *Zh. Eksp. Teor. Fiz, II*, **592**, *J Phys USSR*, **5**, 71.

[3] Landau L D and Lifshitz E M. 1935, Zur Theorie der Dispersion der magnetische Permeabilität der ferromagnetische Körpern, *Physik Zeitschrift fuer Sowjetunion*, **8**(2), 158–164.
[4] Dzyaloshinskii I E and Volovick G E, 1980, Poisson brackets in condensed matter physics, *Ann. Phys.*, **125**(1), 67–97.
[5] Dzyaloshinskii I E and Volovick G E, 1978, On the concept of local invariance in spin glass theory, *J. de. Phys.*, **39**(6), 693–700.
[6] Volovick G E, 1978, Additional localized degrees of freedom in spin glasses, *Zh Eksp Teor Fiz*, **75**(7), 1102–1109.
[7] Martin P C, Paron O and Pershan P S, 1972, Unified hydrodynamic theory for crystals, liquid crystals, and normal fluids. *Phys. Rev. A*, **6**(6): 2401–2420.
[8] Fleming P D and Cohen C, 1976, Hydrodynamics of solids, *Phys. Rev. B*, **13**(2), 500–516.
[9] Lubensky T C, Ramaswamy S, and Toner J, 1985, Hydrodynamics of icosahedral quasicrystals, *Phys. Rev. B*, 32(11), 7444–7452.
[10] Lubensky T C, 1988, Symmetry, elasticity and hydrodynamics of quasiperiodic structures, in Introduction to Quasicrystals, Vol. I, ed. by Jaric, M V, Academic Press, Boston, pp. 199–280.
[11] Rochal S B and Lorman VL, 2002, Minimal model of the phonon-phason dynamics in icosahedral quasicrystals and its application to the problem of internal friction in the i-AlPdMn alloy. *Phys. Rev. B*, **66**(14), 144204.
[12] Khannanov S K, 2002, Dynamics of elastic and phason fields in quasicrystals. *Phys. Met. Metall.*, **93**(5): 397–403.
[13] Coddens G, 2006, On the problem of the relation between phason elasticity and phason dynamics in quasicrystals. *Eur. Phys. J. B*, **54**(1), 37–65.

Index

Q

R

S

www.ingramcontent.com/pod-product-compliance
Lightning Source LLC
LaVergne TN
LVHW010322220926
848484LV00004BB/831
9789811279096